The Developing Brain and Its Connections

The Developing Brain and Its Connections describes the processes of neural development from neural induction through synaptic refinement. Each chapter explores specific mechanisms of development and describes key experiments from invertebrate and vertebrate animal models. By highlighting experimental methods and explaining how hypotheses evolve over time, readers learn essential facts while strengthening their appreciation of the scientific method. Discussions of neurodevelopmental disorders and therapeutic approaches to them bridge basic science discoveries with the clinical aspects of the field. Descriptions of recent work by student researchers and medical residents demonstrate career pathways and options for those interested in pursuing any area neural development. With this distinctive approach, easy-to-follow writing style, and clear illustrations, *The Developing Brain* presents an accessible approach to neural development for undergraduate students.

The Developing Brain and its Connections

Lynne M. Bianchi

CRC Press
Taylor & Francis Group
Boca Raton London New York

CRC Press is an imprint of the
Taylor & Francis Group, an **informa** business

First edition published 2023
by CRC Press
6000 Broken Sound Parkway NW, Suite 300, Boca Raton, FL 33487–2742

and by CRC Press
4 Park Square, Milton Park, Abingdon, Oxon, OX14 4RN

CRC Press is an imprint of Taylor & Francis Group, LLC

Derived from *Developmental Neurobiology* published by Taylor & Francis Group, LLC |
Garland Science 2018

Cover image courtesty of Jason Newbern and Mike Holter.

Library of Congress Cataloging-in-Publication Data
Names: Bianchi, Lynne M. (Professor of neuroscience), author.
Title: The developing brain and its connections / Lynne M. Bianchi.
Description: First edition. | Boca Raton, FL : CRC Press, 2023. | Includes
    bibliographical references and index.
Identifiers: LCCN 2022029030 (print) | LCCN 2022029031 (ebook) |
    ISBN 9780367762438 (hbk) | ISBN 9780367749903 (pbk) |
    ISBN 9781003166078 (ebk)
Subjects: LCSH: Developmental neurobiology. | Neurotransmitters.
Classification: LCC QP363.5 .B59 2023 (print) | LCC QP363.5 (ebook) |
    DDC 612.6/4018—dc23/eng/20220916
LC record available at https://lccn.loc.gov/2022029030
LC ebook record available at https://lccn.loc.gov/2022029031

ISBN: 9780367762438 (hbk)
ISBN: 9780367749903 (pbk)
ISBN: 9781003166078 (ebk)

DOI: 10.1201/9781003166078

Typeset in ITC Leawood
by Apex CoVantage, LLC

To Gene, Aaron, Glia, and Steven

BRIEF CONTENTS

DETAILED CONTENTS

BOXES

AUTHOR BIOGRAPHY

Lynne M. Bianchi is Emerita Professor of Neuroscience at Oberlin College where she taught and mentored undergraduate students for over 20 years. She is currently the Director of Medical Research for the University of Pittsburgh Medical Center Hamot and Gannon University where she helps faculty and clinician-scientists develop their clinical and basic science research projects. Her own research interests focus on the role of nerve growth factors during embryonic development.

PREFACE

There is something inherently fascinating about figuring out how the nervous system forms and makes connections, and it seems nearly everyone wants to know something about how the brain develops. Those who study nervous system development address fundamental questions by drawing from two of the most exciting areas of research—neuroscience and embryogenesis. They use invertebrate and vertebrate animal models and experimental techniques that range from century-old surgical methods to cutting-edge genetic manipulations. With intriguing questions and a multitude of methods, it is not surprising that developmental neurobiology remains one of science's most active areas of research.

Developmental neurobiology is also a great topic for a course, as there are so many interesting areas to discuss—from the initial designation of which tissue has the potential to become neural to the postnatal refinement of synaptic connections. The hardest part for an instructor is deciding what to cover in the time available in the classroom. There is no single approach for what to include, but there are key developmental events that are helpful for students to know something about, and the chapters in this book reflect those events so readers have foundational material on which to build. Chapter topics include neural induction, patterning, proliferation, migration, axonal outgrowth, survival and programmed cell death, and synaptogenesis and synaptic refinement. Each chapter discusses mechanisms governing these stages of neural development and describes experiments that led to our current understanding of those mechanisms. In this way students not only learn the "facts" about a developmental stage, but they also learn how such facts emerged and what it took to generate those facts.

The book was written with the perspectives of both the instructor and student in mind. As instructors know, students usually come to a course about development of the nervous system with very different backgrounds, so providing content that is accessible to a broad audience is always a challenge. The book is designed to be accessible to that broad audience, providing background information where needed so that students taking a first neuroscience course, or those requiring some review, can follow the material as well as more advanced students.

Each chapter includes updated information reflecting recent discoveries in vertebrate and invertebrate animal models, notes areas of ongoing research, and indicates areas in need of further study. Instructors should find this helpful in choosing areas to expand upon given their own interests and the interests and experience of the class. Each chapter also includes historical background to help students appreciate how ideas have evolved, so they can place the latest discoveries into a broader context.

Whether organizing the course or planning a single lecture, an instructor is required to select which content to include, knowing that other material must be set aside. It is never an easy task. The task is equally difficult when choosing what material to include in a book. The examples included in this text highlight experiments that had a major impact on the field or changed how investigators approached a particular question. While these examples illustrate the types of work that has been done, other equally impressive examples are omitted. Readers are encouraged to explore the literature on topics of interest to them and may wish to begin with the references listed under "Further Reading." The name of a lead investigator is indicated for many of the experiments described in each chapter so that students can locate and read original papers. In many instances an investigator's name is listed with the very broad label "and colleagues." In some cases, the colleagues were a few other individuals working on the project in a single lab. In many cases, however, "and colleagues" represents the contributions of several, if not dozens, of researchers over the course of many years or, in some cases, decades. The contributions of these unnamed colleagues, particularly the graduate students and post-doctoral fellows who designed and completed many of the experiments, cannot be underestimated.

A lot has changed since a previous volume of this material was published in 2017. Although global events have slowed research in ways unimaginable five years ago, new discoveries are published weekly, just as in the past. Much of the new knowledge about neural development stems from discoveries that identify, visualize, and track the fate of individual cells, giving us better insight into how neurons form, migrate, and make connections. In 2018 the journal *Science* recognized techniques to track cells during embryonic development as the breakthrough of the year. In 2020, the journal *Nature Methods* named spatially resolved transcriptomics, the mapping of expressed genes, as the method of the year. Such techniques, previously used in only a few large labs, are now commonly used in labs of all sizes throughout the world and have greatly increased our understanding of neural development across animal models. Of course, new technologies alone do not advance the field. Established methods and earlier work, including methods and data from the previous century, are still critical to interpreting current work, a point emphasized through the examples presented in this book.

My goal in creating this book is to provide a helpful and accessible resource for students and instructors, but I also hope that in reading the descriptions of different studies, readers will appreciate that research is a dynamic, creative process that relies on the motivation, ingenuity, and persistence of individuals like them. Reading about the experiments of a Nobel Prize winner or long-established investigator might make science appear like something that only a select few do. However, it takes contributions from researchers at every level to advance a field. Students should also remember that every well-known investigator started out as an inexperienced undergraduate student. Chapters include boxes that highlight work done by undergraduate students, graduate students, medical residents, and physicians and professors at various career stages to help readers recognize that they too can make valuable contributions to understanding how the brain develops and makes its connections.

I would like to thank those who helped make this volume possible. My mentor, Christopher Cohan, who introduced me to the field of developmental biology and always encouraged me to explore my ideas. My students at Oberlin College who initiated great conversations in and out of the classroom, and the many who worked in my lab developing their own projects. In the process of our many interactions, they taught me a lot about how to teach. Thanks to the dedicated staff at the Oberlin College Archives, Alumni Association, and Science Library for finding the photos and books I requested. A big thank you to my colleagues for contributing boxes and for sharing their experiences and educational paths. Special thanks to Larry Zipursky, one of those well-known investigators who started out as an inexperienced undergraduate student, for sharing his story and photo. Thanks to Jason Newbern for generously providing the cover image and other images included in this book and to Elise Marti Gorostiza for providing high resolution images from her 2012 publication on dorsal-ventral patterning. Thanks to Bob Goldstein for his translations of Rita Levi-Montalcini's early papers and his article about her life during World War II

("A Lab of Her Own"; *Nautilus* December 1, 2021) and a special thank you to Aaron Bianchi for reviewing the list of glossary terms.

Many authors acknowledge their families and thank them for their support and patience during the writing process, and with good reason. Writing requires many hours of uninterrupted time, forcing families to adapt to the writer's schedule. I could not have completed this project without the encouragement and support of my husband and son. I also thank them, and our dog, for giving me uninterrupted time each morning. Their willingness to sleep through those early morning writing sessions was integral to the completion of this project.

Books also require many hours of work from editors, illustrators, and publishers. This book would not have been possible without the commitment of Chuck Crumly. I am grateful for his encouragement and support throughout this project. A very special thinks to Kara Roberts and Jordan Wearing for all their help in preparing the final manuscript. The Crumly-Roberts-Wearing team has been a pleasure to work with at every step of the project. This book also benefited greatly from the exceptional work of Nigel Orme, who once again took my crude sketches and turned them into clear and helpful illustrations. Thanks also to Tara Grover Smith and Venkatesh Sundaram for all the work on the final production of the manuscript.

A list of commonly used abbreviations is included here, and a glossary of terms is found after Chapter 10. For consistency in the text, gene names are capitalized and in italics for all species. Protein names are in roman.

FREQUENTLY USED ABBREVIATIONS

Abbreviations used frequently in this book include:

ACh	acetylcholine
AChE	acetylcholine esterase
AChR	acetylcholine receptor
ANB	anterior neural border
ANR	anterior neural ridge
AVE	anterior visceral endoderm
BDNF	brain derived neurotrophic factor
BMP	bone morphogenetic protein
CDK	cyclin dependent kinase
CNS	central nervous system
CNTF	ciliary neurotrophic factor
CSF	cerebrospinal fluid
DBL	dorsal blastopore lip
DRG	dorsal root ganglion
ECM	extracellular matrix
EGL	external granule cell layer (of cerebellum)
Erk	extracellular signal-regulated kinase
FGF	fibroblast growth factor
GDNF	glial-cell-line-derived neurotrophic factor
GEF	guanine nucleotide exchange factor
GMC	ganglion mother cell
GTP	guanosine triphosphate
LGN	lateral geniculate nucleus
LIF	leukemia inhibitory factor
LRP	low-density lipoprotein
MAP	mitogen activated protein
MEK	MAP kinase/Erk Kinase

MHB	midbrain–hindbrain border
MuSK	muscle specific kinase
NCAM	neural cell adhesion molecule
NGF	nerve growth factor
Ngn	neurogenin
NMJ	neuromuscular junction
Nrg	neuregulin
NT-3	neurotrophin 3
NT-4/5	neurotrophin 4/neurotrophin 5
OPC	oligodendrocyte precursor
PCD	programmed cell death
PNS	peripheral nervous system
PSC	perisynaptic Schwann cell
PSD	postsynaptic density
Ptc	Patched
RA	retinoic acid
Raf	rapidly accelerated fibrosarcoma kinase
Ras	rat sarcoma
RG	radial glia
RGC	retinal ganglion cell
SCG	superior cervical ganglia
Sev	sevenless
sFRP	secreted Frizzled-related proteins
Shh	Sonic hedgehog
SOP	sensory organ progenitor
SOS	son of sevenless
SVZ	subventricular zone
TGF	transforming growth factor
Trk	tropomyosin receptor kinase
UPS	ubiquitin proteasome system
VNC	ventral nerve cord
VZ	ventricular zone
ZLI	zona limitans intrathalamica

RESOURCES FOR INSTRUCTORS

The figures for *The Developing Brain and Its Connections* are available to registered instructors in two convenient formats: PowerPoint and JPEG. To gain access to the figure slides, please visit the following link to register as an instructor: https://routledgetextbooks.com/textbooks/instructor_downloads/

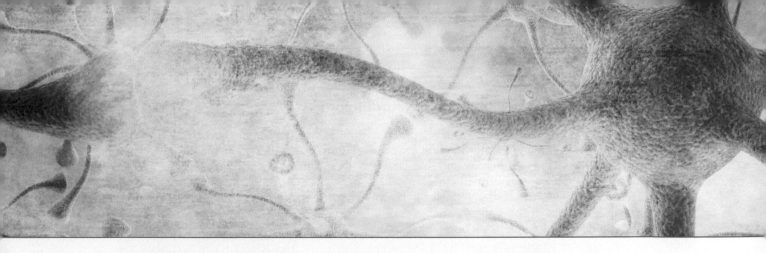

An Introduction to the Field of Developmental Neurobiology

1

The human nervous system is one of the most complex biological systems. Comprised of billions of cells and precisely organized connections, the nervous system governs everything from breathing to thinking. The field of developmental neurobiology, or developmental neuroscience, seeks to understand how such precise cellular organization and wiring arises. Experimentally, developmental neurobiology is one of the most exciting fields to work in, as the questions are addressed using a variety of methods that range from the classical approaches of tissue manipulations to the most sophisticated molecular, genetic, and imaging techniques available today. It is no wonder that developmental neurobiology is a field populated by researchers with backgrounds in fields as diverse as anatomy, biochemistry, cellular and molecular biology, computational sciences, embryology, genetics, medicine, physics, physiology, and psychology (**Box 1.1**).

Fundamental questions about how the nerve cells initially form and extend cellular processes to make the billions of necessary connections have intrigued scientists and non-scientists for centuries (**Figure 1.3**). However, significant advances in understanding how nerve fibers extend from the developing brain and spinal cord to target areas in the body did not occur until the mid- to late nineteenth century, when the microscopic anatomy of neural tissue was described in detail for the first time. These new findings were made possible by several technical advances made during that same period. One important new tool was the microtome, an instrument that cuts tissues into very thin slices. Another was the increasing availability of microscopes with improved optics that allowed for better visualization of these thinner tissue slices. Additionally, scientists continued to test and refine techniques for fixing (preserving) and staining tissues, so that by the end of the nineteenth century, several methods for visualizing the cellular composition of tissues were available. These innovations led to discoveries that were part of the "great age of cellular biology," laying the foundation for many fundamental concepts that we now take for granted.

Several of the first explanations of how the nervous system formed and extended nerve fibers were based on these early microscopic observations. Among the most influential scientists of that period was Santiago Ramón y Cajal, whose work is described in subsequent chapters. What is

DOI: 10.1201/9781003166078-1

Box 1.1 Pathways to Developmental Neurobiology

Investigators have come to the field of developmental neurobiology by following many different career paths. Some, such as Hans Spemann and Rita Levi-Montalcini, began their careers in medicine, but ultimately decided to focus on research instead. Both started research careers in the general area of zoology and gradually, as they undertook one project and then another, started to address questions pertaining to neurodevelopment. Some investigators, such as the biochemist Stanley Cohen, were recruited to help address a particular question in the developing nervous system, and later focused research efforts on topics beyond the nervous system. Spemann's work provided pivotal insights into how neural tissue is first formed in the early stages of embryogenesis (Chapter 2) and Levi-Montalcini and Cohen identified the first protein to promote the survival of developing neurons (Chapter 8).

Researchers also come from a variety of backgrounds. Some who study developmental neurobiology were the first in their families to attend college, whereas others descend from families comprised of several scientists and physicians. Some completed their undergraduate studies at large universities, whereas others began their studies at small colleges. Some came to college expecting to study science, while others began with different majors and uncertain career goals. Roger Sperry, for example, whose early work addressed how neurons extend nerve fibers to contact the correct target cell (Chapter 7), graduated from Oberlin College in 1935 with a major in English. He later earned a master's degree in psychology and studied zoology at Oberlin College prior to beginning his doctoral studies with Paul Weiss at the University of Chicago. In addition to his influential contributions to neuron-target recognition molecules, Sperry won the Nobel Prize for Physiology or Medicine in 1981 for his work on split-brain patients that revealed how the two hemispheres of the brain communicate with one another. Yet, this remarkable career could have easily taken a different path. On his college application, when asked about his future career plans, one of his suggestions was college athletic coach due to his interests and talents in various sports (**Figure 1.1**).

Another Oberlin College graduate, Larry Zipursky, came to college with two clear interests: sports and chemistry. Growing up in Canada, he began playing hockey at age 4 and was on Oberlin's varsity hockey team throughout college. His physician-scientist father first inspired his interest in science and Zipursky always felt the laboratory was a special place. However, like many college students his interests were varied, and he pursued courses in political science

and joined the varsity soccer team in his junior year (**Figure 1.2**). After graduating from Oberlin, he went to Albert Einstein College of Medicine to study the enzymology of DNA replication in bacteria with Jerard Hurwitz. Near the end of his graduate training, Zipursky

Figure 1.1 Roger Sperry as the captain of the basketball team. Like many college students, Roger Sperry was unsure of what career he would pursue. Prior to entering Oberlin College, he expressed interest in science and athletic coaching. As an undergraduate, Sperry majored in English and was captain of the basketball team. After graduating in 1935, he remained at Oberlin to complete a master's degree in psychology (1937), then took additional courses in zoology to prepare for his doctoral studies at the University of Chicago. [Courtesy of the Oberlin College Archives.]

Figure 1.2 Larry Zipursky (Number 17, with ball) during his junior year of college. In addition to pursuing soccer, hockey, political science, and chemistry, Zipursky used his summer breaks to gain research experience. Once convinced that science was the right path for him, he completed his Ph.D. investigating DNA replication enzymes before turning his focus to the nervous system. He began studying the developing nervous system as a postdoctoral fellow in 1981 and has made major contributions to developmental neurobiology ever since. [Photo from Oberlin College 1976 yearbook, courtesy of Larry Zipursky.]

became interested in using genetics and molecular biology to study the brain, which led him to Seymour Benzer's lab (Chapter 6) at the California Institute of Technology (Caltech) in 1981. It was here that he began his developmental studies on the *Drosophila* visual system. In 1985 he joined the faculty of the University of California Los Angeles (UCLA) School of Medicine where is now Distinguished Professor of Biological Chemistry and Investigator of the Howard Hughes Medical Institute. Zipursky's work in developmental neurobiology, begun in his postdoctoral training, has led to major discoveries in many areas, as highlighted in Chapters 6, 7, and 10, including studies

investigating "Sperry-like" neuron-target recognition molecules. The year Zipursky joined the Benzer lab, Roger Sperry whose office was one floor above Benzer's at Caltech, won the Nobel Prize. Like many successful scientists, Sperry and Zipursky appreciated the value of a broad education and the importance of having interests outside the lab.

It is certain that Sperry, Zipursky, and the other scientists whose work is featured throughout this book had no idea as undergraduate students where their careers would take them, what questions they might address, how long their careers would last, or how many other scientists they would influence.

remarkable about the work of Cajal and his contemporaries is that their descriptions of how neurons grew and behaved in an embryonic environment were all formulated based on images of fixed tissues. By careful observation at different stages of development, the researchers were able to propose reasonable hypotheses about how cell growth and movement

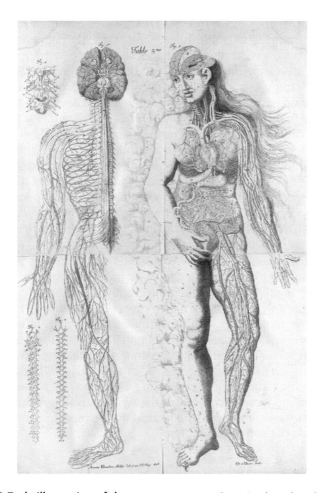

Figure 1.3 Early illustration of the nervous system. Scientists have long been interested in understanding the paths nerve fibers take from the brain and spinal cord to target regions throughout the body. This illustration was completed by the physician Amé Bourdon in 1678, reflecting the artistic style expected for anatomical drawings of that period. [Image courtesy of U.S. National Library of Medicine, Historical Anatomies.]

occurred. While not every idea put forth in the late nineteenth century was accurate, a surprisingly large number were later found to be correct, or very nearly so.

By the early twentieth century scientists had developed a variety of surgical, histological, electrophysiological, and tissue culture techniques that further advanced studies in developmental neurobiology. Major scientific milestones in the field often paralleled advances in other areas.

For example, in the mid-twentieth century, the electron microscope made it possible to view cellular organelles and led to the conclusive identification of the synapse as the site of connection between two nerve cells. Advances in electronics were similarly influential. As recording equipment became more precise, researchers were able to detect the tiny electrical impulses produced by nerve cells. Additionally, instrumentation was developed for making precision microelectrodes to record activity outside of cells, inside of cells, and even on a restricted patch of cell membrane. Because of these technical advances, researchers can measure isolated ionic currents and neural activity and monitor changes that occur as the nervous system develops. Also in the mid-twentieth century, the discovery of **DNA** (**deoxyribonucleic acid**), the various forms of **RNA** (**ribonucleic acid**), amino acid structures, and the genetic code ushered in the entirely new field of molecular biology, which provided insight into the importance of regulated gene and protein expression during neural development. Technological advances continue to be made in imaging, electronics, molecular biology, and genetics. As in the past, researchers commonly combine the available techniques to get a fuller picture of what happens as neurons progress through various developmental stages.

Neural development is studied in a variety of animal models, including flies, worms, frogs, fish, chicks, and mice. These models are used to track normal developmental events as well as manipulate developing systems and evaluate the impact of such changes on neural development. It is now recognized that many developmental mechanisms are highly conserved among species, and scientists working with the fruit fly (*Drosophila melanogaster*), the nematode worm (*Caenorhabditis elegans*), or other invertebrate species often are the first to discover genes and signaling molecules that regulate a particular aspect of nervous system formation in multiple species.

This book describes many of the primary mechanisms by which the nervous system develops, from the initial specification of neural tissue to the refinement of neural connections during early postnatal periods. Each chapter highlights some of the experiments that were key to advancing understanding of a particular stage of neural development. These many experiments highlight the remarkable creativity and insight of the early neurobiologists who made so many major contributions, even without benefit of the more sophisticated techniques available today. One also quickly appreciates how some questions simply could not be answered until suitable technical approaches became available. It is likely that many of the techniques that are considered advanced today will appear crude to scientists in the future. Yet as in the past, the discoveries made today will add to the foundation of knowledge that will be used by future scientists, and together these discoveries will elucidate the mechanisms that govern the formation of the nervous system.

CELLULAR STRUCTURES AND ANATOMICAL REGIONS OF THE NERVOUS SYSTEM

To help orient readers to topics discussed in subsequent chapters, the following sections provide a brief overview of some major cellular, anatomical, and developmental features found in vertebrate and invertebrate animal models and an overview of common experimental techniques discussed in this text. The cellular composition and anatomical organization of key neural structures are described first, followed by information on specific developmental stages documented in the chick, mouse, human, fish, frog, fly, and worm nervous systems. These descriptions focus on the timing of shared developmental milestones, including gastrulation, neural plate and neural tube formation, and early brain segmentation. For more detailed explanations of these topics, refer to the references at the end of the chapter.

The Central and Peripheral Nervous Systems Are Comprised of Neurons and Glia

The vertebrate nervous system is divided into two main regions, the **central nervous system** (**CNS**) and the **peripheral nervous system** (**PNS**). The vertebrate CNS is comprised of the brain and spinal cord, while the PNS consists of collections of neurons called **ganglia** that lie outside of the CNS. The vertebrate PNS includes the neurons of the spinal sensory (dorsal root) ganglia, cranial nerve ganglia, and ganglia of the autonomic nervous system (ANS). The invertebrate nervous system is also divided into CNS and PNS regions; however, different terminology is used for the various CNS and PNS structures, as described in this chapter.

The cells of the CNS and PNS are the **neurons** and **glia**. A vertebrate neuron consists of a cell body and cellular processes called **axons** and **dendrites** (**Figure 1.4**A). Each neuron has only one axon but may have several dendrites. The axon is typically longer than other neural processes, has a uniform diameter, and ends in specialized regions called **axon terminals**. In contrast, the dendrites tend to be shorter, branch extensively,

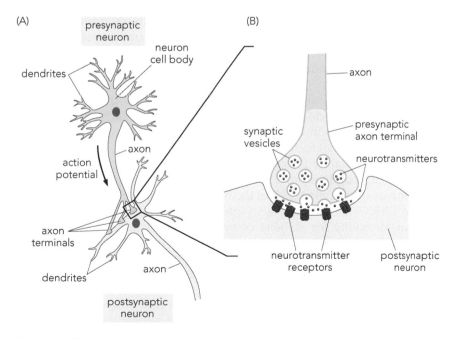

Figure 1.4 Neurons release neurotransmitters to communicate with other cells. (A) Vertebrate neurons consist of a cell body and cellular processes called axons and dendrites. Most vertebrate neurons signal to other cells by conducting the electrical activity of an action potential down the axon to stimulate the release of a neurotransmitter from the axon terminals and, in this example, to the dendrites and cell body of the postsynaptic neuron. (B) The neurotransmitter is released from synaptic vesicles then diffuses across the synaptic cleft to bind to specific receptors on the postsynaptic cell. Depending on the neurotransmitter and receptor pair, the binding will either increase or decrease the likelihood that an action potential will occur in the postsynaptic cell.

and have tapered ends. In some circumstances, the term **neurite** is used to refer to either axons or dendrites. For example, when viewing neuronal processes at early stages of neural development or in tissue culture preparations, it is often difficult to conclusively identify a process as an axon or a dendrite and therefore the term *neurite* is used.

Neurons primarily communicate with one another through electrical signals (**action potentials**) that are conducted along the length of the axon to initiate the release of chemical signals (**neurotransmitters**) from **synaptic vesicles** that accumulate in the axon terminals. The release of neurotransmitter occurs at the **synapse**, a small gap or cleft between the axon terminal of one neuron (the **presynaptic cell**) and the cell body or processes of another (the **postsynaptic cell**). The neurotransmitter diffuses across the **synaptic cleft** to bind to receptors on the postsynaptic cell, which may be another neuron or a muscle cell (Figure 1.4B). Neurotransmitter–receptor pairs that *increase* the likelihood that an action potential will occur are found at **excitatory synapses**. In contrast, neurotransmitter–receptor pairs that *reduce* the likelihood of an action potential firing are found at **inhibitory synapses**. A small percentage of vertebrate neurons and some invertebrate neurons communicate through **gap junctions**—channels that are formed between two cells that are in direct contact with each other. In vertebrates, the chemical synapses and gap junction synapses can work together to enhance neural transmission.

The nervous system is also comprised of several distinct cell types called **glia**. Originally called neuroglia in the mid 1800s, these cells were thought to be connective tissue—the "glue"—needed to support the structures of the nervous system. For over a century and half, glia were thought to be limited to this role. However, it is now clear that glia serve a number of important functions in the nervous system and in some cases participate in cell signaling. The glia in the vertebrate CNS are the oligodendrocytes, astrocytes, microglia, and ependymal cells. **Oligodendrocytes** extend cellular processes that form the myelin around axons in the CNS. Each oligodendrocyte extends processes to wrap around several nearby axons. **Myelin** provides a type of insulation that speeds the propagation of action potentials. Thus, action potentials are conducted faster along myelinated axons than along unmyelinated axons. **Astrocytes** are star-shaped cells that perform many functions in the CNS, such as maintaining the balance of ions in the extracellular fluid surrounding neurons, interacting with cells that form the blood-brain barrier (BBB), secreting signals to influence synaptic formation and stability, and communicating with neurons. **Microglia** are the smallest of the glial cell types and generally function as the immune cells of the brain to remove debris and pathogens in the CNS. Microglia may also interact with signals from the immune system to modify the stability of synaptic connections during development and in neurodegenerative conditions (**Figure 1.5**A). **Ependymal** cells line the **choroid plexus**, the network of cells, capillaries, and connective tissue in the ventricles of the CNS, and produce cerebral spinal fluid (CSF).

In the vertebrate PNS, glial cells consist of the Schwann cells and satellite cells. Most **Schwann cells** function similarly to oligodendrocytes. However, each Schwann cell wraps around only one axon and does not extend processes to nearby axons. The **satellite cells** surround neuronal cell bodies and appear to have functions similar to astrocytes (Figure 1.5B).

It is now recognized that glia in the vertebrate CNS and PNS also release signals that regulate aspects of neural development and similar roles for invertebrate glia have begun to emerge.

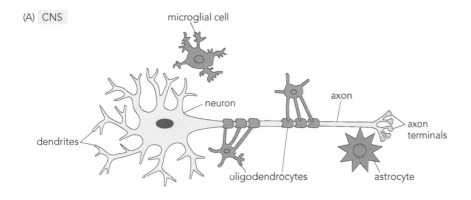

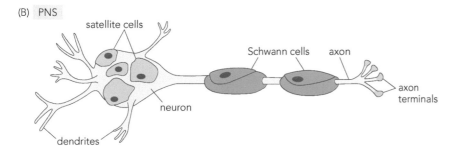

Figure 1.5 The vertebrate nervous system is comprised of neurons and glia. Neurons in the vertebrate nervous system are characterized by a single axon and many dendrites. Axons are generally longer and of uniform diameter, while the dendrites tend to be shorter, with tapered ends. Glial cells surround the neurons and perform diverse functions. (A) Neurons in the central nervous system (CNS) are surrounded by numerous glia, including astrocytes, microglia, and the myelinating oligodendrocytes that wrap around the axons. (B) In the peripheral nervous system (PNS), the cell bodies of neurons are surrounded by glial satellite cells, whereas the axons are wrapped by myelinating Schwann cells.

The Nervous System Is Organized around Three Axes

When describing the location of different anatomical structures in the nervous system, scientists often refer to them relative to other structures along one of three axes. The **dorsal–ventral axis** (also called the **dorsoventral axis**) runs from the back (from the Latin *dorsum*) to the belly (*venter*) side of the animal, and can easily be envisioned in any number of vertebrate species such as mice or humans (**Figure 1.6**A). However, other terms are more easily envisioned in embryos and four-legged animals than in humans. The main body axis of a mouse, for example, is the **rostral–caudal** (or **rostrocaudal**) **axis**. Rostral comes from the Latin word *rostrum*, meaning beak or stiff snout, and caudal from the word *cauda*, meaning tail. In many species, as well as in the early embryonic nervous system, this axis is often called the **anterior–posterior (anteroposterior) axis**, where the terms anterior and posterior substitute for rostral and caudal, respectively. These terms apply to the main body axis as well as the **neuraxis** established by the brain and spinal cord (Figure 1.6B). However, this axis is not as readily envisioned in the adult human nervous system, because the brain and spinal cord (neuraxis) are at a nearly 90-degree angle. For example, along the neuraxis, the cerebellum is caudal (posterior) to the cerebrum. Because of the angle, however, it may at first mistakenly appear that the cerebellum is "dorsal" to portions of the cerebrum (Figure 1.6B). Throughout this book, anterior and posterior refer to the locations along the neuraxis, as shown in Figure 1.6B. Note, however, that when describing location along the adult human torso, the terms

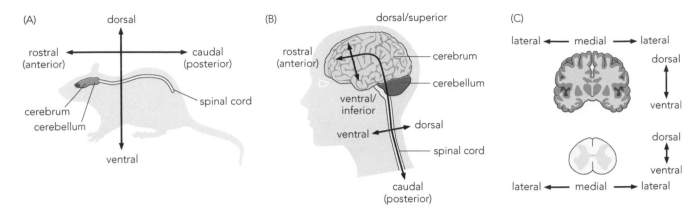

Figure 1.6 The nervous system is organized around three axes. (A) In four-legged animals such as mice, the rostral–caudal or anterior–posterior axis of the nervous system is easily seen as it extends from the region of the snout toward the tail. The dorsal–ventral axis extends from the back to the belly side of the animal. (B) These same axes are present in the adult human nervous system, but the curvature of the brain and spinal cord lead to a corresponding bending of the rostral–caudal (anterior–posterior) axis. (C) In the medial–lateral axis, those structures closest to the midline are called medial, while those further from the midline are designated lateral as shown in these sections from the brain (pink) and spinal cord (yellow).

anterior and posterior are often used differently, corresponding to dorsal and ventral. The **medial–lateral axis** is the third axis used to described structures relative to one another. Structures that are located closer to the midline are said to be medial, while those located further from the midline are called lateral (Figure 1.6C).

ORIGINS OF CNS AND PNS REGIONS

A variety of invertebrate and vertebrate animal models have been used to study neural development, each with its own advantages and disadvantages. Common animal models include fruit flies, worms, frogs, chicks, and mice. Many investigators focus on only one animal model, while some use two or more for comparative studies. Few researchers are fully versed in all of the developmental events of every animal model used, yet having a general idea of how the nervous system forms in different model systems can be extremely useful when reading the literature or when formulating questions to test in another model system.

Among the early structures formed during vertebrate neural development are the blastula, gastrula, neural plate, neural tube, and primary and secondary brain vesicles. Similar structures are found in many invertebrate models. Each structure forms at a specific time during embryogenesis in a given animal model. Because formation of these structures is common across many species, these developmental milestones are often used as a general means for comparing developmental progress in different animal models. Details on the induction of neural tissue and the origins of the neural plate, neural tube, and primary and secondary brain vesicles are provided in Chapters 2, 3, and 4.

The egg cell (**zygote**) begins to divide following fertilization, creating a group of cells called the **blastoderm**. The blastoderm lies above a hollow cavity and together the blastoderm and hollow cavity form a structure called the **blastula**. While the term blastula is often used for all embryos at this stage, more specific terms are used for a given species based on its morphological appearance. For example, blastula is the term used for amphibians, **blastocyst** is used for many mammals, while **blastodisc** is used for birds, fish, and some mammals (**Figure 1.7**). The blastula-stage embryo is organized around the animal and vegetal poles, with the animal pole being the region that gives rise to the nervous system and epidermis

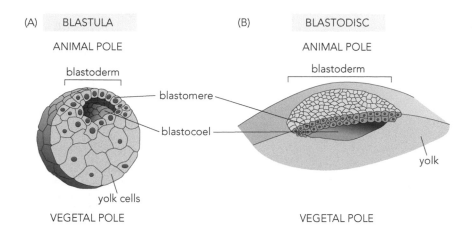

(A)   BLASTULA

ANIMAL POLE

blastoderm

blastomere

blastocoel

yolk cells

VEGETAL POLE

(B)   BLASTODISC

ANIMAL POLE

blastoderm

yolk

VEGETAL POLE

Figure 1.7 The blastula-stage embryo is used in studies of early neural development. Soon after fertilization, the egg cell divides, creating a group of cells that lies above a hollow cavity. The cells are called the blastoderm, while the cells and the hollow cavity together are often referred to as the blastula. However, in different animal models the morphology of these regions varies and more specific terms are applied. For example, in amphibians the ball-shaped structure is called a blastula (A), whereas in birds, fish, and humans, the structure is more flattened and is called a blastodisc (B).

(skin) and the vegetal pole being the site of origin for tissues associated with the gut. Blastula-stage embryos are used in many of the experiments described in Chapter 2.

Gastrulation is the process that begins as the cells of the blastula start to migrate through an indentation that forms on the outer surface of the blastula. As cells migrate though this indentation, the three primary germ layers are formed. The innermost layer becomes **endoderm**, the middle layer forms **mesoderm**, and the outermost layer forms the **ectoderm**. The ectoderm gives rise to both the neural tissue and epidermal (skin) tissue. The vertebrate CNS derives from neural ectoderm along the dorsal surface of the embryo, whereas the invertebrate CNS arises from the ventral ectoderm.

The Vertebrate Neural Plate Gives Rise to Central and Peripheral Structures

The vertebrate neural ectoderm begins as a flattened sheet of cells called the **neural plate** that extends along the anterior–posterior (rostral–caudal) body axis and is wider at the **cephalic** (head) end. Along the length of the neural plate, a central indentation forms called the **neural groove**. The lateral (outermost) edges of the neural plate then begin to curl upward to form the **neural folds**. As the neural folds continue to curve over they eventually contact one another, forming a **neural tube**. Thus, the former lateral edges of the neural plate become the dorsal surface of the neural tube, while the medial section becomes the ventral region. The neural tube lies below overlying epidermal ectoderm. The central lumen of the neural tube will later expand to form the ventricles of the brain and the narrow central canal of the spinal cord, all of which contain cerebral spinal fluid (**Figure 1.8**).

In the zebrafish (*Brachydanio rerio*), another popular animal model for neurodevelopmental studies, the hollow center of the neural tube does not form from the edges of the neural plate curling over. Instead, the neural plate first bends to form the **neural keel** and then the **neural rod**, both of which are solid structures lacking a central lumen. The cells at the center of the rod then migrate, leaving the hollow center of the neural tube.

Many of the neurons and glia of the vertebrate PNS originate from a group of cells that is unique to vertebrates. These cells, called the **neural crest cells** because they originate in the crest of the neural folds, migrate out of the dorsal neural tube, and later coalesce as PNS ganglia (Chapter 4). Other neurons and glia of the PNS form from thickened patches of ectoderm called **placodes** that arise in specified regions of the developing embryo (Chapter 5).

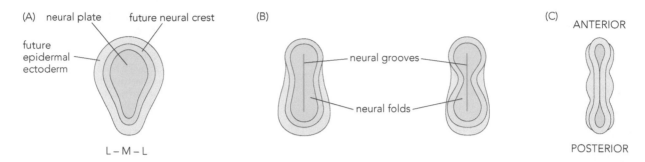

Figure 1.8 The nervous system arises from neural plate ectoderm. (A) The neural plate ectoderm, located on the dorsal surface of the embryo, is wider at the cephalic (head) region. (B) The lateral edges of the neural plate begin to curve upward, leading to the identification of neural folds and a central indentation called the neural groove. (C) The neural folds eventually curl over and contact one another, thus forming the neural tube (blue). Epidermal ectoderm (yellow) surrounds the neural tube. M, medial; L, lateral.

Future Vertebrate CNS Regions Are Identified at Early Stages of Neural Development

Soon after the neural tube closes, the anterior region of the neural tube expands and constricts at specific locations to form three primary brain vesicles. These vesicles are called the **prosencephalon** (forebrain), which is located at the most anterior (rostral) region of the neural tube, the **mesencephalon** (midbrain), and the **rhombencephalon** (hindbrain), which is located just anterior to the developing spinal cord (**Figure 1.9**A). As development continues, five secondary brain vesicles are formed. The prosencephalon forms two vesicles, the **telencephalon** and **diencephalon**, the mesencephalon remains as a single vesicle, and the rhombencephalon is divided into the **metencephalon** and **myelencephalon** (Figure 1.9B).

The five secondary vesicles correspond to the sites of origin for adult CNS structures. The telencephalon gives rise to cerebral cortex, hippocampus, basal ganglia, basal forebrain nuclei, and olfactory bulb. The diencephalon gives rise to structures that include the thalamus, hypothalamus, and the optic cup—the precursor of the retina that contains the sensory cells of the visual system. The mesencephalon gives rise to the midbrain tegmentum, or central gray matter, of the brainstem as well as the tectal regions (the superior and inferior colliculi) that are important relay centers for visual and auditory information, respectively. The metencephalon will ultimately form the cerebellum and pons, while the myelencephalon will form the medulla.

Within each vesicle, neuronal subtypes with specific cellular characteristics begin to develop. Cells begin as neural **precursors** or **progenitors**, meaning they have the capacity to become various neural subtypes. The cell fate adopted by any given cell is determined by the location, development stage, and available intrinsic and extrinsic signals. The signals that coordinate to regulate the formation of the different CNS regions along the anterior–posterior and dorsal–ventral axes are described in Chapters 3 and 4.

The Timing of Developmental Events Is Standardized in Many Vertebrates

The formation of the blastula, gastrula, neural plate, neural tube, and brain vesicles occurs at specific times in embryonic development in each of the animal models studied. While the sequence of developmental events is consistent across all vertebrate species, the actual time that these structures arise varies. Developmental age is reported as the number of hours, days, or weeks post fertilization or by staging criteria established for each species. The stages are based on various morphological criteria, including embryo length and the presence of key developmental features, such as the number of **somites** (the blocks of mesoderm that line either side of the

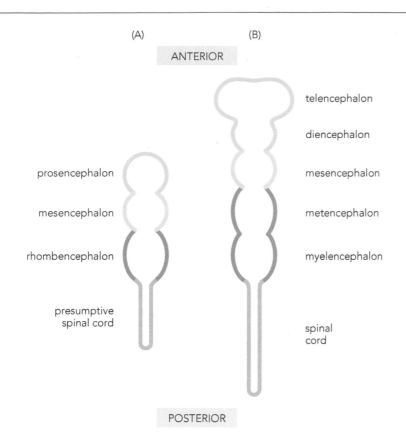

(A)   (B)

ANTERIOR

telencephalon

diencephalon

prosencephalon

mesencephalon

mesencephalon

metencephalon

rhombencephalon

myelencephalon

presumptive
spinal cord

spinal
cord

POSTERIOR

Figure 1.9 The neural tube forms primary and secondary brain vesicles. (A) The early-stage neural tube forms three primary brain vesicles designated the prosencephalon, mesencephalon, and rhombencephalon. (B) The primary vesicles further divide into the five secondary brain vesicles designated the telencephalon, diencephalon, mesencephalon, metencephalon, and myelencephalon. Each of the vesicles is the site of origin for different brain structures.

neuraxis). Such staging corrects for any variations that might arise from genetic or environmental influences.

Human development is often referred to in terms of weeks of gestation or Carnegie stages—stages first defined in the early twentieth century by Franklin Mall and George Streeter, both of whom worked at the Carnegie Institute in Washington, DC. Mice are described in terms of embryonic (E) days or days post coital (d.p.c.) and are often staged using criteria established by Karl Theiler. Development of chick embryos is reported based on the hours or days post fertilization or by the duration of incubation. Chick embryos are staged using criteria published by Viktor Hamburger and Howard Hamilton.

The development of the chick embryo begins *in utero* and continues after the egg is laid (about 20 hours after fertilization). *In utero* development involves the formation of the blastoderm at 10–11 hours after fertilization. The onset of gastrulation begins about the time the egg is laid and subsequent developmental events are easily monitored by cutting a small hole, or window, in the egg. Thus, the chick embryo is an extremely useful model for viewing neural development. Another useful feature of the chick egg is that development of a fertilized egg can be halted for several days if the eggs are maintained at room temperature. Development resumes when the eggs are placed in an incubator at 37.5°C. **Figure 1.10** shows several of the stages first published by Hamburger and Hamilton in 1951. At these stages of development, hours refer to the number of hours the eggs were incubated, rather than the hours post fertilization. The first somites and the head neural folds are visible at Hamburger and Hamilton (HH) stage 7 (23–26 hours of incubation). In the chick, the three primary brain vesicles are detected at HH10 (33–38 hours), and the five secondary vesicles are observed after 40–45 hours of incubation (HH11). The embryo turns to the side beginning at HH13 (48–52 hours) and the enlargement and refinement of the brain vesicles is easily viewed in the translucent embryo from HH14–21 (about 2–3 days after incubation).

Figure 1.10 Chick development is easily viewed throughout embryogenesis. Images of developing chick embryos reveal some of the key events in neural development. Numbers in the corners of the images indicate the stage of development as determined by Hamburger and Hamilton (HH). The neural folds are first identified at 23–26 hours (HH stage 7, arrow). The primary brain vesicles are visible at HH10 (arrow) and the secondary vesicles at HH11 (arrow). The embryo begins to turn to the side at HH13 and further development of the brain regions is observed through the embryo's translucent body until HH21 (embryonic day 4). [From Hamburger V & Hamilton HL [1992] *Dev Dyn* 195:231–272.]

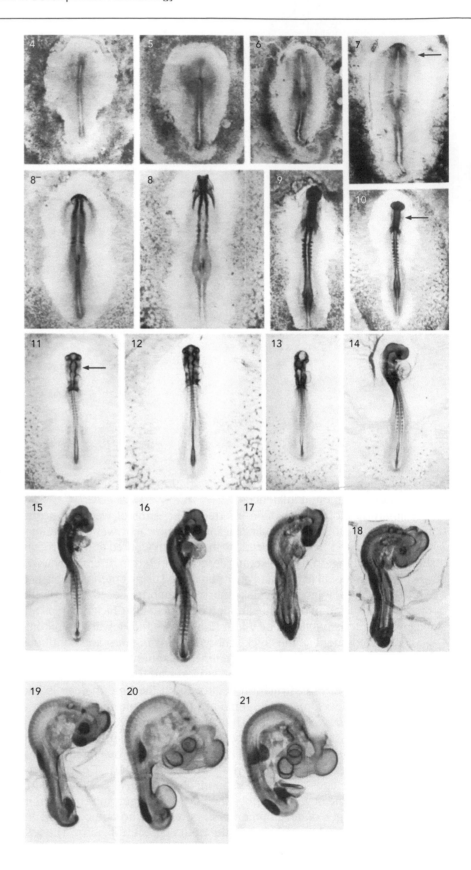

Figure 1.11 compares the development in humans and mice using Carnegie and Theiler criteria. In humans, the blastula is detected in the uterus at 4 days after fertilization (Carnegie stage 3, CS3) and gastrulation begins at day 16 (CS7). The neural plate and neural folds become evident at day 18 (CS8). The three primary vesicles form during the third and fourth

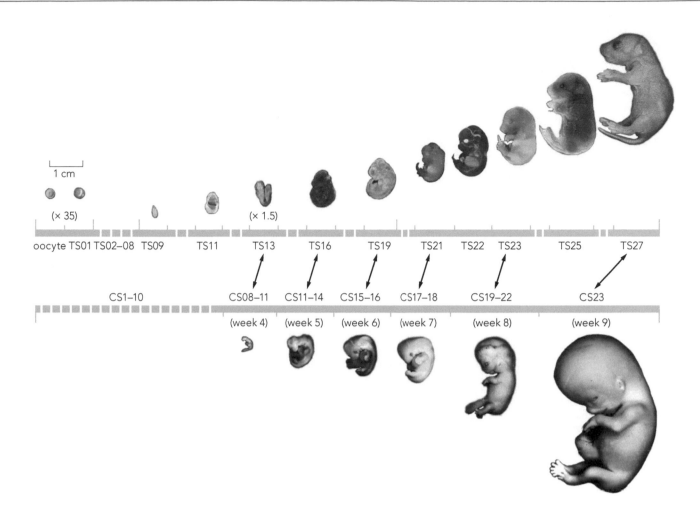

Figure 1.11 Comparison of stages of mouse and human embryonic development. Morphological criteria are used to identify the stages of embryonic development. Mouse development is staged by the criteria of Theiler, whereas human development is marked by Carnegie stages. As shown in the diagram, Theiler stage 13 (TS13) is equivalent to Carnegie stage 11 (CS11). At these stages, the three primary vesicles are observed. TS16 is equivalent to CS11–14, the stages when the five secondary brain vesicles are formed. TS27 is a newborn mouse, which is at a similar stage of development as a 9-week human (CS23). Arrows indicate times at which human and mouse development are at a similar stage. [From Xue L, Cai J-Y, Ma J et al. [2013] *BMC Genomics* 14:568.]

week of gestation (days 20–24; CS8–11) and the five secondary vesicles become visible during the fifth week of gestation (CS14).

In mice, the blastula stage embryo is formed 3 to 4 days post coitus (d.p.c.), corresponding to Theiler stages 4–5 (TS04–05). Gastrulation begins at 6.5–7.5 d.p.c. (TS09) and the neural plate forms at 7.5–8 d.p.c. (TS11). The three primary vesicles are visible at 8–8.5 days of embryogenesis (TS12–13) and the five secondary vesicles are detected at days 9–10 (TS15–16). The development of a newborn mouse (TS27; 19–20 d.p.c.) is similar to that of a 9-week human (CS23). Human development continues for about 38 weeks (9 months).

Zebrafish are another popular vertebrate animal model for studies of neural development. The stages of development in the zebrafish have been documented by Monte Westerfield, Charles Kimmel, and colleagues. The cells of the zygote begin to divide about 40 minutes after fertilization and are easily viewed above the yolk. The zebrafish blastula forms at a little over 2 hours post fertilization, when 128 cells, the blastomeres, are present. The blastula-stage embryo continues to develop through multiple stages during the first 5 hours after fertilization and the blastoderm is identifiable a little over 4.5 hours post fertilization (h.p.f.; **Figure 1.12**). Epiboly and gastrulation, the movement and thinning of cell layers, begins around

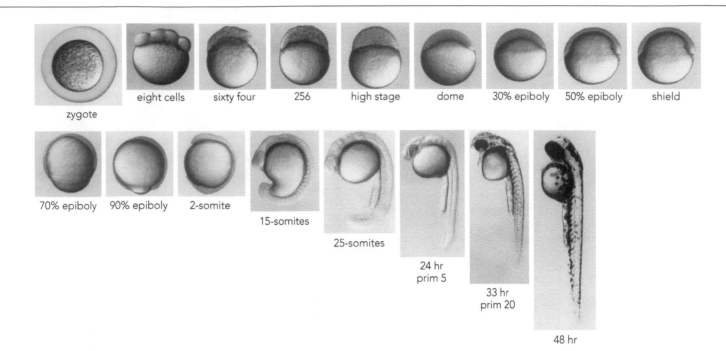

Figure 1.12 Examples of zebrafish development from zygote to hatching. Cells of the zebrafish zygote begin to divide about 40 minutes after fertilization. The resulting blastomeres continue to divide as the blastodisc forms above the yolk, as seen in the eight-cell stage shown in this figure. The blastula stage of development begins at the 128–256 cell stage (2.25–2.5 hours post fertilization, h.p.f.) then progresses through multiple stages. The high stage, for example, indicates the period that the blastodisc is located "high" on the yolk. The blastula stage continues until a little over 4 h.p.f. (dome stage). By about 4.5 hours, epiboly can be measured, indicating the percentage of the yolk surface that is surrounded by the embryo. At 50% epiboly gastrulation begins and at 90% epiboly, the neural plate is present. Somites are first visible by 10 h.p.f., and the divisions of the brain vesicles are first observed at the 14–16 somite stage. The embryo begins to straighten away from the yolk at 24 h.p.f. (prim 5), when development is measured by the myotome number reached by the tip of the primoridum (prim) of the lateral line organ. By 48 h.p.f., the embryo hatches. [From Westerfield M [1993] *The Zebrafish Handbook*, 2nd ed, University of Oregon Press.]

5 h.p.f. At these stages, the embryo begins to curl around the central yolk. Development is measured in terms of the percentage of **epiboly**, indicating the percent of the yolk that is surrounded by the blastoderm. At 50% epiboly, gastrulation begins. At just over 6 hours, the embryonic shield, a key structure in the process of neural induction (Chapter 2), is present. At 90% epiboly, 9 h.p.f., the neural plate is visible. At the completion of epiboly and gastrulation, somites are detected (10 h.p.f.), and by 16 hours, 14 somites and the three primary brain vesicles are observed. At 24 h.p.f., the five secondary brain vesicles are present. At the same time, the embryo begins to straighten away from the yolk sac. The embryos are now measured by indicating which myotome (the segment of the somite that later gives rise to muscle) that the tip of primordium (prim) of the lateral line organ (a sensory organ found in aquatic vertebrates) reaches. Thus, prim 5 indicates that the tip of the primordium of the lateral line reaches the fifth myotome. The embryo hatches at about 48 hours and enters the larval stages by 72 h.p.f. The larval stages last up to 29 days. Juvenile zebrafish form at day 30 and become adults by day 90.

Frogs are also frequently used in studies of neural development, particularly *Xenopus laevis*. These frogs provide many advantages for researchers, including the ability to induce frogs to produce eggs year-round and the one-year cycle needed to complete development from a fertilized egg to an adult frog. The description that follows refers to the timing of developmental events in *Xenopus* based on the time post fertilization and the staging criteria of Pieter Nieuwkoop and Jacob Faber. The timing of these events may be slightly different in other frogs. In *Xenopus*, blastula-stage embryos are observed by 4 hours after fertilization (stage 7). Gastrulation

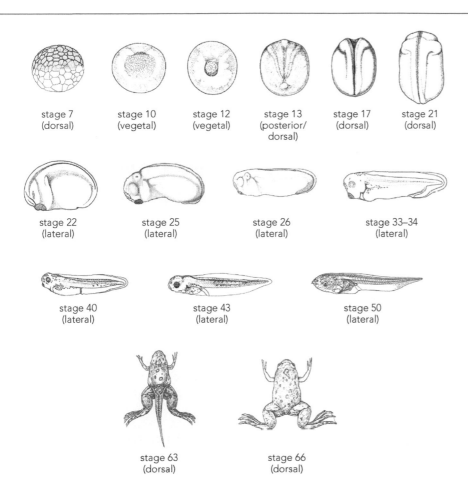

stage 7 (dorsal)

stage 10 (vegetal)

stage 12 (vegetal)

stage 13 (posterior/ dorsal)

stage 17 (dorsal)

stage 21 (dorsal)

stage 22 (lateral)

stage 25 (lateral)

stage 26 (lateral)

stage 33–34 (lateral)

stage 40 (lateral)

stage 43 (lateral)

stage 50 (lateral)

stage 63 (dorsal)

stage 66 (dorsal)

Figure 1.13 Development of the frog from egg through adult stages. Examples from the criteria established by Nieuwkoop and Faber in 1967 identify some of the key stages of development in the frog *Xenopus laevis*. Images show embryos from dorsal, posterior-dorsal, and lateral views. By stage 7 (4 hours after fertilization), the blastula-stage embryo is present. Gastrulation begins 7–10 hours after fertilization (stages 10–12). The neurula-stage embryo, when the neural structures first form, continues from 12 to 22.5 hours after fertilization (stages 13–21). The embryo then enters the tailbud stage at 24–36 hours post fertilization (stages 22–44). The brain vesicles are visible beginning 24 hours after fertilization (stage 22). The embryo develops into a tadpole by 96 hours (stages 45–50), before undergoing metamorphosis (stages 51–65). An adult frog is formed approximately 12 months later (stage 66).

begins approximately 7–10 h.p.f. (stages 10–12) and leads to the formation of the neurula-stage embryo, the stage when the neural tissue begins to form (12 and 13.5 h.p.f.; **Figure 1.13**). The neurula stage (stages 13–21) continues until the early tailbud-stage embryo forms 24–32 h.p.f. (stages 22–28). Primary brain vesicles appear around 24 h.p.f. (stage 22) and the five secondary vesicles about 32 h.p.f. (stage 28). The embryo develops into a tadpole by 96 h.p.f. (stages 45–50), before undergoing metamorphosis (stages 51–65) and reaching the adult stage approximately 12 months later (stage 66).

Anatomical Regions and the Timing of Developmental Events Are Mapped in Invertebrate Nervous Systems

Several invertebrate animal models, particularly the fruit fly *Drosophila melanogaster* and the round worm *Caenorhabditis elegans* (*C. elegans*), were used in pivotal studies described in subsequent chapters and remain important animal models for understanding mechanisms governing neuro-development. *Drosophila* became a popular animal model for research in areas of genetics and developmental biology beginning with the pioneering work of Thomas Hunt Morgan in the early twentieth century. The work of Seymour Benzer and colleagues in the 1960s helped make *Drosophila* an animal model of ongoing interest to developmental neurobiologists. *C. elegans* also became a popular model beginning in the 1960s, largely through the work of Sydney Brenner's lab. These animal models are easily bred, have a short life cycle from the time of fertilization to the adult form, and exhibit naturally occurring and experimentally induced mutations that provide a means to test how specific genes regulate development of the various cells within the nervous system. Like the vertebrate nervous system, the nervous systems of invertebrates arises from the ectoderm. However, there are significant differences in how and where neural structures arise in invertebrate animal models.

The *Drosophila* CNS and PNS Arise from Distinct Areas of Ectoderm

When *Drosophila* are maintained at 25°C, embryogenesis occurs over a period of approximately 22 hours and adult flies are formed within 9–12 days (**Figure 1.14**). This allows for the generation of large numbers of animals in a short period of time. Staging of *Drosophila* embryos is often noted using the criteria of Volker Hartenstein and Jose Campos-Ortega.

In *Drosophila*, the cytoplasm of the fertilized egg contains many nuclei that divide rapidly (stages 1–2) prior to migrating to the outer cortex of the cell to form a syncytial blastoderm (stage 4). Each nucleus is then surrounded by a cell membrane to form the cellular blastoderm (stage 5). Gastrulation (stages 6–7) occurs within 3 hours of fertilization, and by the end of the first day (stages 16–17), the embryo hatches to enter the first larval stage called the first instar. *Drosophila* larvae progress through three instar stages, with each stage lasting 1 to 2 days (Figure 1.14). An epidermal-derived hardened shell called a cuticle surrounds each instar stage larva. At the end of each instar stage, the cuticle sheds to accommodate the growth of the larva and a new, larger cuticle is produced. Following the third instar stage, the cuticle contributes to the extracellular case that surrounds the prepupa. During the pupal stage, metamorphosis occurs. Adult tissues that are derived from ectoderm, such as the nervous system, arise from pockets of epithelium formed during the larval stages. These pockets of tissue, called **imaginal discs**, attach to the inside of the larval epidermis, and later evert during metamorphosis to form adult structures of the head, thorax, legs, and wings. Unlike most other larval-stage organs, the components of the gut and nervous system persist in the adult fly.

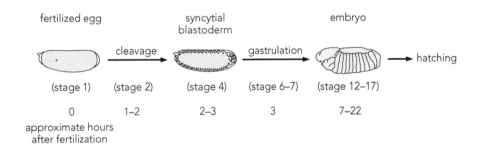

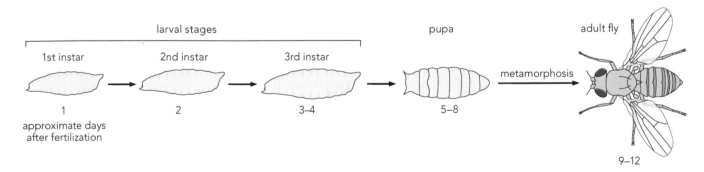

Figure 1.14 Development of the fruit fly *Drosophila melanogaster.* The fertilized *Drosophila* egg (stage 1) undergoes cleavage (stage 2) and forms a syncytial blastoderm 2–3 hours after fertilization followed by a cellular blastoderm (stages 4 and 5, not shown). The embryo then begins gastrulation (stages 6–7) and forms the late-stage embryo 7–22 hours after fertilization (stages 12–17). The embryo enters the first larval stage by 24 hours after fertilization and continues through three larval instar stages before forming a pupa at 5–8 days post fertilization. Metamorphosis takes place and the adult fly emerges 9–12 days after fertilization.

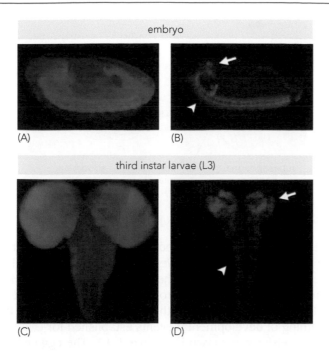

embryo

(A)   (B)

third instar larvae (L3)

(C)   (D)

Figure 1.15 The *Drosophila* central nervous system (CNS) forms in the embryo and continues to develop during the larval stages. The outline of the *Drosophila* embryo (A) and third instar larval stage CNS (C) are shown in blue. (B) CNS structures of the embryo are identified in red. In this panel, the arrow indicates the brain region and the arrowhead indicates the ventral nerve cord. (D) The brain (arrow) and ventral nerve cord (arrowhead) continue to develop and enlarge during the third instar larval stage. In this panel, the areas of red reveal sites of synaptic connections. [From Diaper DC & Hirth F [2014] In *Brain Development Methods and Protocols* [SG Sprecher ed], pp. 3–17. Humana Press.]

In *Drosophila*, the nervous system arises from ventral ectoderm (the ventral neurogenic ectoderm), rather than the dorsal ectoderm as in vertebrates. This ectoderm gives rise to the neuroblasts (beginning at stage 9, about 4 hours after fertilization) that form the adult brain and ventral nerve cord, a structure with functions similar to the vertebrate spinal cord (**Figure 1.15**). During development, **neuroblasts** segregate from the surrounding ectoderm, then move inside the embryo along the anteroposterior axis. As development progresses through the larval stages, the cells of the nervous system proliferate. The cells then begin to differentiate in the pupal stage and the definitive CNS regions form.

In the adult, the anterior-most region of the CNS is divided further, reminiscent of subdivisions found in the mammalian brain (**Figure 1.16**). These anterior brain regions are formed from three pairs of ganglia, with each pair controlling specific functions: the protocerebrum (forebrain, largely associated with visual regions), the deutocerebrum (midbrain, largely associated with sensory information from the antennae), and the tritocerebrum (hindbrain, primarily integrates information from the protocerebrum and deutocerebrum; linked to the ventral nerve cord). The more posterior ganglia of the CNS are part of the ventral nerve cord (Figure 1.16). These include the subesophageal ganglia (associated with head and neck regions), the thoracic ganglia (associated with leg and wing structures), and the abdominal ganglia (associated with abdominal structures).

Many of the neurons of the PNS arise from **sensory organ progenitors (SOPs)** located in the surface ectoderm. The SOPs ultimately develop into the mechanosensory, chemosensory, and chordotonal organs of the fly. PNS neurons generally develop later than the cells of the CNS. Mechanisms regulating the formation of various CNS and PNS structures in *Drosophila* are described in Chapters 2, 3, 4, and 6.

There are four types of glial cells in *Drosophila* that are designated cortex, surface, neuropil, and peripheral glia. The cortex glia are most like vertebrate astrocytes, the neuropil glia are similar to oligodendrocytes, and the peripheral glia function like Schwann cells. The cortex, surface, and neuropil glia also function like the microglia found in the vertebrate CNS.

Figure 1.16 The *Drosophila* central nervous system is comprised of pairs of ganglia. The three pairs of ganglia that make up the *Drosophila* brain are divided into protocerebrum (forebrain), deutocerebrum (midbrain), and tritocerebrum (hindbrain). These ganglia connect to the ventral nerve cord comprised of subesophageal, thoracic, and abdominal ganglia. The nervous system (blue) is shown relative to the digestive (green) and circulatory (yellow) systems. [Adapted from *Agricultural and Life Sciences, General Entomology*, North Carolina State University.]

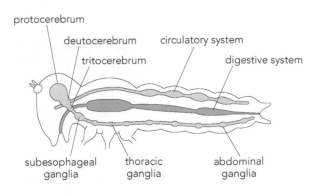

Cell Lineages Can Be Mapped in *C. Elegans*

Most adult *C. elegans* are hermaphrodites, with a smaller percentage being male. The adult hermaphrodite *C. elegans* has a total of 959 cells of which 302 are neurons and 56 are glial. The lineage of each cell has been documented through serial electron micrographs and by following the progeny and fate of individual cells through the translucent body of the tiny worm.

The timing of developmental events established for *C. elegans* maintained at 22° Celsius are shown in **Figure 1.17**. The egg cell is fertilized inside the worm and cells begin to divide in a specific sequence beginning about 40 minutes after fertilization. The eggs are laid at the gastrulation stage, approximately 150 minutes after fertilization, when there are about 26 cells present. Gastrulation is initiated as cells move inward to form the gut and muscle tissues, while the **hypodermis**, the equivalent of ectoderm in vertebrates, remains as the outermost layer. Cells of the hypodermis that subsequently move to the inside of the embryo give rise to most of the neurons. The remaining cells of the hypodermis migrate over the surface of the embryo to form the epidermis. Gastrulation continues until the number of cells increase to 421 (about 5.5 hours after the first cleavage). During the final stages of embryogenesis, the shape of the embryo changes from a spherical structure to the elongated shape of the adult. As elongation continues, the worm begins to fold over first 1.5 times (tadpole stage), then 2 times (plum stage), and eventually 3 times (pretzel stage). Hatching occurs about 14 hours after fertilization, when there are 558 cells. The resulting larva then progresses through the four larval stages (L1–L4) before forming an adult worm. During the larval stages, additional neurons are produced, with the majority born during the late L1 stage. The length of time in each larval stage depends, in part, on environmental conditions such as temperature, food supply, and population density. Under favorable conditions, it takes less than 2.5 days for a *C. elegans* to complete the life cycle from fertilized egg to adult worm.

The cell fate options available to a particular cell in *C. elegans* are established early in development with the asymmetric division of the zygote into the AB and P founder cells. A series of cell divisions then results in a total of six founder cells that are designated AB, P, E, MS, C, and D. Each founder cell gives rise to progeny in a specific pattern. The initial division of the zygote yields an AB cell located at the anterior pole and the P_1 cell at the posterior pole of the embryo (**Figure 1.18**A). Next, the AB cell divides to produce two daughter cells, the anterior AB.a and the posterior AB.p. Shortly after the AB cell divides, the P_1 cell divides, producing the more posterior P_2 cell and the more anterior EMS cell (Figure 1.18A).

As more AB cells are generated, the location of a cell relative to its sister cell is specified along the anterior–posterior axis. Thus, the AB.al cell is the "left-handed" daughter cell of an anterior AB cell, whereas AB.pr is the "right-handed" daughter of the posterior AB daughter cell (Figure 1.18A). The EMS cell divides to produce E and MS cells, whereas P_2 produces C

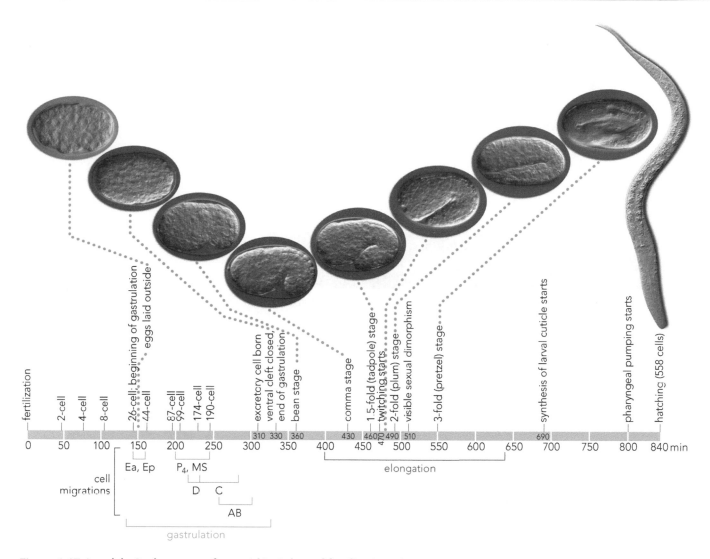

Figure 1.17 An adult *C. elegans* can form within 3 days of fertilization. This timeline of developmental events is shown for *C. elegans* maintained at 22° Celsius. The egg is fertilized and begins to divide in the worm. Gastrulation begins when the egg is laid, about 150 minutes after fertilization, when 26 cells are present. Gastrulation continues until 330 minutes (5.5 hours) after fertilization, when 421 cells are present. During gastrulation, different founder cells migrate, as indicated below the timeline. During elongation, the worm folds over itself 1.5 times (tadpole stage), then 2 times (plum stage), and finally 3 times (pretzel stage) before hatching about 14 hours (840 minutes) after fertilization, when there are 558 cells. After progressing through four larval stages (L1–L4) an adult worm emerges about 56 hours after fertilization. [Adapted from *WormAtlas*. www.wormatlas.org]

and P$_3$. P$_3$ then divides, producing D and P$_4$, the cell that establishes the germ line. Thus, the P$_1$ founder cell functions like a stem cell in that it gives rise to both somatic and germ cells.

Each founder cell produces progeny that go on to contribute to specific tissues. However, not all cells of a given body system arise from a single founder cell type. For example, most of the 302 neurons arise from descendants of the AB founder cell, but some arise from the MS and C cells that descend from the P$_1$ founder cell. In all cases, however, the individual neuronal types always arise from the same precursor and are always found in the same location in the body.

C. elegans has a somatic and pharyngeal nervous system. The somatic nervous system contains 282 neurons found in the head and tail ganglia as well as in the ventral and dorsal nerve cords (Figure 1.18B). These regions contain sensory, motor, and interneurons. The head region also contains numerous sense organs called sensilla that are comprised of free nerve endings and glial sheath and socket cells. The pharyngeal nervous system

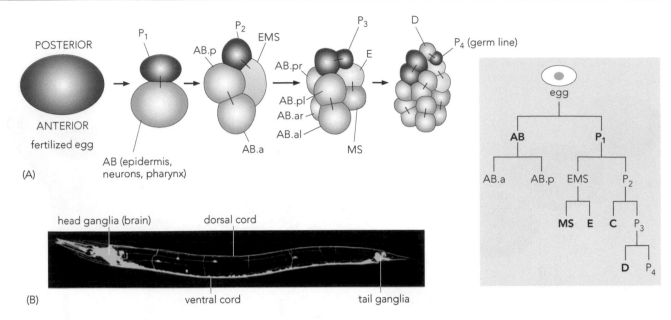

Figure 1.18 Cells in *C. elegans* divide in a precise order. (A) The asymmetric division of the fertilized egg leads to formation of a larger AB founder cell at the anterior pole of the developing embryo and a smaller P_1 founder cell at the posterior pole. The cells continue to divide until a total of six founder cells are produced (AB, P, E, MS, C, and D; bold letters, inset). Most neurons in *C. elegans* arise from the AB founder cell, though some arise from descendants of MS and C founder cells. The AB cell (green) gives rise to the AB.a cell at a more anterior site and the AB.p at a more posterior site. These cells divide to produce additional daughter cells such as AB.al and AB.ar, the left- and right-handed daughter cells of AB.a. The P_1 cell (red) establishes the germ line (red cells) and various somatic cells (yellow, orange, purple, and blue). (B) Regions of the worm nervous system are stained with green fluorescent protein to outline regions of the brain (head ganglia), the tail ganglia, and the dorsal and ventral nerve cords. [(A), Adapted from Alberts B, Johnson A, Lewis J et al. [2008] *Molecular Biology of the Cell*, 5th ed. Garland Science. (B), Courtesy of Harold Hutter.]

contains 20 neurons. The pharynx in *C. elegans* is segregated from the rest of the tissues of the body by a unique basement membrane and functions largely independent of the other parts of the worm.

There are 56 glial cells in *C. elegans* classified into three categories: sheath, socket, and glial-like nerve ring (GLR). The 24 sheath and 26 socket glia are derived from ectoderm, whereas the six GLR cells are derived from mesoderm. While the functions of glia in *C. elegans* are not as well characterized as in other animal models, they appear to assist in synaptic signaling and play roles in the development, maintenance, and activity of their associated synapses. The GLR cells seem to be specifically associated with signaling that regulates movement. Unlike vertebrates and *Drosophila*, axons in *C. elegans* are not myelinated, so glia do not wrap around axons to speed neural conduction.

Despite the variations in anatomy and cellular organization among the different animal models, many of the genes and signaling pathways are conserved across species, allowing discoveries in one animal model to impact discoveries in another. This is particularly helpful when a technique is more readily applied to a simpler invertebrate animal model than a more complex vertebrate model.

GENE REGULATION IN THE DEVELOPING NERVOUS SYSTEM

In all animals, each neuron must selectively produce specific cellular components, such as neurotransmitters, ion channels, cell surface receptors, cytoskeletal elements, and other proteins. The regulated production of these specialized proteins gives individual neurons their unique characteristics and allows them to perform specific functions in the nervous system.

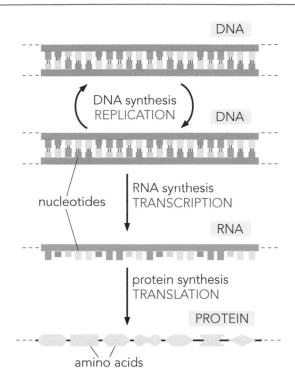

DNA

DNA synthesis
REPLICATION

DNA

nucleotides

RNA synthesis
TRANSCRIPTION

RNA

protein synthesis
TRANSLATION

PROTEIN

amino acids

Figure 1.19 Gene regulation determines which proteins are produced in a cell. DNA synthesis (replication) takes place in the nucleus of the cell, where the template strand of DNA synthesizes a complementary strand of DNA. A DNA template strand can also initiate synthesis (transcription) of a complementary strand of RNA. During protein synthesis (translation) the sequence of nucleotides in a strand of messenger RNA determines the order of amino acids and therefore the resulting protein structure. [Adapted from Alberts B, Johnson A, Lewis J et al. [2015] *Molecular Biology of the Cell*, 6th ed. Garland Science.]

Neurons, like other cells, produce only the proteins required at a particular stage of development. To selectively produce these proteins, individual genes must be turned on (expressed) or turned off (repressed) at the correct stage of development. The process of turning genes on or off is called **gene regulation**.

A **gene** is a segment of DNA, the double-stranded, helical molecule synthesized in the nucleus of all cells. While DNA is the same in every cell of the body, the genes that are expressed in an individual cell at a particular stage of development determine which messenger RNA (mRNA) nucleotides are transcribed from the DNA template and therefore which amino acids are translated into proteins (**Figure 1.19**). Among the numerous proteins a cell produces are **transcription factors**—proteins that bind to specific DNA sequences to enhance or suppress gene expression. **Box 1.2** introduces one family of transcription factors to illustrate an example of the complexity and precision required to construct a nervous system.

In many instances, gene expression and protein production are influenced by extracellular signals. The extracellular signal is typically a **ligand** that binds to a cell surface **receptor** protein to initiate intracellular signal transduction pathways. The structure of each ligand and receptor is unique so that a given ligand only binds to corresponding receptors. This allows for binding specificity. Binding specificity combined with the ability of cells to regulate the expression of the multitude of ligands and receptors ensure that specific signaling pathways are only available to a cell when needed. Thus, cells only respond to required signals and ignore other signals that may be present at that time.

Cell signaling or **signal transduction** is the process by which signals originating outside a cell are conveyed to cytoplasmic components or the nucleus to influence cell behavior. Because the ligand is often thought of as the first messenger in a signal transduction pathway, the subsequent intracellular events are called second messenger pathways. The activation of various signal transduction pathways regulates cellular events such as survival, death, growth, differentiation, movement, and intracellular communication.

Box 1.2 The Many Roles of the Sox Family of Transcription Factors throughout Neural Development

Transcription factors are the proteins responsible for regulating gene activity—they bind to specific sequences of DNA to activate or repress target genes. A given transcription factor may act alone or with binding partners, including other transcription factors. In many cases, transcription factors interact through elaborate networks to determine whether a gene is expressed at a specific developmental stage.

Among the transcription factor families important during neural development are members of the Sox gene family. Sox genes are HMG (high mobility group) box genes related to the founding member *Sry* (sex-determining region of Y). The name Sox is derived from Sry HMB box. In mice and humans, 20 Sox genes have been identified, while eight are known in *Drosophila*. Sox transcription factors are important in nearly every cell throughout the body, from the earliest stages of development through adulthood. Many Sox genes are essential for neural development, regulating everything from cell proliferation, to the specification of a cell as a neural progenitor, to the differentiation of cell-specific characteristics.

In many cases, the sequential expression of Sox transcription factors in the nervous system influences whether a cell continues to proliferate or begins to differentiate into a neuron or glial cell. For example, Sox2 activates genes that promote cell proliferation while at the same time repressing genes that specify a cell to become a neural progenitor—a cell with the potential to become a neuronal or glial cell. Sox3 then promotes the expression of genes associated with neural progenitors, but also suppresses genes that would drive a cell to a specific neural or glial fate too early. Sox11 becomes active when it is time for cell-type specific genes to be expressed. Sox11 therefore activates neuronal genes and suppresses progenitor genes. Later in development, Sox10 promotes the expression of genes needed to drive a cell to become an oligodendrocyte, while repressing astrocyte-specific genes.

However, Sox gene networks are immensely more complex than sequentially turning on and off different family members. Instead, multiple signals converge to influence the activity of a single Sox transcription factor and refine its transcriptional activity. For example, the activity of a given Sox transcription factor may be modified by available binding partners and epigenetic (non-genetic) changes such as methylation that influence DNA accessibility. Sox transcription factors may compete for DNA binding sites or directly inhibit another signal to shape how that signal influences development of a particular region of the nervous system.

Recognizing that the activity of a single member of the Sox gene family can be influenced by numerous overlapping cellular events helps one appreciate the precision needed to generate the many different cell types required in a functional nervous system.

Figure 1.20 outlines an example of a signal transduction pathway, or cascade, where an extracellular signal (ligand) binds to a cell surface receptor. Once the ligand binds to the receptor, subsequent signaling molecules are activated inside the cell. There are often several sequential signaling molecules influenced before the final cellular response is achieved. A signal is said to activate a target **downstream** when it influences the next molecule in the signal transduction pathway. The signal transduction pathway eventually regulates effector proteins that serve a variety of different cellular functions. Common effector proteins are ion channels, metabolic enzymes, cytoskeletal proteins, and gene regulatory proteins (Figure 1.20). Several examples of specific signal transduction pathways important during neural development are detailed in subsequent chapters including those initiated by bone morphogenetic proteins (BMPs), fibroblast growth factors (FGFs), Sonic hedgehog (Shh), Wnts, agrin, and the neurotrophins.

Many signaling pathways initiate what is called a MAP (mitogen activated protein) kinase pathway that, in the nervous system, influences aspects of cell survival, cell differentiation, or synaptic plasticity, the process by which synaptic strength is altered. The MAP kinase pathways involve a series of phosphorylation events that are initiated after the initial binding of the ligand and the recruitment of adaptor proteins to the receptor. Adaptor proteins can activate Ras (rat sarcoma), a small GTPase critical

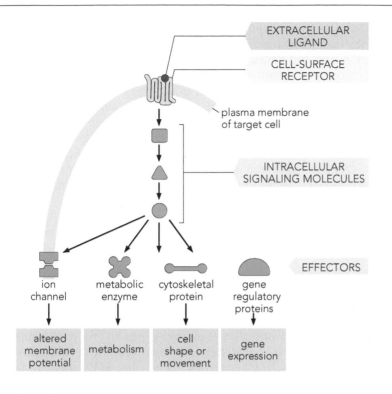

Figure 1.20 Signal transduction pathways transfer extracellular information to the cell. When an extracellular signal (ligand) binds to its corresponding receptor on the cell surface, one of many intracellular signal transduction cascades can be initiated. Each step in the cascade stimulates the next molecule in the pathway until an effector protein is influenced. Examples of effector proteins include ion channels, such as those that alter a neuron's membrane potential, metabolic enzymes that impact cellular metabolism, cytoskeletal proteins that influence cell shape and movement, and gene regulatory proteins, such as transcription factors, that influence whether a gene is expressed or repressed. [Adapted from Luo L [2016] *Principles of Neurobiology*. 1st ed. Garland Science.]

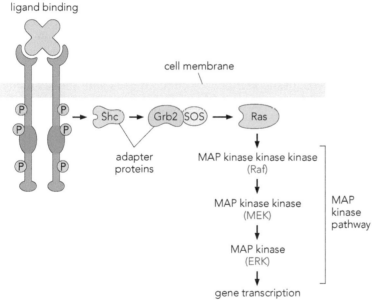

Figure 1.21 MAP kinase pathways involve a series of phosphorylation events. Upon binding of a ligand to a tyrosine kinase receptor, the adaptor proteins Shc, Grb2, and SOS activate the small GTPase called Ras which then phosphorylates and activates a MAP kinase kinase kinase (MAP3K) called Raf. Raf in tun phosphorylates MEK which then phosphorylates Erk. Erk influences gene transcription in the nucleus. Erk, extracellular signal-regulated kinase; Grb2, growth factor receptor-bound protein 2; MAP, mitogen-activated protein; MEK, MAP kinase/Erk kinase; Raf, rapidly accelerated fibrosarcoma kinase; Shc, C-terminal Src-homology-2 domain; SOS, Son of sevenless. [Adapted from Reichardt LF [2006] *Philos Trans R Soc Lond B Biol Sci* 361(1473): 1545–1564.]

in many signal transduction pathways. The first step after Ras activation involves what is called a MAP kinase kinase kinase, abbreviated MAP-KKK or MAP3K that in turn phosphorylates a MAP kinase kinase (MAPKK, MAP2K) that then activates the MAP kinase (MAPK) that phosphorylates transcription factors to a influence gene regulation. For example, when a nerve growth factor binds to a receptor tyrosine kinase, adaptor proteins are recruited that lead to activation of Ras and the subsequent phosphorylation of Raf (rapidly accelerated fibrosarcoma kinase; the MAPKKK in this example) that then leads to phosphorylation of MEK (MAP kinase/Erk, the MAPKK), which then activates a MAPK called Erk (extracellular signal-related kinase). Erk travels to the nucleus to influence the transcription of responsive genes thereby mediating the desired effect, such as cell survival or differentiation (**Figure 1.21**). This is called the MAPK–Erk or Ras–Erk pathway or the Ras/Raf/MEK/Erk pathway.

There are many examples of MAP kinase pathways in this book and other textbooks, and students may wonder why learning about kinase kinase kinases and kinase kinases that activate kinases is necessary. As seen in subsequent chapters, these pathways are critical for normal neural development and any experimental or naturally occurring disruption can cause altered neural function. The importance is also highlighted by what are now often called RASopathies, human neurodevelopmental disorders caused by genetic mutations that interfere with Ras/MAPK signaling. RASopathies include neurofibromatosis 1 (NF-1, see **Box 6.2**), Noonan Syndrome, Costello Syndrome, cardio-facio-cutaneous syndrome, and some forms of epilepsy. While the individual RASopathies are rare, combined they have a significant global impact.

Experimental Techniques Are Used to Label Genes and Proteins in the Developing Nervous System

Because each cell subtype in the nervous system expresses a unique set of genes and proteins, researchers have developed several techniques to identify where and when these molecules are expressed during development and in adulthood. Among the techniques are those that use microscopy to identify the distribution of labeled genes and proteins in tissues or individual cells. These approaches not only reveal the cellular distribution of genes and proteins, but also provide a way to mark particular cells and track them over the course of development. This has been especially helpful because the outward morphological appearance of embryonic neurons is often homogeneous, making it difficult, or impossible, to identify a cell with certainty following any sort of experimental manipulation.

To visualize gene expression in neural tissues, scientists use ***in situ hybridization***. With this technique, mRNA is visualized by incubating whole embryos, tissue sections, or cultured cells with probes made up of a DNA nucleotide sequence of interest. These probes are labeled with a radioactive or fluorescent marker so that when the DNA probe binds to the corresponding mRNA sequence, the cells expressing that mRNA can be visualized. Identifying gene expression patterns often provides insight into the putative function of that gene in the cell, while also providing a labeling method to track changes in gene expression under normal and experimental conditions. Examples of experiments using *in situ* hybridization are found in Figures 3.28 and 4.23.

Scientists use **immunohistochemistry** or **immunocytochemistry** to label proteins in tissues or cells, respectively. These **immunolabeling** methods take advantage of the immune response in which an animal develops antibodies to a foreign substance. For example, one method of generating antibodies is to inject rabbits with a protein of interest. The animals develop antibodies to the protein that are then isolated from the blood serum. The resulting antibodies, called primary antibodies, are then added to tissues or cell cultures, where they bind to the target protein. These antibodies are visualized by either adding an enzymatic reporter molecule, such as horseradish peroxidase, or a fluorescent probe directly to the primary antibody, or by adding a label to a secondary antibody that recognizes and binds the primary antibody. These methods often provide fine details on the distribution of a protein within a cellular region. Advances in microscopy and tissue clearing methods have further enhanced the ability to view protein distribution in three dimensions. Examples include using a labeled neuron-specific antibody to visualize the entire nervous system of an embryonic *Drosophila* (Figure 1.15B) or chick (**Figure 1.22**). Immunolabeling of individual synaptic contacts on a single neuron is shown in Figure 10.1.

Technical advances continue to be made in molecular biology that allow finer resolution of gene expression levels in tissues and individual cells.

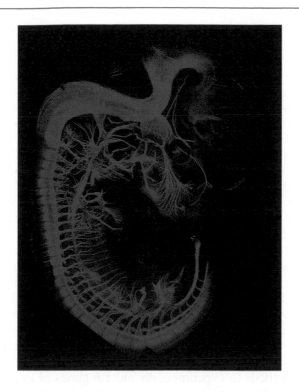

Figure 1.22 Antibodies label nerve fibers in the chick embryo. A fluorescent label (red) is used to detect an antibody bound to the neurofilament protein. In this image from an embryonic day 12 chick embryo, the brain, spinal cord, and nerve fibers for dorsal root and cranial nerve ganglia are seen. [Image courtesy Jason Newbern.]

Researchers can now screen thousands of sequences at once. For example, methods involving what is often called **next-generation sequencing** or **deep sequencing** repeatedly sequence a genomic region hundreds to thousands of times to detect rare material, even from small samples. **RNA-seq** (RNA-sequencing) converts a tissue's RNA to cDNA to indicate the expression levels of transcribed genes in that tissue, while **single-cell RNA-seq** identifies and quantifies mRNA from individual cells. These methods often reveal subtle and specific differences in gene expression in different cell subtypes. These data are used to shape further investigations into the consequences of the observed differences. **Transcriptomics** is the field that identifies and studies the functional differences of expressed RNAs. Examples of applications of transcriptomics in the developing nervous system are described in Chapters 4 and 6.

Altering Development Helps Understand Normal Processes

One common way to assess normal developmental events is to alter some aspect of development and see what happens. This allows researchers to test whether a given tissue, cell, or protein is necessary for normal development to occur. Over the past century and a half, several methods have been used to alter development. Among the approaches used today are techniques to manipulate tissues *in vivo* and *in vitro* and methods to evaluate naturally occurring and experimentally induced genetic mutations.

Tissue manipulations have been used since the earliest neurodevelopmental studies. These methods typically involve surgically removing or rotating a particular region of the developing embryo or grafting extra tissue onto a region of the embryo. Several examples of these types of studies are highlighted in Chapters 2, 4, 7, and 8. Scientists can also observe effects of tissue manipulations in cell culture preparations. In these assays, tissues of interest are surgically dissected from an embryo at a given stage of development and placed into a cell culture dish. The dishes are often coated with substrate molecules that support the attachment and growth of the cells under investigation. The tissues are then covered in a nutrient-containing fluid (cell culture medium). To identify sources of signals that promote the survival, growth, or differentiation of a neural population, the tissues may be grown in the presence of other tissues. In some experiments, specific proteins may be added to test whether they have a direct effect on the developing cells. Examples using these approaches are discussed in Chapters 4, 7, 8, 9, and 10. Cell culture techniques to study neural development were introduced in the 1920s and remain a very popular method for analyzing the development of neural cells. A more recent technique involving **organoids**, three-dimensional cell cultures of pluripotent stems cells that assemble into miniature brain-like structures, is being used to study how neural-like cells assemble and make connections. An advantage of cell culture methods is the ability to test single reagents on a select population of cells. A limitation of the method is that the artificial environment removes other tissue-derived cues that may interact with and alter the effects of the reagent under investigation.

Scientists also observe the effects of additional or missing genes. Such **genetic manipulations** have been instrumental in understanding neural development in both invertebrate and vertebrate animal models. Studies of naturally occurring gene mutations in *Drosophila*, *C. elegans*, and mice have been documented for nearly a century. Many of these spontaneously occurring mutations have provided insight into normal mechanisms of neural development. As detailed in Chapter 6, scientists investigating *Drosophila* initially relied on naturally occurring mutations, but soon developed methods to experimentally mutate genes of interest. Methods for blocking, reducing, or increasing gene expression were also developed for many vertebrate animal models. Examples of mutations induced in frogs are found in Figures 2.10, 2.11, and 6.2, while examples from *Drosophila* are shown in Figures 7.22 and 7.24. A method to experimentally delete, or knock out, individual genes in mice was introduced in the 1980s. The methods for generating **gene knockout mice** greatly advanced studies of mammalian neural development (**Box 1.3**).

Other methods to selectively interfere with gene expression use **short interfering RNAs** (**siRNAs**). Segments of RNA consisting of 20–25 base pairs that are complementary to a gene sequence of interest are introduced to cells by **electroporation**, a method in which an electrical current is used to make cell membranes more permeable. siRNAs can be electroporated into specific regions of an embryo, where they degrade the target mRNA and prevent translation of the protein, thereby providing insight into the normal function of the protein *in vivo*. Examples of this approach are shown in Figure 4.10.

Optogenetics is used to manipulate neuronal activity by selectively stimulating cells genetically modified to express a light-sensitive opsin protein. When light of the correct frequency is presented to the cells, a conformational change occurs in the opsin protein that alters neural activity. Depending on the opsin protein expressed, the light stimulation may activate, inhibit, or modify neural activity. Optogenetic techniques can be applied to *in vitro* and *in vivo* studies. *In vivo* studies often assess behaviors in transgenic mice expressing specific opsin proteins and can be used to map changes in neural connections following activation or inhibition of specific pathways (see Chapter 10).

Researchers continue to refine techniques to selectively alter gene and protein expression in cells at specific stages of development, providing finer resolution of the molecular pathways involved in neural development. There are limits to these approaches, however, and researchers are aware that induced changes represent an artificial environment and that complementary studies are needed to test the role of the molecules during normal development. Despite the inherent limitations of these approaches, tissue and genetic manipulation studies have provided considerable insight into mechanisms underlying normal neural development.

Box 1.3 Knockout Mice

The term "knockout mouse" is now commonly used throughout the scientific literature. The technique has become so widely used and discussed that it may be difficult to imagine what a surprising and significant impact it had when it first emerged in the early 1980s. In fact, when Mario Cappecchi first proposed the technique to a funding agency, the proposal was turned down because reviewers believed the process could not work effectively. The technique relies on the process of **homologous recombination**—the ability of an inserted DNA sequence to line up in the correct orientation and location and replace a specific gene. Homologous recombination takes place naturally and frequently in bacteria, yeast, and viruses, but under normal conditions is rare in mammalian cells, except in germ-line and **embryonic stem (ES) cells**—that is, the cells that have the ability to give rise to all the cells in an organism. Mammalian cells are also capable of the process when foreign genes are intentionally inserted, such as occurs in the process of generating knockout mice.

Figure 1.23 outlines the steps used to generate mice lacking a gene of interest. In the first step, the target gene is removed from a segment of DNA and selector genes are inserted to create a targeting vector. Two commonly used selector genes are the *neomycin resistance* (*neo^R*) gene—a positive selector gene—and the *herpes thymidine kinase* (*tk*) gene—a negative selector gene. The *neo^R* gene is flanked by DNA present in the target gene, while the *tk* gene is located outside the targeted sequence.

In the second step, the target vector is introduced into mouse embryonic stem (ES) cells. Electroporation provides a small electrical charge that opens the cell membranes and permits entry of the DNA. The cells are grown in a culture medium that contains the drugs neomycin and ganciclovir. Cells that have inserted the *neo^R* gene in place of the targeted gene will survive in the medium containing the antibiotic neomycin. Ganciclovir will kill any cells that retain the *tk* gene; thus, the cells that have randomly inserted the targeting vector outside the gene sequence of interest will be eliminated. By using both positive and negative selector genes, all or nearly all the surviving cells in the culture medium will be those with the targeted gene disrupted.

The remaining ES cells are then injected into blastocyst-stage mouse embryos from mice of a particular coat color (Figure 1.23, step 3). The blastocysts are implanted into a surrogate, or foster, female mouse of a different coat color to develop to term. The tissues of the pups from this first litter contain cells that arise from both mice. These chimeras, made up of genetic contributions from the blastocyst and surrogate mouse, can be identified by a coat color that differs from that of the surrogate mother. The male chimeras are then mated with females of another coat color, such as white (step 4). The resulting pups are again selected by coat color to identify those that carry genes from the ES cells (for example, mice that are black). The DNA from these mice is then sequenced and those mice that are heterozygous—that is, the mice that contain one copy of the disrupted gene and one copy of the normal gene—are then mated with littermates that are also heterozygous for the mutation. One quarter of the resulting litter will contain mice that are homozygous for the mutation. These are the knockout mice that contain two copies of the mutated gene and are the ones to be examined for anatomical, physiological, or behavioral deficits. Other mice in the litter will be normal, or wild-type mice, which carry two copies of the normal gene, and the others will be heterozygous.

Depending on the particular gene disrupted, the resulting changes—the phenotype of the mice—can be mild, suggesting that another gene compensates for the loss of the targeted gene, or severe, sometimes even resulting in death of the embryo prior to birth. As scientists have found over the years, the heterozygous mice often have a milder phenotype than the homozygous mice, displaying a gene dosage effect—that is, those with one copy disrupted are impacted less than those with two copies disrupted. Embryonic lethal mutations can sometimes provide information on the role of the gene, depending on the stage when the embryos die. Such severe mutations are often of limited value, however, particularly if the embryo dies prior to the onset of gene function in the cell population of interest.

A refinement to the gene knockout technique was introduced in the late 1980s that allows for researchers to delete a gene in selected tissues or at specific stages of development and therefore overcome the limitations of deleting a gene in every cell of the body. The basic method used to create such **conditional knockout mice** involves inserting loxP (locus of X-over P1) in noncoding regions of the DNA sequence of interest using homologous recombination. These segments are said to be "floxed" (flanked by loxP). The floxed sites of the DNA are recognized by Cre recombinase that mediates the exchange of DNA. The conditional expression of the gene is regulated in one of two ways—namely, Cre recombinase can be linked to tissue-specific promoters or to inducible proteins. Thus, depending on the method used, the gene of

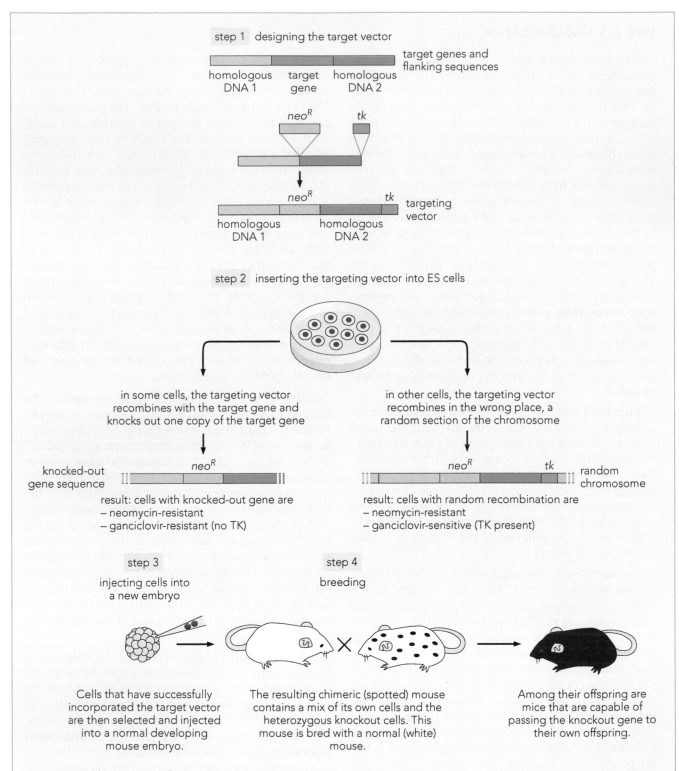

step 1   designing the target vector

target genes and flanking sequences

homologous DNA 1 | target gene | homologous DNA 2

neo^R   tk

neo^R   tk   targeting vector

homologous DNA 1   homologous DNA 2

step 2   inserting the targeting vector into ES cells

in some cells, the targeting vector recombines with the target gene and knocks out one copy of the target gene

in other cells, the targeting vector recombines in the wrong place, a random section of the chromosome

knocked-out gene sequence   neo^R

neo^R   tk   random chromosome

result: cells with knocked-out gene are
– neomycin-resistant
– ganciclovir-resistant (no TK)

result: cells with random recombination are
– neomycin-resistant
– ganciclovir-sensitive (TK present)

step 3
injecting cells into a new embryo

step 4
breeding

Cells that have successfully incorporated the target vector are then selected and injected into a normal developing mouse embryo.

The resulting chimeric (spotted) mouse contains a mix of its own cells and the heterozygous knockout cells. This mouse is bred with a normal (white) mouse.

Among their offspring are mice that are capable of passing the knockout gene to their own offspring.

Figure 1.23 The creation of a knockout mouse. An outline of the steps used to generate mice with the targeted gene disruption.

interest will only be altered in selected tissues or only at the developmental stages in which it is experimentally induced, thus allowing researchers to investigate the role of a gene in a selected cell population or at a specific time in development. Several modifications to these initial knockout and conditional knockout methods have been made since they were first introduced over 25 years ago so that researchers now have the ability to track, delete, or overexpress multiple genes in a single animal.

Using Naturally Occurring Events to Understand Neural Development

Across animal models, naturally occurring genetic mutations contribute to neurodevelopmental deficits. Changes in neural development and maturation also arise from environmental exposures and pathogens. Researchers use these naturally occurring events to better understand normal developmental processes. For example, plant toxins ingested by sheep provided information on the signaling pathway used to separate the two brain hemispheres (see Chapter 4) and a naturally occurring mouse mutant with a reeling gait led to discoveries about how the mammalian cerebral cortex forms (see Chapter 5).

Researchers also adapt toxins and infectious agents to use as research tools. For example, **α bungarotoxin**, a snake venom toxin that causes paralysis and death by blocking acetylcholine receptors on muscle cells, is used in laboratories to investigate acetylcholine signaling mechanisms. It is also modified to carry a visual marker to identify where receptors are located on developing muscle cells (see Chapter 9). **Diphtheria toxin** is a protein synthesis inhibitor secreted by certain bacteria. Diphtheria can cause serious illness in humans and vaccinations are given to prevent disease. Mice, however, do not express the receptors for diphtheria. Researchers use this to their advantage by creating transgenic mice that express subunits of the toxin, but only in specific cells. The cells expressing the toxin are unable to synthesize proteins, and therefore degenerate. Because the surrounding cells do not express the diphtheria receptor, they remain undamaged, even if the toxin is released by dying cells. Selective ablation of cells is helpful in deciphering the normal roles of those cells *in vivo* (see Chapter 4).

Some viruses cause disease and neurological deficits (**Box 1.4**), and some can be modified to selectively deliver genes or gene segments that activate ("knock-in") or repress ("knock-down") target genes of interest. Viruses such as adenoviruses and adeno-associated viruses are commonly used experimental tools. Clinically, viral vectors can deliver genes to treat a disease or condition. An example of this type of gene therapy is the recently approved use of an adeno-associated virus (AAV) to treat infants with spinal muscular atrophy (SMA), a rare condition that leads to loss of motor neurons.

Gene modifications, or editing, can also be accomplished using the clustered regularly interspaced short palindromic repeat (CRISPR) system that is found in bacteria and Archea (single celled microorganisms). CRISPR and CRISPR associated (Cas) proteins edit the genome by selectively breaking DNA strands at a target site, then introducing or altering a gene of interest. The Cas proteins, such as Cas9, are the enzymes that break the DNA strands. The Cas protein is delivered along with an RNA "guide" that identifies the DNA site of interest. As the DNA repairs the induced breaks, a mutation can be removed, or a target gene can be deleted or inserted at the desired location. AAVs and other viruses can deliver the Cas9 and RNA. CRISPR technology has become a valuable experimental tool with potential clinical applications, such as those aimed at stopping the degeneration of photoreceptor cells in some forms of congenital blindness.

Box 1.4 Viruses and the Nervous System: Examples from Chicken Pox, Zika, and SARS-CoV-2

Viruses are very small, non-cellular infectious agents that survive and replicate by invading host cells in another organism. Viruses are much smaller and more numerous than bacteria and are found in animals, plants, and other microorganisms.

The components of a virus are nuclei acid (single or double-stranded DNA or RNA) and a protein coat called a capsid. When outside of a host cell, these particles are called virions. Some viruses form an envelope around the capsid as the virus emerges from the host cell. Thus, the envelope is comprised of host cell membrane. Despite their overall simple composition, viruses have remarkable impacts on all living organisms.

Of the millions of viruses, some infect only a few species while others infect many species. Animal viruses can be transmitted through the air or via contact with infected surfaces, tissues, or organisms. Viruses are often transmitted through blood, sexual contact, and oral-fecal routes. Disease-bearing organisms (vectors) such as insects are another common source for spreading a virus. Some viruses are spread by horizontal transmission, the passing of the virus between individuals; others are spread through vertical transmission, the passing of the virus from mother to fetus. Some viruses are passed through both horizontal and vertical transmission.

Viruses are differentiated by their taxonomy and are often divided into seven categories based on whether they contain single or double-stranded DNA or RNA, have sense or antisense strands, and rely on an intermediate (RNA or DNA) for replication. Examples include double-stranded DNA viruses such as the adenoviruses and herpes viruses; single-stranded RNA sense viruses like the coronaviruses and picornaviruses; and retroviruses, single-stranded sense RNA viruses that require a DNA intermediate.

The impact of viral infection on the host organisms is variable. Some viruses live harmlessly in host cells; in other cases, the host's immune response stops viral replication and eliminates the virus. While the virus remains in the host, the infection may cause mild symptoms, such as those of the common cold, or more severe symptoms and secondary complications. For example, influenza and chickenpox are often mild but can become severe or lethal in some individuals. In contrast, some viruses persist in the host, such as human immunodeficiency virus (HIV), human papilloma virus (HPV), hepatitis B, and hepatitis C. These viruses cause chronic infections and accompanying symptoms. Lethality from acute or chronic viral infections may be directly from the virus, secondary to the host response, or from subsequent infections. Vaccinations can prevent or limit the severity of many viral diseases including influenza, chicken pox, measles, mumps, rabies, poliomyelitis, and cancers caused by HPV infection.

Some Viruses Target Neural Cells

The cells a virus invades determine the initial and latent effects on the host. **Tropism** refers to the cells most vulnerable to invasion of the virus. When discussing viral infections, neurotropic refers to a virus that preferentially enters neurons and glia. For example, the herpes virus known as chickenpox virus (Varicella zoster virus, VZV) that causes a widely distributed itchy skin rash, usually in children, preferentially invades neurons in cranial ganglia, dorsal root ganglia, and autonomic ganglia. After the initial infection, the virus remains dormant in these cells, but can reactivate along a subset of nerve fibers in adulthood. Areas of skin innervated by the nerve fibers with the reactivated virus produce a painful rash commonly known as **shingles**. Some believe the term shingles originated from a distorted pronunciation of *cingulate*, reflecting the girdle or band shape pattern of the rash commonly seen on the torso of many patients.

How the virus becomes reactivated is still unclear. While the rash resolves within two to four weeks, an associated painful burning sensation may persist in about 10–20% of patients leading to a condition called postherpetic neuralgia (PHN). In rare cases, reactivated VZV causes damage to the CNS such as meningoencephalitis (inflammation of meninges and brain) and vasculopathies including stroke, transient ischemic accident (TIA), spinal cord infarction, and cerebral venous thrombosis. Though atypical, these conditions demonstrate the potential for even mild viral infections to cause long-term neurological consequences.

Recent Viral Outbreaks Highlight How Viruses Impact the Developing and Mature Nervous Systems

In the past decade, two viruses have emerged as serious public health concerns: the Zika virus and the SARS-CoV-2 virus. Both viruses impact the nervous system in a variety of ways.

The **Zika virus** reemerged in the early twenty-first century and most recently led to an outbreak associated with increased cases of microcephaly in children exposed *in utero*. The Zika virus is a single strand, sense RNA of the *Flavivirus* genus. Like many other Flaviviruses, Zika virus can be spread to humans through bites from mosquitos. Other mosquito and tick-borne Flaviviruses cause diseases such as yellow fever, dengue fever, and West Nile virus disease. While most Zika infections arise from mosquitos, some cases are transmitted by sexual contact or through vertical transmission from an infected mother to her child *in utero*.

Zika virus was first reported in 1952 stemming from studies begun in rhesus monkeys in the Ziika Forest of Uganda in 1947. The first human disease was reported in 1954 in Nigeria. Until 2007 few human cases were reported. However, outbreaks in the Federated States of Micronesia (2007) and French Polynesia (2013) brought new attention to the virus. Most people presented with fever and rash and had mild symptoms overall. However, symptoms of Guillain-Barre Syndrome, a disease that effects the peripheral nervous system, were reported in the 2013 outbreak, suggesting the virus may be neurotropic. In 2015, a large outbreak in Brazil brought world-wide attention to Zika virus and its long-term impacts on human health as several children of infected mothers were born with microcephaly, or "small brain." The increased incidence of microcephaly led to the discovery that while the virus was primarily transmitted through mosquitos, it was also spread through bodily fluids during sexual contact. Women of childbearing age were therefore advised to take extra precautions to avoid risk of infection from either route. Although many infected individuals remained asymptomatic or had mild cases, the number of serious complications in newborns was surprising, and revealed that mutations in the virus had increased its neurotropism.

With the increase in microcephaly, researchers began to study how the virus could cause such devasting changes in neurodevelopment. *In vitro* and *in vivo* studies found that the Zika virus targets neuronal progenitor cells, disrupting the cell cycle and causing cell death. Studies in mice further suggested the Zika virus disrupts the **adherens junctions** leading to decreased proliferation (see for example, Figure 5.5). In adults, Zika virus infections are associated with defects in the peripheral nervous system including Guillain-Barre syndrome and peripheral neuropathy. Transverse myelitis, a demyelinating disease of the spinal cord has also been reported. Though uncommon, these observations suggest the virus may target myelinating glia cells in the adult peripheral and central nervous systems.

Although the number of cases since the 2015 outbreak has declined, it remains unclear whether long-term neural deficits will emerge in infected patients or if seemingly healthy children born to infected mothers will display as yet unrecognized neurodevelopmental deficits later in childhood. Investigators continue to track the outcomes of patients to better understand the long-term consequences of Zika virus infection on the nervous system.

SARS-CoV-2 (Severe Acute Respiratory Syndrome Coronavirus 2) is the virus that causes Coronavirus disease 2019 (**COVID-19**). **Coronaviruses** are single-stranded RNA viruses that target epithelial cells in the respiratory and gastrointestinal systems. Many are noted for causing respiratory symptoms in mammals and birds. The common cold is an example of a coronavirus that causes mild respiratory disease symptoms. In contrast, other coronaviruses cause diseases that are often severe or lethal. Severe Acute Respiratory Syndrome-associated coronavirus (SARS-CoV),

Middle Eastern Respiratory Syndrome coronavirus (MERS-CoV), and SARS-CoV-2 are examples of coronaviruses that have mild to lethal effects in infected individuals. SARS was first reported in 2002 but no new cases have emerged since 2004. MERS virus first appeared in 2012, with current cases largely restricted to Middle Eastern countries. Both begin with flu-like symptoms that can progress to severe respiratory illness and death.

SARS-CoV-2 emerged in late 2019 and has since spread throughout the world. While many infected individuals remain asymptomatic or have only mild cold- or flu-like symptoms, others develop severe respiratory distress requiring hospitalization. Many people die secondary to effects caused by the host's immune response to the virus. By the end of 2021 over 5 million people had died from COVID-19 disease.

While mainly impacting epithelial cells of the respiratory tract, coronaviruses also target vascular cells. Several cases of COVID-19-associated strokes and blood vessel leakage in the brain have been reported. The receptors SARS-CoV-2 uses to enter host cells include those found in and around neural tissue, suggesting opportunities for direct infection. However, whether neurons or glia are specifically targeted or damaged indirectly remains an ongoing area of research. Thus, unlike the Zika virus, neurotropism of SARS-CoV-2 has not been established.

Among the first symptoms reported by many infected individuals are a loss of smell (anosmia) and taste (ageusia). Normal smell and taste sensations are often disrupted with respiratory infections, largely due to the associated nasal congestion. However, the SARS-CoV-2 virus also appears to directly impact supporting cells and possibly neurons in these systems. In the olfactory system, it seems that the supporting cells in the neuroepithelium are targeted by the virus. When these cells are disrupted, olfactory neuron function becomes disrupted. Most patients recover the sense of smell in a few weeks, corresponding to the time of supporting cell recovery, though others are left with longer-term deficits suggesting direct or indirect damage to olfactory neurons.

The list of potential neurological conditions associated with the SARS-CoV-2 virus continues to increase. While neurotropism is not suspected in all these cases, SARS-CoV-2 infection can ultimately influence neural tissue. Among potential COVID-19-related complications are peripheral neuropathy, transverse myelitis, Guillain-Barre syndrome, acute disseminating encephalopathy, facial nerve palsies, hearing loss, and Parkinson's-like symptoms. Many reports note that neurological symptoms are unrelated to the severity of one's respiratory symptoms. There is also growing concern for long-term neurological complications in those who experienced mild to moderate forms of the disease.

As of the writing of this book, the COVID-19 pandemic continues. New discoveries are reported weekly, and the medical and scientific communities continue efforts to prevent, treat, and stop this highly transmissible virus. An example of one clinician's perspective and experiences during the first 18 months of the pandemic are described next, highlighting how the clinical picture of this novel coronavirus disease rapidly evolved, and continues to change how scientists and clinicians view this disease.

A Physician's Perspective: Presentations of SARS-CoV-2 in an Emergency Department

Erik Iszkula, M.D.

Dr. Iszkula graduated from the University of Pittsburgh. Intending to major in pre-pharmacy he switched to a biology major, then attended medical school at Jefferson Medical College (now the Sidney Kimmel Medical College). He completed his residency in Emergency Medicine at the University of Virginia before returning to the Erie, PA region where he currently practices and trains resident physicians.

As an Emergency Department (ED) physician, I have come to respect and even fear the virus that causes COVID-19. Working in a mid-sized city in the Midwestern United States I have watched the COVID-19 pandemic transition from an event that seemed remote, to one that I deal with daily. Initially, I felt like others who are not involved in the front lines of health care. When the first reports of the illness in Wuhan appeared in early 2020, I remember thinking it funny that Hong Kong shut down public schools 600 miles from an outbreak of what appeared to be nothing more than a flu-like illness.

Yet, soon after, our community began to change too. In March of 2020 our ED saw a drastic drop in volumes as people started to stay at home, concerned they would contract the virus from our hospital. Our volumes dropped nearly 50%, while our caution for COVID-19 dramatically increased. We had virtual briefings daily at 6:30 am. Our residency program went to nearly complete online learning. March 18th was the first confirmed case in our county. We cancelled non-emergent surgeries, shut down schools, and adjusted our lives for an illness that presented with a whisper, rather than the scary bang we expected.

Over the next two months, the number of cases in our region crept upwards, but I still perceived it as more of a number on a newscast, rather than an actual threat. Most hospital workers in our area had not seen a confirmed case. The news bombarded us with stories from larger cities, but it all seemed more theatrical and unrealistic as it was not seen outside our windows. Our cases remained few and far between, hospital volumes were low, and there was an abundance of ventilators at that time.

As June came, our physicians and nursing staff began seeing their first COVID-19 patients. These were mostly mildly symptomatic patients or those concerned they were in contact with someone who had tested positive. Many were tested on an outpatient basis, and we learned of their positive results days later. Again, this disease seemed more flu-like and less SARS-like.

However, my opinion about this disease completely changed in July of 2020. A middle-aged patient with minimal past medical history arrived at our facility after visiting family in another state. We immediately suspected the patient was COVID-19 positive and performed a rapid test, which took approximately an hour. In the meantime, a chest x-ray was performed. The results were startling. This may have been the worst chest x-ray I had seen in 14 years as a physician. I had seen hundreds of pneumonias and influenza patients in my career, but nothing that came close to this. As I spoke with the patient about the presumed diagnosis and received consent for intubation, I saw the fear and confusion on his face. He had been healthy up until now and only began to feel ill earlier that day. Now he was headed for the intensive care unit (ICU) concerned he may never see his family again.

We intubated the patient in the ED before transfer to the ICU. As we were preparing for the procedure the patient began coughing uncontrollably. Although I was fully prepared, in an N-95 mask, gown, and face shield, I felt like I wanted to run out of the room. I have never felt this way before. For me, this patient vividly illustrated what the news stories were trying to convey. I was now horrified by this disease. The sudden reality of what the virus could do inspired me to become more vocal in educating my community about the seriousness of COVID-19 and the precautions needed to limit spread of this highly transmissible and dangerous virus.

Since that encounter in the first summer of the pandemic, I have seen many patients with COVID-19. What began as a rare occurrence is now a common part of my ED practice. In less than two years, the disease has changed from the rare, scary disease seen only on television to one of the most common diagnoses in our department.

The mystery of this disease is how it can affect so many people so differently. I see patients who are asymptomatic and patients with respiratory failure. I see many people who ultimately die from complications of the disease. I have seen babies with no symptoms beyond a fever and unvaccinated 90-year-olds go home with a dry cough. Other days I have to life-flight adolescents by helicopter to children's hospitals where they can receive more advanced treatments for severe respiratory distress. I have seen almost every physical symptom imaginable including severe presentations like blood clots in lungs and legs and strokes and stroke-like symptoms in adults well below the typical age for cerebrovascular disease.

One case that stands out was an otherwise healthy patient in his forties who presented with vertigo (dizziness) and diplopia (double vision). He reported waking up with blurred vision that progressed to double vision and difficulty walking. The patient had rotary and vertical nystagmus (eye movements) and a left eye that tracked slowly and was often misaligned. These symptoms were exacerbated with gaze to the right. I immediately called neurology for a consultation and ordered a special computerized tomography scan (CT scan) to better visualize the brain and blood vessels in his neck. The neurology team was concerned the balance and visual symptoms were due to a posterior cerebral artery stroke. Yet, when the CT came back there were no abnormal findings in the brain or blood supply. However, the scan included the top of the lungs, and I recognized the typical signs of viral pneumonia. Although the patient denied any COVID-19 symptoms or exposure, he tested positive for the disease. This finding surprised him even more than us. The patient's visual disturbances eventually subsided, and he was later discharged with a suspected "MRI-negative stroke." His symptoms are much like other seemingly healthy COVID-19 positive adults presenting with stroke-like symptoms. The patient had no known risk factors for stroke, and his symptoms were likely secondary to COVID-19 infection, though we may never know for certain.

It is still unclear why some patients are more susceptible to neurological complications or what long-term effects might persist in the nervous system of those presenting with mild or moderate disease. What does seem clear is that the virus causing COVID-19 will continue to baffle and amaze us for years to come.

SUMMARY

In the past 130 years, scientists have made remarkable progress in identifying the many mechanisms that lead the human nervous system to develop from a simple region of ectoderm into a complex nervous system comprised of hundreds of billions of neurons and supporting cells. These insights have come from several vertebrate and invertebrate animal models and ongoing studies continue to provide more details about signaling pathways used to form the nervous system.

The development of the entire nervous system, as well as of specific subtypes of neural cells, often occurs via signals that intentionally designate or actively restrict tissues or cells to a particular fate option. Chapter 2 begins by explaining how a specific region of ectoderm is designated to form neural tissue rather than epidermal tissue. Chapters 3 and 4 describe mechanisms that regulate patterning along the anterior–posterior and dorsal–ventral axes of the neural tube. This patterning creates early anatomical boundaries and establishes signaling centers that promote or inhibit subsequent developmental events. Mechanisms that influence cellular migration (Chapter 5) and cellular fate options (Chapter 6) are then described, demonstrating how further cellular specializations are established throughout the nervous system.

The development of the nervous system depends not only on the formation of highly specialized cell types, but also on the formation of precise connections between individual neurons. Chapter 7 explains how axons extend through the developing embryo to establish initial connections. Because the nervous system of vertebrates often over-produces the number of neurons necessary, additional signals are then needed to determine which cells and connections persist. Chapter 8 describes how neurons are selected for survival or death, while Chapters 9 and 10 discuss how synapses are formed and selected connections are lost.

It is thought that some of the same signals that underlie development of the nervous system may also influence repair or regeneration of neural tissues in the adult. Regeneration of many nerves is common in invertebrates; and peripheral nerves regenerate, at least to some extent, in vertebrates. Thus, knowledge gained from studies of developmental neurobiology has the potential to lead to clinical treatments to enhance nerve regeneration and possibly treat neurodegenerative diseases such as multiple sclerosis, spinal cord injuries, amyotrophic lateral sclerosis (ALS), Parkinson's disease, Alzheimer's disease, and others.

FURTHER READING

Alberts B, Johnson A, Lewis J, et al. (2015) *Molecular Biology of the Cell*, 6th ed. Garland Science.

Capecchi M (1994) Targeted gene replacement. *Sci Amer* 270(3):52–59.

Christian KM, Song H & Ming GL (2019) Pathophysiology and Mechanisms of Zika Virus Infection in the Nervous System. *Ann Rev Neurosci* 42:249–269.

Diaper DC & Hirth F (2014) Immunostaining of the developing embryonic and larval *Drosophila* brain. In *Brain Development Methods and Protocols* (Sprecher SG ed), pp. 3–17. Humans Press.

Friedel RH, Wurst W, Wefers B & Kuhn R (2012) Generating conditional knockout mice. In *Transgenic Mouse Methods and Protocols, Methods in Molecular Biology* (Hofker MH & van Deursen JM eds), pp. 205–231. Humana Press.

Gilbert SF (2014) *Developmental Biology*, 10th ed. Sinauer Associates.

Hamburger V & Hamilton HL (1951) A series of normal stages in the development of the chick embryo. *J Morph* 88(1):49–92.

Haretenstein V (1993) *Atlas of Drosophila Development*. Cold Spring Harbor Laboratory Press.

Kimmel CB, Ballard WW, Kimmel SR, et al. (1995) Stages of embryonic development of the zebrafish. *Dev Dyn* 203:253–310.

Luo L (2021) *Principles of Neurobiology*, 2nd ed. CRC Press.

Nieuwkoop PD & Faber J (1967) *Normal Table of Xenopus Laevis*. Elsevier.

Oikonomou G & Shaham S (2011) The glia of *Caenorhabditis elegans*. *Glia* 59:1253–1263.

Sadler TW (2014) *Langman's Medical Embryology*, 13th ed. Wolters Kluwer Health.

Schoenwolf GC, Bleyl SB, Brauer PR & Francis-West PH (2014) *Larsen's Human Embryology*, 5th ed. Churchill Livingstone.

Stern CD & Holand PWH (1993) *Essential Developmental Biology*. IRL Press.

Stevanovic M, Drakulic D, Lazic A, Ninkovic DS, Schwirtlich M & Mojsin M (2021) SOX transcription factors as important regulators of neuronal and glial differentiation during nervous system development and adult neurogenesis. *Front Mol Neurosci* 14:654031.

Stix G (1999) Profile (Mario Capecchi): Of survival and science. *Sci Amer* 281(2):26–27.

Vanderah TW & Gould DJ (2016) *Nolte's The Human Brain: An Introduction to Its Functional Anatomy*, 7th ed. Elsevier.

Westerfield M (1993) *The Zebrafish Handbook*, 2nd ed. University of Oregon Press.

Zhu R, Del Rio-Salgado JM, Garcia-Ojalvo J & Elowitz MB (2022) Synthetic multistability in mammalian cells. *Science (New York, N.Y.)* 375(6578):284.

ONLINE SOURCES FOR ATLASES OF EMBRYONIC DEVELOPMENT

Atlas of *Drosophila* Development:
www.sdbonline.org/sites/fly/atlas/00atlas.htm

C. elegans:
Altun, Z. F. and Hall, D. H. 2005. Introduction to *C. elegans* anatomy. In *WormAtlas*. www.wormatlas.org

eMouseAtlas:
www.emouseatlas.org/emap/ema/theiler_stages/StageDefinition/stagedefinition.html

The Endowment for Human Development:
www.ehd.org/virtual-human-embryo/stage.php

Xenbase:
www.xenbase.org/anatomy/alldev.do

The Zebrafish Information Network:www.zfin.org

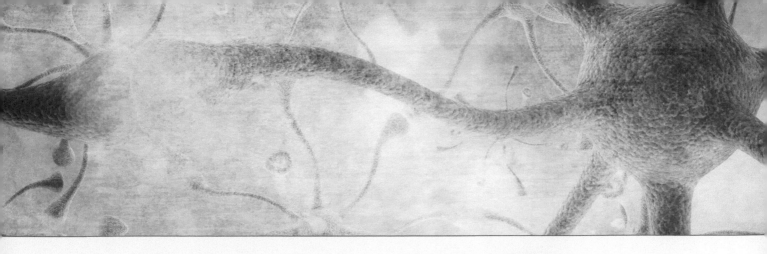

Neural Induction

2

Neural induction, the process by which undifferentiated embryonic tissue is specified to become neural tissue, is the first phase of neural development. Once induced, specialization of the many distinct regions and cells of the nervous system begins. The steps involved in creating a functional nervous system are numerous and complex, and begin very early in development. Segregating neural regions at the earliest stages of development prepares cells to receive the correct signals at the right time so normal development occurs. The following quote from Hans Spemann, a pioneer in the study of neural induction, emphasizes the importance of embryonic organization and segregation in the development of neural tissues:

> We are standing and walking on parts of our body which we could have used for thinking if they had been developed in another position in the embryo.
>
> (Hans Spemann, 1924)

Despite significant differences in nervous system morphology, the signaling molecules that induce neural tissues are largely homologous in vertebrates and invertebrates, demonstrating the evolutionary conservation of this fundamental, early process of neural development. As described in this chapter, neural induction has been the focus of research efforts for well over a century and while many of the general mechanisms are understood, unique species-specific differences continue to be discovered.

NEURAL TISSUE IS DESIGNATED DURING EMBRYOGENESIS

From the earliest stages of development, certain areas of the vertebrate embryo are destined to form neural tissues. Even prior to fertilization, an egg cell is compartmentalized. In egg cells from amphibians, for example, the cytoplasm is segregated into a lighter pigmented **vegetal pole** and more densely pigmented **animal pole** (**Figure 2.1**). The vegetal pole gives rise to gut structures, whereas the animal pole gives rise to the nervous

DOI: 10.1201/9781003166078-2

Figure 2.1 Regions competent to become neural tissue are established in the egg cell. Eggs from amphibians are divided into animal and vegetal poles. The animal pole appears denser due to the accumulation of cytoplasmic lipids and granules. This pole will later give rise to epidermis and neural tissues. The lighter appearing vegetal pole will give rise to structures associated with the gut.

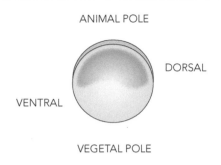

system and the epidermis (the surface layer of skin). In animals with large eggs, such as amphibians, cytoplasmic differences are easily identified prior to fertilization, as are the cytoplasmic rearrangements that occur after fertilization. In other vertebrates, including mammals and birds, the cytoplasmic differences are not as obvious, even after fertilization.

Following fertilization, the egg cell (**zygote**) divides into a number of cells called **blastomeres** that surround a central cavity known as a **blastocoel** (**Figure 2.2**). A group of blastomeres that aggregate above the cavity is called the **blastoderm**. This entire structure—that is, the cells and the hollow cavity they surround—is known by different names, depending on the species: the **blastula** in amphibians, the **blastocyst** in many mammals, and the **blastodisc** in birds, fish, and some mammals. The difference in terminology refers to the morphology of the cells. In birds, for example, the cells form a disc-like structure that is distinct from the cyst-like morphology observed in mice. In many cases, the term "blastula" is used as a generic term for all species.

The amphibian blastula-stage embryo (Figure 2.2A) is spherical, and like the egg itself, the embryo at this stage is characterized by animal and vegetal poles. In birds (Figure 2.2B), the blastodisc lies above the yolk, where it forms two sheets of cells designated the **epiblast** and **hypoblast**—regions that roughly correspond to the animal and vegetal poles, respectively. The epiblast and hypoblast are also found in mammals; however the size of the yolk, a source of nutrients for the embryo, is much greater in birds than in mammals. In fact, in most mammals the size of the yolk is negligible due to maternal sources of nutrients. Yet, despite species differences in blastula shape and yolk size, the next step in early development is comparable across vertebrate species.

Gastrulation Creates New Cell and Tissue Interactions That Influence Neural Induction

Gastrulation is a crucial stage in early embryogenesis—the stage when cells begin to reorganize into the three germ cell layers to form specific

Figure 2.2 Comparison of blastula formation in amphibians and birds. Following fertilization, the egg cell divides repeatedly and the resulting new cells, called blastomeres, surround a hollow, fluid-filled cavity, the blastocoel. The consolidated group of cells above the blastocoel is called the blastoderm. In frogs (A), the blastula is a spherical structure, whereas in birds (B), the blastula, also called the blastodisc, appears as a flattened sheet of cells overlying the yolk, with the blastocoel cavity formed between. The two layers of cells in the blastodisc are termed the epiblast and hypoblast, regions that are roughly equivalent to the animal and vegetal poles. [Adapted from Patten BM [1958] *Foundations of Embryology.* New York: McGraw-Hill.]

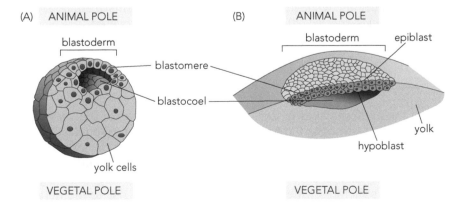

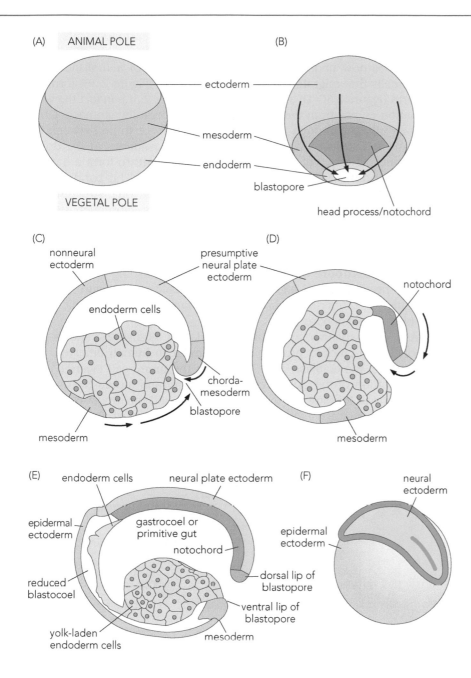

Figure 2.3 Transition from the blastula-to the gastrula-stage embryo in amphibians. (A) A surface view of an early-stage amphibian blastula outlining the general areas of presumptive ectoderm, mesoderm, and endoderm. The animal pole will give rise to ectoderm (blue), the vegetal pole will give rise to endoderm (green), while the mesoderm (light red) will arise from the middle segment. (B) A schematic fate map is shown from the dorsal surface of a blastula-stage amphibian embryo just prior to the onset of gastrulation. The sight of blastopore formation (white) indicates the area through which surface cells will migrate (arrows). The regions fated to become the three germ layers (ectoderm, mesoderm, or endoderm) and the head process/notochord (brown) are also shown. (C–E) show sagittal sections through the center of the blastula to reveal the inward movements of cells and the subsequent arrangement of tissue layers as gastrulation progresses. Dorsal is to the right in all panels. The arrows indicate the direction of cellular movements. As the cells of the presumptive endoderm (green) and mesoderm (light red) push inward through the blastopore (C, D), the primitive gut (gastrocoel) is established (E). At the dorsal surface (C), the chordamesoderm (axial mesoderm) also migrates inward, forming the notochord (brown, D) that extends below the forming neural plate ectoderm (blue, C–E). The cells of the ventral surface ectoderm will give rise to the nonneural, epidermal ectoderm (yellow). The mesoderm-derived dorsal and ventral blastopore lips at the margins of the blastopore are seen in panel E. Signals emanating from the dorsal blastopore lip (shown in gray) are critical for formation of neural ectoderm. (F) Surface view showing regions of neural (blue) and epidermal (yellow) ectoderm. [(B), Adapted from Nagy A, Gertsenstein M, Vinterstan K & Behringer R [2003] *Manipulating the Mouse Embryo*, 3rd ed. Cold Spring Harbor Press; (C–E), adapted from Vogt W [1929] *Gestaltungsanalyse am amphibienkeim mit örtlicher vitalfarbung. II. Teil. Gastrulation und mesodermbildung bei oroden und anuren. Wilhem Roux' Arch* 120:385–706.]

tissues. These layers, the ectoderm, mesoderm, and endoderm, are called the **germ cell layers** or **primary cell layers**, because it is from these three layers that all tissues of the body arise. Gastrulation involves the migration, or invagination, of cells through an indentation on the outer surface of the developing embryo. This site of indentation is called the **blastopore** in amphibians, the **embryonic shield** in zebrafish, and the **primitive streak** in birds and mammals.

As surface cells begin to migrate through the surface indentation, new arrangements of the germ cell layers occur (**Figure 2.3**A, B). In all cases, as a result of gastrulation, the **endoderm** forms as the innermost germ cell layer and gives rise to the gut and organs associated with it. The **mesoderm** forms as the middle layer and gives rise to muscle, bone, connective tissues, and the cardiovascular and urogenital systems. The outermost layer, the **ectoderm**, remains at the embryo's surface (animal pole or epiblast) to give rise to the epidermis and nervous system. In all these animal models, during a limited period in the late gastrula-stage embryo, the ectodermal cells have the potential to become either epidermal cells or neural cells.

In 1929 Walter Vogt illustrated how cellular rearrangements occur during amphibian gastrulation and many of the structures identified in his drawings are now understood to be key regulators of neural induction. The first cells that migrate through the blastopore (Figure 2.3B) form the endoderm (Figure 2.3C, D) and primitive gut (Figure 2.3E). In fact the word gastrulation comes from the Greek word *gaster* meaning stomach. Other cells then migrate through the blastopore and come to lie between the endoderm and surface ectoderm, forming the mesoderm. At the dorsal side, a specified group of these cells migrates inward to form a band of **chordamesoderm** (also called the axial mesoderm). This band of mesoderm gives rise to the notochord, a transient embryonic structure that lies below the surface ectoderm and is important for subsequent aspects of neural development (Figure 2.3D, E). Ectoderm remains at the surface of the embryo. At the opening of the blastopore, the margins form the mesoderm-derived dorsal and ventral blastopore lips. As described later, the **dorsal blastopore lip** (**DBL**) is critical in establishing which areas of ectoderm become neural and which become epidermal (Figure 2.3E, F). Gastrulation in zebrafish is similar to amphibians, as the embryonic shield resembles the blastopore.

As a comparison to amphibian gastrulation, **Figure 2.4** shows gastrulation in the chick, a process that is similar in human embryos. The outer surface, the epiblast, gives rise to all the future cell layers (Figure 2.4A). A portion of the cells becomes migratory and moves through the primitive streak forming the middle mesoderm and innermost endoderm layers. As the endoderm forms, the hypoblast that lies just above the yolk

Figure 2.4 Gastrulation in chick embryos. (A) A surface view of a chick blastodisc at approximately 16 hours of development. The surface cells (epiblast) give rise to the future germ layers. The primitive streak, which forms the longitudinal axis of the embryo, establishes the site where cells migrate inward during gastrulation. Hensen's node, located at the anterior end of the primitive streak, is the equivalent of the amphibian dorsal blastopore lip. The area of the prechordal plate, where ectoderm and endoderm adhere, is outlined anterior to Hensen's node. (B) A cross-section of a chick embryo demonstrates how the surface cells of the epiblast (gray) migrate medially into the opening of the primitive streak, then migrate laterally to establish the mesoderm and endoderm layers. The hypoblast, lying above the yolk, is displaced as the endoderm forms. The hypoblast later gives rise to extraembryonic tissues. (C) Cells migrating inward along the midline give rise to the notochord, which lies below the future neural plate (not shown). As more cells migrate inward, the primitive streak regresses. (D) The cells of the epiblast that remain at the surface form epidermal (yellow) and neural (blue) ectoderm, with the neural tissue forming anterior to the primitive streak. (E) A sagittal section through the embryo reveals the orientation of the neural tissue and primitive streak as well as the prechordal and notochord regions. The notochord lies beneath the neural plate tissue. The prechordal plate is formed where the surface ectoderm and underlying endoderm form a tight junction. Gastrulation is similar in humans and other mammals that form a blastodisc. [(A), Adapted from Patten BM [1958] *Foundations of Embryology*. McGraw-Hill; (C, E), adapted from Balinsky BI [1975] *An Introduction to Embryology*, 4th ed. Saunders.]

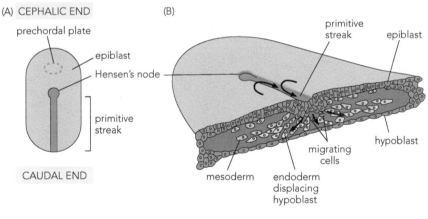

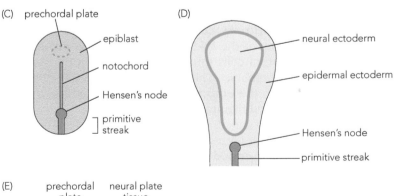

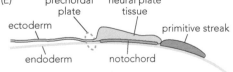

is displaced (Figure 2.4B). The hypoblast cells give rise to **extraembryonic tissues**—tissues that are not part of the embryo proper but contribute to the maintenance of the embryo. As in an amphibian embryo, a specified group of cells migrates inward along the midline to form the notochord. As the notochord elongates anteriorly and the germ layers are established, the primitive streak regresses (Figure 2.4C, E). At the anterior (cephalic) region of the epiblast, the **prechordal plate** (Figure 2.4A, C, E), an area where the surface ectoderm adheres to the underlying endoderm, forms. This tight association prevents cells of the notochord from migrating further anteriorly (Figure 2.4E). The epiblast cells remaining at the surface form neural ectoderm or epidermal ectoderm. The neural ectoderm forms a sheet of cells that rises and extends from the anterior portion of the primitive streak.

Like the dorsal blastopore lip, **Hensen's node**, a transient group of cells at the anterior (cephalic) end of the primitive streak, provides signals important for the development of neural ectoderm (Figure 2.4D, E), as does the **node**, the equivalent transient cell population in mammals.

Gastrulation in mice is a bit more difficult to envision because the epiblast is found inside the blastocyst, a cuplike structure (**Figure 2.5**). If one imagines the blastocyst cut open (dashed line, Figure 2.5A) and flattened out, the similarities between gastrulation in a blastodisc and blastocyst are more apparent. The mouse blastocyst is surrounded by **visceral endoderm** (VE). Like cells of the blastodisc, a subset of epiblast cells migrates through the primitive streak to form the mesoderm and endoderm layers between the epiblast and VE (Figure 2.5A, B). The VE does not become part of the definitive endoderm and is similar to the hypoblast in that it becomes displaced by the forming endoderm germ layer. A portion of the VE, called **anterior visceral endoderm** (AVE), secretes factors that keep some epiblast cells from migrating through the primitive streak, thereby designating cells that will form ectoderm. The AVE comes to lie below the anterior portion of the neural ectoderm and, as described in

Figure 2.5 Gastrulation in mouse embryos is similar to that in the chick. (A) The blastocyst with part of the posterior region cut away to reveal the orientation of the forming cell layers. If the blastocyst were cut open along the dashed lines and flattened out, the epiblast would be on the surface and the migration of cells (arrows) would be appear like the those of the chick blastodisc. (B) A sagittal section through the cup-shaped blastocyst of the mouse embryo reveals the epiblast in the inner region. The cells of the epiblast migrate through the primitive streak that begins at the junction of the extraembryonic and embryonic regions. The migrating epiblast cells begin to form germ layers between the epiblast and visceral endoderm. The visceral endoderm is an extraembryonic tissue that is replaced by cells that form the endoderm layer, similar to the hypoblast in the chick. The node is located at the anterior end of the primitive streak. (C) A section through the blastocyst at a later stage of development reveals the orientation of the neural ectoderm, epidermal ectoderm, and notochord. The anterior visceral endoderm underlies the anterior portion of the neural tissue. [Adapted from Nagy A, Gertsenstein M, Vinterstan K & Behringer R [2003] *Manipulating the Mouse Embryo*, 3rd ed. Cold Spring Harbor Press.]

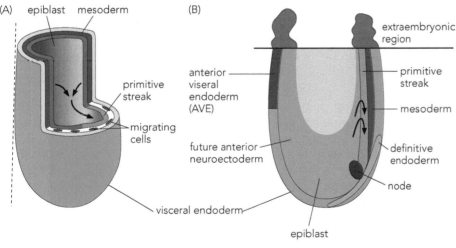

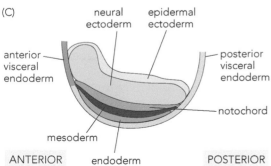

Chapter 3, subsequently plays a role in defining regions of the developing neural tube (Figure 2.5C).

NEURAL INDUCTION: EARLY DISCOVERIES

The ectoderm of the late gastrula-stage vertebrate embryo has the potential to become either neural or epidermal tissues. By the early twentieth century, embryologists had established that dorsal ectoderm forms neural plate tissue, whereas ventral ectoderm forms epidermal tissues. What remained unclear was how the ectoderm became specialized as one tissue or the other.

Amphibian Models Were Used in Early Neuroembryology Research and Remain Popular Today

Many of the first studies of neural induction were conducted using amphibian embryos, particularly those from salamanders and frogs. Frogs remain a popular animal model today, in part because of the comparatively large size of their eggs and early-stage embryos. For example, the frog egg is about 2–3 mm in diameter compared to the 0.1–0.2 mm diameter of a human egg. This larger size makes cellular regions easier to identify and experimentally manipulate. Yet, despite the advantages of larger embryos, these early experiments required considerable creativity (for example, designing tiny glass needles to use as surgical knives and using a baby's hair to separate blastomeres), tremendous patience, and very steady hands. The technical accomplishments of these first experiments, now almost a century old, are nearly as impressive as the scientific hypotheses that resulted from the studies.

Among the technical achievements was the grafting of tissue harvested from one region of an early gastrula-stage amphibian embryo (the donor) to another region of a second embryo at the same developmental stage (the host). As the host embryo continued to develop, the effects of the donor and host tissues on subsequent developmental events were evaluated.

A Region of the Dorsal Blastopore Lip Organizes the Amphibian Body Axis and Induces the Formation of Neural Tissue

In 1924 Hans Spemann and Hilde Mangold published what has become one of the most cited studies in developmental neurobiology. In this study, grafting experiments were done with two different species of newts, one species that was pigmented and one that was not. The differences in pigmentation allowed them to visualize the fate of the donor and host tissues. Spemann and Mangold grafted the dorsal blastopore lip (DBL) of a pigmented, early gastrula-stage embryo to the ventral side of a nonpigmented embryo (the host) at the same stage of development (**Figure 2.6**).

Figure 2.6 The classic grafting experiment performed by Hans Spemann and Hilde Mangold. The dorsal region of a blastopore lip from a pigmented donor embryo was grafted onto a nonpigmented host embryo's ventral surface, an area of ectoderm that does not give rise to neural tissue. As development proceeded, a neural plate and body axis formed along the dorsal surface of the host embryo, as expected. In addition, a second neural plate and body axis formed from the ventral tissue of the host. This result indicated that the donor dorsal blastopore lip produced signals that induced host ventral ectoderm to form an entirely new body axis complete with neural tissue.

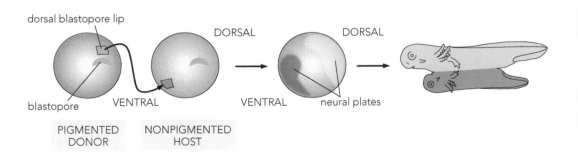

dorsal blastopore lip

DORSAL   DORSAL

blastopore   VENTRAL   VENTRAL   neural plates

PIGMENTED   NONPIGMENTED
DONOR   HOST

Although the ventral side of the embryo normally gives rise to epidermal tissue, at this early stage of gastrulation the regions of epidermal and neural ectoderm have not yet been specified. Therefore, the investigators could evaluate any inductive effects of the DBL.

Spemann and Mangold discovered that the host embryos not only continued to develop but also formed a second body axis, complete with neural tissue. The key observation from this experiment was that the new body axis formed primarily from the nonpigmented host embryo. This indicated that the ventral tissue of the host—the ectoderm that normally becomes epidermal—was induced to form neural tissue when provided with a signal originating in the DBL. Thus, the second body axis did not result by a simple expansion of the pigmented donor dorsal lip tissue, as was proposed previously by other investigators.

The band of DBL that induced the formation of new neural tissue came to be called **Spemann's organizer** because it organized the entire body axis. Spemann and his students, as well as investigators in many other labs, continued to explore the mechanisms that governed neural induction. Spemann ultimately received the Nobel Prize in Physiology or Medicine in 1935 for his work. Tragically, Hilde Mangold died in 1924 at the age of 26 following a household oven explosion and was unable to witness the continuing impact of her work. Her husband, Otto Mangold, also a student of Spemann's, made additional contributions regarding the mechanisms governing cell determination and induction of embryonic tissues.

The intriguing discovery of a neural inducer originating in the organizer—that is, the DBL of amphibians or the presumptive equivalent in other species, led to numerous studies investigating the nature of this signal. Additional grafting experiments in the 1930s noted that the signal arising from the organizer was not species-specific. For example, when grafts of organizer tissue (Hensen's node) from duck were transplanted to chick, neural tissues formed. Similarly, when grafts of chick organizer tissue were transplanted to either duck or rabbit, neural tissues were induced. Thus, the signal appeared to be universal in its ability to induce the formation of neural tissue from early-stage ectoderm. Other grafting experiments in the 1930s revealed that the medial section of the amphibian DBL preferentially gave rise to head structures while the lateral portion induced tail structures. This suggested the possibility of multiple signals available to direct the formation of specific neural regions along the anterior–posterior body axis of the embryo (see Chapter 3).

The Search for the Neural Inducer Took Decades of Research

For almost 100 years, scientists have investigated the signals governing neural induction, using an ever-expanding array of experimental approaches. It is hard to capture in just a few paragraphs the extensive efforts made and the number of scientific debates and alternative hypotheses put forth during the search for putative neural inducers.

One serious obstacle to identifying neural inducers is the very small size of the organizer region. Because the small tissue size prevented direct isolation of the signaling molecule, some researchers screened larger tissues for more plentiful sources of inductive activity while others tested various substances for inductive properties.

Through the many studies, it was recognized that the neural inducer was diffusible and that, surprisingly, any number of biological and nonbiological substances could induce competent undifferentiated ectoderm to form neural tissue. For example, tissues that were first boiled to denature active proteins still induced neural tissue, as did dyes and other chemicals known to be toxic to cells. For several years scientists struggled to

Figure 2.7 *In vitro* assays show the effect of the mesoderm on the fate of cultured animal caps. (A) When intact *Xenopus* animal caps without any mesoderm tissue are removed and placed in culture, epidermal tissue forms. (B) When intact animal caps include mesoderm tissue, neural tissue forms. (DBL, dorsal blastopore lip.)

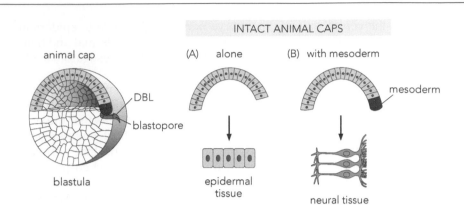

distinguish true neural inducers from "artificial activators," as these non-specific substances came to be called.

New Tissue Culture Methods and Cell-Specific Markers Advanced the Search for Neural Inducers

One experimental technique developed during the early era of neuroembryological research proved essential for distinguishing true neural inducers. Scientists identified ways to isolate animal caps from blastula and early gastrula-stage amphibian embryos and maintain them in various tissue culture environments. This technique—first developed in the 1920s, and later refined in the 1960s—has been employed to assess interactions among cells from the animal pole, the vegetal pole, and the band of mesoderm that forms between the poles.

Cells from these early stages of development are not morphologically distinct from one another, so identifying whether embryonic cells in tissue culture are endodermal, mesodermal, epidermal, or neural is often difficult. Methods for identifying the different cell types were needed to sort out whether a given experimental treatment caused direct or indirect induction of a tissue type. For example, a given treatment might induce the formation of neural cells directly or might first induce the formation of organizer-like mesoderm that would then induce neural cells. As techniques were developed in the latter part of the twentieth century for labeling sequences of DNA or RNA by *in situ* **hybridization** and specific proteins by **immunocytochemistry**, it became possible to identify early cell types based on the expression of these **cell markers** (see Chapter 1). For example, **neural cell adhesion molecule** (**NCAM**) is found only in neural cells and provides a means of identifying these cells in a mixed cell population.

Examples of some of the first discoveries made using the animal cap assay and labeling techniques are shown in **Figure 2.7**. When amphibian animal caps (that is, ectoderm) were cultured alone, epidermal tissues formed. However, when animal caps were cultured in the presence of mesoderm, the animal cap cells became neural (Figure 2.7A, B). As the DBL is formed from a band of cells that includes mesoderm tissue, these results were consistent with the notion that signals from the organizer induce neural tissue from unspecified ectoderm.

NEURAL INDUCTION: THE NEXT PHASE OF DISCOVERIES

Prior to the 1980s, it was generally assumed that neural inducers influenced ectoderm directly, instructing select cells of that germ layer to become neural rather than epidermal tissue. However, more than 50 years after the first description of the organizer, several experiments carried out in different labs changed this fundamental assumption.

Some alternative hypotheses arose after recognizing limitations of previously used techniques. For example, it was found that animal cap cells from some amphibians can "auto-neuralize"—that is, become neural tissue in the absence of other tissues or added factors. This explained, in part, why so many different substances appeared to induce the formation of neural tissues; animal cap cells from these amphibians would form neural tissues no matter what substance was added. It was also observed that animal cap cells of the *Xenopus laevis* embryo do not auto-neuralize to the same extent as those from other frogs. Therefore, researchers relied on these amphibians to design experiments to directly test for neural inducers.

Studies Suggest Neural Induction Might Require Removal of Animal Cap-Derived Signals

As noted previously, mesoderm forms between the vegetal and animal poles. Scientists in the 1970s had already determined that mesoderm formation required interaction with one or both poles, and by the 1980s various groups sought to identify mesoderm-inducing signals emanating from either pole. Mesoderm formation was also of interest to scientists studying neural induction because the organizer is a mesoderm-derived structure.

In the search for mesoderm inducers, growth factors from the **fibroblast growth factor** (**FGF**) and **transforming growth factor β** (**TGFβ**) families were added to cultures of dissociated animal cap cells. In these experiments, isolated animal caps were first treated with enzymes to break apart cell contacts. These dissociated cells were then placed in cell cultures and treated with different growth factors. The treated cells reaggregated in the cell culture dishes and formed mesoderm tissue (**Figure 2.8**A). Notably, the investigators observed that untreated, control cultures (those lacking any additional growth factors) reaggregated as either epidermal or neural cells depending on how long they remained dissociated. When cells were kept dissociated for only short periods prior to reaggregation, they became epidermal. However, if cells remained dissociated for over 3 hours prior to reaggregation, they became neural (Figure 2.8B, C). These observations led to the suggestion that during the longer dissociation period presumptive contact-mediated or secreted signals arising from the animal cap cells were no longer at a concentration sufficient to prevent neural formation. Thus, without this signal, neural tissue formed instead of epidermal tissue. Other explanations were also offered, but the idea that neural tissue could form when a specific signal was *absent* received support from other studies conducted around the same time.

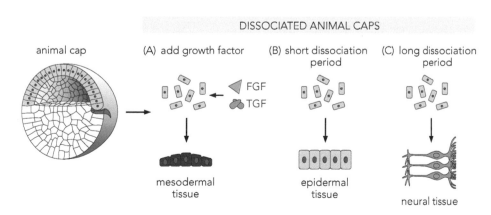

Figure 2.8 *In vitro* assays reveal how growth factors and dissociation time influence the fate of dissociated animal cap cells. (A) If *Xenopus* animal caps without any mesoderm tissue are dissociated into single cells and then cultured in the presence of a growth factor, such as fibroblast growth factor (FGF) or transforming growth factor (TGF), the cells reaggregate to form mesodermal tissue. (B) If cells are kept dissociated for a brief period without additional growth factors, the cells reaggregate to form epidermal tissue. (C) If the untreated animal cap cells are kept dissociated for longer periods (>3 h), the cells reaggregate to form neural tissue.

Mutation of the Activin Receptor Prevents the Formation of Ectoderm and Mesoderm but Induces Neural Tissue

Another set of experiments investigating mesoderm formation focused on the role of the TGFβ signaling pathway. Several TGF-related proteins, including one called **activin**, were identified as candidate mesoderm inducers. Activin and its associated serine/threonine kinase receptors were noted to be endogenous to blastula-stage animal cap regions. Investigators therefore disrupted the activin signaling pathway to assess its role in mesoderm formation.

To interfere with normal signaling, large amounts of synthetic RNA coding for a truncated (and thus inactive) version of the activin receptor were injected into animal cap cells. The truncated receptors lacked only the kinase domain of the receptor subunits; the extracellular binding region remained. Therefore, activin could bind to the extracellular regions, but the ligand was unable to transduce intracellular signals. The truncated receptors were injected in excess so the mutated receptors would sequester the majority of activin or activin-like ligands and prevent signal transduction through any endogenous activin receptors. This is a type of **dominant negative** approach in which a mutation is great enough to interfere with a normal signaling event. RNA that did not interfere with receptor signaling was used as a control.

Control or experimental RNA was injected into the two-cell stage animal pole of the frog *Xenopus* (**Figure 2.9**A, B). After developing to the blastula stage, the animal caps were dissected and placed in cell culture for further experimentation. When animal cap cells with normal (control) receptor activity were placed in culture with increasing concentrations of activin, different mesoderm derivatives, including blood, muscle, notochord, and heart cells, formed (Figure 2.9C). These animal caps also gave rise to neural tissues, presumed to be induced by the newly formed

Figure 2.9 Blocking activin signaling disrupts mesoderm formation but leads to the formation of neural tissue. (A) Signal transduction through activin receptors is similar to other members of the TGF receptor family. The receptors are present as dimers that come into close association following ligand binding. As the type II and type I receptor subunits come together, the type II receptor phosphorylates the intracellular region of the type I receptor, leading to signal transduction. (B) When the intracellular portion of the type I receptor is removed, ligand binding cannot initiate signal transduction because there is no site for phosphorylation to occur. Thus, these truncated receptors sequester the ligand but are nonfunctional (dominant negative approach). (C) When intact animal caps from embryos expressing functional activin receptors were cultured in the presence of activin, mesoderm formed, suggesting a role for activin signaling in mesoderm induction. Neural tissue also formed under these conditions, presumably in response to signals derived from the mesoderm. (D) When intact animal caps from embryos expressing the truncated activin receptor were cultured in the presence of activin, mesoderm was not induced, further supporting the hypothesis that activin signaling is needed for mesoderm formation. Surprisingly, however, neural tissue formed, rather than epidermal tissue, which normally arises from intact animal cap cells lacking mesoderm signals (see Figure 2.7A). Thus, the disruption of activin signaling led to neural tissue formation, suggesting neural tissue may be the default state, with epidermis requiring an inductive signal.

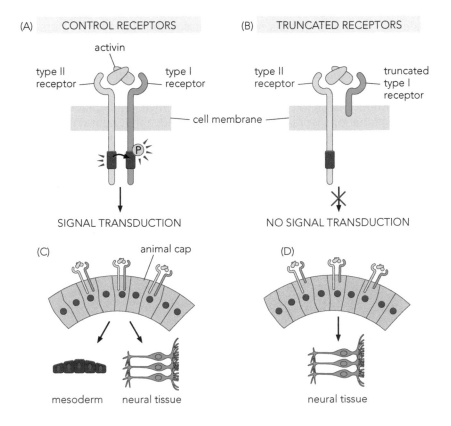

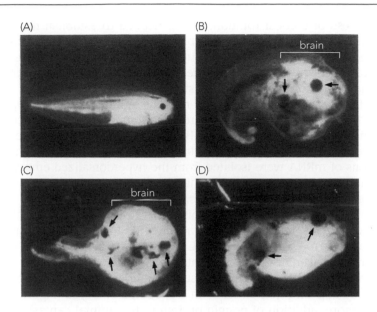

(A)

(B)

brain

(C)

brain

(D)

Figure 2.10 Disruption of activin signaling in whole embryos leads to increased neural tissue. *Xenopus* embryos at the two-cell stage were injected with mRNA for globin (control condition) or for the truncated activin receptor (experimental condition). (A) Embryos developed normally when injected with globin mRNA. (B–D) Embryos formed greater amounts of neural tissue, including extra eyes (arrows) and expanded brain (brackets, B, C) when activin signaling was disrupted. [Adapted from Hemmati-Brivanlou A & Melton DA [1994] *Cell* 77:273–281.]

mesoderm. In contrast, the animal caps expressing the truncated receptor did *not* form mesoderm when treated with activin. Thus, as hypothesized, interfering with activin signaling prevented mesoderm induction.

What was not anticipated however, was that these animal caps expressed neural cell markers and formed only neural tissue (Figure 2.9D). This observation was surprising because in previous experiments, intact animal caps cultured in the absence of activin or mesoderm formed only epidermal tissue (see Figure 2.7A). This suggested that functional activin signaling was normally needed not only for mesoderm induction, but also for the formation of the epidermal ectoderm. Because blocking activin signaling led to the formation of neural tissue rather than epidermal tissue, it appeared that neural tissue formed by default if epidermal formation was inhibited.

To explore this possibility further, additional studies were done in which truncated activin receptors were again injected into embryos at the two-cell animal cap stage, but this time the embryos were left to develop until the tail began to emerge (tail bud stage). These embryos had expanded regions of neural tissue, further suggesting that inhibiting a signaling pathway allowed the formation of neural tissues (**Figure 2.10**).

At first these results may appear to contradict earlier work. The tissue culture studies showed that neural tissue formed when activin-mediated mesoderm induction was prevented, suggesting mesoderm was not important for neural induction. Yet, it was also well established that the mesodermal DBL induced neural tissue (see Figure 2.7B). Fortunately, at the same time these studies were being conducted, additional labs were investigating mesoderm and ectoderm formation. Together the results from those labs helped revise the concept of neural induction and revealed how mesoderm blocks epidermal formation and allows neural tissue to form.

Modern Molecular Methods Led to the Identification of Three Neural Inducers

One influential and elaborate set of experiments in *Xenopus* led to the first successful isolation of a neural inducer. This research utilized **ventralized embryos** that lack a nervous system and other dorsal structures and **hyperdorsalized embryos** that have a larger than normal nervous system. Treatment of embryos with lithium chloride leads to the expansion of dorsal regions, including the nervous system. In contrast, ventralized embryos are created by exposure to ultraviolet (UV) light, which disrupts the process of cortical rotation, a step needed to establish the dorsal–ventral body axis in *Xenopus*. Notably, the ventralized embryos could be rescued to develop normally by grafting DBL tissue from a normal embryo. As a result, this preparation became a useful assay to identify molecules that induce the formation of neural structures.

Hypothesizing that the larger nervous systems found in hyperdorsalized embryos would provide a greater amount of neural-inducing signals than normal embryos, Richard Harland and William Smith used these embryos as a source of material to isolate candidate neural inducers. Fractions of mRNA were isolated from the hyperdorsalized embryos and screened for the ability to rescue ventralized embryos. Smaller and smaller fractions of active mRNA extracts were tested until eventually only a single active molecule remained. When this final extract was injected into ventralized embryos, a normal embryo resulted (**Figure 2.11**A), reminiscent of what occurred with grafts of DBL tissue. Through this process a novel protein they called **noggin** was identified.

Purified noggin protein was subsequently found to induce a normal body axis and neural tissue in UV treated embryos (Figure 2.11B). Furthermore, addition of noggin protein to the animal cap assay led to the formation of neural cells rather than epidermal cells, even in the absence of mesoderm tissue. This suggested that noggin was a direct inducer of neural tissue. Significantly, noggin expression was also found in the organizer regions, beginning in the late blastula stage. More than half a century after Spemann and Mangold revealed the importance of the organizer, noggin was the first novel neural inducer to be conclusively identified.

A second neural inducer was soon identified by Ali Hemmati-Brivanlou, Douglas Melton, and colleagues. In other developing tissues, it had already been shown that the reproductive hormone **follistatin** could bind to and inactivate activin. Subsequent studies noted that follistatin was expressed in the *Xenopus* gastrula-stage organizer region, suggesting a possible role in neural induction. Like noggin, follistatin induces expression of neural

Figure 2.11 The neural inducer noggin was discovered in studies using ventralized and hyperdorsalized embryos. (A) Experimental treatment of early-stage amphibian embryos leads to expansion or loss of neural regions. (Top) When lithium chloride (LiCl) was added to the 32-cell stage embryo to inhibit dorsal–ventral axis specification, a hyperdorsalized embryo with expanded dorsal regions formed. (Bottom) In embryos treated with UV light for 20–30 min following fertilization, dorsal structures failed to form because cortical rotation was disrupted. However, when RNA extracts from hyperdorsalized embryos were injected into the ventralized embryos, dorsal structures formed and the embryos were rescued. By testing smaller and smaller fractions of the RNA extracts, the protein noggin was ultimately identified. (B) Images from the 1992 study of Smith and Harland compare control embryos to those treated with UV light and those treated with UV light then rescued with injection of noggin protein. Injection of noggin was able to rescue many of the UV treated, ventralized embryos so that they developed dorsal structures comparable to controls. [Adapted from Smith WC & Harland RM [1992] *Cell* 70:833–840.]

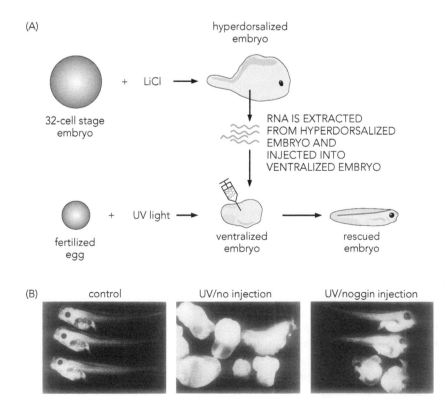

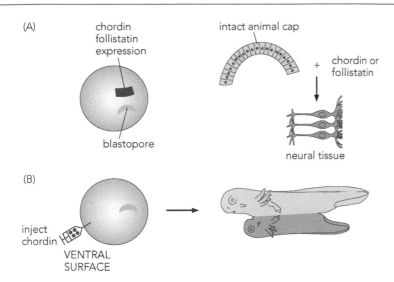

Figure 2.12 Chordin and follistatin also induce neural tissue. (A) Both chordin and follistatin are expressed in the dorsal blastopore lip (the organizer), and both molecules induce intact animal caps to form neural tissue rather than epidermal tissue. (B) Injection of chordin into the ventral (nonneural) surface of the embryo induces a second body axis, similar to the results of the original grafting experiment by Spemann and Mangold.

markers when injected into animal cap cells, yet mesoderm tissue does not form (**Figure 2.12**A). Thus, follistatin was proposed to be a direct neural inducer that might act by inhibiting activin.

During this same period, a third candidate neural inducer, **chordin**, was identified by Edward De Robertis and colleagues while screening for gene transcripts expressed specifically in the organizer region. When chordin mRNA was injected into the animal pole or when soluble chordin protein was added to animal cap cultures, neural tissue formed, but mesoderm did not (Figure 2.12A). When chordin was injected in the ventral regions of a frog embryo, it also induced a second body axis much like the grafting experiment by Spemann and Mangold (Figure 2.12B). Thus, within a span of about five years, three unrelated candidate neural inducers were identified using methods that were unavailable, and likely unimaginable, to earlier generations of neuroembryologists.

NOGGIN, FOLLISTATIN, AND CHORDIN PREVENT EPIDERMAL INDUCTION

The newly discovered neural inducers seemed to act directly, without the need for mesoderm tissue. However, other studies had found that blocking activin, a molecule used in mesoderm formation, induced neural tissues. Research efforts in multiple areas resolved this apparent contradiction.

Studies of Epidermal Induction Revealed the Mechanism for Neural Induction

One important discovery was that **bone morphogenetic proteins (BMPs)** induced epidermal markers while suppressing neural markers. The BMPs are a large family of molecules and are part of the larger TGFβ superfamily. BMPs influence many developmental events; despite their name, BMP functions are not limited to bone formation. BMP2 and BMP4 are expressed throughout most ventral regions of the frog embryo, but are ultimately restricted to regions of ectoderm associated with epidermal tissue. In cultures of animal cap cells dissociated for 3 hours or more, treatment with BMP2 or BMP4 led to the formation of epidermal rather than neural cells (**Figure 2.13**A, B). The BMP-treated cells expressed epidermal markers such as keratin but did not express any neural markers. Other studies using intact animal cap assays found that BMP4, at a high enough concentration, inhibited the ability of chordin, noggin, or follistatin to form

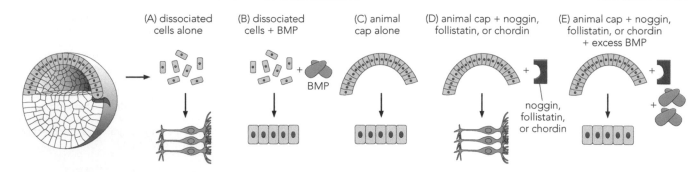

Figure 2.13 Animal cap assays demonstrate that bone morphogenetic proteins (BMPs) induce epidermal tissue. (A) Animal cap cells that remain dissociated for >3 hours form neural tissue, as previously seen (Figure 2.8C). (B) Dissociated cells treated with BMP form epidermal tissues. (C) Intact animal caps that are untreated also form epidermal tissue (see also Figure 2.7). (D) Intact animal cap cells treated with noggin, chordin, or follistatin form neural tissue. (E) The addition of an excess concentration of BMP to the treated cultures induces epidermal tissue because the concentration of BMP is sufficient to prevent the actions of the neural inducers (noggin, chordin, or follistatin).

neural cells, and instead induced epidermal tissue, as occurred in cultures of animal cap tissue alone (Figure 2.13C–E). In contrast, in experiments that blocked BMP signaling, including those using truncated activin receptors, neural tissue, rather than epidermal tissue, formed (see Figure 2.9D). Together these observations suggested that BMP signaling was required for the formation of epidermal tissue, and that blocking BMP signaling was necessary to form neural tissue.

The Discovery of Neural Inducers in the Fruit Fly *Drosophila* Led to a New Model for Epidermal and Neural Induction

Emerging data from experiments in the fruit fly species, *Drosophila melanogaster*, helped link the roles of neural inducers and BMPs in the formation of neural and epidermal tissues (**Box 2.1**). Scientists discovered that like their *Drosophila* homologs, chordin, as well as noggin and follistatin, interacted with BMP molecules to block BMP signaling. Experiments with

---

Box 2.1 Neural Induction: Lessons from Invertebrates

Studies in the fruit fly *Drosophila melanogaster* helped elucidate how changes in bone morphogenetic protein (BMP) signaling established epidermal versus neural fate in vertebrate animal models. Despite major structural and developmental differences between *Drosophila* and the vertebrates, there is a remarkable conservation of signaling pathways across species. Studies in *Drosophila* first revealed how genes expressed on the opposite sides of the embryo could differentially induce epidermal or neural tissues. Unlike vertebrate animal models, *Drosophila* epidermis is formed on the dorsal side of the embryo, whereas neural tissue is formed on the ventral side (see also Chapter 1).

Fruit fly embryogenesis is quite different from that of vertebrates. Following fertilization of the ovoid-shaped *Drosophila* egg, nuclei divide rapidly within the cytoplasm prior to migrating to the cortex of the cell, where they form a **syncytial blastoderm**. The

nuclei are then surrounded by cell membrane, creating individual cells of the **cellular blastoderm** (**Figure 2.14**).

Beginning during the syncytial blastoderm stage, gradients of a protein called Dorsal are established with the highest concentrations ventrally and the lowest concentrations dorsally. At the dorsal surface, the lower concentration of Dorsal protein induces expression of the gene *Decapentaplegic* (*Dpp*), the homolog of vertebrate BMP. Intermediate levels of Dorsal protein lead to the expression of the gene *Short of gastrulation* (*Sog*), the chordin homolog, in ventrolateral regions. In the ventral-most region, the highest concentrations of Doral induce *Twist*, a gene encoding a basic Helix-Loop-Helix (bHLH) transcription factor important in mesoderm induction (**Figure 2.15**).

The gradients of Dorsal protein thus create gradients of the secreted proteins, Dpp and Sog, with higher concentrations of Dpp located dorsally and higher

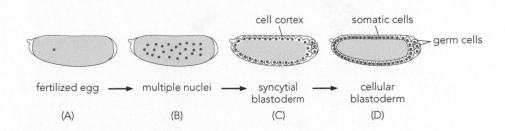

cell cortex

somatic cells

germ cells

fertilized egg → multiple nuclei → syncytial blastoderm → cellular blastoderm

(A)   (B)   (C)   (D)

Figure 2.14 Formation of the blastoderm in *Drosophila*. After the ovoid-shaped egg cell of a *Drosophila* is fertilized (A), nuclei begin to divide rapidly within the egg cytoplasm (B). The nuclei then migrate to the cell surface (cortex), forming the syncytial blastoderm (C). The nuclei become surrounded by cell membranes and form the cellular blastoderm that contains both the somatic and germ cells (D).

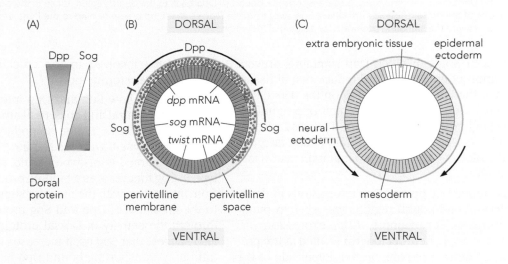

(A)   (B)   DORSAL   (C)   DORSAL

Dpp Sog

Dpp

extra embryonic tissue   epidermal ectoderm

dpp mRNA

sog mRNA

twist mRNA

Sog   Sog

neural ectoderm

Dorsal protein

perivitelline membrane   perivitelline space

mesoderm

VENTRAL   VENTRAL

Figure 2.15 Signals direct the formation of epidermal tissue at the dorsal surface of the *Drosophila* embryo. (A) The protein Dorsal is produced in a gradient, with the highest levels produced ventrally and the lowest levels produced dorsally. The concentration of Dorsal differentially regulates gene expression for the proteins Decapentaplegic (Dpp) and Short of gastrulation (Sog). (B–C) Cross-sections through the *Drosophila* embryo reveal the origins of the signals that induce epidermal and neural tissues. The lower concentrations of Dorsal protein in the dorsal-most region of the embryo induce expression of *Dpp* mRNA. Intermediate levels of Dorsal protein lead to the expression of *Sog* mRNA in ventrolateral regions, while the highest concentrations of Dorsal protein in the ventral-most region induce expression of mRNA for *Twist*. The protein products that result from *Dpp* and *Sog* expression are set up as gradients within the perivitelline space that surrounds the embryo such that Dpp is found in the highest concentration in dorsal regions of the embryo, whereas Sog is found in higher concentrations at progressively more ventral regions. Like the homologous vertebrate neural inducer chordin, Sog inhibits the activity of Dpp—the *Drosophila* homolog of BMP—to block the induction epidermal ectoderm in the ventrolateral regions that normally give rise to neural ectoderm. (C) The high concentrations of Dpp in the dorsal region induce epidermal ectoderm and extraembryonic tissue. Higher concentrations of Sog in the more ventral regions block the activity of Dpp to prevent the induction of epidermal ectoderm in the regions that give rise to neural ectoderm. Mesoderm, induced by Twist, is located to the ventral-most region prior to gastrulation, but will later invaginate to form the middle germ cell layer, while the neural ectoderm will extend more ventrally (arrows).

concentrations of Sog ventrolaterally. The high concentration of Dpp on the dorsal side of the embryo induces extraembryonic (amnioserosa) and epidermal

tissue, whereas the high concentration of Sog ventrolaterally induces neural tissue. Sog also antagonizes Dpp. On the ventrolateral sides of the embryo, the

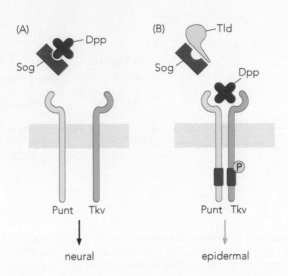

Figure 2.16 Sog and Dpp interactions regulate formation of neural and epidermal tissue. (A) To block epidermal formation and induce neural tissue, Sog binds to Dpp to prevent it from activating the serine/threonine kinase receptor subunits Punt and Thick veins (Tkv). (B) Under normal conditions, Sog transports the bound to Dpp back to the dorsal region. Once in the dorsal region, local concentrations of the protein Tolloid (Tld) cleave Sog away from the Dpp so that Dpp is able to bind to the Punt and Tkv receptor subunits and activate the signaling pathways for epidermal induction.

high levels of Sog can block Dpp signaling, prevent the formation of epidermis, and induce neural tissue. In contrast, the lower levels of Sog on the dorsal side of the embryo are insufficient to inhibit Dpp signaling, thereby allowing epidermis to form. Thus, in *Drosophila*, Sog blocks Dpp signaling to retain neural fate, just as chordin blocks BMP activity to maintain neural fate in vertebrates.

The importance of Dpp and Sog interactions in defining epidermal and neural regions was seen in multiple experiments. For example, if *Dpp* expression was eliminated, ectopic neuroectoderm formed in the presumptive epidermal regions on the dorsal side of the fly embryo. Conversely, experimental overexpression of *Dpp* converted regions of presumptive neuroectoderm to epidermis.

Evidence for cross-species conservation of these signaling pathways was demonstrated when *Drosophila Sog* mRNA was injected into *Xenopus* embryos, inducing the formation of a second body axis, much like that observed in the early transplantation studies of Spemann and Mangold. Thus, studies in the fruit fly revealed the evolutionary conservation of signaling

mechanisms involved in the crucial early steps of nervous system formation.

The details of how Sog and Dpp interact to differentially induce neural and epidermal structures continue to be explored. Like BMPs, homodimers of Dpp proteins bind to type II and type I serine/threonine kinase receptors named Punt (type II) and Thick veins (Tkv; type I). Sog functions as a binding protein to block Dpp from interacting with the receptor subunits. In addition to the gradients of Dpp and Sog proteins established through the activity of Dorsal protein, investigations also suggest that Sog itself increases the Dpp concentration by transporting bound Dpp back to the dorsal region. In the dorsal region, the Sog is cleaved from Dpp by the protein Tolloid (Tld), thus releasing Dpp so it can again bind to local receptors (**Figure 2.16**). Other proteins related to Dpp and Sog are also differentially expressed in the fly embryo, where they contribute to the signaling pathways that induce the formation of neural, epidermal, and extraembryonic tissues. The mechanisms by which these proteins interact, and how homologs of the proteins function in vertebrates, continue to be explored.

noggin or chordin revealed specific binding between each of these inducers and BMP2, BMP4, and BMP7 (**Figure 2.17**A). Together, the accumulated data pointed to a mechanism by which ectoderm becomes epidermis when the BMP signaling pathway is active. However, if the BMP signal is blocked by one of the candidate neural inducers, neural tissue forms. As this model indicates, the elusive neural inducers sought by scientists for so many years were actually the inhibitors of epidermis-inducing signals. The

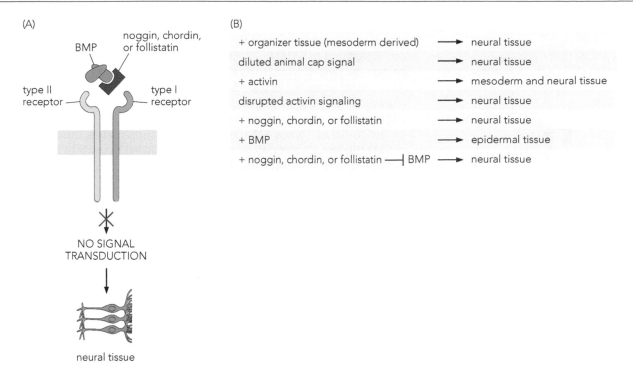

Figure 2.17 Blocking the BMP signal transduction pathway allows neural tissue formation. (A) The neural inducers, (noggin, chordin, and follistatin) bind to BMP and prevent it from activating the type II and type I BMP receptors. Because this signal transduction pathway is blocked, neural tissue forms by default. (B) Among the key discoveries that led to the realization that neural tissue forms by default and BMPs actively induce epidermal tissue were: (1) The organizer, a mesoderm-derived tissue, induced neural tissue. (2) Diluting a signal normally present in animal cap cells induced neural tissue. (3) The protein activin, a member of the TGFβ superfamily, induced mesoderm and neural tissues. (4) Disrupting activin signaling also induced neural tissue. (5) Three proteins in the organizer induced neural tissue. (6) BMP proteins, also members of the TGFβ superfamily, induced epidermal tissue. (7) Neural inducers blocked BMP signal transduction to prevent epidermal induction and allow neural tissue to form.

DBL or node secretes the BMP antagonists noggin, follistatin, and chordin that then diffuse into the dorsal regions of the embryo where they inhibit endogenous BMP signals. Thus, the ectoderm closest to the node receives the least BMP signaling and therefore forms neural tissue. In contrast, BMP signaling in the ectoderm further from the node is not blocked, because the ventral ectoderm is not exposed to sufficient levels of inducers. Thus, BMP is able to bind to the endogenous BMP receptors and induce epidermal tissue. After nearly a century of work and the contributions of dozens of labs, the basic mechanisms for neural induction were beginning to be understood (Figure 2.17B).

Interestingly, activin itself is not part of the neural induction pathway. It was later determined that the truncated activin receptor used in the original experiments was not specific for activin, but recognized many ligands of the TGFβ family, including the BMPs. The fortuitous use of this nonspecific receptor contributed greatly to our understanding of both epidermal and neural inducers.

BMP Signaling Pathways Are Regulated by SMADs

BMPs, like other members of the TGFβ superfamily bind to heterodimeric serine/threonine kinase receptors. The interaction of the two receptor subunits leads to phosphorylation of a serine or threonine on the type I subunit. The phosphorylated type I subunit then phosphorylates certain transcription factors within the cytoplasm that translocate to the nucleus to activate target genes (**Figure 2.18**).

Among the transcription factors phosphorylated following BMP receptor binding are those of the SMAD family. The *Smad* genes are homologs of

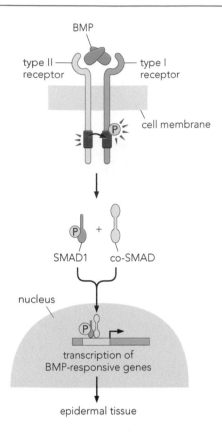

Figure 2.18 SMAD transcription factors activate BMP-responsive genes. BMP binds to heterodimeric serine/threonine kinase receptors leading to phosphorylation of the type I receptor. The now-activated type I receptor phosphorylates C-terminus serines on SMAD1, one of the receptor-regulated SMAD proteins (R-SMADs). SMAD1 then associates with co-SMAD and both are transported to the nucleus to activate BMP-responsive genes that induce epidermal tissue.

the *Drosophila Mad* and *Caenorhabditis elegans Sma* genes (see Chapter 4, Box 4.1). In mammals, eight SMAD family members have been identified and are designated as SMAD1, -2, -3, -4, -5, -6, -7, and -8/9. SMAD8/9 is a single protein, but was referred to as either SMAD8 or SMAD9 in earlier studies.

SMAD signaling is transduced by a trimer comprised of two receptor-regulated SMADs (R-SMADs) and one common-mediator SMAD (co-SMAD). R-SMADs include SMAD1, -2, -3, -5, and -8/9. SMAD4 functions as the co-SMAD, while SMAD6 and SMAD7 function as inhibitory SMADs (I-SMADS; see Chapter 4). SMAD1, the R-SMAD important for epidermal induction, is phosphorylated at its C-terminus serine residues, and once phosphorylated, joins the co-SMAD to travel to the nucleus to activate transcription of the BMP-responsive genes that lead to formation of epidermal tissue (Figure 2.18). Because chordin, noggin, and follistatin interfere with BMP signaling, SMAD1 is never phosphorylated and therefore the transcription of BMP-responsive, epidermal inducing genes does not occur. Thus, those regions of ectoderm remain neural.

Additional Signaling Pathways May Influence Neural Induction in Some Contexts

As the default pathway for neural tissue formation was established and the general mechanisms by which BMPs induced epidermal tissue were identified, it appeared that the process of neural induction was at last understood. However, the signaling pathways responsible for designating ectoderm as neural or epidermal are more complex and in some cases, species specific. Further, several findings suggest that the neural inducers are used mainly to designate anterior brain structures, at least in some species.

Additional Neural Induction Pathways May Be Used in Some Species

Research on neural induction in other vertebrates has revealed interesting differences and discrepancies. For example, the expression of chordin, noggin, and follistatin in the node of the chick does not entirely coincide with the timing of neural induction. Furthermore, chordin alone is unable to induce neural tissue in chick, but if the ectoderm is transiently exposed to a graft of node tissue, neural induction occurs, suggesting that other signals are present in the node at earlier stages. Further, in genetically altered mice that lack one or more of the neural inducers, neural tissue was still detected, though not always in its entirety. Thus, while the node tissues of chick and mice play an important role in neural induction, other signals, including those expressed prior to gastrulation, may be needed in these species.

Among the signaling molecules thought to contribute to neural induction are FGFs and Wnts (homologs of the *Wingless* genes of *Drosophila*). FGFs have multiple roles at different stages of neural induction. For example, in at least some species, FGFs in the hypoblast regulate which cells remain at the epiblast surface during gastrulation. At later stages, FGFs expressed in the organizer help influence which cells become neural. There are 22 FGF ligands in vertebrates that bind with high affinity to specific FGF receptors. The four vertebrate FGF receptors are tyrosine kinase receptors that dimerize and cross-phosphorylate upon ligand binding (**Figure 2.19**). FGF dimers and their receptors contain a heparan sulfate proteoglycan (HSPG) binding region that allows HSPGs in the extracellular space to facilitate ligand binding and signal transduction. The phosphorylation of the FGF receptors leads to the activation of various intracellular pathways, including the Ras–Erk pathway (see Chapters 1 and 3).

FGF inhibits BMP signaling by phosphorylating a specific region of the SMAD1 transcription factor, called the linker region. Phosphorylation of

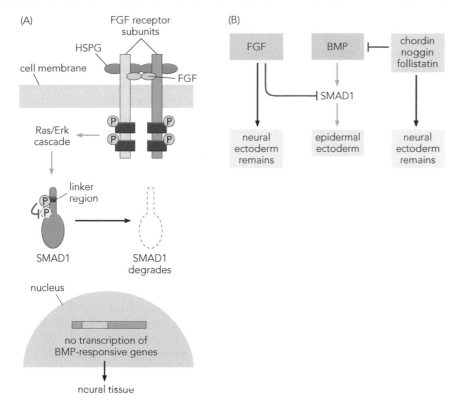

Figure 2.19 Fibroblast growth factors interfere with SMAD1 phosphorylation to block epidermal formation. (A) Fibroblast growth factors (FGFs) form dimers that bind to tyrosine kinase receptors. Ligand binding causes the receptor subunits to dimerize and cross-phosphorylate one another. Binding to the receptors is enhanced by heparan sulfate proteoglycan (HSPG) localized to extracellular spaces. Phosphorylated FGF receptors activate various intracellular pathways, including the Ras–Erk cascade (see Chapter 3). Activation of the Ras–Erk pathway leads to phosphorylation of the linker region of SMAD1. This inhibits the BMP-induced phosphorylation on the C-terminus serines and causes degradation of SMAD1 in the cytoplasm. Without SMAD1, BMP-responsive genes cannot be transcribed, so the default neural tissue remains. (B) Induction of epidermal ectoderm requires BMPs signaling (green arrows) whereas signals that interfere with BMP signal transduction (red bars) permit the default neural ectoderm to remain (black arrows).

the linker region stops the phosphorylation of the SMAD1 C-terminus serines that occurs following BMP receptor binding (Figure 2.18). Phosphorylation at the linker region leads to degradation of SMAD1, ending the BMP-induced gene transcription pathway needed for epidermal formation and instead allowing the formation of neural tissue (Figure 2.19). Thus, BMP signaling can be inhibited by neural inducers that prevent receptor binding, or by FGFs that interfere with SMAD1 activity (Figure 2.19). Whether and how these pathways integrate to influence neural induction in different species is a question that remains under investigation.

Additional proteins, such as Wnt family members, may further influence FGF and BMP signaling to determine whether neural or epidermal tissue forms. Wnt proteins are important for numerous developmental events from early gastrulation through adult neurogenesis. Some studies suggest Wnt signaling promotes neural induction by increasing expression of BMP antagonists, others suggest Wnt signals inhibit BMP expression so FGF can induce neural tissue. Still others suggest high levels of Wnt interfere with FGF signaling thereby allowing BMPs to induce epidermal tissue. The conflicting roles proposed for Wnt in neural and epidermal induction may be due to differences in the species examined, the stage of development studied (late blastula, early gastrula, or late gastrula), or gene expression levels at the time of experimentation.

Multiple signaling pathways likely interact for Wnt to exert any effects on neural or epidermal induction. For example, studies in chick and *Xenopus* found that Wnt proteins upregulate BMP signaling by inhibiting glycogen synthase kinase-3β (GSK-3β). In the absence of Wnt, GSK-3β, like FGF, phosphorylates the linker region of SMAD1, thus terminating the activity of SMAD1 and preventing the transcription of BMP-responsive genes. When Wnt signal is present however, GSK-3β is inhibited, retaining the phosphorylation of SMAD1 C-terminus serines needed for epidermal tissue. Recent studies further suggest that FGF and Wnt signals influence epiblast cells prior to neural induction to establish which cells will form the future spinal cord regions.

SUMMARY

Surface ectoderm forms epidermis (skin) through an instructive or active process, whereas neural tissue forms through a permissive or passive process. The ectoderm itself secretes the signals necessary to induce epidermal formation (for example, BMPs), and the underlying mesoderm tissue (the organizer) secretes molecules (for example, noggin, chordin, and follistatin) to block epidermal induction. By inhibiting the pathway for epidermal formation, the neural tissues remain. Thus, the spatial and temporal expression pattern of the neural inducers (that is, the epidermal blockers) permits that ectoderm to maintain a neural identity. The regions of ectoderm not exposed to the signals from the organizer are not inhibited from transducing BMP signals and therefore generate epidermal tissue. Although Spemann had no way of knowing that neural inducers functioned in this way, his quote from 1924 still holds true. If the epidermis on the soles of our feet had been located in another region of the embryo, it may have been exposed to an inhibitor of BMP signaling and induced to form neural tissue instead.

While the basic mechanisms of neural induction are well defined, details regarding how different signaling pathways interact to influence neural induction across species continues to be investigated. Several investigators note that the concept of a single organizer may need to be expanded, at least for non-amphibian species because the DBL does not have a single, homologous structure in other vertebrates such as mice or chicks. Thus, it may be that organizer tissues are dispersed in space or time in some species.

FURTHER READING

Appel B (2000) Zebrafish neural induction and patterning. *Dev Dyn* 219(2):155–168.

Attisano L & Lee-Hoeflich ST (2001) The smads. *Genome Biol* 2(8).

Brafman D & Willert K (2017) Wnt/β-catenin signaling during early vertebrate neural development. *Dev Neurobiol* 77(11):1239–1259.

Dorey K & Amaya E (2010) FGF signalling: Diverse roles during early vertebrate embryogenesis. *Development* 137(22):3731–3742.

Eivers E, Fuentealba LC & De Robertis EM (2008) Integrating positional information at the level of Smad1/5/8. *Curr Opin Genet Dev* 18(4):304–310.

Hemmati-Brivanlou A, Kelly OG & Melton DA (1994) Follistatin, an antagonist of activin, is expressed in the Spemann organizer and displays direct neuralizing activity. *Cell* 77(2):283–295.

Hemmati-Brivanlou A & Melton DA (1994) Inhibition of activin receptor signaling promotes neuralization in *Xenopus. Cell* 77(2):273–281.

Holley SA, Jackson PD, Sasai Y, et al. (1995) A conserved system for dorsal–ventral patterning in insects and vertebrates involving sog and chordin. *Nature* 376(6537): 249–253.

Huang CT, Tao Y, Lu J, et al. (2016) Time-course gene expression profiling reveals a novel role of non-canonical WNT signaling during neural induction. *Sci Rep* 6:32600.

Kaneda T & Motoki JY (2012) Gastrulation and pre-gastrulation morphogenesis, inductions, and gene expression: Similarities and dissimilarities between urodelean and anuran embryos. *Dev Biol* 369(1):1–18.

Lamb TM, Knecht AK, Smith WC, et al. (1993) Neural induction by the secreted polypeptide noggin. *Science* 262(5134):713–718.

Martinez Arias A & Steventon B (2018) On the nature and function of organizers. *Development (Cambridge, England)* 145(5):dev159525.

Mulligan KA & Cheyette BN (2012) Wnt signaling in vertebrate neural development and function. *J Neuroimmune Pharmacol* 7(4):774–787.

Nieto MA (1999) Reorganizing the organizer 75 years on. *Cell* 98(4):417–425.

Sasai N, Kadoya M & Ong Lee Chen A (2021) Neural induction: Historical views and application to pluripotent stem cells. *Dev Growth Differ* 63(1):26–37.

Sasai Y, Lu B, Steinbeisser H, et al. (1994) *Xenopus* chordin: A novel dorsalizing factor activated by organizer-specific homeobox genes. *Cell* 79(5):779–790.

Shimmi O, Umulis D, Othmer H & O'Connor MB (2005) Facilitated transport of a Dpp/Scw heterodimer by Sog/Tsg leads to robust patterning of the *Drosophila* blastoderm embryo. *Cell* 120(6):873–886.

Shparberg RA, Glover HJ & Morris MB (2019) Modeling mammalian commitment to the neural lineage using embryos and embryonic stem cells. *Front Physiol* 10:705.

Smith WC & Harland RM (1992) Expression cloning of noggin, a new dorsalizing factor localized to the Spemann organizer in *Xenopus* embryos. *Cell* 70(5):829–840.

Smith WC, Knecht AK, Wu M & Harland RM (1993) Secreted noggin protein mimics the Spemann organizer in dorsalizing *Xenopus* mesoderm. *Nature* 361(6412):547–549.

Spemann H & Mangold H (2001) Historical article: Induction of embryonic primordia by implantation of organizers from a different species. 1924. *Int J Dev Biol* 45(1):13–38.

Stern CD (2005) Neural induction: Old problem, new findings, yet more questions. *Development* 132(9):2007–2021.

Stern CD & Downs KM (2012) The hypoblast (visceral endoderm): An evo-devo perspective. *Development* 139:1059–1069.

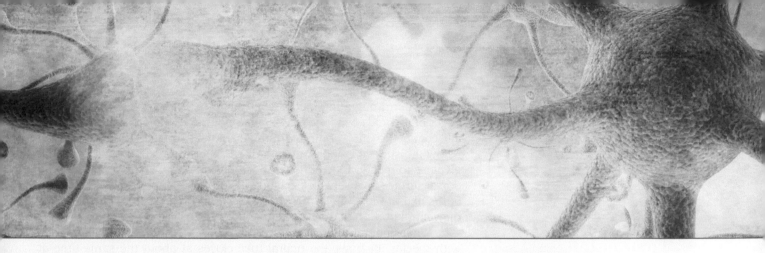

Segmentation of the Anterior–Posterior Axis

3

I n vertebrates, shortly after neural induction, the **neural plate** begins to form. The neural plate forms as a flattened structure along the surface ectoderm but soon begins to bend and curve to create the **neural tube**. Subsequent expansions and constrictions along the length of the neural tube create segments and boundaries that designate future anatomical regions of the central nervous system (CNS).

In the latter part of the twentieth century, as techniques for investigating the genetic and molecular aspects of cellular development evolved, various mechanisms underlying anterior–posterior (A/P) segmentation were discovered. Methods for manipulating gene expression in *Drosophila melanogaster* and chick embryos and generating transgenic mice, combined with classic methods of tissue grafting and cell culture, revealed many complex and often shared mechanisms governing segmentation of the neural tube. It is now clear that several signaling molecules regulate boundary formation along the A/P axis and many of these signals cross-inhibit one another to limit the activity of each signal to the correct segment at the correct time in development.

In several animal models, neural tissue is first induced with characteristics of the anterior-most region of the neural tube with progressively more posterior regions developing in response to signals arising from adjacent tissues. These signals transform the anterior-like neural tissue into progressively more posterior structures. This chapter describes how the different segments along the vertebrate A/P axis are established through the interaction of various inductive, transforming, and cross-inhibitory signals and discusses recent evidence that suggests that posterior, spinal cord regions of the neural tube require separate, species-specific cues for proper development.

NEURAL TUBE FORMATION

Among the first morphological changes noted in the developing neural plate are a longitudinal central indentation called the **neural groove** and upward flexures near the lateral edges that create the **neural folds** (**Figure 3.1**). In most vertebrates, the folds are pushed upward by cellular

DOI: 10.1201/9781003166078-3

contractions and forces within the neural plate, as well as by lateral epidermal ectoderm migrating medially, pushing the folds together. The lateral edges of the folds continue to wrap around the central groove and eventually connect to one another, forming the dorsal surface of the neural tube below the surface ectoderm (epidermis). Later, the neural crest cells that give rise to much of the peripheral nervous system will emigrate from the dorsal surface as a separate cell population (see Chapter 5). Upon closure of the neural tube, the former medial section of the neural plate becomes the ventral portion of the neural tube (**Figure 3.1**).

Where neural tube closure begins along the A/P axis varies slightly with species. In frogs, the neural tube closes at about the same time at all A/P levels. In chicks, the anterior neural tube closes first, followed by progressively more posterior regions. In mammals, the neural tube first closes at the mid-anterior region, then progresses to the more anterior (rostral) and posterior (caudal) regions (Figure 3.1E). Failure of the neural tube to close results in human birth defects that can have mild to fatal outcomes, depending on which region along the A/P axis is affected. Thus, the process of neural tube formation and closure is studied to reveal the basic cellular and molecular mechanisms that govern these events and to understand how **neural tube defects** (**NTDs**) can be prevented in children (**Box 3.1**).

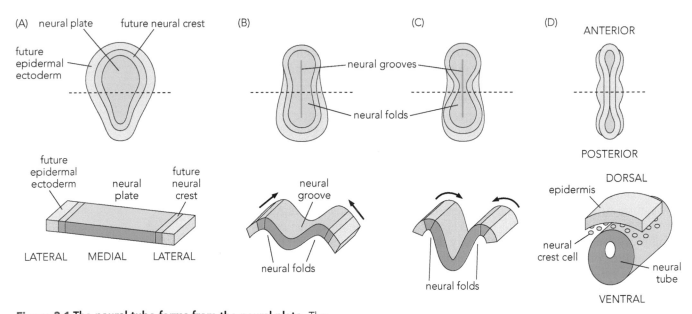

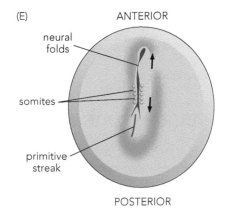

Figure 3.1 The neural tube forms from the neural plate. The general process of neural tube formation is similar across vertebrates. The stages shown in this figure occur between approximately embryonic days 7.5–10.5 in mice, gestation days 18–28 in humans, and 23–33 hours post fertilization in the chick. In A–D, the top panels reveal surface views of the neural plate and forming neural tube. Dashed lines indicate the plane of section shown in the lower panels. (A) The specified area of ectoderm that forms the neural plate (blue) indents to form the neural groove (B) while the edges curve upward to form neural folds (B, C). As the lateral margins of the neural plate continue to curve over (arrows in B and C), the neural tube closes, forming the dorsal surface of the neural tube where the neural crest cells arise (D, lower panel). The remaining ectoderm originating at the lateral margins of the neural plate form the epidermis overlying the neural tube (D). (E) The neural tube is elevated above the surface ectoderm. In mammals, the neural tube first closes in the mid-anterior regions then progresses in more anterior and posterior regions as indicated by the arrows. The somites and primitive streak are also labeled.

Box 3.1 Adequate Intake of Folic Acid Can Reduce Incidence of Neural Tube Defects

In the United States, about one pregnancy per 2000 results in a neural tube defect (NTD), with anencephaly and spina bifida the most prevalent. When the cranial region fails to close, the result is **anencephaly**, a condition in which the anterior portions of the brain and back of the skull fail to develop and close properly (**Figure 3.2**). While pregnancy can continue through full term, the condition is fatal. In contrast, **spina bifida** arises from incomplete closure of the posterior (caudal) region of the neural tube, leaving the spinal cord exposed through the opened vertebral column (**Figure 3.3**). Spina bifida occurs in many forms that result in mild to severe disabilities.

The incidence of NTDs can be dramatically decreased with proper diet. **Folic acid** (folate, vitamin B_9) was first discovered in 1941; by the 1960s, it was recognized to have potentially important nutritional effects during pregnancy. In the 1980s, folic acid was specifically noted to decrease the occurrence of NTDs. Studies found that women who included folic acid in their daily diets during the first four weeks of pregnancy decreased the risk of neural tube defects by 50–70%. In 1996, the United States Preventive Services Task Force recommended that women capable of conception take folic acid supplements, and the United States Food and Drug Administration published regulations to require the addition of folic acid to enriched breads, cereals, flours, and other grains. It is now recommended that women of child-bearing age take 400 micrograms of folic acid daily.

The neural tube closes between 21 and 28 days of gestation, so women are unlikely to know they are pregnant at the time of neural tube closure. The average

Figure 3.2 An infant with anencephaly. Anencephaly occurs when the neural tube fails to close in the anterior-most regions, leaving the skull open and the remaining brain tissue exposed. While pregnancy may continue full term, the condition is fatal.

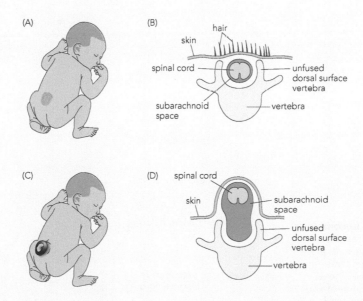

Figure 3.3 Forms of spina bifida. (A) An infant with a mild form of spina bifida in which the vertebrae fail to fuse. (B) The spinal cord develops normally but is only protected by a thin patch of skin that is often covered by a patch of hair. (C) A more severe form of spina bifida results when the spinal cord bulges outward through the opened vertebral column (D).

U.S. diet is estimated to include only 200 micrograms of folic acid. Therefore, to ensure adequate intake, women are encouraged to add folic acid supplements to their diets and eat foods rich in folate, such as spinach, orange juice, and fortified foods, to reduce the risk of NTDs.

The precise mechanisms by which folic acid prevents NTDs are still not entirely clear, but several clues have been identified in recent years. In some mothers whose children were born with NTDs, autoantibodies that target the folate receptor were detected, revealing one pathway by which folate uptake can be limited. In several animal models, reduced nucleotide synthesis has been associated with NTDs. Because folic acid is involved in the synthesis of nucleotides, it may prevent NTDs by improving otherwise reduced rates of synthesis. Folic acid is also involved in the methylation pathway, an important epigenetic modification that can influence gene expression. In mice in which the methylation cycle was inhibited, NTDs developed. Further, hypomethylated DNA was found in the brains of mice with NTDs. Thus, folic acid can influence numerous cellular events and multiple genetic and environmental factors may be involved in determining the likelihood of a NTD occurring. Whatever the precise mechanisms involved, taking the recommended amount of folic acid daily can substantially reduce the risk of this common birth defect.

Early Segmentation of the Neural Tube Establishes Subsequent Organization

As the neural tube closes, cells continue to proliferate and the length of the neural tube extends at the posterior end. The neural tube also constricts, expands, and bends at specific sites to establish early anatomical and cellular boundaries for the future CNS. Such changes in appearance are first noted shortly after neural tube closure and are soon readily identified as the **prosencephalon** (future forebrain), **mesencephalon** (future midbrain), and **rhombencephalon** (future hindbrain; **Figure 3.4**A). As development continues, these three primary brain vesicles become the five secondary brain vesicles. The prosencephalon is divided into the

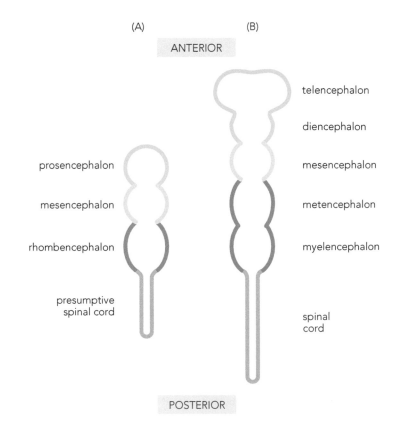

(A)   (B)

ANTERIOR

telencephalon

diencephalon

prosencephalon   mesencephalon

mesencephalon   metencephalon

rhombencephalon   myelencephalon

presumptive
spinal cord   spinal
cord

POSTERIOR

Figure 3.4 The neural tube becomes segmented into brain vesicles. (A) A dorsal view of the neural tube reveals that shortly after neural tube closure, three primary brain vesicles are visible. These are the prosencephalon (future forebrain), mesencephalon (future midbrain), and rhombencephalon (future hindbrain). (B) As development progresses, the five secondary brain vesicles are formed as the prosencephalon is divided into the telencephalon and diencephalon and the rhombencephalon is divided into the metencephalon and myelencephalon.

telencephalon and diencephalon, while the rhombencephalon is divided into the metencephalon and myelencephalon. In contrast to the segmentation noted in the developing brain, the spinal cord region remains largely uniform in outward appearance (Figure 3.4B). However, differences in spinal cord patterning are more easily observed along the dorsal–ventral axis (see Chapter 4).

The early expansions along the A/P axis of the neural tube are often described as metameric segments, or **neuromeres**, enlargements that provide regionally restricted areas for further development. In the developing forebrain region, these expansions are called **prosomeres**, and in the hindbrain, **rhombomeres (Figure 3.5A)**. Although the neuromeres correspond to general anatomical areas found in the adult nervous system, the purpose of segmentation at such early stages is not to designate rigid boundaries for adult structures. Instead, the transient boundaries serve important functions, such as limiting the movement of cells until they have been exposed to signals critical to their subsequent development and specialization.

Many of the signals necessary for A/P patterning originate in tissues adjacent to neuromere segments. For example, the **anterior neural ridge (ANR)** forms the anterior-most signaling center to influence development of the **telencephalon**. More posteriorly, the **zona limitans intrathalamica (ZLI)**, located at the junction of the prechordal and notochordal regions, provides signals that influence development of the **diencephalon**. The **midbrain–hindbrain border (MHB)**, or **isthmus,** located between the mesencephalon and metencephalon is the site of the **isthmic organizer (IsO)** that influences development of the future midbrain, pons, and cerebellum. In addition, the **notochord** extends along much of the neuraxis and is an important source of signals for patterning hindbrain and spinal cord regions. The notochord extends as far anteriorly as the mesencephalon but is prevented from extending further by the prechordal plate, the area where ectoderm and mesoderm tightly adhere to one another (Figure 3.5B; see also Figure 2.4). How molecules produced by these signaling centers influence A/P development is detailed later in the chapter.

Temporal–Spatial Differences in Organizer-Derived Signals Induce Head and Tail Structures

Shortly after the discovery that the dorsal blastopore lip (DBL) of the gastrula stage amphibian embryo (Spemann's organizer) was a source of

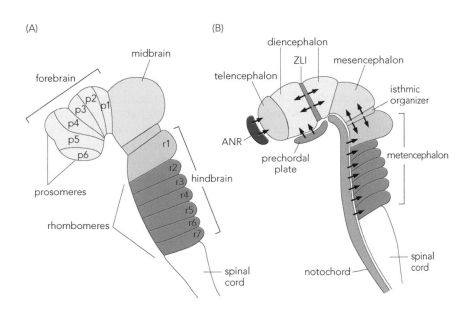

Figure 3.5 Early boundary formation along the anterior–posterior axis. (A) Areas of expansions called neuromeres form along the anterior–posterior (A/P) axis. In the forebrain, these expansions are called prosomeres and are numbered from 1–6 (p1–p6) beginning at the junction of the forebrain (prosencephalon) and midbrain. In the hindbrain, the expansions form rhombomeres that are numbered sequentially beginning at the midbrain–hindbrain junction (r1–r6). (B) Signaling centers that provide local development cues (arrows) are located along the A/P axis. The anterior neural ridge (ANR) lies at the anterior-most end of the neural tube (telencephalon) to provide signals that influence development of the forebrain. Signals influencing forebrain development are also generated by the prechordal plate and the zona limitans intrathalamica (ZLI) in the diencephalon. The isthmic organizer is found at the boundary between the mesencephalon (midbrain) and metencephalon (hindbrain), where signals are generated to influence development of these regions of the developing nervous system. The notochord extends from the mesencephalon to the spinal cord and provides signals that influence the development of cells in these more posterior regions of the neural tube.

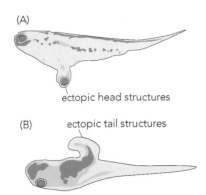

Figure 3.6 Induction of head or tail structures is determined by the age of the organizer. Grafting studies by Otto Mangold and others demonstrated that when younger (anterior) organizer tissue was grafted into a host embryo, extra ectopic head structures were induced (A). In contrast, grafts of older (posterior) organizer tissue induced ectopic tail-like structures in the host embryos (B). [Adapted from Stern CD [2001] *Nat Rev Neurosci* 2:92–98. With permission from Macmillan Publishers Ltd.]

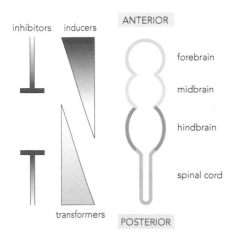

Figure 3.7 The traditional model of how inducers, transformers, and inhibitors pattern A/P segments. Cells along the A/P axis respond best to a given concentration of a morphogen. Inducers are thought to form primarily forebrain-like regions, while transforming signals modify the effects of the inducers to influence development of more posterior-like regions. Various inhibitory signals limit the spread of inducers and transformers to help maintain the proper boundaries of the morphogen signals.

neural-inducing signals, other studies found evidence of additional signals that directed further development along the neuraxis. For example, grafting studies conducted independently by Hans Spemann and Otto Mangold demonstrated that the age of the organizer tissue determined whether anterior or posterior regions of the body axis formed (**Figure 3.6**). Younger organizer tissues induced anterior structures, whereas slightly older tissues induced posterior regions. If the DBL from an early-stage gastrula or a section of anterior (younger) chordamesoderm were grafted into a host embryo, extra ectopic head structures formed. Conversely, if slightly older DBL or posterior (older) chordamesoderm were grafted in the host embryo, extra tail structures formed. Together, these results suggested that the organizer generated different signals at different times to direct development along the A/P axis.

Over the years, many scientists concluded that the anterior (forebrain-like) region of the neural tube is a "default" state for neural tissue, with additional signals required to transform neural tissues into more posterior regions. In the 1950s, Pieter Nieuwkoop expanded on the ideas of Conrad Waddington and others to develop the activation-transformation model. The model proposes that the signal for neural induction, the "activator," induces competent ectoderm to form anterior brain regions, with progressively higher concentrations of "transforming" signals converting the tissue into hindbrain and spinal cord. Studies have largely supported this model in amphibians and in the forebrain through hindbrain regions of other species. However, in at least some species, spinal cord regions may develop using separate mechanisms that do not rely on transforming signals.

Activating, Transforming, and Inhibitory Signals Interact to Pattern the A/P Axis

To establish forebrain, midbrain, and hindbrain regions a complex and elegant balance of activating, transforming, and antagonizing signals is required. In many animal models, after neural inducers generate anterior-like tissue in the neural plate, region-specific signals emerge to refine each area along the A/P axis of the neural tube. Some of the signals transform previously induced areas into new regions, while other signals inhibit (antagonize) those signals to prevent transformations (**Figure 3.7**). The signals often overlap and interact through concentration gradients established by **morphogens**—diffusible molecules that act on cells at a distance. In general, cells within different regions along the A/P axis respond best to specific concentrations of each signal and the interactions of these signals, in combination with other region-specific cues, establish anatomical specializations along the neuraxis.

Although anterior brain regions are generally specified prior to more posterior regions, the signals used to pattern each A/P level are not simply produced in a wave from anterior to posterior. Instead, signals are produced at several locations and at various times to pattern nearby areas while restricting the influence of molecules from adjacent territories. For example, signals that specify midbrain structures must be prevented from reaching more anterior or more posterior regions. Additionally, midbrain structures must not be converted by signals originating in the forebrain or hindbrain regions. As will be seen, the expression of regionally specific signals and the repression of adjacent signals are equally important in patterning the A/P axis.

Many of the genes and signal transduction pathways for A/P axis patterning are conserved across invertebrate and vertebrate species, although some differences in signaling mechanisms have been noted in the various animal models that may reflect, in part, the differences in brain anatomy.

Further, different members of a given gene family are often more influential in one species than in others. The examples that follow focus primarily on the gene families involved in patterning the different regions of the vertebrate neuraxis, rather than species-specific family members.

SPECIFICATION OF FOREBRAIN REGIONS

The developing forebrain has traditionally been divided into six different prosomeres beginning with prosomere 1 (p1) adjacent to the mesencephalon and progressing anteriorly to prosomere 6 (p6) in the telencephalon (**Figure 3.8**). Signaling molecules first establish anterior regions then regulate gene expression within each prosomere to specify the telencephalon and diencephalon regions of the forebrain.

Signals from Extraembryonic Tissues Pattern Forebrain Areas

After gastrulation, the **anterior visceral endoderm** (**AVE**) in mice serves as a signaling center important for head and forebrain development (see Chapter 2, Figure 2.5). The hypoblast in the chick serves a similar function. The AVE—the population of extraembryonic cells that comes to lie beneath the anterior neural plate—does not induce neural tissue but rather influences development of the anterior regions of the neural tube. For example, in mice lacking the AVE, neural tissue is induced, but anterior brain regions do not form.

Cells of the AVE and hypoblast also produce signals that specify the most anterior limit of the neural ectoderm, the area called the **anterior neural ridge** (**ANR**, mouse and chick) or the **anterior neural border** (**ANB**, fish). In chick, hypoblast-derived fibroblast growth factor (FGF) specifies the ANR which then serves as a signaling center to regulate expression of forebrain-specific genes. Similarly, in mid-late gastrula stage zebrafish, ANB cells help establish telencephalon gene expression patterns. If ANB cells are transplanted to posterior regions, genes normally restricted to the telencephalon are expressed. The AVE and hypoblast may further regulate normal anterior development by inhibiting posteriorizing signals originating in the developing midbrain.

Forebrain Segments Are Characterized by Different Patterns of Gene Expression

Regional expression of genes, including members of the *Otx* (orthodenticle homeobox), *Pax* (paired homeobox), and *Six* (sineoculus homeobox) families, are detected within boundaries of the prosomeres (Figure 3.8). These genes play direct roles in forebrain patterning and are homologs of *Drosophila* genes that often have similar patterning functions during fly development. For example, experiments have demonstrated the necessity of *Otx* genes in forebrain development across species. In vertebrates, *Otx* genes are expressed widely in the regions anterior to the isthmus/MHB. In mice lacking *Otx* genes, head structures do not form. However, if *Otd* (orthodenticle), the homologous gene in *Drosophila*, replaces the missing *Otx* genes, the head structures are rescued. Conversely, human *Otx* genes can rescue *Drosophila Otd* mutants, further demonstrating the highly conserved nature of these genes and their functions in patterning forebrain and head structures.

Genes of the *Pax* family are also expressed widely in the developing neural tube. *Pax6*, for example, is expressed in the telencephalon through diencephalon regions, where it is required for forebrain development and eye and ear formation. Loss of *Pax* genes leads to small or absent eyes and ears in mice and humans.

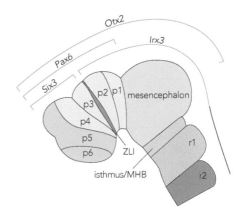

Figure 3.8 Restricted patterns of gene expression regulate development of forebrain regions. The prosomeres (p1–p6) are located in the forebrain anterior to the mesencephalon. Gene expression within many of these prosomeres establishes boundaries for further neural development. *Otx2* expressed in p1–p6 regulates development of the forebrain, as well as the mesencephalon and other head structures. *Pax6* expression in p1–p6 is needed for normal forebrain, eye, and ear formation. *Six3*, which is localized to p3–p6, opposes expression of *Irx3* in p1, p2.

Like *Pax6*, *Six3*, a member of the Six family of homeodomain transcription factors, is needed to direct development of forebrain regions (Figure 3.8). Balancing the role of Six3 in the normal embryo is the Irx family of transcription factors expressed in the posterior diencephalon and midbrain (see Figure 3.8). Beginning at the neural plate stage, the ZLI is bordered anteriorly by *Six3* and posteriorly by *Irx3*, a member of the Iroquois homeobox gene family. The restricted expression domains of the Six and Irx family members are maintained because these two proteins suppress one another. *Irx3* expression is also influenced by Wnt. Loss of Wnt causes a loss of *Irx3* expression, whereas increasing Wnt levels leads to expansion of *Irx3* expression into more anterior regions. Thus, coordinated expression of Wnt and Irx3 are needed to pattern posterior diencephalon regions and delineate them from more anterior *Six3*-expressing regions of the forebrain.

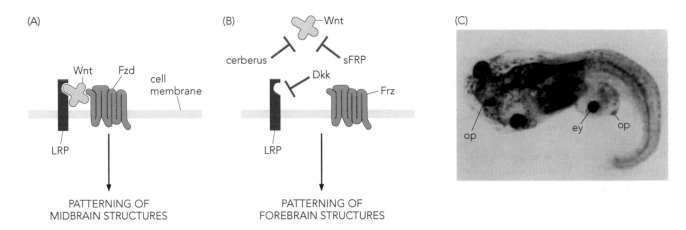

Figure 3.9 Wnt signaling is normally inhibited in forebrain regions. (A) Wnt can signal through a receptor complex formed by a transmembrane protein of the Frizzled family (Fzd) and a low-density lipoprotein receptor (LRP5 or LRP6). (B) Several forebrain-derived signals inhibit the activity of Wnt. Cerberus and secreted Frizzled-related proteins (sFRPs) prevent Wnt from binding to the receptor complex. Dickkopf binds to subunits of LRP5/LRP6 to inhibit Wnt signaling. (C) Endogenous Wnt signaling was blocked in *Xenopus* embryos injected with excess Dickkopf, leading to larger head and brain regions that extended into the midbrain. op, olfactory placode; ey, eye. [(C), From Bouwmeester T, Kim S, Lu B & DeRobertis EM [1996] *Nature* 382:595–601. With permission from Macmillan Publishers Ltd.]

Figure 3.10 Signaling molecules along the A/P axis. (A) Signals interact to influence boundary formation by inducing (green arrow) or blocking (red bars) the expression of adjacent genes. For example, Cerberus, Dickkopf, and secreted Frizzled-related proteins (sFRPs) inhibit Wnt signals to prevent posteriorization of anterior regions. (B) *Six3* in the forebrain and *Irx3* in the posterior diencephalon/midbrain region antagonize the expression of each other, thus limiting expression of each gene to a certain region. Wnt located at the isthmus (midbrain–hindbrain border, MHB) regulates expression of Irx3 levels to ensure sufficient Irx3 is present to limit expansion of the forebrain into these more posterior regions.

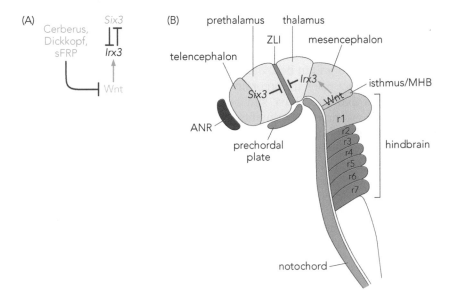

Signals Prevent Wnt Activity in Forebrain Regions

The specification of forebrain versus midbrain regions also depends on the levels of neural inducers and Wnt signaling. Some Wnt signaling pathways use a receptor complex comprised of a transmembrane protein of the Frizzled family of receptors (Fzd, also abbreviated Frz or Fz) and a low-density lipoprotein receptor (LRP5 or LRP6). Both receptor components interact with Wnt (**Figure 3.9**A). If Wnt is overexpressed in the presence of a neural inducer (for example, chordin, follistatin, or noggin), genes and structures associated with anterior brain regions are suppressed. In contrast, if endogenous Wnt expression is inhibited in midbrain regions, anterior brain regions expand posteriorly (**Figure 3.10**).

Wnt-inhibiting molecules, including Cerberus, soluble members of the Frizzled family, and Dickkopf are initially expressed in the organizer and later found in the AVE, the two signaling centers shown to be important for patterning the anterior neuraxis.

Cerberus is a secreted protein that binds to both Wnt and BMP and thus blocks the activity of both (Figure 3.9B). When Cerberus mRNA was injected into early-stage *Xenopus* embryos, one or more incomplete ectopic heads formed. Based on this finding, the molecule was named Cerberus after the mythical three-headed dog that guards the gates of Hades. Although the *Xenopus* embryos injected with the excess mRNA did not develop three complete heads, they clearly had extra head structures.

A second group of proteins associated with anterior regionalization is the Frizzled family. In addition to the Frizzled (Fzd) transmembrane proteins that serve as receptors for Wnt, there are additional small, soluble molecules of the Frizzled family called **secreted Frizzled-related proteins (sFRPs)**. The sFRPs sequester Wnt and thus block signaling through Fzd receptors (Figure 3.9B). Injection of sFRPs into *Xenopus* embryos resulted in formation of larger heads, and sFRP overexpression in zebrafish caused an expansion of the telencephalon.

Another secreted molecule discovered shortly after Cerberus was **Dickkopf (Dkk)**, which binds directly to subunits of the Wnt receptor complex. By binding to the LRP5 or LRP6 subunits, Dkk interferes with Wnt signaling and prevents formation of midbrain structures in more anterior regions. The importance of Dkk signaling in formation of anterior regions of the nervous system is seen in experimental manipulations of *Xenopus* embryos. Truncated (inactive) BMP receptors had previously been shown to induce neural tissue (see Chapter 2). When these truncated receptors and Dkk were co-injected into *Xenopus* embryos, a larger, expanded head and brain region formed (Figure 3.9C). The resulting phenotype of an expanded head and brain seen in the Dkk-treated embryos led to the name of the gene. Dickkopf is German for "big head" or "stubborn." Conversely, if antibodies raised against Dkk were also injected, the Dkk-induced expansion of head structures was inhibited resulting in a smaller head and incomplete brain development (microcephaly). Similarly, mice lacking the *Dkk1* gene developed with a smaller head and brain. Recent studies in zebrafish suggest FGF signaling induces expression of chordin and maintains expression of the zebrafish homologue of Dkk (*Dkk1b*) to induce head structures.

Other proteins that inactivate or interfere with Wnt are expressed early in development to ensure Wnt signaling is prevented in anterior regions. For example, the metalloprotease Tiki is expressed in the organizer and the hydrolase Notum is expressed in the neural plate. Thus, multiple signals coordinate to limit the activity of midbrain-associated Wnt to the proper locations along the A/P axis.

REGIONALIZATION OF THE MESENCEPHALON AND METENCEPHALON REGIONS

Posterior to the diencephalon is the mesencephalon—the future midbrain region of the vertebrate brain. The mesencephalon forms dorsal tectal and ventral tegmental structures and establishes a border with the metencephalon, the region that gives rise to the pons and cerebellum. The metencephalon is typically described as extending posteriorly to rhombomeres 1 and 2 (r1 and r2), the anterior-most rhombomeres in the hindbrain.

Intrinsic Signals Pattern the Midbrain–Anterior Hindbrain

As noted, the mesencephalon and metencephalon regions are separated by a narrow constriction called the midbrain–hindbrain border (MHB) or isthmus. This anatomical constriction also serves as one of the intrinsic signaling centers in the developing nervous system known as the isthmic organizer (IsO; Figure 3.5). The IsO produces multiple molecules that pattern the midbrain and anterior hindbrain regions, as well as signals that prevent the spread of signals originating in the forebrain and posterior hindbrain.

Signaling centers intrinsic to the mesencephalon and metencephalon were identified using various experimental approaches, including transplantation studies in quail and chick embryos. The quail–chick transplantation method was selected because the cells originating from the quail donor could be distinguished from those of the chick host with the Feulgen nucleolar stain, or with specific antibodies that label the condensed heterochromatin characteristic of quail cells (**Figure 3.11**A). When quail mesencephalon or metencephalon was grafted to corresponding regions of chick embryo, the chick embryos developed correct midbrain or hindbrain structures. These results indicated that signals were present in the donor tissue to induce formation of these brain regions. The results also demonstrated that the mesencephalon- and metencephalon-derived signals were not species specific, an observation that proved helpful in determining the cellular origins of different signals in other experiments, including transplant experiments between mouse and chick.

The mesencephalon also produced signals needed to form the cerebellum and induce adjacent midbrain tissue. For example, Salvador Martinez and colleagues found that when quail metencephalon was rotated 180 degrees and grafted into certain areas of the developing chick forebrain, an ectopic cerebellum formed and adjacent midbrain structures were induced in an orientation consistent with the rotated graft (Figure 3.11B, C).

Figure 3.11 Midbrain tissue induces ectopic structures in the forebrain. (A) Cells originating from a quail donor embryo (left side of panel) can be distinguished from those of the chick host (right side of panel) with the Feulgen nucleolar stain that reveals the condensed heterochromatin (arrow) that is characteristic of quail but not chick cells. (B) Grafts of quail mesencephalon tissue were rotated 180 degrees and placed in the forebrain (telencephalon) region of chick embryos. (C) The graft of mesencephalon tissue led to the formation of an extra ectopic midbrain and induced the formation of an ectopic cerebellum. When viewed from the dorsal surface, the new structures were observed to form in an opposite, or mirror image, orientation (top panel) compared to the normal cerebellar/midbrain region (bottom panel). Tel, telencephalon; cb, cerebellum; mb, midbrain. [(A), From Le Douarin NM [2004] *Mech Dev* 121:1089–1102. With permission from Elsevier Inc.]

(A)

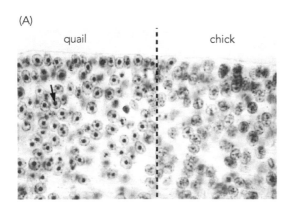

quail | chick

(B)

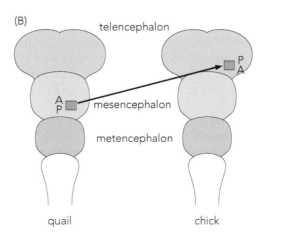

telencephalon

mesencephalon

metencephalon

quail   chick

(C)

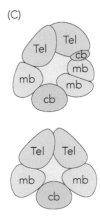

However, a new cerebellum and midbrain were only found if the transplant occurred posterior to the ZLI. The expression of *Six3* anterior to the ZLI may have prevented the induction of midbrain structures in the more anterior region.

Multiple Signals Interact to Pattern Structures Anterior and Posterior to the Isthmus

Like forebrain regions, signals originating in the isthmus/MHB coordinate to maintain gene expression patterns in certain regions while repressing the activity of other molecules in adjacent structures. Many of the genes for these signals are detected during gastrulation or early neural plate stages indicating that the steps to establish A/P polarity begin very early in development. For example, a division between anterior expression of *Otx* and posterior expression of *Gbx* is first observed in the gastrula-stage embryo and continues through A/P patterning (**Figure 3.12**).

Interactions between *Otx* and *Gbx* were demonstrated in several studies. In mice lacking *Otx* genes, the areas anterior to the isthmus that normally express *Otx* are respecified and take on more posterior-like characteristics. Similarly, if *Gbx* is missing, midbrain areas extend more posteriorly leading to midbrain-like characteristics in regions posterior to the isthmus. Thus, with the loss of either gene, the adjacent area is able to dominate and expand to re-pattern neighboring regions of the neural tube. Later in development, additional signals are employed to restrict *Otx* to regions anterior to the isthmus and *Gbx* to areas posterior to the isthmus.

Two well-characterized gene products that participate in midbrain–anterior hindbrain patterning are FGF8 and Wnt. There are 22 known members of the FGF family. FGF8, FGF17, and FGF18 are expressed in the isthmus region, but the role of FGF8 is the most fully characterized to date. Wnt is initially expressed throughout much of the midbrain but eventually becomes restricted to a narrow band in the isthmus, immediately anterior to FGF8 expression. Numerous experimental findings suggest a direct role for FGF8 in cerebellar induction and an indirect role in the patterning of adjacent regions of the nervous system. In contrast, Wnt does not directly induce midbrain–hindbrain structures, but instead interacts with other signaling molecules, such as Otx and Gbx to regulate development in these areas.

FGF Is Required for Development of the Cerebellum

A role for FGF8 in cerebellar induction was demonstrated in studies using experimental beads soaked in FGF8 protein. FGF8 remained attached to the beads that were then implanted into embryonic chick brains at the

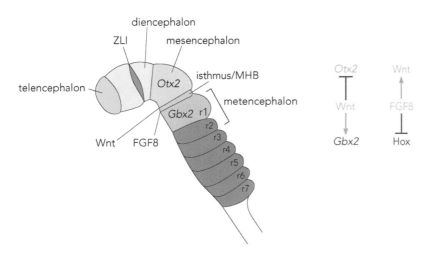

Figure 3.12 The midbrain–anterior hindbrain border (MHB) is a source of FGF8 and Wnt. A narrow constriction at the MHB, the isthmus, expresses FGF8. Wnt is initially expressed throughout much of the midbrain but becomes restricted to a band just anterior to FGF8. These signals influence the expression of other patterning genes. For example, Wnt inhibits activity of *Otx2* anteriorly and activates *Gbx2* posteriorly, whereas FGF8 activates Wnt expression in the region of the isthmus, but inhibits expression of *Hox* genes in rhombomere 1 (r1).

three primary vesicle stage. When beads were implanted into the posterior segment of the chick forebrain (near the prosomere 1 and mesencephalon border; **Figure 3.13**A), the FGF8-coated beads induced ectopic cerebellar and midbrain structures in these more anterior regions (Figure 3.13B). In several of the host embryos, the orientation of the cerebellum and midbrain were opposite of that found in control conditions, similar to the results obtained when rotated metencephalon tissue was transplanted into the forebrain (Figure 3.13D).

In the posterior regions of the forebrain, the FGF8 beads also altered gene expression patterns. The cells adjacent to a bead now failed to express forebrain *Otx2*, but instead expressed *Engrailed 2 (En2)*, *FGF8*, and *Wnt*, genes normally found in midbrain–anterior hindbrain regions. Additionally, like the transplant studies, beads placed in more anterior forebrain regions were unable to induce midbrain structures, suggesting the anterior forebrain-derived signals inhibited midbrain formation.

FGF Isoforms and Intracellular Signaling Pathways Influence Cerebellar and Midbrain Development

In studies of mice with decreased levels of FGF8, defects in midbrain and cerebellum formation were observed. Other studies found the concentration of FGF8 influences whether midbrain or cerebellum forms. Lower concentrations of FGF8 appear to impact the development of the midbrain while higher concentrations primarily influence the cerebellum. Different isoforms of FGF8 may also affect development in this region. For example, FGF8a appears to be the isoform important for midbrain induction, whereas FGF8b, the isoform used in the bead implantation experiments, appears necessary for cerebellar induction.

The different effects of these two isoforms may result from differences in the strength of binding to the four FGF receptor tyrosine kinases. FGF8 binds with highest affinity to FGFR3 and FGFR4. The addition of 11 amino acids in FGF8b may allow this isoform to bind to these FGF receptors more strongly than the FGF8a isoform. Such enhanced binding is thought to increase the strength of the FGF signal, leading to the induction of cerebellar tissue.

Experiments manipulating intracellular signaling cascades support the importance of FGF signaling in cerebellar formation. Under normal conditions, the Ras/Raf/MEK/Erk pathway (see Chapter 1) is one of the signaling cascades initiated when FGF binds to its receptor. This pathway is activated when the tyrosine kinase domains of the FGFR become phosphorylated, causing subsequent phosphorylation of the docking protein FRS2 (FGFR substrate 2) and recruitment of the adaptor protein Grb2

Figure 3.13 Exogenous FGF8 induces ectopic midbrain and anterior hindbrain structures at the juncture of the midbrain and forebrain. When beads coated with FGF8 were implanted into the posterior forebrain at the mesencephalon–prosomere 1 boundary (arrowhead, A), ectopic midbrain and cerebellar tissues were induced. Side views of the brain show extra midbrain regions (asterisks, B) and an ectopic cerebellar-like outgrowth (arrow, B) in the FGF8-treated embryos compared to untreated controls (C). Sagittal sections through FGF8-treated (D) and untreated control (E) brains show that the two ectopic midbrain regions (asterisks, D) formed in an opposite orientation (arrows, D). Double asterisks indicate ectopic regions that are anterior to those marked by a single asterisk. Tel, telencephalon; Cb, cerebellum; Mb, midbrain; Tc, tectum; Di, diencephalon; Is, Isthmus; ic, isthmic constriction; mes, mesencephalon; v4, fourth ventricle. [From Martinez S, Crossley PH, Cobos I et al. [1999] *Development* 126:1189–1200. With permission from The Company of Biologists Ltd.]

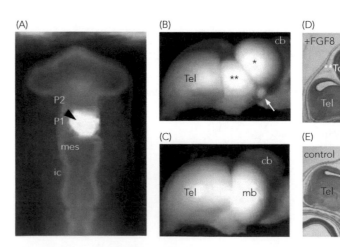

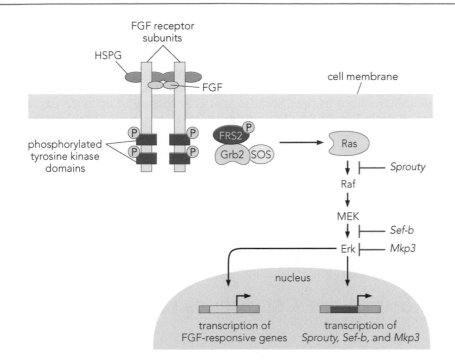

Figure 3.14 FGF regulates transcription of genes important for cerebellar development. Fibroblast growth factor (FGF) dimers bind to tyrosine kinase receptors. Binding is enhanced by heparan sulfate proteoglycan (HSPG) localized to extracellular spaces. Ligand binding causes the receptor subunits to dimerize, cross-phosphorylate one another, then phosphorylate the docking protein FRS2 (FGFR substrate 2), leading to recruitment of the adaptor protein Grb 2 (growth factor receptor-bound protein 2). Grb2 then interacts with SOS (Son of sevenless) to activate Ras (rat sarcoma). Activation of Ras initiates the sequential phosphorylation of Raf (rapidly accelerated fibrosarcoma kinase), MEK (MAP kinase/Erk kinase), and Erk (extracellular signal-regulated kinase). Phosphorylated Erk then enters the nucleus, where it regulates transcription of FGF-responsive genes. FGF also induces the expression of the genes *Sprouty*, *Sef-b*, and *MKp3*. The protein products of each inhibit the Ras–Erk pathway at specific sites (pink bars) to regulate FGF expression.

(growth factor receptor-bound protein 2). Grb2 then interacts with SOS (Son of sevenless) to activate Ras (rat sarcoma). Activation of Ras leads to the sequential phosphorylation of Raf (rapidly accelerated fibrosarcoma kinase), MEK (MAP kinase/Erk kinase), and Erk (extracellular signal-regulated kinase). Phosphorylated Erk then enters the nucleus, where it regulates transcription of FGF-responsive genes (**Figure 3.14**). Experiments in zebrafish revealed that when this pathway was inhibited, the expression of *Gbx2* was repressed in the metencephalon while *Otx2* expression was induced in its place. This led to the formation of more anterior structures in areas that normally give rise to the cerebellum. Thus, FGF8 normally induces *Gbx2* expression via the Ras–Erk pathway.

FGF also provides a negative regulator of the Ras–Erk pathway to prevent expansion of metencephalon structures too far anteriorly. This occurs because FGF induces expression of the genes *Sprouty*, *Sef-b*, and *MKp3* in the isthmus, all of which inhibit the Ras–Erk pathway at specific sites (Figure 3.14). By inducing expression of these genes, FGF can regulate its own activity.

FGF and Wnt Interact to Pattern the A/P Axis

As noted in the bead implantation experiments, FGF8 not only directly induced formation of cerebellum, but also influenced expression of genes used to pattern the midbrain region. Further, in addition to maintaining the expression of Wnt and En in the regions immediately anterior to FGF8 expression, FGF8 also inhibits the expression of *Hox* genes in the first rhombomere (**Figure 3.15**). As detailed later, *Hox* genes are necessary for the formation of the more posterior rhombomere segments of the hindbrain. However, suppression of *Hox* genes in rhombomere 1 (r1) permits the formation of cerebellar tissues from this segment, rather than rhombomere-like structures that would arise following the activation of *Hox* genes. Thus, FGF8 is a critical signaling molecule for cerebellar formation and the regulation of genes that pattern midbrain regions.

Wnt also contributes to normal midbrain and cerebellar development, as noted by the absence of midbrain and cerebellar regions in mice lacking Wnt. However, unlike FGF8, Wnt does not induce midbrain–anterior hindbrain regions directly. Instead, the coordinated expression of Wnt,

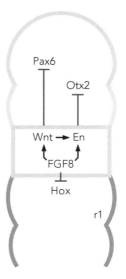

Figure 3.15 FGF8 influences expression of genes in the midbrain and hindbrain. FGF8 maintains expression of *Wnt* and *En* in the regions immediately anterior to FGF8 expression. Wnt and En then inhibit expression of *Otx2* in the midbrain and forebrain and *Pax6* in the forebrain. FGF8 also inhibits expression of *Hox* genes in the first rhombomere (r1) so that the cerebellum, rather than *Hox*-dependent hindbrain structures, can form in this region.

FGF8, and En1/En2 enables the formation and maintenance of boundaries within the midbrain–anterior hindbrain region (Figure 3.15). Loss of any one of these molecules not only causes defects, but also impacts the expression of the other two signals. For example, in mice lacking the transcription factors En1 and En2, the midbrain and cerebellar regions are dramatically reduced in size. Because Wnt is necessary to maintain expression of En, a loss of Wnt likewise disrupts development of the midbrain and anterior hindbrain. Wnt, FGF8, and En2 also work together to repress the expression of genes associated with more anterior regions, including *Pax6* and *Otx2*.

The various signaling molecules in the midbrain–anterior hindbrain region likely serve multiple purposes in segregating cell types, limiting migration of cells into adjacent regions, and maintaining gene expression and signal transduction pathways that impact final cell fate. Scientists continue to uncover how these signaling centers integrate and complement signals arising from more anterior and posterior regions.

RHOMBOMERES: SEGMENTS OF THE HINDBRAIN

In terms of A/P segmentation along the developing neural tube, the rhombomeres of the hindbrain are perhaps the most studied and best characterized to date. The rhombomeres are numbered in an anterior-to-posterior sequence beginning with rhombomere 1 (r1) in the anterior hindbrain and extending to the junction of the spinal cord (**Figure 3.16**). The exact number of rhombomeres varies from seven to nine, depending on the species and the criteria used to designate the segments.

Each rhombomere expresses a unique set of proteins to regulate proliferation, differentiation, and axonal growth of the developing hindbrain cells. For example, the formation of several of the cranial nerves is impacted by early rhombomere boundaries. Figure 3.16 provides an example from the mouse of the rhombomere origin of various cranial motor neurons in r2–r7. Any loss of rhombomere-specific signals leads to defects in hindbrain patterning and associated changes in cranial nerve development.

Many of the proteins that establish cellular identities in each rhombomere are regulated by signals and transcription factors found in other areas of the neural tube. Expression of genes critical for hindbrain formation often begins in late blastula or early gastrula stage embryos, setting up subsequent spatial and temporal interactions required for rhombomere-specific development.

Cells Usually Do Not Migrate between Adjacent Rhombomeres

One mechanism used to restrict cells to specific rhombomeres is to limit cell migration between adjacent rhombomeres. It is estimated that under normal conditions, fewer than 10% of the cells migrate across rhombomere boundaries during development. Limiting cellular migration into adjacent rhombomeres at early stages ensures that cells are exposed to rhombomere-specific developmental cues that direct subsequent cell fate. Thus, sharp boundaries are critical at early stages of rhombomere formation. At later stages, when the cells need to migrate to another region of the nervous system, the transient rhombomere boundaries are no longer present.

Cell surface molecules differentially expressed in even- versus odd-numbered rhombomeres are among the cues used to ensure proper cellular

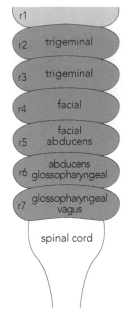

Figure 3.16 Segmentation of the hindbrain. Rhombomeres form segments that extend from the junction with the midbrain to the junction with the spinal cord. Rhombomeres are transient swellings that provide restricted regions for early neural development along the hindbrain. Many cell types, such as cranial motor neurons associated with the trigeminal, facial, abducens, vagus, and glossopharyngeal nerves, originate in specific rhombomeres as shown in rhombomere 2 (r2) through rhombomere 7 (r7). The largest rhombomere, r1, contributes to formation of the cerebellum.

groupings in the hindbrain. For example, a series of experiments by Andrew Lumsden and colleagues revealed that cells from even- or odd-numbered rhombomeres preferentially adhere to cells from other even- or odd-numbered rhombomeres. In one *in vitro* study, dissociated cells from odd-numbered rhombomeres reaggregated with cells from the same rhombomere or those from another odd-numbered rhombomere, but not with cells from even-numbered rhombomeres. Similarly, cells of even-numbered rhombomeres preferentially reaggregated with cells from other even-numbered rhombomeres. Moreover, grafting studies in chick embryos confirmed that boundaries are normally established between adjacent rhombomeres. In one series of experiments, rhombomere segments were removed from one side of a donor and host embryo of the same developmental stage (**Figure 3.17**A). The donor segments were then grafted into the region of missing segments of the host embryo and allowed to develop for another 1.5–2.5 days. When segments of odd- and even-numbered rhombomeres were grafted adjacent to one another in the host embryo, a rhombomere boundary remained and the migration of cells between the two segments was restricted, as on the untreated side of the embryo (Figure 3.17B). However, surgically recombining two even-numbered segments allowed intermixing of cells from these different segments (Figure 3.17C). Similar results were obtained if two odd-numbered segments were grafted next to one another. Together, *in vitro* and *in vivo* studies suggested that adhesive cues group cells together within a rhombomere segment, while inhibitory cues limit migration to adjacent segments.

Subsequent studies revealed that the migration of cells between even- and odd-numbered rhombomeres is inhibited by proteins of the Eph family of tyrosine kinase receptors (Ephs) and their associated cell surface ligands (ephrins). The interaction of membrane-bound **ephrin** ligands with their corresponding **Eph** receptors leads to bidirectional signaling. In other words, a signal is transduced through both the receptor and the ligand (**Figure 3.18**A). In the hindbrain, the Eph receptors and ephrin ligands are often present in alternating patterns. For example, in r3 and r5, the receptors EphA4, EphB2, and EphB3 are highly expressed while one or more of

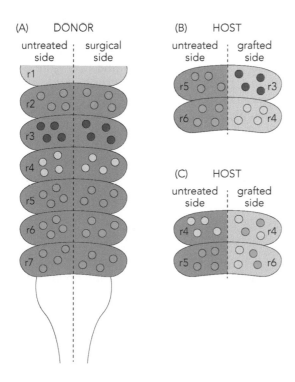

Figure 3.17 **Inhibitory cues restrict migration of cells between even- and odd-numbered rhombomeres.** (A) Under normal conditions, cells from adjacent rhombomeres do not intermingle. In this example, the cells in each rhombomere are shown as a single color to illustrate the segregation of cells within rhombomere boundaries. To test whether rhombomere segments prevent intermingling, segments were removed from one side of a donor chick embryo (surgical side, A) and grafted to a host embryo at the same stage of development (grafted side, B, C). When grafts of rhombomeres from adjacent odd and even rhombomeres (r3 and r4) were transplanted at a different axial level in the host embryo, cell migration was still inhibited across rhombomere boundaries (B). However, intermingling across rhombomere boundaries was observed when two even-numbered rhombomere segments (r4 and r6) were grafted adjacent to one another in the host embryo, even though cells on the untreated side remained segregated (C).

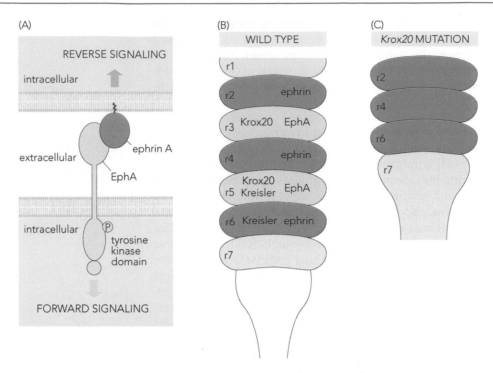

Figure 3.18 Multiple signals interact to regulate the formation of rhombomere boundaries. (A) EphA receptors and ephrin A ligands initiate forward and reverse signaling pathways in cells located in adjacent rhombomeres. An ephrin A ligand binds to an EphA receptor to initiate tyrosine phosphorylation (P) and forward signaling in an adjacent cell. The EphA receptor also initiates signal transduction and reverse signaling through the membrane-attached ephrin A ligand. The ephrin A ligand interacts with co-receptors (not shown) to initiate signal transduction in the ligand-bearing cell. (B) EphA receptors are present on cells in odd-numbered rhombomeres (r3 and r5), while the corresponding ephrin ligands are present on cells of even-numbered rhombomeres (r2 and r4). The resulting bidirectional signaling limits migration between adjacent rhombomeres. The transcription factor Krox20 is also present in r3 and r5 and regulates expression of these Eph receptors. Another transcription factor, Kreisler, is localized to r5 and r6. Both Krox20 and Kreisler are needed to maintain rhombomere boundaries. (C) In mice lacking *Krox20*, r3 and r5 are not maintained and the remaining even-numbered rhombomeres, r2–r6, are joined together, creating a shortened hindbrain.

the corresponding ligands (ephrin A2, ephrin B1, and ephrin B3) are highly expressed in r2, r4, and r6. In this segment of the hindbrain (r2–r6), the Eph/ephrin signaling is repulsive, inhibiting the mixing of cells between adjacent rhombomeres.

EphA4 receptor expression in r3 and r5 is regulated by the transcription factor Krox20, also expressed in r3 and r5 (Figure 3.18B). Although r3 and r5 begin to form in mice lacking *Krox20*, these rhombomeres fail to develop further and are ultimately lost. As a consequence, the even-numbered rhombomeres adjacent to r3 and r5 are brought together creating a shortened hindbrain (Figure 3.18C). The neural structures normally associated with r3 and r5 are also lost and the growth of axons from cranial nerve neurons originating in r2 (trigeminal nerve), r4 (facial nerve), and r6 (abducens nerve) become rerouted within the shortened hindbrain structure.

Multiple Signals Interact to Regulate Krox20 and EphA4 Expression in r3 and r5

Krox20 expression levels are carefully regulated by multiple transcription factors and morphogens. Although species and timing differences exist, a general mechanism for regulating *Krox20* gene expression involves activation of *Pax6*. *Pax6* expression in the hindbrain begins after *Pax6* forebrain expression. FGF in adjacent hindbrain segments helps regulate *Pax6*. Initially, *Pax6* is highest in r3 and r5 before being expressed in all rhombomeres. In most species *Pax6* expression remains most prominent in r3 and r5 where it indirectly limits *Krox20* expression levels via the transcription factor Nab1 (NGFI-A binding protein 1) that is also expressed in r3 and r5. Nab1 represses *Krox20* in these segments while Krox20 positively regulates Nab1, creating a negative feedback loop to ensure proper *Krox20* expression levels are maintained (**Figure 3.19**).

FGF signaling can also influence *Krox20* expression via the transcription factors vHNF1 (variant of hepatocyte nuclear factor 1) and MafB. MafB is part of the larger Maf (Musculo-aponeurotic fibrosarcoma) family. The *MafB* gene is also called *Kreisler* in mice and *Valentino* in zebrafish.

Hindbrain-derived FGF induces vHNF1 which then induces *Krox20* expression in r3 (Figure 3.19). vHNF1 also upregulates MafB which in turn induces *Krox20* in r5. Experimental manipulations in multiple species, including zebrafish, mice, chick, rat, and *Xenopus* revealed that altering any step of the signaling pathway impacts the expression of *Krox20* and *EphA4* in r3 and r5, further demonstrating the importance of these signals for normal hindbrain development. For example, in chick, overexpression of *Pax6* leads to a decrease in *EphA4* expression due to the increased inhibition of *Krox20*. Conversely, when *Pax6* is mutated or experimentally downregulated, *Krox20* and *EphA4* territories expand. Similarly, if FGF is inhibited, *Pax6* expression is downregulated allowing *Krox20* expression to expand to adjacent rhombomeres.

Krox20 also influences the expression of rhombomere-specific *Hox* genes, as explained in the next section.

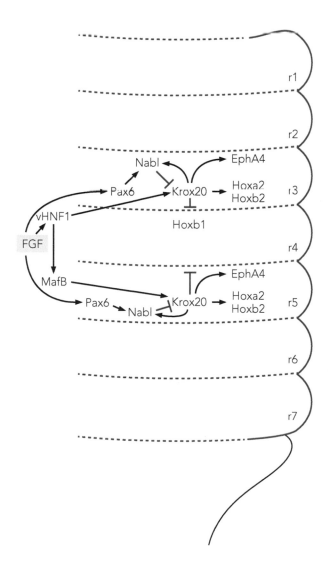

Figure 3.19 *Krox20* expression in rhombomeres (r) 3 and 5 is regulated by several transcription factors. *Pax6* expression, regulated by FGF from adjacent hindbrain segments, indirectly limits *Krox20* expression through the Nab1 transcription factor expressed in r3 and r5. Nab1 represses Krox20 expression. At the same time, Krox20 activates Nab1, creating a negative feedback loop to ensure the proper levels of *Krox20* are maintained in r3 and r5. FGF also influence *Krox20* expression via the transcription factors vHNF1 in r3 and MafB (Kreisler, Valentino) in r5. Levels of Krox20 must be maintained so the expression of other rhombomere-specific genes are expressed in the proper location. For example, Krox20 activates expression of *EphA4*, *Hoxa2*, and *Hoxb2* in r3 and r5 while restricting expression of *Hoxb1* to r4. Nab1, NGFI-A binding protein 1.

HOX GENES REGULATE HINDBRAIN SEGMENTATION

Among the most important signals necessary for segmentation of the hindbrain are the homeodomain-containing transcription factors encoded by various *Hox* genes. In the hindbrain, specific *Hox* genes are expressed in distinct, yet often overlapping patterns within each rhombomere. The combination of *Hox* genes expressed in every hindbrain segment is different. The specific combination of genes expressed has traditionally been referred to as the **Hox code**, which is thought to impact the patterning and development of cells that arise in each rhombomere. Therefore, the expression of individual *Hox* genes must be carefully regulated so that the right *Hox* genes are activated at the right time in the correct rhombomere. Much of our understanding of how *Hox* genes provide region-specific developmental cues for hindbrain segmentation stems from studies in *Drosophila*.

The Body Plan of *Drosophila* Is a Valuable Model for Studying Segmentation Genes

The fruit fly has proven an exceptionally helpful model for investigating genes that regulate segmentation of both the main body axis and the nervous system. The wild-type *Drosophila* body plan consists of three easily identifiable segments—the head, thorax, and abdomen—with each segment contributing to the formation of specific body parts or regions (**Figure 3.20**A). X-ray or chemical exposure mutates single genes allowing

Figure 3.20 Homeotic genes are conserved across species. (A) The fruit fly body plan is organized into distinct segments for the head, thorax, and abdomen. (B) The HOM-C genes (also called *Hox* genes) that regulate this body segmentation are clustered in the antennapedia and bithorax complexes found on chromosome 3. These genes are arranged in order from the 3′ end to the 5′ end of the chromosome so that anterior segments develop in response to the genes expressed closer to the 3′ end, while progressively more posterior regions develop in response to the genes expressed closer to the 5′ end. This is known as the principal of co-linearity, where the relative position of a gene along the chromosome corresponds to the relative position along the A/P axis. lab, labial; pb, proboscipedia; Dfd, deformed; Antp, antennapedia; Ubx, ultrabithorax; AbdA, abdominal A; AbdB, abdominal B; Scr, sex combs reduced. (C) Segmentation of the body plan of mice and other mammals is regulated by homologs of the HOM-C genes, the *Hox* genes. These *Hox* genes are found in four clusters (A–D) on four different chromosomes (chromosomes 6, 11, 15, and 2). Each cluster contains a subset of the 13 *Hox* gene subfamilies. As with *Drosophila*, the relative position of a *Hox* gene from the 3′ to 5′ end of the cluster corresponds to the relative position along the A/P body axis. [Adapted from Alberts B, Johnson A, Lewis J et al. [2015] *Molecular Biology of the Cell*, 6th ed. Garland Science.]

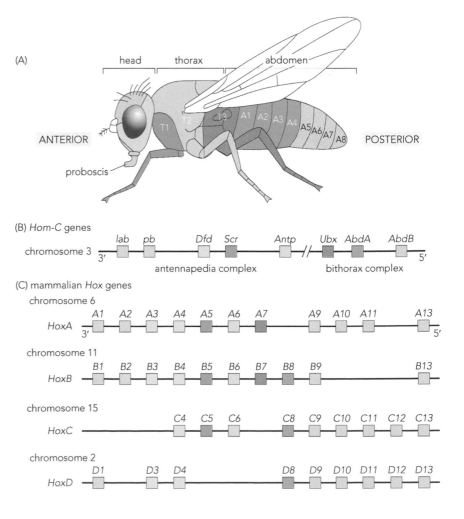

researchers to observe the impact of a gene mutation on normal development. Using these approaches, gene mutations that altered the body plan were identified. Some mutations led to flies with missing or misplaced body parts or, in extreme cases, bodies with no observable body segmentation. As genetic and molecular biology techniques advanced, investigators identified specific genes necessary for body plan patterning.

Among the genes identified were the **segmentation genes**, a group that includes the gap, pair-rule, and segment polarity gene classes. Each class of genes works in sequence to divide the body into smaller and smaller segments along the A/P axis. The **gap genes** are the first class to be active and establish the larger boundaries of the head, thorax, and abdomen. Many of these genes, including *Caudal*, *Hunchback*, *Krüppel*, and *Orthodenticle*, contribute to multiple aspects of neural development. Combinations of gap genes then control the expression of the **pair-rule genes** that divide the three segments into smaller units. Pair-rule genes include *Paired*, *Even-skipped*, *Hairy*, and *Odd-paired*. These genes in turn regulate expression of the **segment polarity genes** that establish smaller boundaries within each existing segment (**Figure 3.21**). The segment polarity genes are also important for establishing characteristics of cells restricted to a given segment. Examples of segment polarity genes include *Hedgehog*, *Wingless*, and *Engrailed*. Like the gap genes, the pair-rule and segment polarity genes play additional roles at other stages of neural development. Mammalian homologs of common segmentation genes important for neural development are listed in **Table 3.1**.

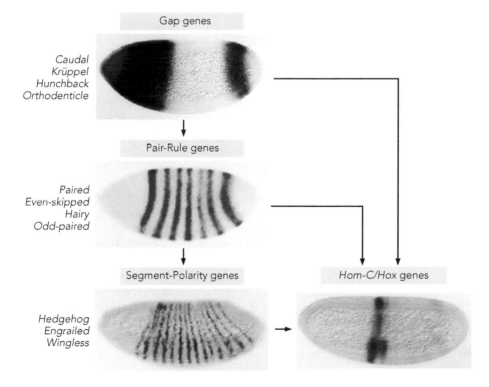

Figure 3.21 Segmentation genes work in sequence to pattern progressively smaller segments along the fruit fly body axis. (A) Gap genes are the first group of segmentation genes to be active and are needed to establish the boundaries between the head, thorax, and abdomen. Gap genes include *Caudal*, *Hunchback*, *Krüppel*, and *Orthodenticle*. (B) The three main body segments are then divided into smaller segments by the activity of pair-rule genes whose expression is regulated by the gap genes. Pair-rule genes include *Paired*, *Even-skipped*, *Hairy*, and *Odd-paired*. (C) Segment polarity genes, including *Hedgehog*, *Wingless*, and *Engrailed*, are then expressed in response to the pair-rule genes, leading to further segmentation of the boundaries along the A/P axis. (D) Gap and pair-rule genes also regulate the expression of the homeotic complex (*HOM-C*)/*Hox* (homeobox) genes that code for a group of helix-turn-helix transcription factors critical for further patterning of regions along the A/P axis. [Adapted from Alberts B, Johnson A, Lewis J et al. [2015] *Molecular Biology of the Cell*, 6th ed. Garland Science.]

Table 3.1 Examples of Segmentation Genes.

	Drosophila	**Mammalian homolog**
Gap Genes	*Caudal*	*Cdx* (caudal type homeobox)
	Hunchback	*Ikaros*
	Krüppel	*Krüppel-like*
	Orthodenticle	*Otx* (orthodenticle homeobox)
Pair-Rule Genes	*Even-skipped*	*Evx-1* (even-skipped homeobox-1)
	Hairy	*Hes1* (hairy and enhancer of split 1)
	Odd-paired	*Zic* (zinc finger of the cerebellum)
	Paired	*Pax* (paired box)
Segment Polarity Genes	*Wingless*	*Wnt* (wingless-related integration)
	Hedgehog	*Shh* (Sonic hedgehog)
	Engrailed	*En* (Engrailed)

The Homeotic Genes That Establish Segment Identity Are Conserved across Species

The gap and pair-rule genes also regulate expression of a group of homeotic genes, originally called the *homeotic complex* (*HOM-C*) genes that code for a group of helix-turn-helix transcription factors. The *HOM-C* genes are the homologs of vertebrate *Hox* (Homeobox) cluster genes that function by activating or repressing downstream genes. These transcription factors contain a 60-amino-acid DNA binding domain (the **homeodomain**) that is encoded by a specific 180-base-pair region of DNA termed the **homeobox**. The prefix "homeo" refers to similarity or sameness; mutations in homeotic genes caused one segment of the fruit fly body to become similar to another.

Although termed *HOM-C* genes in *Drosophila*, these genes are now typically referred to as *Hox* genes because they all arise from the same ancestral gene complex. The remarkable conservation across species is highlighted in studies in which experimental substitution of a mouse *Hox* gene restored function in a fly mutant lacking the homologous *Drosophila* gene.

In *Drosophila*, the *Hox* genes are comprised of the antennapedia and bithorax complexes located on chromosome 3 (Figure 3.20B). The antennapedia complex includes the *Labial, Proboscipedia, Deformed, Sex combs reduced*, and *Anttennapedia* genes. The bithorax complex includes the *Ultrabithorax, Abdominal A*, and *Abdominal B* genes.

Hox genes in both *Drosophila* and vertebrates are arranged on chromosomes in a linear fashion from the 3′ end to the 5′ end. Each gene's relative position in its cluster corresponds to its position of expression along the A/P axis. This is known as the principle of **co-linearity**. *Hox* genes located toward the 3′ end of a cluster are expressed earliest and at the anterior end of the embryo, while those located toward the cluster's 5′ end are expressed later and at the posterior region.

The vertebrate *Hox* genes are not identical to *Drosophila* genes, however. For example, over the course of evolution, gene duplications led to 39 mammalian *Hox* genes. There are 13 subfamilies of the mammalian *Hox* genes that are organized into four *Hox* gene clusters located on four different chromosomes. Each *Hox* cluster contains 9–11 of the 13 *Hox* gene subfamilies. The *Hox* genes are designated *HOXA–D* with the corresponding number (for example, *HOXA1*). **Paralogous groups** of genes, which share homology due to gene duplication, are located at the same relative position in the cluster. For example, *HoxA1*, *HoxB1*, and *HoxD1* are paralogs (Figure 3.20C). Most other vertebrates also have four *Hox* clusters, though some fish, including zebrafish, have seven clusters. Despite the extra clusters, hindbrain expression patterns are similar to those in mice (**Figure 3.22**).

Just as mutations in the *Drosophila* homeotic genes lead to altered body segmentation, a loss of *Hox* genes in mice can lead to altered formation and patterning of rhombomere segments. Some *Hox* gene mutations result in an absence of rhombomere boundaries, causing adjacent rhombomeres to merge and the neural progenitor cells to take on characteristics of neurons found in the adjacent rhombomeres. Although some *Hox* mutations in mice are embryonic lethal—that is, the embryos do not develop to term—in many cases the severity of a single *Hox* gene mutation is not as dramatic as seen with a single homeotic gene mutation in *Drosophila*. The decreased severity in *Hox* mutations is thought to be due in part to the overlap of *Hox* gene expression within a given rhombomere. Overlapping *Hox* gene expression would allow remaining *Hox* genes to partially compensate for the loss of a single *Hox* gene. The ability of *Hox* gene **enhancers** to act over long distances may also protect against deficits arising from a single gene mutation (**Box 3.2**).

Hox expression patterns are not fixed, however; temporal and spatial changes in *Hox* gene expression patterns are noted during hindbrain development, and these changes lead to further specialization of the cells that originate in each segment. For example, *Hoxa1* is expressed in the presumptive r4–r7 regions of the mouse embryo, but the expression is lost by the time the rhombomeres are fully formed. Other *Hox* genes, such as *Hoxb5*, *Hoxb6*, and *Hoxb8*, are initially expressed in spinal cord regions, but extend their expression anteriorly to rhombomeres as

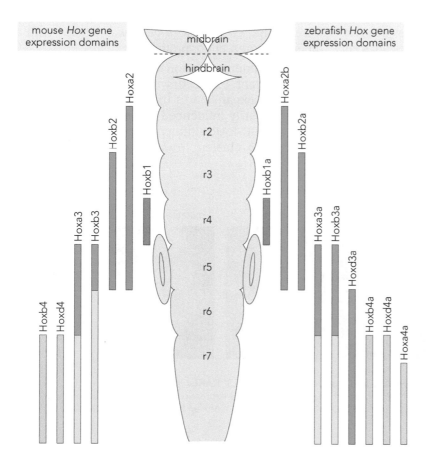

Figure 3.22 Rhombomere segments are distinguished by the expression of different Hox genes. Each rhombomere expresses specific *Hox* genes at relatively higher (darker shading) or lower levels (lighter shading). The combination of genes expressed in any given rhombomere impacts development of cells arising in that segment. The expression of specific *Hox* genes varies between species as shown in mice (left) and zebrafish (right), and at different times in development. [Adapted from Moens CB & Prince V E [2002] *Dev Dyn* 224:1–17. With permission from John Wiley & Sons, Inc.]

Box 3.2 Developing Neuroscientists: Additional Insights into *Hox* Gene Regulation

Hillary Mullan, M.D.

Hillary Mullan, graduated from Oberlin College in 2014 where she majored in neuroscience and biology. Following graduation, Hillary was a research associate in a neurology lab prior to beginning medical school in 2016. She is currently completing a residency in neurology. During the summer of 2012 she worked in the laboratory of Robb Krumlauf at the Stowers Institute for Medical Research in Kansas City, Missouri. Here she describes the project she worked on during that period.

As described in this chapter, *Hox* genes are a group of genes that are critical for normal embryogenesis in many species. These genes are activated during gastrulation and produce transcription factors that bind to enhancer regions on DNA and, in turn, initiate a number of downstream cascades. Within the mammalian genome, the 39 *Hox* genes are organized into four clusters (A, B, C, and D). An interesting characteristic of these genes lies in the relationship between their structural organization along the chromosome and the order in which they are spatially and temporally activated during development. Genes located closer to the 3′ end of the cluster are expressed earlier and more anteriorly along the body axis than those located toward the 5′ end of the cluster. *Hox* genes at the 3′ end are activated by lower levels of retinoic acid (RA) than those at the 5′ end, which require a higher concentration of RA to be activated. As also noted in this chapter, RA binds to the retinoic acid response element (RARE) in the promoter region of the gene to regulate *Hox* expression. These RAREs are associated with sequences adjacent to the corresponding *Hox* gene. However, the deletion or mutation of specific RAREs does not result in the same morphological deficits observed with loss of the *Hox* gene itself. This suggested that other enhancers may be available to regulate *Hox* gene expression and compensate for the loss of specific RAREs.

Other experimental observations also suggested additional enhancers are available to regulate RA-dependent *Hox* gene expression. In the spinal cord, a subset of *Hox* genes exhibits an expression pattern known as rostral expansion. The expression of these *Hox* genes expands anteriorly from their initial site in the spinal cord to include sites in the hindbrain. For example, in mice, expression of *Hoxb5*, *Hoxb6*, and *Hoxb8* is first detected in the spinal cord at embryonic day 9.5 (E9.5), but by E11.5 is found in the hindbrain as well. Although this shift in expression has been associated with the presence of retinoic acid, there are no RAREs near these 5′ *Hoxb* genes. However, RAREs are located in the 3′ region of the *Hoxb* genes. This suggested the possibility that 3′ RAREs function as enhancers to influence transcription of 5′ *Hox* genes and influence their rostral expansion.

Candidates for this long-distance regulation of 5′ *Hox* gene transcription include the RA-dependent enhancers early neural enhancer (ENE) and distal enhancer (DE) located near the 3′ end of the *Hox* cluster, where they activate expression of *Hoxb4* and *Hoxb5* in the hindbrain. ENE is located 3′ to *Hoxb4* and DE is located 3′ to *Hoxb5*.

We tested whether these 3′ enhancers regulate the rostral expansion of 5′ *Hoxb* genes. Our lab produced mice with mutations in either one or both RAREs and observed the effect of these mutations on the expression of different *Hoxb* genes. Gene expression patterns were visualized by genetically engineering the mice so that each 5′ *Hoxb* gene was labeled with a different fluorescent or protein tag.

The experiments demonstrated that while both RA-dependent enhancers are needed to regulate the rostral expansion of the *Hoxb* genes, the ENE-RARE primarily influenced the nearby *Hoxb4* gene while the DE-RARE influenced all the 5′ members of the *Hoxb* cluster, *Hoxb5–Hoxb9* (**Figure 3.23**).

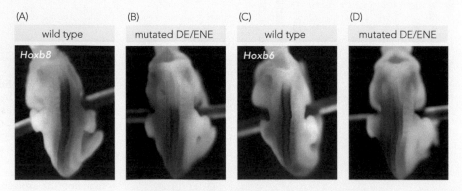

Figure 3.23 The normal rostral expansions of Hoxb8 and Hoxb6 are lost when 3′ RAREs are mutated. *Hoxb8* (A) and *Hoxb6* (C) are seen in the hindbrain regions of wild-type mice at embryonic day 11.5. This expression was lost when both the early neural enhancer (ENE) and distal enhancer (DE) were mutated (B, D). [Adapted from Ahn Y, Mullan H & Krumlauf R [2014] *Dev Biol* 388:134–144.]

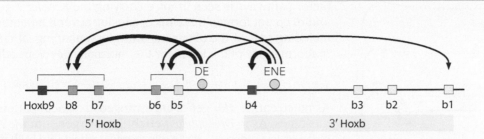

Figure 3.24 Schematic representation of the shared long-range retinoic acid response elements (RAREs). The RAREs DE and ENE are both located close to 3' *Hoxb* genes (*Hoxb1–Hoxb4*). However, as indicated by the arrows, the enhancers can also act over a longer distance to influence rostral expansion of the 5' *Hoxb* genes (*Hoxb5–Hoxb9*). [Adapted from Ahn Y, Mullan H & Krumlauf R [2014] *Dev Biol* 388:134–144.]

Thus, DE-RARE functions over a long distance to influence multiple *Hox* genes and regulate the rostral expansion of 5' *Hoxb* genes (**Figure 3.24**). The research also supports previous studies suggesting enhancers that work at a distance and influence multiple *Hox* genes provide a means of compensating for a single genetic mutation. *Hox* genes may have remained clustered over the course of evolution to ensure the different *Hox* genes are within reach of the various enhancers.

development progresses (see **Box 3.2**). Thus, *Hox* gene expression patterns shift as development progresses, often in response to specific transcription factors.

Transcription Factors Regulate *Hox* Gene Expression and Rhombomere Identity

Current research efforts focus on identifying how and when *Hox* gene expression is regulated in different species. Recent discoveries suggest various homeodomain and non-homeodomain transcription factors interact to influence *Hox* gene expression in each rhombomere. Although species-specific differences are noted in some cases, several common themes are beginning to emerge.

One observation is that early in development, before the neural plate is fully formed, homeodomain transcription factors of the TALE family (three amino-acid loop extension) activate anterior *Hox* genes (paralog groups 1–4). Combinations of TALE and *Hox* proteins then interact to transcribe target genes including other *Hox* genes and *Krox20*. For example, the TALE transcription factors Meis (mouse ectopic integration site) and Pbx (pre-B-cell leukemia) work synergistically to activate a *Krox20* enhancer element in r3 of chick and mice.

Further, at early stages of hindbrain formation, Hoxa1 binds to enhancers of *Meis 2* and *Meis 3* in mice and zebrafish to regulate their expression. As development progresses, combinations of two or three Meis/Pbx/Hox proteins work together to regulate other *Hox* gene expression patterns in the hindbrain.

The non-homeodomain transcription factors, vHNF1, Krox20, and MafB/Kreisler/Valentino also regulate early *Hox* gene expression. For example, vHNF1 is required for expression of *Mafb/Kresiler/Valentino* in r5 and r6.

In zebrafish, vHNF1 induces *Mafb/Kresiler/Valentino* and *Krox20* expression in r5 and r6 and these in turn activate rhombomere-specific *Hox* genes. At the same time, Krox20 keeps *Hoxb1* expression limited to r4 by repressing its expression in r5 (**Figure 3.25**). Similar mechanisms are used in mice to regulate expression of equivalent r5 and r6-specific *Hox* genes (see Figure 3.19). The functional importance of these transcription

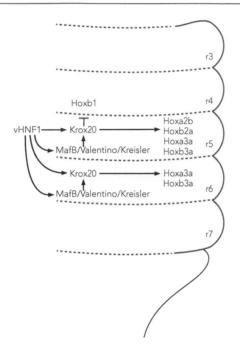

Figure 3.25 Transcription factors regulate Hox gene expression and rhombomere identity. In this example from zebrafish, vHNF1 regulates expression of the transcription factors Krox20 and Mafb/Kreisler/Valentino in rhombomeres 5 and 6 (r5, r6). These transcription factors then activate expression of r5- and r6-specific *Hox* genes (*Hoxa2b*, *Hoxb2a*, *Hoxa3a*, and *Hoxb3a* in r5; *Hoxa3a* and *Hoxb3a* in r6). In r5, Krox20 also inhibits expression of the *Hoxb1* gene that is specific for r4. Similar mechanisms are used to restrict *Hox* expression in mice (see also Figure 3.19).

Figure 3.26 Hox genes are differentially affected by retinoic acid. The concentration of retinoic acid (RA) regulates expression of the *Hox* genes. The RA concentration is lowest in the anterior hindbrain, where it influences expression of the 3′ *Hox* genes but increases as it approaches the junction with the spinal cord. RA is found in an opposing concentration gradient in the spinal cord, where it influences development cells along the dorsal–ventral axis (see Chapter 4). (B) Retinoic acid (RA) binds to a receptor to act as a transcription factor for *Hox* gene transcription. RA released from RA-generating tissues adjacent to the neural tube enters cells of the hindbrain, where it is then translocated to the nucleus. Translocation may be aided by association with cellular RA-binding proteins (CRABPs). Once in the nucleus, RA binds to a retinoic acid receptor (RAR), which subsequently associates with the RA-response element (RARE) of a *Hox* gene's promoter to activate gene transcription.

factor pathways is seen in mice carrying the *Kreisler 1* gene mutation. R5 and r6 do not form in these mice, causing several hindbrain abnormalities, including loss of the abducens and glossopharyngeal cranial nerves and malformations of the inner ear that normally develops adjacent to r5.

Retinoic Acid Regulates *Hox* Gene Expression

As noted earlier, *Hox* genes are arranged in a linear pattern along the chromosomes. As in *Drosophila*, vertebrate *Hox* genes at the 3′ end (for example, paralog groups 1–4) are expressed earliest and influence development of the more anterior structures, in this case the rhombomeres 2–7 in the hindbrain (**Figure 3.26**A), while those at the 5′ end (for example, paralog groups 5–13) are expressed later and influence development of the more posterior structures (for example, the spinal cord).

During normal development, *Hox* genes are not expressed in r1 or regions anterior to it. However, if r1 is transplanted to a more posterior region of the hindbrain, the grafted r1 will begin to express the *Hox* genes associated with that region. Similarly, misexpression of *Hox* genes in r1 leads to formation of neural cells with characteristics of those normally found in the corresponding *Hox*-expressing hindbrain region. Thus, under normal developmental conditions, *Hox* gene expression is established within a segment of the hindbrain and the resulting gene expression patterns establish which structures will form at that A/P level.

The expression of the different *Hox* genes is regulated by the coordinated efforts of multiple interacting signaling pathways. **Retinoic acid (RA)**, a vitamin A derivative, is one of the critical modulators. RA directly activates *Hox* gene expression by binding to receptors located in the nuclei of responsive cells. RA is taken up by the cell and translocated to the nucleus, often aided by cellular RA-binding proteins (CRABPs). The RA

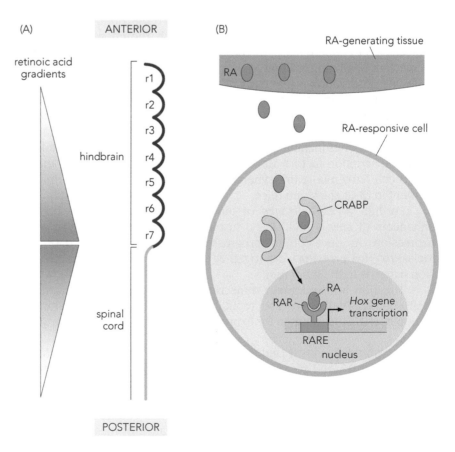

then associates with one of the retinoic acid receptors (RARs) to act as a transcription factor that binds to a retinoic acid response element (RARE) in the promoter region of the target *Hox* gene. In this way, RA can regulate expression of the individual *Hox* genes (Figure 3.26B).

In the developing chick, mouse, and zebrafish, RA is expressed beginning in the gastrula stage embryo and is later secreted by the mesoderm adjacent to the neural tube. RA is synthesized by the enzyme retinaldehyde dehydrogenase 2 (Raldh2, also called aldehyde dehydrogenase 1A2) that is present in the paraxial mesoderm (chordamesoderm) and becomes localized adjacent to the posterior hindbrain. RA is found in two opposing gradients anterior and posterior to the junction of the presumptive hindbrain and spinal cord. Both gradients increase as they approach the spinal cord junction (Figure 3.26A) and contribute to cellular specializations along the A/P as well as the dorsal–ventral (D/V) axis (see Chapter 4).

In the hindbrain, each *Hox* gene responds preferentially to a given RA concentration. *Hox* genes located at the 3′ end of the chromosome are activated by lower concentrations of RA than the *Hox* genes located nearer the 5′ end. Thus, because *Hox* gene location on the chromosome corresponds to the development of structures along the A/P axis, structures in the more anterior regions of the hindbrain develop in response to lower RA concentrations, whereas the more posterior hindbrain structures develop following exposure to higher concentrations of RA (Figure 3.26).

Early investigations of RA–*Hox* gene interactions in the hindbrain noted that experimentally increasing the RA concentration to which embryos were exposed led to the expansion of more posterior structures, while limiting the formation of anterior structures. Additionally, when RA was added to embryonic stem cells, low concentrations led to expression of *Hox* genes that are active in anterior hindbrain regions, whereas higher RA concentrations led the cells to express *Hox* genes that are active in progressively more posterior hindbrain segments. Thus, several lines of evidence supported the role of RA concentration differences in regulating *Hox* gene expression and influencing A/P regionalization.

Although there is a clear link between RA concentration and *Hox* gene expression in the hindbrain, current research indicates that the regulation of these interactions is more complex than originally envisioned. As explained next, the levels of RA are regulated by multiple signals and the concentration requirements at specific A/P regions appear to change as development progresses, at least in some animal models. Additionally, RA levels regulate the expression of other genes, which together with the *Hox* genes pattern the hindbrain segments and prevent segmentation of the spinal cord.

The RA-Degrading Enzyme Cyp26 Helps Regulate *Hox* Gene Activity in the Hindbrain

Just as *Hox* gene expression must be carefully regulated, so too must RA levels. The Cyp26 family of proteins is a group of enzymes that degrade RA. Cyp26 family members are thus a primary modulator of RA levels and one of the essential signals required for hindbrain development. There are three or four members of the Cyp26 family detected in most vertebrate species. Specific Cyp26 enzymes exhibit overlapping and unique functions in different animal models. The following description is a general summary of the role of Cyp26 enzymes as a group rather than the individual role of each enzyme in the different vertebrate animal models.

Cyp26 enzymes are expressed in a concentration gradient along the hindbrain and spinal cord. In the hindbrain, higher concentrations of Cyp26 are found in the more anterior regions and are therefore able to degrade more RA. This results in the lower concentrations of RA needed for activation of *Hox* genes in the anterior hindbrain (**Figure 3.27**A). In

Figure 3.27 The RA concentration gradient in the hindbrain is regulated by gradients of Cyp26 enzymes. (A) Higher concentrations of Cyp26 family members, the enzymes that degrade RA, are found in the anterior hindbrain segments, where lower concentrations of RA are needed to activate 3′ *Hox* genes. Progressively lower concentrations of Cyp26 are found approaching the posterior hindbrain, where higher RA concentrations are needed to induce *Hox* genes. (B) Cyp26 enzymes are required for maintaining the RA gradient during normal developmental conditions (wild-type). When Cyp26 was experimentally inhibited in mice or zebrafish (–Cyp26), RA levels increased throughout the hindbrain, leading to the posteriorization of anterior hindbrain segments.

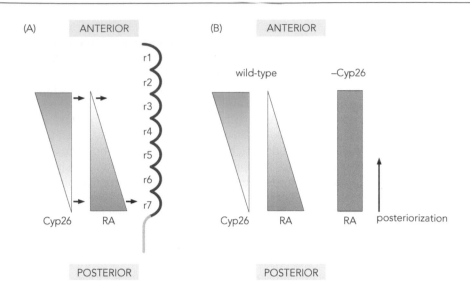

contrast, lower concentrations of Cyp26 are found at more posterior rhombomere segments leading to less RA degradation and thus the higher RA concentrations needed for activation of *Hox* genes necessary for posterior rhombomeres development.

If Cyp26 enzymes fail to degrade sufficient RA in the anterior regions, RA levels increase, leading to expression of the *Hox* genes that normally regulate the development of posterior regions. The importance of tightly regulating Cyp26 was noted in studies of mice and zebrafish, where loss of Cyp26 led to increases in anterior RA levels and posteriorization of otherwise anterior hindbrain segments (Figure 3.27B). Thus, Cyp26 enzymes may be the primary signal utilized to generate the RA gradient necessary for patterning the anterior hindbrain.

RA can also influence Cyp26 levels to some extent. RA levels naturally fluctuate in response to levels of vitamin A available to the embryo. Experiments in zebrafish found that as RA synthesis fluctuates in the future hindbrain regions, the Cyp26 levels also change. Further, when RA synthesis was experimentally increased, Cyp26 activity increased. Conversely, when RA synthesis was decreased, Cyp26 activity decreased. The coordinated activity of RA and Cyp26 allows for a consistent level of RA in each region of the hindbrain, and therefore activation of the correct of hindbrain-associated genes throughout embryogenesis.

RA and FGF Differentially Pattern Posterior Rhombomeres and Spinal Cord

In addition to roles in anterior patterning, members of the FGF family also influence patterning of the posterior regions of the neural tube. In the spinal cord, FGF is present in a gradient that progressively increases toward more posterior regions, where it regulates transcription of posterior *Hox* genes and influences the levels of RA synthesis. The mechanisms by which FGF regulates RA levels are not entirely clear. There appear to be variations in signaling mechanisms utilized by the different animal models, as well as differences in mechanisms used in different organ systems of the same species. Several studies found that FGF can influence the synthesis of RA either directly or indirectly via Wnt signaling. Increases in RA levels, in turn, inhibit FGF signaling in the hindbrain (**Figure 3.28A**). Thus, by increasing RA synthesis, FGF restricts its effects to the more posterior regions.

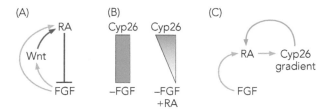

Figure 3.28 FGF interacts with RA to pattern the posterior hindbrain and spinal cord. (A) Under different conditions, members of the fibroblast growth factor (FGF) family influence levels of RA synthesis either directly or via Wnt. RA, in turn, inhibits FGF signaling in the hindbrain. Thus, by regulating RA levels, FGF can restrict its activity to the spinal cord. (B) In the absence of FGF, the normal Cyp26 gradient is lost. Addition of RA restores this gradient, suggesting that FGF regulates RA levels so RA can modify the existing Cyp26 gradient. The Cyp26 gradient, in turn, influences RA levels (C).

FGF also appears to influence the shape of the Cyp26 gradient in the hindbrain. Studies reveal that a loss of FGF leads to broader expression of Cyp26 (Figure 3.28B). In zebrafish lacking FGF, the addition of RA restored the Cyp26 gradient, suggesting that FGF may normally have an indirect effect on Cyp26 by regulating RA synthesis, which in turn refines the existing Cyp26 gradient (Figure 3.28C). Precisely how FGF and RA interact and how they influence Cyp26 expression patterns in different animal models remains under investigation. Recent studies further suggest that FGF and Wnt together prevent upregulation of *Cyp26* in the posterior-most hindbrain thereby allowing the higher levels of RA needed for posterior hindbrain regions. Thus, FGF and Wnt may work together to both increase RA synthesis and prevent upregulation of *Cyp26* expression in posterior regions of hindbrain.

Cdx Transcription Factors Are Needed to Regulate *Hox* Gene Expression in the Spinal Cord

FGF also regulates *Hox* gene transcription by interacting with the Cdx family of transcription factors that bind to *Hox* gene enhancers. Cdx (caudal-related homeobox) transcription factors are related to *Hox* genes and are homologs of the *Drosophila Caudal* (*Cad*) genes. In *Drosophila*, the Caudal transcription factors directly activate several different segmental genes. Studies in mice, *Xenopus*, zebrafish, and chick demonstrated the importance of FGF and Cdx interactions in regulating *Hox* gene expression and patterning in posterior regions of the neural tube. Although the specific number of *Cdx* genes present in these species varies, the general requirement for *Cdx* genes in spinal cord patterning is consistent across species. In the zebrafish, for example, a loss of *Cdx1* and *Cdx4*—genes expressed in posterior regions of the embryo—leads to a loss of the *Hox* genes normally expressed in the spinal cord. Further, in the absence of these *Cdx* genes, the expression of hindbrain *Hox* genes expands through spinal cord regions (**Figure 3.29**). These results show that the Cdx transcription factors are needed for the expression of spinal cord *Hox* genes, but not for hindbrain *Hox* genes.

Studies in early chick embryos reveal that RA and FGF differentially regulate expression of 3′ (hindbrain) and 5′ (spinal cord) *Hox B* genes. For example, in chick, *Hoxb1*, *Hoxb3*, *Hoxb4*, and *Hoxb5* are expressed in the developing hindbrain. These 3′ *Hox* genes were influenced by RA, but not responsive to FGF. More posteriorly, in the future spinal cord region, *Hoxb6*, *Hoxb7*, *Hoxb8*, and *Hoxb9* are expressed. In contrast to 3′ *Hox* genes, these 5′ *Hox* genes responded to FGF, but not to RA (**Figure 3.30**A). Researchers also confirmed that the effects of FGF on *Hoxb9* activation are mediated by Cdx transcription factors. By this stage of development, *Cdx* gene expression is limited to the spinal cord region of the developing neural tube and does not extend anteriorly into the hindbrain regions. However, when *Fgf* and *Cdx* were both ectopically expressed in hindbrain regions, *Hoxb9* expression was induced in these more anterior regions. When *Cdx* expression alone was increased in the hindbrain, *Hoxb4* expression was

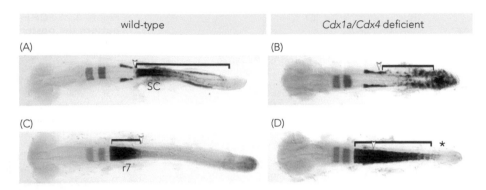

Figure 3.29 Loss of Cdx results in a loss of spinal cord Hox genes with an expansion of hindbrain Hox genes. (A) The expression of a spinal cord *Hox* gene (*Hoxb8a*, bracketed region) was lost in zebrafish lacking *Cdx1a* and *Cdx4* (B, bracket). (C) In the absence of the *Cdx* genes, the expression of the *Hox* gene normally found in rhombomere 7 (*Hoxb1a*, bracket) expanded throughout spinal cord levels (D, bracket), except for the posterior-most tip (asterisk). The transcription factor Krox20 (red) is labeled in the figures to indicate the levels of rhombomeres 3 and 5. The white arrow points to the junction of the hindbrain and spinal cord located at the level of the third somite in wild-type embryos (SC, spinal cord). [From Skromne et al. [2007] *Development* 134:2147–2158. With permission from The Company of Biologists.]

now induced in response to FGF, but no longer in response to RA. These findings support the idea that Cdx is required in FGF, but not RA, signaling.

Under normal conditions, *Cdx* expression may be restricted to the spinal cord region by RA. Thus, RA not only activates 3' *Hoxb* genes, but also restricts the influence of FGF by preventing Cdx activity in the hindbrain. The mechanisms by which RA can mediate both effects are not fully understood, but RAREs have been detected adjacent to the *Cdxl* gene, suggesting a likely means by which RA could both activate 3' *Hoxb* gene expression and repress *Cdx* expression in the hindbrain.

Intricate and carefully balanced interactions between signals associated with hindbrain and spinal cord formation appear to be necessary to pattern these areas of the nervous system. Tightly controlled levels of RA, Cyp26, FGF, and Cdx combine to regulate which *Hox* genes are expressed in hindbrain and spinal cord regions of the A/P axis (Figure 3.30B).

The Activation-Transformation Model Is Being Revised

Recent studies suggest that in some species, the anterior most (forebrain, midbrain, hindbrain) and posterior most (spinal cord) region of the neuraxis arise from two different cell lineages. Thus, some aspects of regionalization may occur even before the onset of neural induction. Evidence now suggests that anterior neural regions originate from cells in the anterior epiblast while cells of the spinal cord arise from cells in the posterior-lateral epiblast. This posterior (caudal) portion of the epiblast also gives rise to the neuromesodermal progenitors (NMPs) found in paraxial mesoderm and spinal cord regions.

Signals associated with spinal cord patterning appear necessary to induce NMPs and direct their development toward a spinal cord cell fate. For example, when FGF and Wnt were added to mouse embryonic stem cells, NMPs were induced. Transient exposure to FGF and Wnt also induced expression of the Cdx proteins and caused the NMPs to differentiate into spinal cord progenitor cells. In contrast, the embryonic stem cells not exposed to FGF and Wnt developed characteristics of hindbrain and the anterior-most spinal cord cells.

These findings are forcing researchers to re-assess the activation-transformation model across species. In at least some animal models, the cells of the anterior epiblast develop in a manner consistent with the activation-transformation model, while the posterior-lateral epiblast cells develop independent of transformation signals. It has been suggested that

(A)

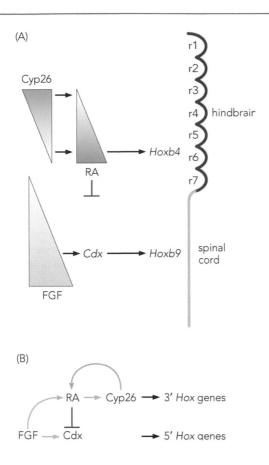

r1
r2
r3
r4 } hindbrain
r5
r6
r7

Cyp26

Hoxb4

RA

Cdx ⟶ Hoxb9   spinal cord

FGF

(B)

RA ⟶ Cyp26 ⟶ 3' Hox genes

FGF ⟶ Cdx ⟶ 5' Hox genes

Figure 3.30 RA and FGF differentially influence 3' and 5' Hox genes in the posterior hindbrain and spinal cord. (A) In the hindbrain, the gradient of Cyp26 influences the gradient of RA. In the area of rhombomere r6/r7, the lower levels of Cyp26 enzymes lead to the higher RA concentration that induces expression of the 3' *Hox* gene *Hoxb4*. In the spinal cord, on the other hand, the expression of the 5' *Hox* gene, *Hoxb9*, is regulated by a concentration gradient of FGF. Unlike RA, the effects of FGF are mediated by Cdx transcription factors that are only expressed in the spinal cord. RA helps restrict Cdx to the spinal cord, thereby preventing its activity in the hindbrain. (B) The roles of RA, Cyp26, Cdx, and FGF are interconnected. FGF influences RA synthesis, while levels of RA and Cyp26 are refined by one another. RA inhibits the expression of Cdx transcription factors, likely by associating with a RARE located adjacent to the *Cdx* gene. Thus, in the hindbrain, RA can activate transcription of 3' *Hox* genes while also repressing transcription of *Cdx*. FGF limits its own expression to the spinal cord by influencing the synthesis of RA.

different cell linages for anterior versus posterior regions may reflect the evolutionary origins of brain and spinal cord regions.

SUMMARY

The A/P axis of the neural tube is established at early stages of neural development. From the developing forebrain through hindbrain region, many structural landmarks including expansions and constrictions set up early boundaries to limit the migration of cells and provide local signals to induce specific regions of the forming nervous system. Through multiple signaling pathways, these segments are further delineated allowing for cellular specialization within each region.

As discussed in Chapter 2, neural inducers first set up the general axis of the neural plate. As seen in this chapter, gradients of signals arising from forebrain, midbrain, hindbrain, and spinal cord regions, as well as antagonists to these signals, interact to pattern the structures along the A/P axis. For example, *Hox* genes are expressed to pattern the hindbrain region of the nervous system, but they are blocked from the midbrain by the activity of FGFs, particularly FGF8. Wnt family members are also important for patterning the midbrain region, while multiple signals from more anterior regions, such as Cerberus, Dickkopf, and sFRP, inhibit Wnt activity from extending to the forebrain. Other gene families act in opposition to one another to further refine boundaries. Members of the *Six* and *Irx* gene families establish boundaries between midbrain and forebrain structures, *Otx* and *Gbx* family members help define midbrain and anterior hindbrain segments, while RA and FGF signals establish the boundary between hindbrain and spinal cord regions.

Although the steps involved in patterning the A/P axis of the neural tube are most often described as individual events, they often overlap both temporally and spatially. The level of detail that is currently understood and the relatively short time frame in which much of this information has been discovered are quite remarkable. At the same time, much remains to be discovered, and uncovering the intricate mechanisms that regulate regionalization along the A/P axis remains an active area of research.

FURTHER READING

Ahn Y, Mullan H & Krumlauf R (2014) Long-range regulation by shared retinoic acid response elements modulate dynamic expression of posterior *Hoxb* genes in CNS development. *Dev Biol* 388:134–144.

Bally-Cuif L, Alvarado-Mallart RM, Darnell DK & Wassef M (1992) Relationship between Wnt-1 and En-2 expression domains during early development of normal and ectopic met-mesencephalon. *Development* 115(4):999–1009.

Bel-Vialar S, Itasaki N & Krumlauf R (2002) Initiating *Hox* gene expression: In the early chick neural tube differential sensitivity to FGF and RA signaling subdivides the *HoxB* genes in two distinct groups. *Development* 129(22): 5103–5115.

Bouwmeester T, Kim S, Sasai Y, et al. (1996) Cerberus is a head-inducing secreted factor expressed in the anterior endoderm of Spemann's organizer. *Nature* 382:595–601.

Cambronero F, Ariza-McNaughton L, Wiedemann LM & Krumlauf R (2020) Inter-rhombomeric interactions reveal roles for fibroblast growth factors signaling in segmental regulation of EphA4 expression. *Dev Dyn* 249(3):354–368.

Crossley PH, Martinez S & Martin GR (1996) Midbrain development induced by FGF8 in the chick embryo. *Nature* 380(6569):66–68.

Duester G (2008) Retinoic acid synthesis and signaling during early organogenesis. *Cell* 134:921–931.

Dupe V & Lumsden A (2001) Hindbrain patterning involves graded responses to retinoic acid signalling. *Development* 128(12):2199–2208.

Fongang B, Kong F, Negi S, et al. (2016) A conserved structural signature of the homeobox coding DNA in *HOX* genes. *Sci Rep* 6:35415.

Fossat N, Jones V, Garcia-Garcia MJ & Tam PPL (2012) Modulation of WNT signaling activity is key to the formation of the embryonic head. *Cell Cycle* 11(1):26–32.

Frank D & Sela-Donenfeld D (2019) Hindbrain induction and patterning during early vertebrate development. *Cell Mol Life Sci* 76(5):941–960.

Glinka A, Wu W, Delius H, et al. (1998) Dickkopf-1 is a member of a new family of secreted proteins and functions in head induction. *Nature* 391:357–362.

Green DG, Whitener AE, Mohanty S, et al. (2020) Wnt signaling regulates neural plate patterning in distinct temporal phases with dynamic transcriptional outputs. *Dev Biol* 462(2):152–164.

Grinblat Y, Gamse J, Patel M & Sive H (1998) Determination of the zebrafish forebrain: Induction and patterning. *Development* 125(22):4403–4416.

Guo Q, Li K, Sunmonu NA & Li JY (2010) Fgf8b-containing spliceforms, but not Fgf8a, are essential for Fgf8 function during development of the midbrain and cerebellum. *Dev Biol* 338(2):183–192.

Guthrie S & Lumsden A (1991) Formation and regeneration of rhombomere boundaries in the developing chick hindbrain. *Development* 112(1):221–229.

Imbard A, Benoist JF & Blom HJ (2013) Neural tube defects, folic acid and methylation. *Int J Environ Res Public Health* 10(9):4352–4389.

Janssens S, Denayer T, Deroo T, et al. (2010) Direct control of *Hoxd1* and *Irx3* expression by Wnt β-catenin signaling during anteroposterior patterning of the neural axis in *Xenopus*. *Int J Dev Biol* 54(10):1435–1442.

Kawano Y & Kypta R (2003) Secreted antagonists of the Wnt signalling pathway. *J Cell Sci* 116:2627–2634.

Kayam G, Kohl A, Magen Z, et al. (2013) A novel role for Pax6 in the segmental organization of the hindbrain. *Development* 140(10):2190–2202.

Keenan ID, Sharrard RM & Isaacs HV (2006) FGF signal transduction and the regulation of Cdx gene expression. *Dev Biol* 299(2):478–488.

Kudoh T, Wilson SW & Dawid IB (2002) Distinct roles for Fgf, Wnt and retinoic acid in posteriorizing the neural ectoderm. *Development* 129(18):4335–4346.

Leung B & Shimeld SM (2019) Evolution of vertebrate spinal cord patterning. *Dev Dyn* 248(11):1028–1043.

Liu A & Joyner AL (2001) Early anterior/posterior patterning of the midbrain and cerebellum. *Annu Rev Neurosci* 24:869–896.

Maden M, Horton C, Graham A, et al. (1992) Domains of cellular retinoic acid-binding protein I (CRABP I) expression in the hindbrain and neural crest of the mouse embryo. *Mech Dev* 37(1–2):13–23.

Marshall H, Nonchev S, Sham MH, et al. (1992) Retinoic acid alters hindbrain *Hox* code and induces transformation of rhombomeres 2/3 into a 4/5 identity. *Nature* 360(6406):737–741.

Martinez S, Crossley PH, Cobos I, et al. (1999) FGF8 induces formation of an ectopic isthmic organizer and isthmocerebellar development via a repressive effect on Otx2 expression. *Development* 126(6):1189–1200.

Martinez S, Wassef M & Alvarado-Mallart RM (1991) Induction of a mesencephalic phenotype in the 2-day-old chick prosencephalon is preceded by the early expression of the homeobox gene en. *Neuron* 6(6):971–981.

Metzis V, Steinhauser S, Pakanavicius E, et al. (2018) Nervous system regionalization entails axial allocation before neural differentiation. *Cell* 175(4):1105–1118.e17.

Moens CB & Prince VE (2002) Constructing the hindbrain: Insights from the zebrafish. *Dev Dyn* 224(1):1–17.

Nakayama Y, Kikuta H, Kanai M, et al. (2013) Gbx2 functions as a transcriptional repressor to regulate the specification and morphogenesis of the mid-hindbrain junction in a dosage- and stage-dependent manner. *Mech Dev* 130(11–12):532–552.

Rhinn M & Dolle P (2012) Retinoic acid signalling during development. *Development* 139(5):843–858.

Ribes V, Wang Z, Dolle P & Niederreither K (2006) Retinaldehyde dehydrogenase 2 (RALDH2)-mediated retinoic acid synthesis regulates early mouse embryonic forebrain development by controlling FGF and sonic hedgehog signaling. *Development* 133(2):351–361.

Sanchez-Arrones L, Stern CD, Bovolenta P & Puelles L (2012) Sharpening of the anterior neural border in the chick by rostral endoderm signalling. *Development* 139(5):1034–1044.

Sasai N, Toriyama M & Kondo T (2019) Hedgehog signal and genetic disorders. *Front Genet* 10:1103.

Shimizu T, Bae YK & Hibi M (2006) Cdx-*Hox* code controls competence for responding to Fgfs and retinoic acid in zebrafish neural tissue. *Development* 133(23):4709–4719.

Simeone A, Puelles E & Acampora D (2002) The Otx family. *Curr Opin Genet Dev* 12(4):409–415.

Skromne I, Thorsen D, Hale M, et al. (2007) Repression of the hindbrain developmental program by Cdx factors is required for the specification of the vertebrate spinal cord. *Development* 134(11):2147–2158.

Stern CD & Downs KM (2012) The hypoblast (visceral) endoderm: An evo-devo perspective. *Development* 139:1059–1069.

Stower MJ & Srinivas S (2014) Heading forwards: Anterior visceral endoderm migration in patterning the mouse embryo. *Philos Trans R Soc Lond B Biol Sci* 369(1657).

Sunmonu NA, Li K & Li JY (2011) Numerous isoforms of Fgf8 reflect its multiple roles in the developing brain. *J Cell Physiol* 226(7):1722–1726.

Tanaka S, Hosokawa H, Weinberg ES & Maegawa S (2017) Chordin and dickkopf-1b are essential for the formation of head structures through activation of the FGF signaling pathway in zebrafish. *Dev Biol* 424(2):189–197.

White RJ & Schilling TF (2008) How degrading: Cyp26s in hindbrain development. *Dev Dyn* 237(10):2775–2790.

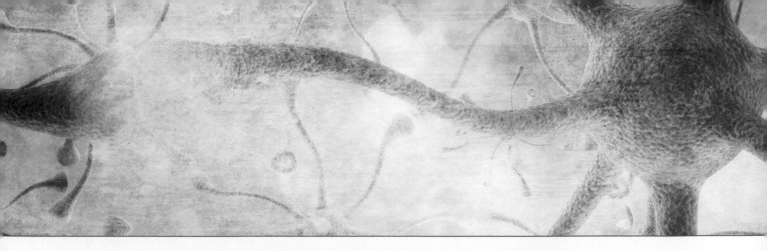

Patterning along the Dorsal–Ventral Axis

4

Macroscopic changes along the **dorsal–ventral (D/V) axis** of the developing vertebrate neural tube are not as obvious as those observed along the anterior–posterior (A/P) axis. However, segmentation and patterning of cells along this axis are equally important to the development of the neural tube and the various neuronal subtypes.

In vertebrates, signals are present in nonneural tissues located at or near the dorsal and ventral halves of the neural tube. These signals regulate the expression of specific transcription factors in neuronal progenitors located at various distances from the signal source. **Neural progenitors**, also called **precursors,** are cells that have the capacity to develop into a restricted number of neural cell types but have not yet taken on all the characteristics of a final, specialized cell. Each progenitor developing along the D/V axis is exposed to a particular concentration of signals that helps establish a unique **transcription factor code** for each cell. The expression of different transcription factors regulates, in turn, the expression or repression of genes that establish the cell's morphological and behavioral characteristics, a process called **neuronal specification**. Much more is discussed about cell fate, differentiation, and specification in Chapter 6. Here, the focus is on the initial patterning of progenitor cells along the D/V axis and how scientists identify the progenitor populations that give rise to different neuronal subtypes. However, by exploring D/V axis organization, principles of gene regulation and neuronal specificity are introduced.

This chapter begins with an overview of the anatomical features found along the D/V axis of the vertebrate nervous system, followed by a description of signals known to influence the development of specific D/V cell types. Homologous signaling mechanisms utilized by *Drosophila* are also discussed.

ANATOMICAL LANDMARKS AND SIGNALING CENTERS IN THE POSTERIOR VERTEBRATE NEURAL TUBE

The early segregation of sensory and motor regions in the vertebrate neural tube is perhaps the most distinctive feature of D/V axis patterning. As described in Chapter 3 and reviewed in **Figure 4.1**, the neural tube forms

DOI: 10.1201/9781003166078-4

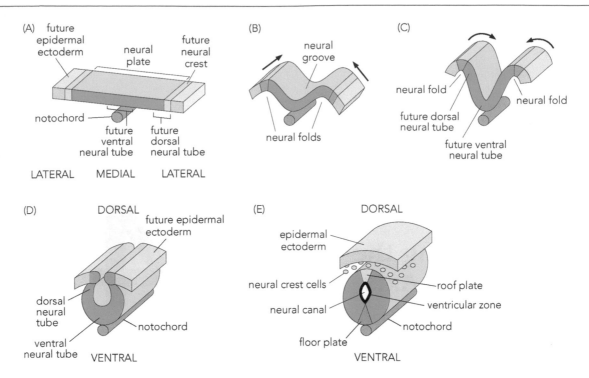

Figure 4.1 Neural tube formation leads to specializations along the dorsal–ventral axis. (A) At the neural plate stage, the lateral edges mark where structures associated with the dorsal surface of the neural tube will arise. The lateral-most edges will ultimately form epidermal ectoderm (yellow), while the adjacent regions will form neural crest cells (green) and the dorsal region of the neural tube (blue). Future ventral neural tube structures will arise in the medial region of the neural plate above the notochord—the mesodermal structure running along the length of the caudal neural tube. (B) The lateral edges of the neural plate begin to curve upward, creating the neural folds. The bending of the neural plate also creates the neural groove that lies above the notochord. (C, D) The lateral edges continue to curve upward and approach one another as the neural tube (blue) begins to take shape, thereby creating distinct dorsal and ventral neural tube regions. (E) As the edges of the neural tube come together dorsally, a strip of specialized cells—the roof plate—is identified. The epidermal ectoderm is now separated from the neural tube, and neural crest cells begin to migrate away from the dorsal surface. At the ventral surface, another strip of specialized cells—the floor plate—is seen above the notochord. The closure of the neural tube also results in the formation of the centrally located neural canal.

from the neural plate. The lateral edges of the neural plate curve upward, forming the neural folds. The neural folds then come together to form the dorsal surface of the neural tube (Figure 4.1A–D). At this dorsal junction, a wedge-shaped region of specialized glial and neuroepithelial cells called the **roof plate** develops. As the neural tube closes, the former medial region of the neural plate becomes the ventral surface of the neural tube and is marked by another region of glial cells called the **floor plate** (Figure 4.1E). The floor plate lies just above the **notochord**, the mesodermal structure that runs the length of the main body axis. Along the lumen of the neural tube, new neurons are generated in the ventricular zone (VZ). These neurons then migrate laterally and differentiate into specific cell types. The lumen of the neural tube later forms the ventricles of the brain and central canal of the spinal cord.

The Sulcus Limitans Is an Anatomical Landmark That Separates Sensory and Motor Regions

At the early neural tube stage, a longitudinal groove along the inner surface of the lateral walls serves as an anatomical landmark that indicates where sensory and motor structures will arise. This groove, called the **sulcus limitans**, is most apparent in the spinal cord and hindbrain regions, but also extends through the midbrain (**Figure 4.2**A). In the spinal cord region of the developing neural tube, the alar plate is identified between the sulcus limitans and the roof plate, while the basal plate is found between the sulcus limitans and the floor plate. The alar plate contains

(A) EMBRYONIC NERVOUS SYSTEM

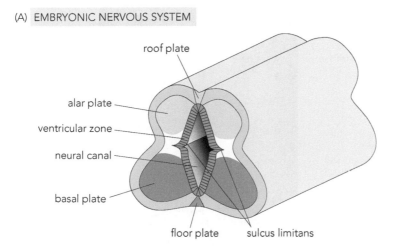

(B) ADULT NERVOUS SYSTEM

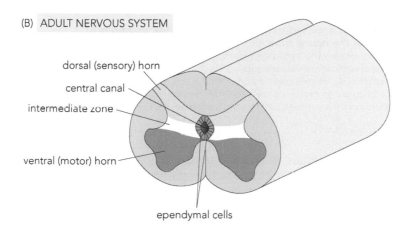

Figure 4.2 The sulcus limitans marks the boundary between sensory and motor neurons of the developing nervous system. (A) In the posterior (caudal) regions of the embryonic nervous system, a longitudinal groove—the sulcus limitans—is readily observed along the inner surface of neural tube. This anatomical landmark separates the alar plate from the basal plate. The alar plate gives rise to sensory interneurons while the basal plate gives rise to motor neurons and associated interneurons. Newly formed neurons arise in the ventricular zone (VZ) adjacent to the neural canal. (B) In the mature spinal cord, the division between sensory and motor neurons remains with sensory interneurons located in the dorsal horn and motor neurons and associated interneurons located in the ventral horn. An intermediate zone containing a mixture of sensory and motor interneurons is also noted at some spinal cord levels. The former neural canal becomes the much smaller central canal lined with ependymal cells.

sensory interneuron progenitor cells, whereas the basal plate contains the progenitors of motor neurons and motor interneurons.

In the adult spinal cord, the division of the sensory and motor neurons in the gray matter remains. Sensory neurons are found in the dorsal horn, whereas motor neurons are in the ventral horn. An intermediate zone containing a mixture of sensory and motor interneurons is also visible at some spinal cord levels. The former neural canal persists as the much smaller central canal, and the former ventricular zone is replaced by a lining of **ependymal cells**, another type of specialized glial cell (Figure 4.2B). The functional significance of the sulcus limitans is unknown, but it provides a visible landmark for developmental neurobiologists to easily distinguish early patterning along the D/V axis. As the nervous system matures, the sulcus limitans is no longer visible in most regions as it becomes obscured by the expanding neuronal populations.

Labeling Techniques Identify Cell Types along the D/V Axis

Although the roof plate and floor plate are relatively easy to identify in the early neural tube, the morphological appearance of individual cells along the D/V axis is remarkably homogeneous, making it difficult to identify different cell types. Fortunately, by the late twentieth century techniques for identifying cell-specific proteins and mRNA were available. Immunohistochemistry reveals protein expression patterns, while *in situ* hybridization and single cell mRNA methods localize differentially expressed mRNA.

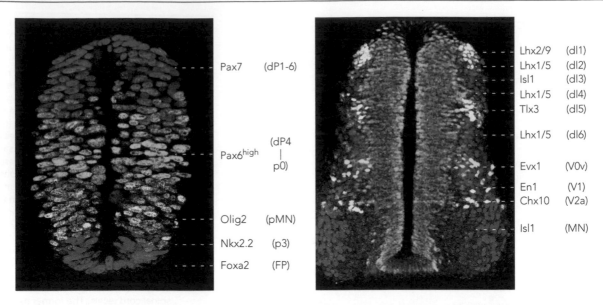

Figure 4.3 Transcription factor expression patterns are used to distinguish different populations of developing cells along the dorsal-ventral axis of the neural tube. (A) In early-stage chick embryos (Hamburger and Hamilton stage 16, HH16), the dorsal most cells express the transcription factor Pax7, intermediate cells express Pax6, future motor neurons express Olgi2, and a subset of ventral interneurons express Nkx2.2. The floorplate is identified at the ventral midline by Foxa2. (B) Later in development (HH24), as cells begin to differentiate and migrate laterally, additional transcription factors distinguish subsets of dorsal interneuron progenitor cells (dl1–dl6), ventral motor interneurons (V0–V2), and motor neurons (MN). Green indicates the ventricular zone where proliferating cells arise. [Le Dréau G & Martí E [2012] *Dev Neurobiol* 72(12):1471–1481.]

With these techniques, scientists can now readily distinguish various dorsal and ventral progenitors and differentiated neurons *in vivo* and *in vitro*. These methods not only identify different cell types during normal development, but also reveal how experimental manipulations influence development of cell populations along the D/V axis.

Examples of cell markers used to identify progenitors or differentiating neurons in the chick spinal cord are shown in **Figure 4.3**. Early in development, Pax7 marks dorsal regions, Pax6 indicates intermediate regions, while Olig2 and Nkx2.2 mark ventral regions and Foxa2 (HNF3β) labels the floor plate (Figure 4.3A). Later in development, after progenitor cells have migrated laterally, additional transcription factors delineate territories for specific dorsal interneurons (dI), ventral interneurons (V), and the motor neurons (MN; Figure 4.3B). Because the discovery of D/V cell markers occurred over the course of many years, some of the experiments described later in this chapter took place before the associated genes or proteins were identified.

The Roof Plate and Floor Plate Produce Signals That Influence D/V Patterning

Signals originating in the epidermal ectoderm and roof plate help pattern the dorsal half of the neural tube. The epidermal ectoderm ultimately separates from the neural tube, as does a population of cells originating in the neural folds—the neural crest cells (see Figure 4.1). Individual or small clusters of neural crest cells take defined migratory routes throughout the developing embryo to form specific peripheral neurons, including the sensory neurons found in the dorsal root ganglia that come to lie along the dorsal region of the spinal cord (see Chapter 5). The roof plate cells, initially in contact with the surface ectoderm, become a separate population of glia-like cells after neural crest cell emigration.

The dorsal epidermal ectoderm provides signals to influence development of the roof plate, which in response produces signals for patterning the dorsal horn sensory interneurons, the group of neurons that receive and process sensory information from the body and connect with motor neurons in the spinal cord and brain.

The floor plate located at the ventral surface of the neural tube is similarly influenced by signals originating in the nearby notochord (Figure 4.3). The notochord is ultimately incorporated into the vertebral column in most vertebrates, but first serves multiple functions important for neural development, including the production of proteins that induce formation of the floor plate. Once induced, the floor plate produces signals that regulate the specialization of motor neurons and their associated interneurons in the ventral half of the neural tube. Thus, along the D/V axis, dorsal sensory and ventral motor specializations develop in response to signals provided by the nonneural cells of the roof plate and floor plate.

Roof Plate and Floor Plate Signals Influence Gene Expression Patterns along the D/V Axis of the Neural Tube

It is now known that the signals originating in the dorsal epidermis and roof plate coordinate, often through opposing actions, with signals originating in the notochord and floor plate. As described in the following sections, this knowledge emerged as more was learned about the unique gene expression patterns in the various progenitor and neural subpopulations of the nervous system.

It is important to emphasize that while the expression of different markers is very helpful as an experimental tool, these regulated expression patterns are required to establish the ultimate function of each cell. Thus, the dorsal and ventral signaling centers influence the expression of specific transcription factors in each cell type, and the resulting transcription factor code regulates which genes are expressed or repressed, thereby determining the final characteristics of each neuronal cell type along the D/V axis. The importance of the transcription factor code in D/V patterning is highlighted by experiments in which expression levels were experimentally manipulated and corresponding changes in cell development were observed. Other examples of temporally regulated transcription factor expression patterns are provided in Chapters 5 and 6.

VENTRAL SIGNALS AND MOTOR NEURON PATTERNING IN THE POSTERIOR NEURAL TUBE

In the adult spinal cord, motor neurons and interneuron subpopulations are found on either side of the ventral midline. The motor neurons control muscle movements along the body axis, while the various interneurons modify motor output. Embryonically, motor neurons and interneurons arise from distinct progenitor pools located in discrete regions of the ventral neural tube. Each progenitor pool receives specific developmental cues based on its location.

The Notochord Is Required to Specify Ventral Structures

Experiments testing how the notochord might influence neural tube development were among the first to describe mechanisms underlying D/V patterning. In 1939, for example, the classic work of Johannes Holtfreter demonstrated that surgically removing the notochord in amphibians led to an absence of ventral structures, including the floor plate (**Figure 4.4**). Later studies found that surgically inhibiting or disrupting notochord formation caused the cells of the ventral neural tube to appear similar to those

Figure 4.4 The notochord induces ventral specializations, including the floor plate and motor neurons. (A) Under normal developmental conditions, the notochord provides signals (large arrows) that can induce the formation of the floor plate and adjacent motor neurons. Once formed, the floor plate becomes the primary source of signals (small arrows) to induce motor neurons. (B) Following surgical removal of the notochord, neither floor plate nor motor neurons form. (C) When notochord tissue is grafted to lateral regions of the neural tube, the signals that arise from a notochord graft induce the formation of an ectopic floor plate with motor neurons on either side.

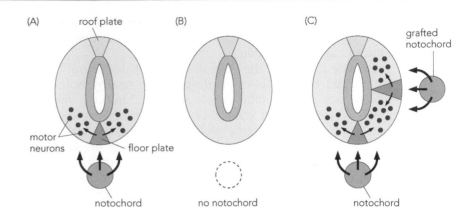

of the dorsal neural tube. Expansion of dorsal-like cells into ventral regions was confirmed once markers for different neuronal types were identified. For example, in the absence of notochord, the expression of *Pax3* and *Pax7*, genes normally restricted to dorsal neural tube regions, expanded into ventral regions. Because dorsal-like patterning occurred in the absence of notochord tissue, dorsal characteristics were considered the default state for the neural tube.

In another series of experiments in amphibians, grafting extra (supernumerary) notochord tissue at the lateral plate regions of the developing neural tube induced floor plate tissue. Subsequent studies in chicks, mice, and zebrafish confirmed this finding and further noted that **ectopic** motor neurons formed on each side of these misplaced floor plate regions (Figure 4.4C). Thus, lateral plate regions of the neural tube that normally do not give rise to floor plate or motor neurons could form these ventral specializations if provided with a signal from the notochord.

Other experiments observed that when the size of two grafted notochords differed, the larger graft induced a larger floor plate and more motor neurons than the smaller graft. Thus, the amount of signal produced by the notochord determined the size of the induced tissues. Additional studies suggested that contact with the notochord was necessary for formation of the floor plate. When another tissue, such as mesenchyme, was placed between a grafted notochord and the neural tube, the floor plate tissue was no longer induced, presumably because the notochord signal was unable to penetrate the mesenchymal tissue.

Cell culture assays further demonstrated the role of the notochord in inducing ventral patterning. When segments of neural tube were cultured in contact with a segment of notochord, the floor plate marker *FoxA2* (*HNF3β*) and the motor neuron markers *Isl1* and *Isl2* were identified in the neural tube region adjacent to the notochord. Thus, multiple studies from different species revealed that the notochord is essential for ventral neural tube patterning.

Sonic Hedgehog (Shh) Is Necessary for Floor Plate and Motor Neuron Induction

The accumulated evidence showing the necessity of the notochord for inducing ventral specializations in the neural tube led several research groups to focus efforts on identifying the notochord-derived signal.

The first studies describing a specific notochord-derived signal were published in the early and mid-1990s. During this period, several labs reported roles for **Sonic hedgehog** (**Shh**) in the development of vertebrate limb and neural tissues. *Shh* is one of three vertebrate genes related to the *Drosophila Hedgehog* (*hh*) gene (**Box 4.1**). Studies in the developing neural tube revealed that *Shh* was first expressed in the notochord then

Box 4.1 How Genes Are Named and Why They Are Sometimes Renamed

You will no doubt notice some unusual names for genes. In some cases it is easy to figure out why a gene was given a particular name. In other cases, the origin of the name is not as readily apparent to someone outside the lab that first discovered and named the gene. Many gene names discussed in this book originated in *Drosophila* and were often intended to help scientists remember the names of the numerous genes being rapidly discovered.

Several *Drosophila* genes are named based on the appearance or behavior of the flies. An example of a gene name that ties to the mutation is the *Sevenless* gene, described in Chapter 6. In flies with a mutation in this gene, the R7 photoreceptor cell is missing in the fly eye. Thus, without the gene there is no R7 cell and the fly is therefore "sevenless." Here the gene name describes what happens when the gene is missing, and the origin of the name is easy to follow.

Other gene names are not as intuitive. For example, the vertebrate *Smad* genes described in this chapter were named by combining the names of the homologous genes first discovered in *C. elegans* and *Drosophila*. In *C. elegans*, the reduced size of worms lacking the gene led to the gene name Small (*Sma*), the first part of the vertebrate *Smad* gene name. In *Drosophila*, the gene was named *Mad*, the second half of the vertebrate gene name. *Mad* stands for "mothers against decapentaplegic." A mutation in *Mad* represses the gene *Decapentaplegic*, which is required for the formation of the 15 (decapenta) imaginal discs of the fly used to form several tissues, including limbs. "Mothers against decapentaplegic" was derived from the campaign in the United States and Canada called Mothers Against Drunk Driving (MADD). So, while the name *Smad* may be easy to remember and its origin makes sense when explained, it is not a gene name one can easily relate to its function.

The long tradition of creative gene naming has recently come under scrutiny by many groups. When a gene first discovered in *Drosophila* has a human homolog that is later found to be associated with a particular disease, the clever name may create unexpected problems. In recent years a number of articles have discussed the problem of gene names and scientific panels have been created to review and rename genes if needed.

In some cases, the name of a gene can seem insensitive to families. For example, families may feel that the severity of a condition is not being recognized when they are told the condition is caused by "Sonic hedgehog." The origin of the gene name *Sonic hedgehog* (*Shh*) was based on the original *Drosophila* gene *Hedgehog* (hh), which was named after the physical appearance of the flies. Flies lacking the gene were covered with small, spiky projections that reminded the scientists of a hedgehog. Later three vertebrate homologs were discovered. *Indian hedgehog* (*Ihh*) and *Desert hedgehog* (*Dhh*) were named after types of hedgehogs and the third was named *Sonic hedgehog* (*Shh*) after the video game character. None of these names seemed offensive until it was discovered that some human birth defects, including holoprosencephaly (HPE), arise from disruption in the *Shh* signaling pathway. Some parents felt doctors and researchers were not taking the condition seriously when they found the gene that causes this tragic defect was named after a cartoon video character.

In some cases, genes have been renamed after threats of lawsuits. For example, in 2005 a mammalian cancer proto-oncogene was discovered and originally named *Pokémon* after characters in a popular game. *Pokémon* stood for POK erythroid myeloid ontogenic. While the gene name may be easy to remember, the company that makes Pokémon did not appreciate having a popular product associated with a cancer-causing gene. The scientists renamed that gene *ZBTB7*, which is a member of the POK (POZ and *Krüppel*)/ZBTB (zinc finger and BTB) protein family. One can see how the original name *Pokémon* was easier to remember, but one can also understand why the company was concerned about the negative associations that could be made with its product.

How to name genes, and when and if to rename them, will continue to be an issue for scientists. Today scientists find they must balance creativity and humor with sensitivity when naming newly discovered genes. For further exploration of the topic, some suggested readings are as follows:

Simonite T (2005) Pokémon blocks gene name. *Nature* 438:897.

Schwartz J (2006) Rename That Gene "Sonic Hedgehog" Sounded Funny, at First. *The New York Times* (www.nytimes.com/2006/11/12/weekinreview/12schwartz.html).

Krulwich, R. (2009) Fruit fly scientists swatted down over "Cheap Date." National Public Radio, All Things Considered, (www.npr.org/templates/story/story.php?storyId=100468532).

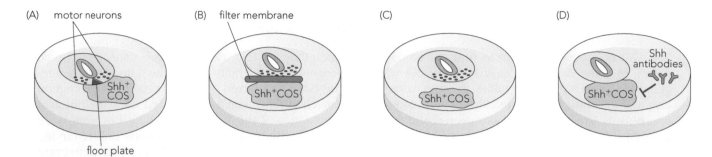

(A) motor neurons   (B) filter membrane   (C)   (D)

floor plate

Figure 4.5 In vitro assays revealed that Sonic hedgehog (Shh) mimics the activity of notochord tissues. (A) When COS cells expressing Shh protein were cultured in contact with neural tube explants, both floor plate and motor neurons formed. However, only motor neurons formed when contact between the neural tube explants and Shh-producing COS cells was prevented by placing a filter membrane between the explant and COS cells (B) or placing the tissues apart from each other in the culture dish (C). Together, these findings indicated Shh functions as a diffusible factor. (D) The specificity of Shh was confirmed when antibodies to Shh were added with cultures of Shh-expressing COS cells. Under these conditions, neither floor plate nor motor neurons formed.

subsequently expressed in the floor plate. The location and timing of *Shh* expression made it a plausible candidate for regulating ventral neural tube specializations.

In vitro experiments demonstrating a critical role for *Shh* in ventral patterning are outlined in **Figure 4.5**. These experiments, based on work by Henk Roelink and colleagues, employed the immortalized COS cell line to express Shh protein. An immortalized cell line is a group of cells carrying a mutation that permits continuous proliferation. The COS cell line originated from a monkey kidney cell line called CV-1 that was immortalized with the SV40 virus (CV-1 in Origin SV40). COS cells are often used experimentally to express a protein of interest to test the putative functions of that protein. Normally, COS cells alone have no influence on neural tube development. However, COS cells transfected with the *Shh* gene produced the Shh protein and were able to induce floor plate and motor neurons in cultures of neural tube segments (Figure 4.5A). These studies also noted that different modes of Shh delivery resulted in the induction of different ventral cell types. When Shh-expressing COS cells were grown in contact with the neural tube cultures (Figure 4.5A), floor plate tissue formed first, as indicated by the expression of the transcription factor Foxa2 (HNF3β). The formation of the floor plate was followed by the induction of motor neurons. In contrast, if the cultured neural tube tissues were physically separated from the COS cells, only motor neurons formed, as revealed by *Isl1* and *Isl2* expression. The neural tube segments and COS cells were separated either by placing a filter membrane between the COS cells and the neural tube segment or by placing the two tissues on opposite sides of the cell culture dish (Figure 4.5B, C). These results indicated that this Shh was a diffusible signal because it was able to act through a filter membrane and at a distance. The results also implied that the floor plate formed in response to the higher Shh concentrations provided by direct contact with the COS cells, whereas motor neurons formed in response to lower concentrations provided by the diffusible Shh signal. Other studies soon supported the hypothesis that notochord-derived Shh was necessary for floor plate induction while floor-plate-derived Shh contributed to motor neuron development.

Additional experiments confirmed that COS cells alone did not influence ventral neural tube development and further demonstrated the specificity of Shh signaling in inducing floor plate and motor neurons in the ventral neural tube (Figure 4.5D). When antibodies to Shh protein were added to the cultures of neural tube segments and Shh-expressing COS cells, ventral structures failed to form in the neural tube segments.

In vivo experiments further confirmed the importance of Shh in inducing ventral structures. For example, mice lacking the *Shh* gene failed to develop several ventral cell types, including the floor plate, motor neurons, and the ventral-most interneurons. In fact, when *Shh* was eliminated, the dorsal interneurons and intermediate cell types expanded into the ventral region. Additional support for the role of Shh was seen in transgenic mice in which *Shh* was misexpressed in the dorsolateral regions of the neural tube. Under these conditions, ectopic floor plate and motor neurons arose, similar to what was observed when supernumerary notochords were transplanted in these regions. Thus, the presence of Shh promoted ventral neural tube differentiation, whereas blocking production of Shh *in vitro* or *in vivo* inhibited the formation of ventral cell types and led instead to the expansion of dorsal cell types.

Shh Concentration Differences Regulate Induction of Ventral Neuron Subtypes

Different concentrations of Shh are necessary and sufficient to instruct cellular identity of most ventral cell types in the neural tube. The highest concentrations of Shh induce floor plate, lower Shh concentrations induce motor neurons, and progressively lower concentrations induce the different ventral interneuron subtypes. Ventral interneurons arise from distinct progenitor cell populations (p3, p2, p1, p0) that migrate to specific sites in the ventral half of the neural tube to form the V3, V2, V1, and V0 populations of the adult spinal cord (**Figure 4.6**).

In vivo the Shh concentrations needed to induce the floor plate and the various ventral neurons arise from cleavage of the Shh protein into amino (Shh-N) and carboxyl (Shh-C) fragments. Shh uses autoproteolysis to generate these two fragments and only the Shh-N fragment is used for ventral cell patterning. Shh-N is further modified by the addition of a cholesterol molecule that allows tethering of Shh to the surface of notochord cells. Thus, much of the Shh-N remains at the cell surface, providing the higher concentration needed for induction of the floor plate. The remaining Shh-N diffuses away from the sites of production, first from the notochord

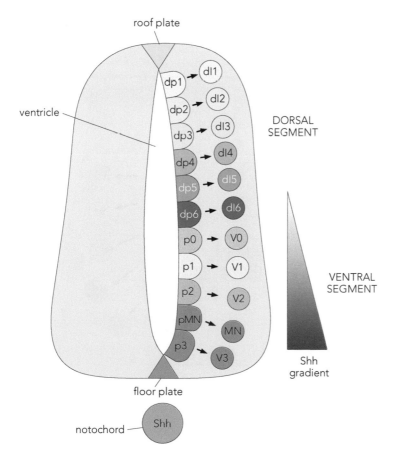

Figure 4.6 Neurons in the ventral neural tube arise in response to graded concentrations of Shh. In the ventral segment, the floor plate is located at the midline, above the notochord. Motor neurons (MN) and interneurons (V3, V2, V1, and V0) lie on either side of the midline. These neurons arise from specific progenitor cell groups located in a ventral-to-dorsal progression (p3–p0). The concentration of Shh is greatest near the ventral midline (dark pink), where it is released by notochord and floor plate. As Shh diffuses from these regions, the concentration gradually decreases.

and later from the floor plate, to provide the progressively lower concentrations needed for induction of the various subtypes of ventral neurons (Figure 4.6). The graded concentrations of Shh therefore lead to the generation of specific neuronal subtypes within the ventral neural tube, with those located most ventrally (p3 and pMN) requiring higher concentrations than those located closer to the intermediate zone (p2, p1, p0). Thus, in the ventral neural tube, Shh functions as both an inducer to floor plate tissue and as a morphogen signaling gradient to cells along the D/V axis.

Genes Are Activated or Repressed by the Shh Gradient

A given concentration of Shh either activates or represses certain genes in the progenitor cells of the ventral neural tube (**Figure 4.7**). Among the genes regulated by Shh are those that code for class I and class II transcription factors. Class I transcription factor genes are repressed by Shh whereas class II transcription factor genes are activated by Shh. Class I transcription factors include the homeodomain proteins Dbx1, Dbx2, Irx3, and Pax6. Class II transcription factors include homeodomain proteins of the Nkx family and the basic helix-loop-helix (bHLH) protein Olig2.

Specialization of the different ventral progenitors occurs because class I and class II transcription factors cannot be active in the same cell at the same time. To maintain the boundaries between progenitor cell types, cross-repression of pairs of class I and class II transcription factors occurs (Figure 4.7).

The class I transcription factors, though initially expressed in the ventral segment, are ultimately repressed by Shh in a gradient-dependent manner. For the dorsal-most progenitors, those found in p1 and p0, low

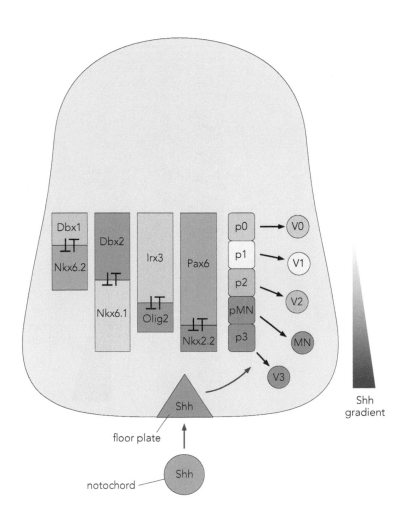

Figure 4.7 Shh levels differentially regulate expression of class I and class II transcription factors. Shh either activates or represses transcription factor genes to specify the cell types in the ventral neural tube. Genes for the class II transcription factors (for example, Nkx6.2, Nkx6.1, Olig2, and Nkx2.2) are activated by Shh, whereas genes for the class I transcription factors (for example, Dbx1, Dbx2, Irx3, and Pax6) are repressed by Shh. These classes of transcription factors cross-repress each other (black bars) so that only one is activated in a cell at any time. [Adapted from Dessaud E, McMahon AP & Briscoe [2008] *Development* 135:2489–2503. With permission from The Company of Biologists.]

levels of Shh are sufficient to repress these genes. In contrast, at more ventral regions closer to the floor plate, higher Shh concentrations are required to repress class I transcription factor genes (Figure 4.7).

Unlike the class I transcription factors, the genes for class II transcription factors, such as *Nkx2.2*, *Nkx6.1*, *Nkx6.2*, and *Olig2* are activated by the higher concentrations of Shh produced closest to the floor plate (Figure 4.7).

As an example of how the cross-repression of class I and class II transcription factors occurs, consider the paired transcription factors Nkx2.2 and Pax6. As the Shh concentration increases above a certain threshold, *Nkx2.2* in the ventral-most progenitors (p3) becomes activated, while *Pax6* expression in that same progenitor region is repressed. If expression of either gene is altered, subsequent patterning shifts. For example, *in vitro* studies of neural tube explants found that when antibodies that inhibit Shh were added, *Nkx2.2* expression was lost and only *Pax6* was expressed. Conversely, in mice lacking *Pax6*, *Nkx2.2* expression extended further dorsally so the area devoted to V3 interneurons expanded while the area available for the more dorsal motor neurons (pMN) decreased.

Similar cross-repression takes place between the other pairs of class I and class II transcription factors. For example, Olig2 pairs with Irx3, while Nkx6.1 pairs with Dbx2, and Nkx6.2 pairs with Dbx1 (Figure 4.7). Thus, because different concentrations of Shh regulate the repression or activation of the genes for these transcription factor pairs, the resulting neural subtypes (V3, MN, V2, V1, or V0) are patterned along the D/V axis. Under normal conditions, the gradient of Shh throughout the ventral neural tube is sufficient to induce each cell type.

The gradient of Shh also determines a corresponding intracellular gradient of GLI (glioma-associated) zinc finger transcription factors. The GLI family plays important roles in regulating transcription factor expression, including the expression of class I and class II transcription factors. GLI proteins are transcriptional effectors that function as transcription activators or repressors. Whether activator or repressor forms predominate depends on whether the Shh ligand is bound to the receptor Patched (Ptc). Similar mechanisms are used in *Drosophila* where *Hh* binding status influences cubitus interruptus proteins—homologs of vertebrate GLI proteins.

Shh Binds to and Regulates Patched Receptor Expression

When Shh binds to Ptc receptors, an intracellular signaling complex is initiated that leads to the transcription of target genes. In contrast, in the absence of Shh binding, the intracellular signaling complex represses the transcription of target genes.

In the absence of Shh, the Shh signaling complex is associated with microtubules of the primary cilium, a slim extension of cell membrane with unique cytoplasmic components (**Box 4.2**). Further, Ptc that is not bound by Shh inhibits the activity of a transmembrane protein called Smoothened (Smo), likely by sequestering it in intracellular vesicles. Though the mechanism by which Ptc suppresses Smo activity is still not fully understood, it is clear that inhibiting Smo activity allows intracellular protein kinase activity and proteolytic cleavage of GLI proteins associated with SuFu, a cytoplasmic protein that interacts with the GLI transcription factors.

With Smo inhibited and GLI cleavage allowed, the transcriptional repressor form of GLI (GLI-R) takes over. GLI-R then travels to the nucleus to block the transcription of Shh target genes. Thus, in the absence of Shh binding, the repressor forms of GLI predominate so Shh-target genes are repressed (**Figure 4.8**A).

In contrast, when Shh binds to Ptc, Ptc is internalized and therefore no longer inhibits Smo. Smo then travels to the cilium, where it is phosphorylated by protein kinase A (PKA) and other kinases. Phosphorylation

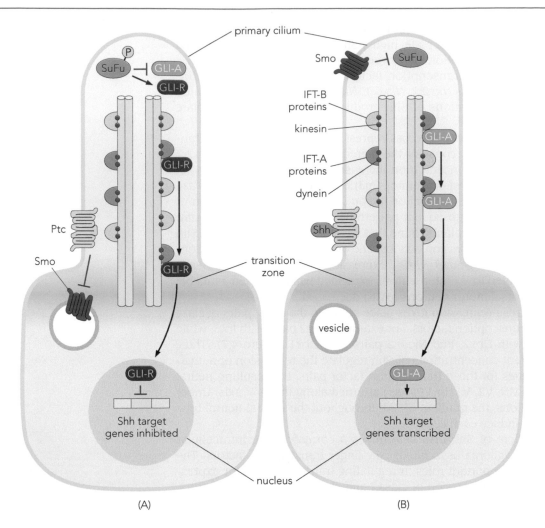

(A)   (B)

Figure 4.8 The binding state of Patched (Ptc) receptors determines whether target genes are repressed or activated. (A) In the unbound state, Ptc is localized near the base of the cilium where it inhibits the activity of the transmembrane protein Smoothened (Smo), likely by keeping it sequestered in intracellular vesicles. With Smo inhibited, intracellular kinase activity activates the microtubule associated Shh signaling complex, comprised of suppressor of fused (SuFu) and GLI proteins, in the primary cilium. Phosphorylation (P) of SuFu prevents GLI-activator forms (GLI-A) by cleaving GLI into the transcriptional repressor form (GLI-R) that is then transported to the nucleus to inhibit Shh-target genes. (B) Binding of Shh to Ptc stops the inhibition of Smo, allowing Smo to travel to the primary cilium where it is phosphorylated (P), undergoes a conformational change, and inhibits SuFu. With SuFu inactivated, GLI is not cleaved to the repressor form, so the activator form (GLI-A) travels to the nucleus to initiate transcription of Shh-target genes.

leads to a conformational change in the Smo protein that allows it to interact with and inhibit SuFu. Inactive SuFu is unable to cleave GLI into repressor forms, so the activator forms of GLI (GLI-A) remain. GLI-A travels to the nucleus, where it initiates transcription of Shh-responsive genes (Figure 4.8B).

Vertebrates have three known GLI proteins (GLI1, GLI2, and GLI3). In the absence of Shh binding, GLI3 predominates and functions as the repressor form. In some species, including mammals, GLI2 is the primary transcriptional activator. In other vertebrates, such as zebrafish, GLI1 is the primary activator. In all cases, Shh binding allows activator forms of GLI to override the effects of the repressor forms.

The necessity of balancing the activator and repressor forms of GLI proteins is seen in mice lacking genes for both Shh and GLI3. *Shh* $^{-/-}$ mice have a loss of ventral cell types, particularly those closest to the floor plate. However, if *GLI3* is also absent, then the repressor activity of *GLI3* is lost

Box 4.2 The Primary Cilium

Most vertebrate cells contain a primary, non-motile cilium that extends from the cell membrane. In the nervous system, primary cilia are found on progenitor cells, neurons, and astrocytes. Although long observed on neurons by electron microscopy, only since the turn of the twenty-first century have scientists recognized the importance of the cilium in normal cellular functions including developmental events in the spinal cord, cerebral cortex, and cerebellum. The primary cilium is associated with crucial signaling pathways including TGFβ, Wnt, and Shh. The importance of proper cilia development and function is noted in genetic disorders, called ciliopathies, that lead to defects in several neural tissues, including the retina.

The cilium is a separate compartment represented as a slim extension of cell membrane. Although continuous with the cell membrane, the cytoplasm of the cilium is separated from the rest of the cell by a selective pore-like transition zone at its base. As a separate compartment, the cilium has distinct proteins that are selectively transported along the microtubule core (axoneme) via intraflagellar transport (IFT) proteins. IFT-A proteins transport elements of signal transduction pathways away from the tip via retrograde motor proteins such as dynein. In contrast, IFT-B proteins move proteins toward the tip of the cilium via anterograde motor proteins such as members of the kinesin family (**Figure 4.9**). An example of protein transport in the Shh-Ptc signaling pathway is seen in Figure 4.8.

Any changes in IFT transport disrupts localization of signaling components and consequently interferes with normal development, such as ventral patterning. For example, cilia are lost and Shh signaling is significantly decreased if IFT-B is mutated. Although low levels of Shh activity persist, the signaling pathways are not sufficient to regulate proper ventral patterning.

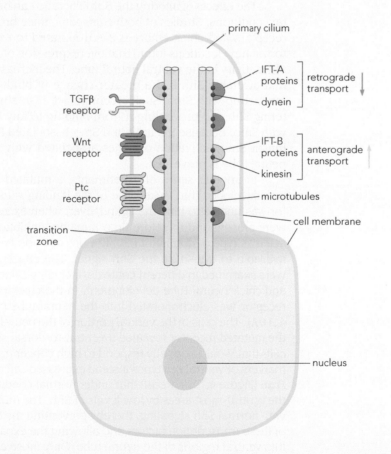

Figure 4.9 The primary cilium is involved in many developmental signaling pathways including those initiated by TGFβ, Wnt, and Ptc receptors. The primary cilium is an extension of the cell membrane, but the cytoplasm of the cilium is separated from that of the cell body by a transition zone. Elements of signaling pathways are transported by intraflagellar transport (IFT) proteins along microtubules that extend to the tip of the cilium. Retrograde transport of proteins away from the tip and toward the cell body is regulated by IFT-A proteins and the motor protein dynein. Anterograde transport from the cell body toward the tip of the cilium occurs via IFT-B proteins and motor proteins of the kinesin family.

and some of the ventral neurons remain, particularly those furthest from the floor plate. These results indicate that under normal conditions, Shh is not needed to specify all ventral neuronal precursors. Instead, the gradient of Shh influences the gradient of GLI that establishes a balance between activator and repressor forms of the protein. The pattern of ventral subtypes along the D/V axis is thus influenced by the type, level, and duration of GLI proteins available.

The importance of Shh signaling to patterning ventral cell types was also seen when Smo activity was experimentally manipulated. For example, a constitutively active form of the *Smo* gene was introduced into mice or chick embryos. As discussed, activated Smo normally leads to the induction of genes that typify ventral neuron subtypes. In both the mouse and chick preparations, regions of ventral neuronal cell types expanded so the areas of the neural tube that would normally give rise to dorsal precursor cell types now produced ventral cell types.

Among its many roles, Shh regulates *Ptc* expression through a feedback loop, so the relative concentrations of Shh and Ptc are carefully balanced to ensure the proper levels of Shh signaling in each cell. Under normal conditions, *Ptc* is upregulated in cells that respond to Shh. For example, *Ptc* expression is greatest near the floor plate and gradually decreases dorsally, in the same pattern as the Shh gradient. A direct link between *Shh* and *Ptc* expression is seen in mice that lack *Shh*; these mice also have lower levels of *Ptc* expression.

The effects of altering the Shh/Ptc ratio can be seen under experimental conditions. Studies of both transgenic mice bred to produce excess Ptc receptors and chick embryos electroporated to express excess Ptc receptors in new locations found that overexpression of *Ptc* led to a repression of Shh activity in the ventral neural tube. The increase in Ptc receptors meant that available Shh had a greater chance of binding to a Ptc receptor not interacting with Smo. Such receptors act as a "sink" for Shh, thus sequestering Shh and preventing it from binding to any Ptc receptors associated with Smo. Because the unbound Smo-associated Ptc receptors continue to inhibit Smo, activation of genes associated with the ventralization of the neural tube is prevented.

In another series of experiments, a mutated form of the Ptc receptor was developed that lacked one of the binding sites for Shh/Hh. The receptor was unable to bind the ligand, even when excess concentrations of Shh were present and therefore Smo remained inhibited. The mutated receptor acted as type of dominant inhibitor because too few normal receptors were available to transduce the Shh signal. The effects of the mutated receptor were examined in different contexts, including *Drosophila* wing development and chick neural tube development. In the chick experiments, the mutated receptor was electroporated into the neural tube of chick embryos (**Figure 4.10**A). The cells in the ventral portion of the neural tube that now expressed the mutated receptor revealed a ventral-to-dorsal shift in cell identity. Those cells that would normally respond to high concentrations of Shh and express markers of ventral cell types instead expressed cell markers such as *Pax7* and *Pax6* (Figure 4.10B). Recall that under normal conditions *Pax6* is repressed in the ventral-most areas by low levels of Shh. The mutated receptors interfered with normal Shh signaling, thereby preventing the Shh-mediated repression of these transcription factors and allowing the expansion of dorsal cell types into ventral regions of the neural tube. Only those cells that incorporated the mutated gene shifted cell fate, indicating that the receptor functioned in a **cell-autonomous** manner, that is, only the cells that carried the gene mutation had an altered phenotype. This supported the hypothesis that Shh acts directly on the cells of the ventral neural tube.

Many of the mechanisms regulating Shh-mediated transcriptional repressors and activators are unclear. However, differences in both the

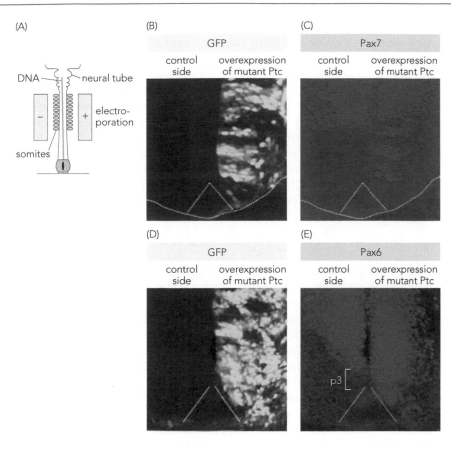

(A)

DNA — neural tube

— / + electro-poration

somites

(B) GFP

control side | overexpression of mutant Ptc

(C) Pax7

control side | overexpression of mutant Ptc

(D) GFP

control side | overexpression of mutant Ptc

(E) Pax6

control side | overexpression of mutant Ptc

p3 [

Figure 4.10 Mutated Ptc receptors interfere with Shh signaling. (A) DNA for a mutated Ptc receptor lacking a Shh binding site was introduced to one side of a chick neural tube by electroporation (+). The mutated receptors were overexpressed so they functioned as a dominant inhibitor. (B, D) Cells overexpressing the mutant Ptc receptors are labeled with GFP (green). (C, E) On the control side where normal Shh signaling continues, the expression of Pax7 and Pax6 (red) remains restricted to more dorsal regions. However, Pax7 and Pax6 expression expands into ventral regions on the side expressing the mutant Ptc receptors. Thus, without sufficient Shh signaling on the treated side of the embryo, *Pax6* and *Pax7* were no longer repressed in ventral regions. The bracket in E indicates the region where p3 ventral progenitor cells are found. The triangle outlines the region of the floor plate. GFP, green fluorescent protein. [Adapted from Briscoe J, Chen Y, Jessell TM, Struhl G. [2001] *Molecular Cell* 7:1279–1291. With permission from Elsevier.]

concentration and length of Shh exposure appear to be important. For example, *in vitro* experiments revealed that in the absence of Shh, ventral progenitors express the class I transcription factors associated with more dorsal fates. Adding Shh activated the class II transcription factors Nkx6.1 and Olig2, inducing the corresponding p2 and pMN progenitors. However, sustained Shh exposure at the same concentration led to a corresponding increase in GLI activator activity, which in turn increased expression of Nkx2.2, and the development of p3 progenitors, the ventral-most interneurons. Thus, both the concentration and duration of the Shh signal might regulate the transcriptional activity that directs progenitor fate in the ventral neural tube.

Shh Signals Interact to Influence Gene Expression and Ventral Patterning

While much is now understood about the developmental signals that establish ventral regions of the spinal cord, new discoveries continue to be made. For example, recent evidence suggests that the sclerotome segment of the somite is an additional source of Shh for patterning motor neuron progenitors. Experimental reduction of Shh levels in the sclerotome led to a reduction in motor neurons in the ventral neural tube. How the notochord, floor plate, and sclerotome may interact to regulate ventral patterning is not yet known.

Additional genes regulated by Shh continue to be defined. For example, Shh influences *Prdm* gene expression. The Pdrm family (named for two proteins first identified as sharing a conserved N-terminal domain, Positive-regulatory domain-I-binding factor and Retinoblastoma protein-binding zinc finger protein) is implicated in the differentiation of neural, somatic, and muscle cells. There are three *Prdm* genes in *Drosophila*, four in *C. elegans* and up to 17 genes in vertebrates. The number in vertebrates varies by species due to the duplication or loss of specific *Prdm* genes.

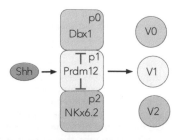

Figure 4.11 Shh and Prdm12 influences ventral patterning. Shh increases *Prdm12* expression in p1. Prdm12 then regulates the expression of transcription factors necessary for development of V1 interneurons. In at least some animal models, Prdm12 represses Dbx1 and Nkx6.2 in p1. This prevents p1 cells from adopting fates associated with the adjacent p0 and p2 progenitor pools.

The *Prdm* genes appear to first specify progenitor domains, then influence differentiation by regulating the transcription of cell-specific genes. In the neural tube, various *Prdm* genes are differentially expressed in ventral and dorsal progenitor domains where they activate or repress other transcription factors, especially bHLH transcription factors. In ventral regions, Prdm8 is found in p1, p2 and pMN. Prdm 14 is found in a subset of cells in pMN while Prdm12 is in p1. An example of how Shh and *Prdm* genes interact was seen in zebrafish where the inhibition of Shh signaling reduced *Prdm12* expression in p1 and altered the subsequent development of V1 interneurons. Prdm12 may directly influence p1 development by activating V1-associated transcription factors. Prdm12 in p1 can also repress the expression of transcription factors necessary for development of adjacent progenitor pools, such as Dbx1 and Nkx6.2 (**Figure 4.11**).

RA and FGF Signals Are Also Used in Ventral Patterning

Accumulating evidence indicates that Shh is only one of the signals used to pattern the posterior ventral neural tube. For example, in zebrafish, floor plate cells still form in the absence of notochord tissue, the source of Shh. Other studies suggest that only lateral floor plate cells require Shh, while the medial floor plate cells also require nodal, a member of the transforming growth factor β (TGFβ) superfamily. Additional studies suggest Shh and nodal cooperate to regulate early and late stages of floor plate induction in both chick and zebrafish.

Further, signals associated with A/P patterning, such as fibroblast growth factor (FGF) and retinoic acid (RA), may refine cell specializations in the D/V axis. FGFs appear to suppress the expression of transcription factors in progenitor cell populations to ensure the cells remain in the progenitor pool long enough to produce a sufficient number of cells. Once FGF signals cease, progenitors begin to express class I and class II transcription factors. The activity of these classes of the transcription factors is subsequently regulated by the concentrations of Shh, and possibly RA, they encounter (**Figure 4.12**). For example, a gradient of RA produced by the paraxial mesoderm (see Chapter 3) appears to be important for patterning intermediate and ventral cell types by promoting expression of the class I transcription factors Pax6, Dbx1, Dbx2, and Irx3. In one *in vitro* study using lateral (intermediate) spinal cord segments, the addition of RA induced expression *Dbx1*, *Dbx2*, and *Irx3* while repressing the expression of the dorsal marker *Pax7*. Conversely, disruption of RA signaling or production prevented expression of these class I transcription factors. For example, chick embryos electroporated with a dominant negative form of a retinoic acid receptor (RAR) had decreased expression of *Pax6*, *Irx3*, *Dbx1*, and *Dbx2*. In mice lacking the RA-synthesizing enzyme Radlh2 (*Radlh2⁻/⁻*), Pax6, Irx3, Nkx6.2, and Olig2 were downregulated. Similar results are observed in vitamin-A-deficient quails that lack RA. Thus, several studies in different animal models demonstrate the need for RA to help pattern the intermediate and ventral regions of the neural tube. This effect may be due to a direct action of RA on the promoter regions of the transcription factors. RA may also help repress the actions of Shh at progressively more dorsal

Figure 4.12 FGF inhibits expression of class I and II transcription factors in ventral progenitor cells. (A) The presence of FGF prevents progenitor cells in the ventral region of the neural tube from expressing class I or class II transcription factors. (B) When FGF production stops, the progenitor cells are able to express both class I and class II transcription factors. (C) The gradient of Shh then selectively activates either class II or class I transcription factors, with class II transcription factors activated by the higher levels of Shh.

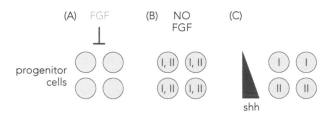

regions of the ventral segment. Thus, coordinated signaling between Shh, RA, and FGF is likely needed to shape the differential expression of genes in ventral or dorsal regions.

DORSAL PATTERNING IN THE POSTERIOR NEURAL TUBE

As noted previously, the dorsal region of the neural tube gives rise to the midline roof plate, dorsal sensory interneurons, and neural crest cells. The roof plate cells form a wedge-shaped stripe along the dorsal midline of the spinal cord that serves as a signaling center during early spinal cord development.

Within the dorsal segment of the spinal cord, six distinct regions of progenitor cells are identified (Figure 4.6). As with the ventral region, the development of these different dorsal interneuron subtypes is influenced by midline signals that regulate transcription factor expression. The signals that pattern the dorsal interneurons are distinct from those in ventral regions however, and are often antagonistic to ventral-derived signals.

TGFβ-Related Molecules Help Pattern the Dorsal Neural Tube

The dorsal neural tube appears to be the default state for the neural tube. Many cell markers ultimately associated with dorsal regions are first expressed throughout the entire neural plate. For example, Pax3 and Pax7 are initially widely distributed, but as the neural tube begins to close, the floor-plate-derived signals block their expression in the ventral segment. However, ventral signals cannot convert the dorsal-most neural tube regions to ventral structures. For example, although notochord grafted to lateral regions of the neural tube led to ectopic formation of a floor plate and motor neurons (see Figure 4.4), grafts to the dorsal neural tube did not have the same effect. When the notochord was transplanted to the dorsal surface, the roof plate cells persisted (**Figure 4.13**A), though the expression of some dorsal genes was inhibited. Similarly, if a transplanted neural tube was rotated 180° so that the host notochord was now adjacent to the neural tube's dorsal surface (Figure 4.13B), a roof plate still formed above the notochord (Figure 4.13C). Thus, the roof plate could not be converted to a floor plate by notochord signals. Therefore scientists began to search for signals that induce the roof plate and specify dorsal cell types.

By the late 1990s, members of the transforming growth factor β (TGFβ) superfamily were identified as candidate molecules for dorsal specializations. Recall that during neural induction, TGFβ-related molecules of the bone morphogenetic protein (BMP) family induce the nonneural,

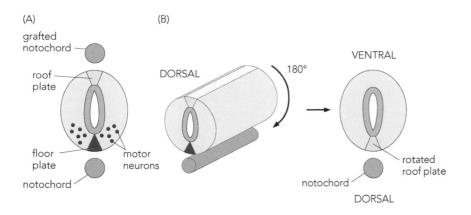

Figure 4.13 Notochord does not induce floor plate in the dorsal neural tube. (A) Notochord grafted to dorsal regions did not induce floor plate or motor neurons and was unable to convert the roof plate to a floor plate. (B) Rotation of the neural tube 180° to align the dorsal surface with the notochord also failed to induce floor plate tissue in the dorsal region. The rotated dorsal surface, now located above the notochord, continued to produce the roof plate. Together, these experiments show the roof plate forms in response to signals other than those associated with the notochord.

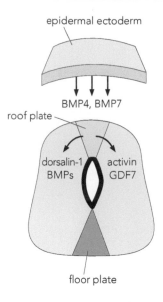

Figure 4.14 TGFβ-related molecules help pattern the dorsal neural tube. Epidermal ectoderm overlying the dorsal surface of the neural tube secretes BMP4 and BMP7. The roof plate becomes a secondary source of signals, releasing dorsalin-1, BMPs, activin, and growth differentiation factor 7 (GDF7), all members of the TGFβ superfamily.

epidermal ectoderm by inhibiting formation of neural tissue (see Chapter 2). At later stages of development, however, BMPs are used to specify distinct neuronal cell types, including the neural crest cells, the roof plate cells, and the six subtypes of dorsal interneurons.

Experiments revealed that expression of BMPs in the epidermal ectoderm overlying the neural tube induced formation of the roof plate, much like the notochord induces the floor plate (**Figure 4.14**). The epidermal BMPs induce the expression of the LIM homeodomain transcription factor Lmx1a in the roof plate cells; thus Lmx1a is used as a roof plate marker. Similar to the floor plate in the ventral neural tube, the roof plate appears to serve as a secondary, local source of signals to pattern cell types unique to the dorsal segment of the neural tube. Investigators found that the roof plate becomes a secondary source of BMPs as well as other TGFβ-related molecules, including dorsalin-1 (Dsl-1; also called BMP-9), growth differentiation factor 7 (GDF7), and activin (Figure 4.14). The BMPs expressed varies by species, as not all species express the same BMPs. In rodents, BMP6, BMP7, and GDF7 are present, whereas BMP4, BMP5, BMP7, and BMP-9 (dorsalin-1) are expressed in chick.

Roof Plate Signals Pattern a Subset of Dorsal Interneurons

In cell culture experiments, roof plate tissue induced formation of interneurons in the dorsal segment of the spinal cord (**Figure 4.15**A). Similarly, BMPs and activin induced formation of dorsal interneurons (Figure 4.15B). Further evidence to support the role of TGFβ-like molecules was seen with the addition of BMP inhibitors such as noggin or follistatin. As hypothesized, inhibiting BMP signaling blocked the expression of dorsal transcription factors and therefore the formation of the dorsal horn interneurons (Figure 4.15C).

In vivo evidence for the necessity of the roof plate in patterning dorsal interneurons came from studies in a naturally occurring mouse mutant known as the *Dreher* mouse. Among the developmental abnormalities detected in these mice are cerebellar defects, an absence of roof plate tissue, and a loss of the dorsal-most interneurons. Mice with this mutation have several neurological deficits, including an ataxic gait caused by altered axonal projections from the dorsal spinal cord to the cerebellum.

BMP-Related Signals Pattern Class A Interneurons

Based on studies of roof-plate-derived signals, scientists ultimately distinguished two classes of progenitor cells (class A and class B) that give rise to the six subtypes of dorsal interneurons. The class A progenitors (dp1–dp3, see Figure 4.6) depend on signals from the roof plate and are the most studied to date. These populations go on to form the dorsal interneurons 1–3 (dI1–dI3; (**Figure 4.16**). In literature prior to 2002, dorsal interneurons 1–3 were most often designated as D1, D3A, and D2, respectively. In contrast to class A neurons, class B progenitor populations (dp4–dp6)

Figure 4.15 *In vitro* studies confirmed the importance of roof-plate-derived signals for dorsal neural tube patterning. (A) When roof plate tissue was cultured with explants of neural tube, dorsal interneurons formed (green). (B) In the absence of roof plate tissue, the addition of BMP, dorsalin-1, or activin was sufficient to induce dorsal interneurons. (C) Adding the BMP inhibitors noggin or follistatin to the BMP culture media prevented formation of dorsal cell types, thus demonstrating the specificity of BMP signals for developing dorsal interneurons.

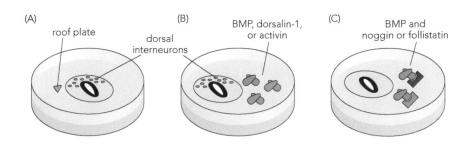

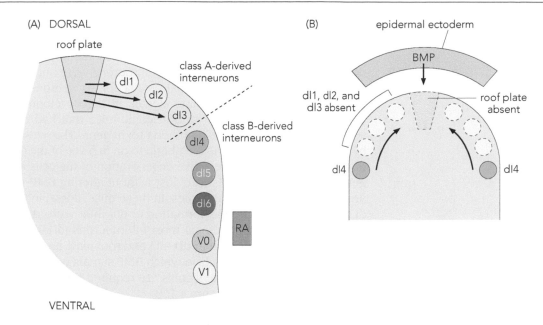

Figure 4.16 Class A and class B progenitor cells in the dorsal neural tube respond differently to roof-plate-derived signals. (A) Progenitor populations for interneurons dI1, dI2, and dI3 (class A) depend on roof-plate-derived signals for development. In contrast, progenitor populations for interneurons dI4, dI5, and dI6 (class B), which are located between the class A interneurons and the dorsal-most ventral interneurons (V0 and V1), do not require roof-plate-derived signals for differentiation. However, retinoic acid (RA) located in paraxial mesoderm appears to contribute to the development of class B dorsal interneurons and the V0 and V1 ventral interneurons. (B) The roof plate fails to form in mice genetically engineered to express the diphtheria toxin gene in place of *Gdf7*, a roof-plate-specific gene. The toxin causes neuronal degeneration of roof plate cells, but does not interfere with BMP signals arising from the epidermal ectoderm. In the absence of the roof plate, the dI1, dI2, and dI3 interneurons fail to form. Because dI4 interneurons are not influenced by roof plate signals they develop normally and expand into the more dorsal regions normally occupied by dI1–dI3 interneurons (arrows).

are largely independent of roof-plate-derived signals and appear to rely on non-BMP-related signaling molecules for their specialization. Class B progenitors give rise to dorsal interneurons 4–6 (dI4–dI6; formerly D3–D5). As in the ventral regions, cross-repression ensures each progenitor region develops correct neuronal characteristics. For example, *Olig3* represses expression of *Lbx1* in dp1–3. Because Lbx1 is necessary for differentiation of dI4–6, *Olig3* prevents dp1–3 cells from adopting the more ventral fates.

BMP influences dp1–dp3 development. If BMP levels are decreased at the stages these progenitor cells are present, the final number of dI1–dI3 interneurons is decreased. In addition, the class B (dI4–dI6) interneurons expand to occupy the regions typically populated by the dI1–dI3 interneurons.

Specification of the dI1 and dI2 interneurons is regulated in part by concentration differences and the timing of exposure to BMP-related molecules. In experiments manipulating available levels of BMP, higher concentrations induced primarily to dI1 interneurons, while lower concentrations induced only dI2 interneurons. In contrast, the more ventral dI5 and dI6 subtypes were unaffected by changes in available BMP levels.

The *Dreher* mouse also reveals the importance of roof-plate-derived signals in mediating the expression of the transcription factors needed to specify dI1 neurons. Investigations of these mutant mice determined that the phenotype arises from a mutation in the LIM homeodomain transcription factor gene *Lmx1a* that is expressed in roof plate cells. BMP from the epidermis normally signals to the roof plate to induce expression of *Lmx1a* and specify the roof plate cells. However, due to the mutation in the *Lmx1a* gene, the BMP signal is unable to induce roof plate cells. Thus, in the absence of *Lmx1a*, the roof plate does not form, BMP signaling is lost in the spinal cord, and dI1 interneurons do not develop normally. These mice have fewer dI1 neurons and those remaining migrate abnormally.

GDF7 is another BMP-related signal present in roof plate cells, but not other regions of the spinal cord or in the overlying epidermis. Mice lacking *Gdf7* also revealed a loss of dI1 interneurons. In these mutant mice, no other subpopulations of dorsal interneurons were affected, suggesting roof-plate-derived GDF7 is a signal specific for dI1 neurons.

The necessity of roof plate signals in the development of dI1–dI3 interneurons was also seen in transgenic mice that had roof plate cells selectively destroyed by the diphtheria toxin gene. This was accomplished by inserting the gene for the diphtheria toxin in place of the *Gdf7* gene. The diphtheria toxin caused selective degeneration of the cells expressing the gene, so all roof plate cells were lost without altering BMP-related signals arising from the epidermal tissues. In these mice, dorsal interneurons dI1–dI3 failed to form, whereas formation of dI4 interneurons was not inhibited. With the loss of the dorsal-most interneurons (dI1–dI3), the dI4 cells expanded dorsally into former dI1–dI3 and roof plate areas similar to what was observed in experiments in which BMP signals were decreased. Thus, roof-plate-derived BMP-like signals are required for normal development of dI1–dI3 interneurons (Figure 4.16B).

Activin, another TGFβ-related molecule, appears to specifically influence development of the dI3 subpopulation. When chick embryos were electroporated with a constitutively active from of an activin receptor, the dI3 population was induced independent of BMP signaling. Whereas the level of BMP signaling influences which cell populations develop, the level of activin receptor activation only affected dI3 interneurons. This result suggested that activin and BMP signaling pathways function independently of one another in the dorsal neural tube.

BMP-Signaling May Influence Dorsal Cell Specification in Multiple Ways

Unlike Shh, BMP may not function as a traditional morphogen gradient. Instead, the duration of BMP exposure may be more critical for establishing dorsal interneuron identities. Although the results to date are inconsistent, some studies suggest sustained BMP signaling induces dp1, dp3, and possibly dp5 while more limited exposure to BMP induces dp2, dp4, and dp6 populations. How the duration of BMP signaling would selectively influence dp1/3/5 and dp2/4/6 remains to be determined.

BMP signaling also regulates the expression of genes differentially expressed along the D/V axis. For example, *Prdm13* is expressed in dorsal progenitor domains where its expression appears to be directly or indirectly influenced by BMP signaling, just as Shh influences ventrally expressed *Prdm* genes. The Prdm13 protein is currently thought to act as a transcriptional repressor to prevent or limit activity of ventral-specific genes such as *Olig1*, *Olig2*, and *Prdm12* in dorsal cell types. Prmd13 is also thought to limit the expression levels of the proneural bHLH transcription factors neurogenin 1 and neurogenin 2 that are normally highly expressed in ventral regions. Prdm13 represses the activity of these bHLH transcription activators in dp2–dp6, preventing them from adopting a ventral fate. An example of the importance of Prdm13 is seen in mice lacking *Prdm13*. In the absence of *Prdm13*, Prdm12 expression expands to the dp2–dp6 domains. In addition, levels of neurogenin 1 and neurogenin 2 increase as Prdm13 is no longer available to suppress their activity, suggesting Prdm12 and neurogenins normally cooperate to drive ventral fates (**Figure 4.17**).

Prdm13 is also implicated in differential development of inhibitory (d4) interneurons over excitatory (d1–d3, d5) interneurons by repressing *Tlx1* and *Tlx3*, genes associated with excitatory neurons. Prdm13 can directly repress *Tlx* expression or prevent activity of the transcriptional activators that drive *Tlx* gene expression. This is just one example of how a

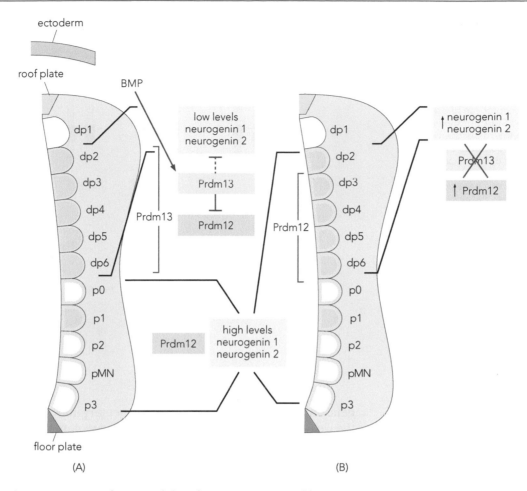

Figure 4.17 Prdm13 is necessary for normal dorsal patterning. (A) In wild-type mice, Prdm 13 (blue) is expressed in dl2–dl6. Prdm13 limits the expression of neurogenin 1 and neurogenin 2 (yellow) in these dorsal domains and prevents expression of Prdm12 (pink) from extending to dorsal regions. (B) In the absence of Prdm13, dorsal levels of neurogenin 1 and neurogenin 2 increase and Prdm12 expression extends into dl2–dl6 regions.

single gene product can differentially regulate the transcription factor code in different cells along the D/V axis.

BMP-Like Signaling Pathways Are Regulated by SMADS

As described in Chapter 2 (see Figure 2.20), BMPs, like other members of the TGFβ superfamily, bind to heterodimeric serine/threonine kinase receptors. BMP binds to the type II subunit of the receptor, which then associates with the type I subunit. The interaction of the two subunits leads to phosphorylation of a serine or threonine on the type I subunit. The phosphorylated type I subunit then phosphorylates certain SMAD proteins within the cytoplasm that translocate to the nucleus to activate specific genes (**Figure 4.18**). The *Smad* genes are homologs of the *Drosophila Mad* and *C. elegans Sma* genes (see **Box 4.1**).

SMADs not only activate gene transcription, but two SMAD proteins (SMAD6 and SMAD7) function as inhibitory SMADs (I-SMADS). I-SMADs block the BMP signaling cascade by inhibiting phosphorylation of R-SMADs (receptor-regulated SMADs), thereby preventing translocation to the nucleus and the activation of BMP-responsive genes (Figure 4.18).

The importance of BMP signal transduction can be explored by manipulating parts of the SMAD pathway. For example, chick embryos were electroporated with **short interfering RNA (siRNA)** against the co-SMAD, SMAD4. siRNAs are synthetic double-stranded RNAs that transiently block

Figure 4.18 BMP signal transduction is regulated by SMAD proteins. BMP and other TGFβ superfamily members bind to heterodimeric serine/threonine kinase receptors. The type II receptor binds BMP, then associates with the type I receptor, leading to phosphorylation. The now-activated type I receptor can phosphorylate receptor-regulated SMAD proteins (R-SMADs), which then associate with co-SMAD and are transported to the nucleus to activate or inhibit BMP-responsive genes. However, inhibitory SMAD proteins (I-SMADs) can block this gene activation pathway by preventing the phosphorylation of R-SMADs.

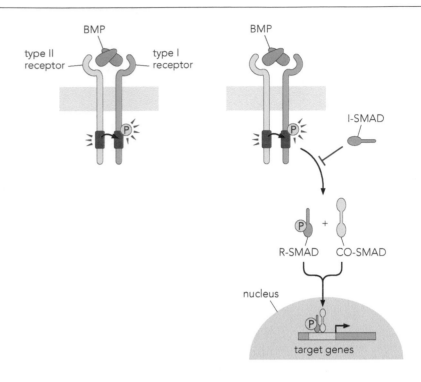

gene transcription by binding to complementary mRNA sequences, causing the cleavage and degradation of the targeted region. When siRNAs against SMAD4 were selectively introduced into dorsal neural tube cells, a specific loss of dI1–dI3 interneurons was noted. This indicated a cell-autonomous role for SMAD signaling in specifying the dorsal-most interneurons. In the areas normally occupied by dI1–dI3 interneurons, markers for dI4–dI6 interneurons were now detected. These results were similar to previous studies showing a need for BMP-related, roof-plate-derived signals in regulating the expression of the transcription factors expressed in dI1–dI3 interneurons. Together, the investigations indicate that roof-plate-derived signals utilize the SMAD pathway to selectively express the genes associated with the dI1–dI3 interneurons.

Although several molecules have been identified that impact class A progenitors, the signals used to induce specification of class B progenitors are still under investigation. RA is one plausible candidate, because it also impacts development of the ventral interneuron populations V0 and V1 that lie closest to the class B interneurons in the dorsal segment of the neural tube (see Figure 4.16A).

Wnt Signaling through the β-Catenin Pathway Influences Development in the Dorsal Neural Tube

Other signals help pattern the dorsal regions of the neural tube. For example, Wnt family members have been implicated as roof-plate-derived signals that complement the actions of TGFβ superfamily members. In Chapter 3, Wnt (also called Wnt1) was noted to pattern midbrain regions and prevent the expansion of forebrain regions into more posterior territories. Similarly, in the dorsal neural tube, Wnt might prevent dorsal expansion of dp4–6 neurons by regulating the expression of *Olig3*.

Along the D/V axis, Wnt signaling, acting downstream of BMP, may regulate cell cycle control or specification of interneuron subtypes. This hypothesis is supported by studies showing that a loss of BMP or BMP-related signals in the dorsal neural tube altered Wnt expression patterns. BMPs may activate Wnt signaling in dorsal regions and inhibit Wnt signaling in ventral regions using mechanisms that intersect with Shh pathways originating in the ventral neural tube.

The Wnt family of ligands includes up to 19 members, depending on the species. Among the Wnt ligands important for D/V patterning in the nervous system of mice and chick are Wnt1 and Wnt3a. The specific role of these ligands in neural patterning along this axis is still unclear, but it appears that Wnts in the dorsal regions of the neural tube influence the proliferation of cells as well as modify the effects of BMP and Shh signaling on the gene expression patterns in dorsal interneurons.

The most common intracellular signaling cascade initiated by the binding of a Wnt ligand to its receptor complex is the β-catenin or **canonical pathway** introduced in Chapter 3. This pathway involves Wnt binding to the receptor complex comprised of Frizzled and LRP4 or LRP6. This leads, in turn, to recruitment of a scaffolding protein called Dishevelled (Dvl), phosphorylation of LRP4/6 by the protein kinases GSK3 (glycogen synthase kinase 3) and CKI (casein kinase 1), and the recruitment of the scaffolding protein axin and the protein APC (adenomatous polyposis coli). The association of these proteins with the Wnt-bound receptor complex allows cytoplasmic β-catenin to travel to the nucleus, where it displaces a co-repressor protein called Groucho and interacts with LEF1/TCF (lymphoid enhancer factor 1/T-cell factor) to initiate transcription of Wnt-target genes (**Figure 4.19**A). In the absence of Wnt binding, β-catenin is degraded in the cytoplasm by a degradation complex comprised of axin, APC, GSK3, and CK1. Active GSK3 and CK1 phosphorylate β-catenin, leading to its degradation. Thus, β-catenin is unavailable to translocate to the nucleus and initiate transcription of Wnt-responsive genes (Figure 4.19B). Instead, the Wnt-target genes are repressed by the binding of the co-repressor Groucho to LEF1/TCF. Wnt binding and the stabilization of β-catenin therefore changes the actions of LEF1/TCF from interacting with Groucho and acting as a transcriptional repressor to interacting with β-catenin and functioning as a transcriptional activator.

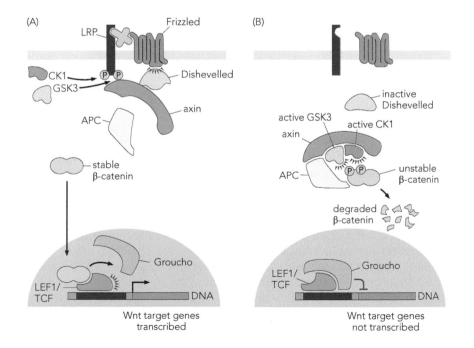

Figure 4.19 The Wnt/β-catenin pathway activates Wnt-target genes. The canonical Wnt/β-catenin pathway is the most common Wnt signaling pathway. (A) When Wnt binds to a receptor complex comprised of Frizzled receptors and LRP4 or LRP6, a series of events allows β-catenin to travel to the nucleus and activate Wnt-target genes. Binding of Wnt leads to the phosphorylation of the LRP subunit by protein kinases CK1 (casein kinase 1) and GSK3 (glycogen kinase 3) and the recruitment of a complex that includes the scaffolding proteins Dishevelled and axin, as well as the protein APC (adenomatous polyposis coli). With the recruitment of these to the Wnt receptor complex, the β-catenin is free to travel to the nucleus, where it displaces Groucho to interact with LEF1/TCF (lymphoid enhancer factor 1/T-cell factor) and initiate transcription of target genes. (B) In the absence of Wnt binding, the scaffolding protein Dishevelled is inactive and the other proteins associate as a degradation complex, which phosphorylates β-catenin. This causes β-catenin to become unstable and degrade in the cytoplasm. Because β-catenin is unavailable to travel to the nucleus, Groucho remains associated with LEF1/TCF and together they function as a transcriptional repressor to prevent Wnt-target gene transcription.

Figure 4.20 Wnt and Shh pathways interact with GLI proteins to pattern dorsal and ventral regions of the neural tube. Wnt1 and Wnt3a in the dorsal neural tube bind to the enhancer region of GLI3, the predominant repressor form of the GLI proteins (GLI3-R). Because the Shh concentration in the dorsal region is too low to activate Ptc-induced GLI-A activity, GLI3-R dominates and prevents Shh gene transcription in dorsal interneurons. However, in the ventral neural tube, Shh is present in a high enough concentration to bind to Ptc receptors and initiate the activator form of GLI (GLI-A) thereby promoting Shh gene transcription and ventral cell patterning.

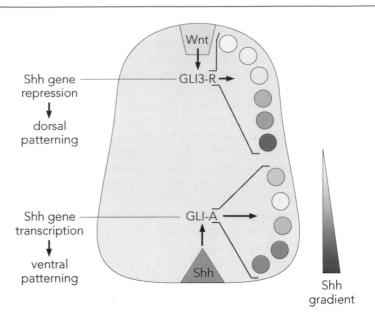

Wnt/β-catenin signaling appears to regulate the proliferation of neuronal precursor cells by upregulating the specific enzymes needed to drive cellular proliferation (cyclin-dependent kinases, CDKs; see Chapter 5).

Due to this effect on proliferation, it was originally thought that Wnt influences D/V patterning indirectly by increasing cell numbers to expand the size of dorsal neural tube regions, rather than directly influencing the transcription factor code and gene expression patterns. However, several studies indicate that the Wnt/β-catenin pathway intersects with the Shh–GLI pathway to directly influence patterning. Wnt1 and Wnt3a appear to induce activity of GLI3—the predominant repressor form of GLI—in the dorsal regions of the neural tube (**Figure 4.20**). When sufficient levels of Wnt signaling are present, the stabilized β-catenin is able to bind to enhancer regions of the *GLI3* gene. Because the Shh concentration in the dorsal neural tube is too low to activate sufficient Ptc receptor signaling and induce transcription of the activator forms of GLI, the Wnt-induced GLI3-R form represses *Shh* gene transcription. Thus, Wnt signaling increases GLI3-R activity and prevents *Shh* gene transcription in the dorsal neural tube, allowing for the expression of transcription factors associated with dorsal interneuron cell fates.

BMP and Shh Antagonize Each Other to Form D/V Regions of the Neural Tube

Mutually antagonistic signaling by BMP and Shh helps establish the boundaries between dorsal and ventral structures in the vertebrate neural tube, similar to homologous signals in *Drosophila* (**Box 4.3**). BMPs upregulate genes in the dorsal neural tube, such as *Pax3* and *Pax7*. In contrast, these same genes are repressed in the ventral neural tube by Shh. Experimentally manipulating the concentration of one signal relative to the other helped demonstrate how BMP and Shh interact to specify whether dorsal or ventral cell types form. For example, when levels of BMP were increased relative to Shh, ventral structures were absent. The higher concentrations of BMP inhibited the ventralizing capabilities of Shh so that only dorsal structures formed. In contrast, when BMP antagonists were also included, only ventral structures formed, even when Shh was present at low concentrations that would not usually

Box 4.3 Signaling Mechanisms That Regulate D/V Patterning in the *Drosophila* Embryo

Much like the role of signaling molecules in vertebrates, signals originating in dorsal and ventral regions of the *Drosophila* embryo differentially regulate the expression and repression of genes characteristic of different neural cell types along the dorsal–ventral (D/V) axis of the developing nervous system. As described in Box 2.1, the *Drosophila* nervous system arises from the ventrolateral region of the ectoderm and ultimately forms on the ventral side of the embryo. The epidermal ectoderm arises from the ectoderm located on the dorsal side of the embryo, opposite the orientation in the vertebrate embryo (**Figure 4.21**A). The ventrolateral neuroectoderm extends to the ventral side as the mesoderm begins to invaginate (Figure 4.21B, C). The cells of the mesoderm and neuroectoderm are ultimately surrounded by the epidermis (Figure 4.21D). Initially, the neural region is comprised of three rows of undifferentiated neural cells (neuroblasts) that form as individual cells delaminate from the neuroectoderm region. The cells in each row are identified by the expression of different homeodomain transcription factor genes: *Msh*, muscle segment homeobox; *Ind*, intermediate neuroblasts defective; and *Vnd*, ventral nervous system defective. Theses genes are homologs of the vertebrate genes *Msx1/2*, muscle segment homeobox; *Gsh*, genomic screen homeobox; and *Nkx2.2*, NK homeobox 2.2, respectively.

The specific neural genes expressed in the three primary rows of neuroblasts are determined by the actions of the opposing gradients originating from the dorsal and ventral signaling centers. Like the vertebrate BMP signals, the *Drosophila* Dpp signals in the dorsal region first repress the expression of neural genes in cells destined to become epidermal (see Box 2.1) and later regulate expression of different neural genes in a dosage-dependent manner established by the Dpp gradients. Neuronal cell types are further influenced by the gradient of Dorsal protein originating on the ventral side of the blastoderm stage embryo. In the fly, the mechanisms that govern neural gene expression along the D/V axis are said to be ventral dominant—that is, the signals originating from the ventral region of the embryo are the primary signals that first shape neuronal gene expression patterns. Signals from the dorsal region of the embryo then modify the gene expression patterns via a threshold-dependent repression of neural genes. The expression of these genes begins in the neuroectoderm regions of the blastoderm. In addition, the genes expressed more ventrally can repress the expression of the other genes. Thus, *Vnd* represses *Ind* and *Msh*, while *Ind* represses *Msh* (**Figure 4.22**).

The cells that form from the neuroectoderm closest to the ventral region express *Vnd*, those in the central layer express *Ind*, and those located closest to the dorsal side express *Msh*. Ventrally derived Dorsal protein influences the expression of these genes in a gradient-dependent manner. The higher concentration of Dorsal at the ventral region induces expression of *Vnd*, while progressively lower concentrations induce expression of *Ind* and *Msh*. The expression of *Vnd* in the ventral-most region also represses *Ind* and *Msh* expression in the ventral region, further contributing to the boundaries of gene expression found in the future neuroblasts. The higher concentration of Dorsal protein in the

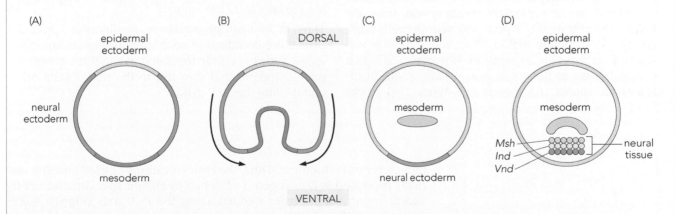

Figure 4.21 Neural ectoderm arises from the ventrolateral regions of the Drosophila embryo. (A) Neural ectoderm is found along the lateral sides of the *Drosophila* embryo. (B) As mesoderm at the ventral side invaginates, the neuroectoderm extends further ventrally (arrows) until it lies on the ventral side of the embryo (C). Individual cells of the neuroectoderm delaminate to form three rows of neuroblasts (D). Each row expresses a different transcription factor gene. The ventral-most cells express *Vnd*, the cells in the middle row express *Ind*, and those of the dorsal-most row express *Msh*.

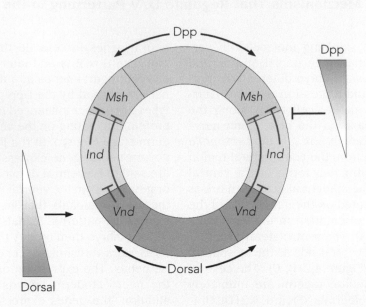

Figure 4.22 Signals arising from dorsal and ventral regions organize the transcription factor expression in neuroblasts along the D/V axis. The protein Dorsal on the ventral side of the embryo regulates the expression of each transcription factor in a concentration-dependent manner. Higher levels of Dorsal are able to induce expression of *Vnd*, whereas progressively lower concentrations induce expression of *Ind* and *Msh*. The boundaries between these cell regions are further specified by the repression of adjacent genes. *Vnd* is able to repress expression of *Ind* in the ventral-most regions to maintain a boundary between the *Vnd*- and *Ind*-expressing cells. The gradient of Dpp that originates on the dorsal side of the embryo represses neural gene expression. The higher concentration at the dorsal-most region is insufficient to influence *Msh* expression. However, the lower concentration at the mid-regions of the neuroectoderm can repress *Ind* expression. With *Ind* repressed, the inhibitory effect of *Ind* on *Msh* is lost and therefore *Msh* is expressed in the most dorsal regions of the neuroectoderm. In this way Dpp helps establish the boundary between *Ind* and Msh. [Adapted from Mizutani CM, Meyer N, Roelink H & Bier E [2006] *PLoS Biol* 4(10):e313. With permission from *PLoS*.]

ventral region is thought to be particularly important for establishing the border between *Vnd*- and *Ind*-expressing cells (Figure 4.22, left).

The Dpp originating on the dorsal side further shapes the D/V axis by repressing the expression of neural genes. Dpp can repress *Ind* expression to a greater extent than it can repress *Msh* expression. Thus, the lower concentration of Dpp found near the cells originating in the middle region of neuroectoderm is sufficient to repress *Ind* expression. However, the higher concentration at the more dorsal side is still insufficient to repress *Msh* expression. When Dpp blocks

expression of *Ind*, it is no longer available to repress the more dorsally located *Msh*, so this gene is able to be expressed in the dorsal-most region of neuroectoderm (Figure 4.22, right). The boundary between *Msh*- and *Ind*-expressing cells is believed to be influenced primarily by the gradient of Dpp originating at the dorsal side of the embryo.

Thus, in the *Drosophila* embryo, gradients of Dpp and Dorsal are coordinated so that expression of the different genes is tightly regulated in the three areas of neuroectoderm that give rise to the neuroblasts oriented along the D/V axis.

induce ventral structures. Thus, the concentrations of BMP and Shh normally produced *in vivo* appear to interact to ensure specialization of the various neuronal subtypes needed along the D/V axis (**Figure 4.23**). Opposing BMP and Shh signals may similarly act to maintain boundaries between forebrain regions along the A/P axis, such as proposed in the zona limitans intrathalamica (ZLI) region (see Chapter 3). These observations further demonstrate that patterning of the A/P and D/V axes do not require two entirely separate and distinct groups of signaling molecules.

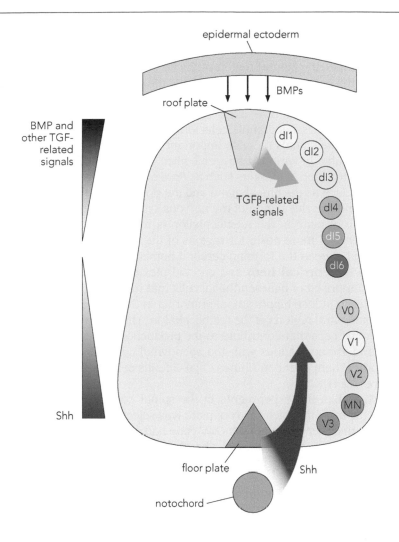

Figure 4.23. BMP and Shh interact to induce specific cell types along the dorsal–ventral axis. The epidermal ectoderm releases BMPs to induce formation of the roof plate. The roof plate then releases BMPs and other TGFβ-related molecules to induce formation of dorsal sensory interneurons (dl1–dl3). The timing, identity, and concentration of BMPs affect which cell types form from the progenitor cells. The roof plate signals are not required for the induction of dl4–dl6. The notochord releases Shh to induce the floor plate, which then also releases Shh. Together these sources of Shh induce motor neurons (MN) and interneurons (V3, V2, V1, and V0) in the ventral segment in a concentration-dependent manner. D/V patterning is further established because BMP and Shh antagonize each other, causing genes upregulated by one to be repressed by the other.

D/V PATTERNING IN THE ANTERIOR NEURAL TUBE

The patterning mechanisms employed in more anterior regions of the vertebrate neural tube are less well defined than those used in posterior regions. This is due in part to fewer identified cell markers for dorsal and ventral neuron subtypes and the more complex anatomical organization of the anterior neural tube. Despite these complications, progress in understanding D/V patterning in anterior segments is being made. In some cases, the same molecules used to pattern the spinal cord also participate in patterning the D/V axis at more anterior regions. For example, in the hindbrain, a gradient of Shh induces different motor neuron populations. Studies have shown that the concentrations of Shh required to induce the more ventrally located visceral motor neurons are higher than the Shh concentrations required to induce the more dorsally located somatic motor neurons. Thus, levels of available Shh regulate the types of motor neurons that develop in ventral regions of the hindbrain, similar to the process that occurs in the spinal cord.

Roof Plate Signals Interact with the Shh Signaling Pathway in the Cerebellum, Diencephalon, and Telencephalon

As noted in Chapter 3, the cerebellum arises from the dorsal region of rhombomere 1 (r1) in the hindbrain. Signals from the roof plate influence cerebellar development. When the roof plate is reduced in size, such as occurs in *Dreher* mice, the resulting cerebellum also is much smaller and lacks a vermis, the central region of the cerebellum. The roof plate in this region of the neural tube expresses BMP6, BMP7, and GDF7 and cerebellar granule cells (see Chapter 5), the dorsal-most cells to arise in the cerebellum, appear to rely on BMP-like signals. For example, when BMP is added to cultures of r1 tissue, granule cells form.

Roof plate signals are also important in the diencephalon portion of the neural tube. In various transgenic mice in which roof plate cells have been depleted, dorsal markers such as *Pax6* are downregulated and dorsal structures such as the pineal gland and the posterior commissure, a major axonal bundle in this region of the nervous system, are lost.

The roof plate in the telencephalon is morphologically quite different than that of more posterior regions of the neural tube. The roof plate is located between the forming cerebral hemispheres and later differentiates into the **cortical hem** and choroid plexus epithelium. The cortical hem is comprised of neuroepithelial cells that function as an early signaling center to induce hippocampal primordia in dorsal regions and choroid plexus in ventral regions of the telencephalon. The choroid plexus consists of epithelial cells that contribute to the production of cerebral spinal fluid (CSF) and are continuous with the ependymal cells lining the ventricles. The cortical hem later contributes Cajal–Retzius cells in the cerebral cortex (**Figure 4.24**).

Much like the experiments in the spinal cord, the telecephalon of transgenic mice in which the roof plate was selectively destroyed by diphtheria toxin under control of the *Gdf7* gene had defects in dorsal structures. In these mice, the roof plate was absent, *Lhx2* expression was decreased, and the number of cells in the cortical hem and choroid plexus were significantly reduced.

Both the cortical hem and choroid plexus express several BMP-related molecules and Wnt family members. The importance of BMP signals for normal development of the choroid plexus was also demonstrated in transgenic mice. When BMP receptor signaling was disrupted, choroid plexus cells were lost. In contrast, constitutively activated receptor signaling caused the entire alar plate region to become choroid plexus. However, the ventral region in these mice was not altered, even though BMP receptors are expressed throughout both dorsal and ventral regions of the telencephalon.

Wnt expression is widely distributed in the dorsal telencephalon, where it induces expression of dorsal markers such as *Pax6*. The Wnt signals also block the ability of ventral signals to convert dorsal structures to a ventral fate, likely by interacting with Shh signaling pathways. Wnt in the cortical hem normally upregulates GLI3, the repressor form that suppresses Shh signaling along A/P axis. If *GLI3* is mutated, cortical hem patterning is abnormal. Wnt also appears necessary for hippocampal induction as a loss of Wnt signaling causes truncation of the hippocampus.

Figure 4.24 Roof plate formation in the telencephalon. (A) In the anterior-most region of the neural tube, the roof plate (RP) forms along the dorsal surface similar to that observed at more posterior regions. As the telencephalon begins to invaginate (B), the roof plate comes to lie between the forming cerebral hemispheres of the neocortex (neo). (C) The roof plate cells give rise to choroid plexus epithelium (CP), which produces cerebral spinal fluid in the ventricles. The cortical hem cells (CH) later contribute cells to the outermost layer of the cerebral cortex. During development, these cells function as a signaling center to influence the development of dorsal regions of the telencephalon. (D) Image of a mouse embryo at embryonic day 12.5. The expression of *Lmx1a* (dark staining), a roof plate marker, is revealed by *in situ* hybridization. [Adapted from Chizhikov VV & Millen KJ [2005] *Developmental Biology* 27:287–295. With permission from Elsevier.]

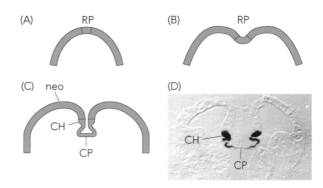

FGFs originating in anterior signaling centers, such as the ANR, may act antagonistically with BMP and Wnt signals to pattern the midline dorsal telencephalon. For example, when FGF was overexpressed, the cortical hem decreased in size.

Zic Mediates D/V Axis Specification by Integrating Dorsal and Ventral Signaling Pathways

The integration of dorsal and ventral signaling pathways in the midbrain and hindbrain is also influenced by the expression of various *Zic* (zinc finger of the cerebellum) genes, the vertebrate homologs of the *Drosophila Odd-paired* genes. Vertebrates have five *Zic* genes that are expressed in different patterns throughout the developing embryo and regulate a variety of functions, including those related to neural and limb development. In the nervous system, *Zic* genes are expressed during different stages of neural development and serve important roles during neural induction, neural crest development, and D/V patterning. *Zic* expression is regulated by other signals including BMPs, Wnts, and Shh. Elucidating the mechanisms by which these signals interact to pattern the D/V axis is currently an active area of research. During neural induction, BMPs appear to repress *Zic* genes, thereby preventing a neural fate in BMP-responsive cells that become ectoderm. In contrast, the neural inducers chordin, noggin, and follistatin interfere with BMP signaling to allow *Zic* gene expression and promote neural fate. However, later in development when *Zic* genes are expressed in dorsal regions of the neural tube, BMP signals promote *Zic* expression. Wnt signals also promote expression of *Zic* genes in dorsal areas of the neural tube. Zic proteins subsequently promote the expression of genes such as *Pax3* and *Wnt* that are needed for normal development of dorsal structures, particularly in the midbrain and hindbrain. Further patterning along the D/V axis occurs because Shh from ventral regions prevents expansion of *Zic* gene expression to the ventral areas of the neural tube.

The exact mechanisms that regulate the different effects of signaling on *Zic* gene expression patterns at different stages of development are still unclear. However, the ability of multiple signals to regulate *Zic* expression may occur because the signals intersect at specific DNA binding domains to influence the transcription of various genes associated with dorsal or ventral regions. For example, the actions of the Zic proteins associated with dorsal cell fates and the GLI-A proteins activated in ventral cells may inhibit one another by binding to each other's zinc finger domains (**Figure 4.25**). In this way, the different levels of Zic and GLI-A proteins available along the D/V axis ensure that Shh signaling occurs only in ventral regions.

The Location of Cells along the A/P Axis Influences Their Response to Ventral Shh Signals

An important point to keep in mind when thinking about structures differentiating along the neural tube is that A/P and D/V patterning often occur at the same time, through a process described as analogous to a Cartesian grid. In this way cells at each point along the two axes process multiple signals that combine to regulate differentiation of specific cell types at each segment of the nervous system. Thus, the location of signals along the A/P axis often influences the response of cells along the D/V axis. For example, the types of neurons that develop in response to Shh depend, in part, on the location of the cells along the A/P axis. In the hindbrain, Shh from the notochord and floor plate regulate the development of serotonergic neurons, whereas in the midbrain Shh regulates the development of dopaminergic neurons. This was also observed *in vitro*. Addition of Shh induced dopaminergic neurons in cultures of midbrain explants, but addition of Shh

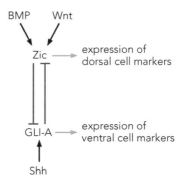

Figure 4.25 Zic proteins integrate dorsal and ventral signaling pathways along the D/V axis. The expression of *Zic* (zinc finger of the cerebellum) genes is regulated by the BMP and Wnt signals arising in dorsal regions of the neural tube. Shh in the ventral neural tube activates GLI-A proteins that interact with Zic proteins. The Zic and GLI proteins inhibit the zinc finger domains of one another, contributing to the repression of Shh signaling in the dorsal neural tube and the repression of BMP and Wnt signaling in the ventral neural tube.

Figure 4.26 Shh differentially regulates ventral cell types in the midbrain and hindbrain. (A) The addition of Shh to cultures of midbrain neural tube segments induces dopaminergic neurons (DA). However, when Shh is added to cultures of hindbrain segments (B), no DA neurons form.

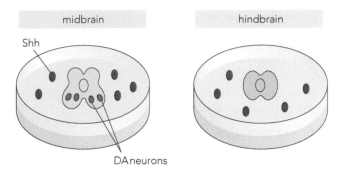

could not induce dopaminergic neurons in cultures of hindbrain explants (**Figure 4.26**).

Shh also plays a role in patterning more anterior regions. Shh and nodal released by the prechordal plate appear to influence forebrain patterning, including the formation of the two symmetrical cerebral hemispheres and the segregation of the eye fields. In experiments in which the prechordal plate was removed, a fused cerebral hemisphere and a single eye were observed, indicating the necessity of prechordal-plate-derived signals such as Shh in patterning these forebrain structures. Together, studies in the hindbrain, midbrain, and forebrain demonstrate that the outcomes of Shh signaling depend on its location along the A/P axis.

Analysis of Birth Defects Reveals Roles of D/V Patterning Molecules in Normal Development

The failure of the two hemispheres to separate is called **holoprosencephaly** (**HPE**), a birth defect that typically impacts both brain and craniofacial development. There are several subtypes of HPE that produce neurological deficits that range from mild to severe. One severe form of HPE occurs when the forebrain regions fail to separate. This leads to a single undivided hemisphere and often a single midline eye (cyclopia). This is a relatively common birth defect in humans, occurring in 1 out of 250 conceptions and over 1 in 10,000 live births, though in the most severe cases the infant dies shortly after birth. HPE also occurs in farm animals, and studies of those animals led to our understanding of the importance of Shh signaling in preventing this form of HPE.

Cyclopia in farm animals has been observed since the 1950s. It was found that when pregnant sheep or goats grazed on corn lily plants (*Veratrum californicum*) during the first two weeks of gestation, up to one quarter of the offspring were born with cyclopia—the observable manifestation of this form of HPE (**Figure 4.27**A, B). Later, scientists realized that the plants contain a naturally occurring alkaloid, now termed cyclopamine. It was subsequently determined that cyclopamine inhibits the action of Shh. This inhibitory effect was confirmed by *in vitro* experiments in which neural tube explants were cultured in the presence of Shh and cyclopamine. In contrast to control cultures that only contained Shh, the cyclopamine-treated explants failed to form either floor plate or motor neurons (Figure 4.26C). Thus, a naturally occurring birth defect was confirmed to result from disruption of the Shh signaling pathway.

Another typically less severe form of HPE, the middle interhemispheric (MIH) form, is caused by disruption in BMP signaling from the roof plate. In most cases, disruption of roof plate signals in the dorsal region of the anterior neural tube causes mild craniofacial deformities and defects in the posterior frontal and the parietal lobes. These developmental changes typically result in mild-to-moderate developmental delays, learning difficulties, and cerebral palsy.

(A) 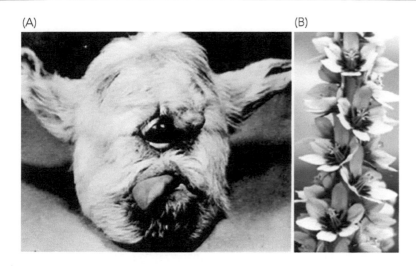 (B)

(C)
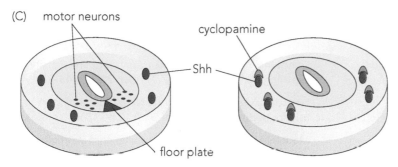

Figure 4.27 Inhibition of Shh by cyclopamine leads to congenital malformations in the forebrain. (A) Farmers discovered that sheep grazing on the corn lily plant often give birth to offspring with a single eye (cyclopia). Cyclopia is one characteristic of holoprosencephaly (HPE), a birth defect that is characterized by fused cerebral hemispheres and craniofacial defects. It was later determined that this birth defect occurs because the plant (B) contains a naturally occurring alkaloid, cyclopamine, that inhibits Shh signaling. (C) The effects of cyclopamine on Shh signaling were confirmed *in vitro*. When explants of neural tube were cultured in the presence of Shh, the floor plate and motor neurons formed. When cyclopamine was added to the Shh treated cultures, no floor plate or motor neurons were induced. [(A and B), Courtesy Philip Beachy/HHMI.]

These descriptions of congenital malformations in farm animals and humans illustrate that disruption of ventral and dorsal patterning molecules can lead to birth defects. Thus, by focusing on the basic science underlying the cellular and molecular pathways utilized in neural tube patterning, scientists discover important information relevant to veterinary and human medicine.

SUMMARY

Opposing gradients of signaling molecules help pattern many cell types along the D/V axis. These gradients coordinate to influence the expression of distinct transcription factors that, in turn, establish gene expression patterns and cellular specialization along the D/V axis. It is important to keep in mind that differentiation of both the A/P and D/V axes often occurs at the same time. Thus, cells must recognize and respond to multiple signals simultaneously. Scientists continue to identify new gene expression patterns along both axes and to unravel how signals are integrated in each cell population. It is now clear that multiple signals direct and refine the characteristics of each neuronal subtype. In some regions, transcription factors cross repress one another. In other populations, transcriptional activators are broadly expressed throughout all cells, but specific repressor proteins are expressed in subsets of progenitors. It also appears that for many cells, fluctuations in signaling molecule concentrations are integrated over time, and the sum of the effect leads to activation or repression of different genes. As information continues to accumulate, our understanding of the early stages of neural tube segmentation and boundary formation will also expand.

FURTHER READING

Andrews MG, Kong J, Novitch BG & Butler SJ (2019) New perspectives on the mechanisms establishing the dorsal-ventral axis of the spinal cord. *Curr Top Dev Biol* 132:417–450.

Attisano L & Lee-Hoeflich ST (2001) The smads. *Genome Biol* 2(8):1–8.

Briscoe J & Small S (2015) Morphogen rules: Design principles of gradient-mediated embryo patterning. *Development* 142(23):3996–4009.

Briscoe J & Therond PP (2013) The mechanisms of Hedgehog signalling and its roles in development and disease. *Nat Rev Mol Cell Biol* 14(7):416–429.

Caspary T, Larkins CE & Anderson KV (2007) The graded response to Sonic Hedgehog depends on cilia architecture. *Dev Cell* 12(5):767–778.

Cave C & Sockanathan S (2018) Transcription factor mechanisms guiding motor neuron differentiation and diversification. *Curr Opin Neurobiol* 53:1–7.

Chang C & Hemmati-Brivanlou A (1998) Cell fate determination in embryonic ectoderm. *J Neurobiol* 36(2): 128–151.

Chesnutt C, Burrus LW, Brown AM & Niswander L (2004) Coordinate regulation of neural tube patterning and proliferation by TGFβ and WNT activity. *Dev Biol* 274(2): 334–347.

Delile J, Rayon T, Melchionda M, et al. (2019) Single cell transcriptomics reveals spatial and temporal dynamics of gene expression in the developing mouse spinal cord. *Development* 146(12):dev173807.

Dessaud E, McMahon AP & Briscoe J (2008) Pattern formation in the vertebrate neural tube: A sonic hedgehog morphogen-regulated transcriptional network. *Development* 135(15):2489–2503.

Goodrich LV, Jung D, Higgins KM & Scott MP (1999) Overexpression of ptc1 inhibits induction of Shh target genes and prevents normal patterning in the neural tube. *Dev Biol* 211(2):323–334.

Gross MK, Dottori M & Goulding M (2002) Lbx1 specifies somatosensory association interneurons in the dorsal spinal cord. *Neuron* 34(4):535–549.

Hynes M, Ye W, Wang K, et al. (2000) The seven-transmembrane receptor smoothened cell-autonomously induces multiple ventral cell types. *Nat Neurosci* 3(1):41–46.

Ille F, Atanasoski S, Falk S, et al. (2007) Wnt/BMP signal integration regulates the balance between proliferation and differentiation of neuroepithelial cells in the dorsal spinal cord. *Dev Biol* 304(1):394–408.

Iulianella A & Stanton-Turcotte D (2019) The Hedgehog receptor Patched1 regulates proliferation, neurogenesis, and axon guidance in the embryonic spinal cord. *Mech Dev* 160:103577.

Kahane N & Kalcheim C (2020) Neural tube development depends on notochord-derived sonic hedgehog released into the sclerotome. *Development* 147(10).

Kim JJ, Gill PS, Rotin L, et al. (2011) Suppressor of fused controls mid-hindbrain patterning and cerebellar morphogenesis via Gli3 repressor. *J Neurosci* 31(5): 1825–1836.

Le Dreau G & Marti E (2012) Dorsal–ventral patterning of the neural tube: A tale of three signals. *Dev Neurobiol* 72(12):1471–1481.

Lee KJ, Dietrich P & Jessell TM (2000) Genetic ablation reveals that the roof plate is essential for dorsal interneuron specification. *Nature* 403(6771):734–740.

Liem KF Jr, Tremml G & Jessell TM (1997) A role for the roof plate and its resident TGFβ-related proteins in neuronal patterning in the dorsal spinal cord. *Cell* 91(1):127–138.

Litingtung Y & Chiang C (2000) Control of Shh activity and signaling in the neural tube. *Dev Dyn* 219(2):143–154.

Lumsden A & Graham A (1995) Neural patterning: A forward role for hedgehog. *Curr Biol* 5(12):1347–1350.

Maden M (2006) Retinoids and spinal cord development. *J Neurobiol* 66(7):726–738.

Megason SG & McMahon AP (2002) A mitogen gradient of dorsal midline Wnts organizes growth in the CNS. *Development* 129(9):2087–2098.

Millen KJ, Millonig JH & Hatten ME (2004) Roof plate and dorsal spinal cord dl1 interneuron development in the dreher mutant mouse. *Dev Biol* 270(2):382–392.

Mizutani CM, Meyer N, Roelink H & Bier E (2006) Threshold-dependent BMP-mediated repression: A model for a conserved mechanism that patterns the neuroectoderm. *PLoS Biol* 4(10):e313.

Mona B, Uruena A, Kollipara RK, et al. (2017) Repression by PRDM13 is critical for generating precision in neuronal identity. *eLife* 6:e25787.

Monsoro-Burq AH, Bontoux M, Vincent C & Le Douarin NM (1995) The developmental relationships of the neural tube and the notochord: Short and long term effects of the notochord on the dorsal spinal cord. *Mech Dev* 53(2): 157–170.

Moore SA & Iulianella A (2021) Development of the mammalian cortical hem and its derivatives: The choroid plexus, Cajal-Retzius cells and hippocampus. *Open Biol* 11(5):210042.

Pachikara A, Dolson DK, Martinu L, et al. (2007) Activation of class I transcription factors by low level Sonic hedgehog signaling is mediated by Gli2-dependent and independent mechanisms. *Dev Biol* 305(1):52–62.

Patten I, Kulesa P, Shen MM, et al. (2003) Distinct modes of floor plate induction in the chick embryo. *Development* 130(20):4809–4821.

Persson M, Stamataki D, Welscher P, et al. (2002) Dorsal–ventral patterning of the spinal cord requires Gli3 transcriptional repressor activity. *Genes Dev* 16(22):2865–2878.

Roelink H, Porter JA, Chiang C, et al. (1995) Floor plate and motor neuron induction by different concentrations of the amino-terminal cleavage product of sonic hedgehog autoproteolysis. *Cell* 81(3):445–455.

Sagner A & Briscoe J (2019) Establishing neuronal diversity in the spinal cord: A time and a place. *Development* 146(22):dev182154.

Stasiulewicz M, Gray SD, Mastromina I, et al. (2015) A conserved role for Notch signaling in priming the cellular response to Shh through ciliary localisation of the key Shh transducer Smo. *Development* 142(13): 2291–2303.

Timmer J, Chesnutt C & Niswander L (2005) The activin signaling pathway promotes differentiation of dl3 interneurons in the spinal neural tube. *Dev Biol* 285(1):1–10.

Tozer S, Le Dreau G, Marti E & Briscoe J (2013) Temporal control of BMP signalling determines neuronal subtype identity in the dorsal neural tube. *Development* 140(7): 1467–1474.

Van Kampen KR & Ellis LC (1972) Prolonged gestation in ewes ingesting *Veratrum californicum:* Morphological changes and steroid biosynthesis in the endocrine organs of cyclopic lambs. *J Endocrinol* 52(3):549–560.

Wilson L & Maden M (2005) The mechanisms of dorsoventral patterning in the vertebrate neural tube. *Dev Biol* 282(1):1–13.

Wilson SW & Houart C (2004) Early steps in the development of the forebrain. *Dev Cell* 6(2):167–181.

Yamada T, Placzek M, Tanaka H, et al. (1991) Control of cell pattern in the developing nervous system: Polarizing activity of the floor plate and notochord. *Cell* 64(3):635–647.

Zannino DA & Sagerström CG (2015) An emerging role for prdm family genes in dorsoventral patterning of the vertebrate nervous system. *Neural Dev* 10:24.

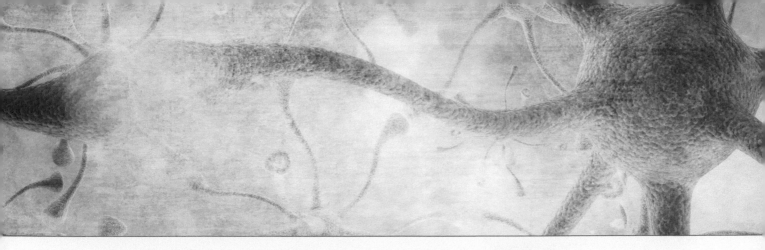

Proliferation and Migration of Neurons

5

As the neural tube forms and begins to segregate into distinct regions along the anterior–posterior (A/P) and dorsal–ventral (D/V) axes, neuroepithelial cells at the ventricular surface divide to produce the progenitors of central nervous system (CNS) neurons and glia. Early in nervous system development, the neuroepithelial cells must divide rapidly to produce the necessary complement of neurons (**neurogenesis**) and glial cells (**gliogenesis**). While the earliest born cells can form either neurons or glia, as development continues, cells stop dividing and begin to express the genes that limit them to a neural or glial fate. Thus, some cells in the neuroepithelium begin to express proneural genes that set them on the path to become neurons. Others begin to express genes that lead them to a glial cell fate. Similar mechanisms are used by the neural crest cells that give rise to neurons and glia in the peripheral nervous system (PNS). As introduced in Chapters 3 and 4, various intrinsic and extrinsic signals activate or repress genes that determine the type of cell that forms. This is the process of **cell fate determination**. Once fate is established, the type of cell that forms can no longer be altered by further development or experimental manipulation. However, during the subsequent, gradual process of **cellular differentiation**, the unique cellular characteristics associated with different neuronal and glial subtypes are established.

This chapter explores how vertebrate neuroepithelial cells begin to form different cell types and how the migration of cells from the cerebral cortex, cerebellum, and neural crest influences subsequent cell fate options. In Chapter 6, examples of cell determination and differentiation in these and other CNS and PNS populations from vertebrate and invertebrate animal models are described. As becomes apparent in reviewing these developmental events, aspects of segmentation, neurogenesis, migration, and cell fate take place concurrently within the developing nervous system.

NEUROGENESIS AND GLIOGENESIS

It is now understood that newly formed neuroepithelial and neural crest cells are exposed to a variety of spatially and temporally regulated signals that influence cell proliferation, migration, and differentiation. However, it

DOI: 10.1201/9781003166078-5

took many decades of research to understand the basic mechanisms that govern such processes. In the late nineteenth century, for example, scientists were unsure if CNS neurons and glia arose from a single cell population or two separate populations. Twentieth-century scientists began to sort out how proliferation and migration could influence cell fate options, while current efforts focus on deciphering the intricate signaling pathways required by individual cell populations.

Scientists Debated Whether Neurons and Glia Arise from Two Separate Cell Populations

Published reports by early neurobiologists such as Wilhelm His (1886, 1889) and Santiago Ramón y Cajal (1894) provided detailed descriptions of the cellular composition of the vertebrate embryonic nervous system. Both produced elaborate drawings of cells arranged in layers near the lumen of the neural tube (that is, the ventricular surface). His and Cajal correctly identified these cell layers to be the precursor, or germinal, cells of the nervous system, though they interpreted the organization differently.

Based on the appearance of the cells under the microscope, His hypothesized that neural and glial cells form from two different cell populations (**Figure 5.1**A). He proposed that the glial cells arose from the "spongioblasts," a layer that remained along the ventricular surface as a **syncytium**, a connected network of cells. The idea that cells formed this sort of continuous network was popular at the time and was described in many other tissues as well. The appearance of the cells as they were prepared by His was consistent with the concept of a spongioblast network.

Cajal's observations confirmed much of what His stated, but based on his own preparations, Cajal described the glia as a separate population of individual cells that were *not* connected as a syncytial network. Subsequent reports from other scientists ultimately supported Cajal's conclusion that the ventricular surface was lined with individual cells and that glial cells did not form a connected network (Figure 5.1B).

However, in contrast to Cajal's interpretation of two separate cell populations, Alfred Schaper (1897) proposed that a single population of uncommitted precursor cells formed along the ventricular surface then

Figure 5.1 Evolving views on the origin of neurons and glia. (A) In 1889, Wilhelm His proposed that neurons and glia arise from two different populations located at the ventricular surface of the neural tube. He proposed that neurons (blue) arose from one cell population and that glia (yellow) arose from another population of interconnected cells he called spongioblasts. (B) In 1894, Ramón y Cajal agreed with much of His's hypothesis, but suggested that neurons and glia each arose from individual cells and that the glia were not connected by a syncytium. (C) In 1897, Alfred Schaper suggested that neurons and glia arise from the same population of cells (gray). He described the cells as forming at the ventricular surface and migrating toward the pial surface, where they would become neurons, or would divide again before giving rise to a neuron (blue) and glial cell (yellow). [Adapted from Levitt P, Cooper ML & Rakic P [1981] *J Neurosci* 1:27–39.]

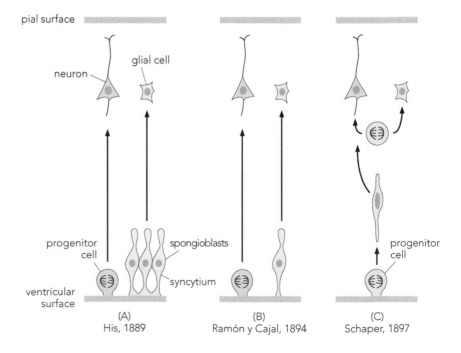

migrated out of the ventricular zone to become neurons or divided again outside the ventricular zone to form neurons and glia (Figure 5.1C). Because this idea was inconsistent with the two cell populations described by Cajal and His, Schaper's idea remained unpopular at the time. However, Schaper's interpretation that a single cell population gives rise to neurons and glia was ultimately shown to be correct.

As detailed later in this chapter, neuroepithelial cells self-renew and also produce **radial glial** (RG) cells. RG, in turn, serve as both **multipotent** progenitor cells and as a scaffold for new neurons to migrate upon. Thus, the RG cells act like stem cells to generate more neurons, then give rise to glia. Both neuroepithelial cells and RG cells are therefore considered neural **stem cells**. The term "radial glia" persists even though it is now recognized that the cells give rise to both neurons and glia.

Neuroepithelial Cell Nuclei Travel between the Apical and Basal Surfaces

Nineteenth-century scientists had also observed neuroepithelial cell bodies located at different distances from the apical (ventricular) to basal (pial) surfaces in the neural tube (**Figure 5.2**A). The morphological changes that occur during neural tube formation explain the origins of the terminology used to describe the apical and basal regions. These surfaces are first identified in the neural plate stage (Figure 5.2B). As the neural plate curves over to form the neural tube, the former apical surface is located surrounding the lumen (the future ventricles), while the former basal surface becomes the outer surface of the neural tube located just below the pial membrane.

Many early investigators believed that the cell bodies represented different populations of neuroblasts settled in different layers. In contrast, Schaper reported that the cell bodies observed at the apical surface and those seen closer to the basal surface were the same cells viewed at different phases of the cell cycle. Schaper died in 1905 at the age of 42, 30 years before experimental evidence reported by Fred C. Sauer confirmed his interpretation.

Figure 5.2 During the cell cycle, cells undergo interkinetic nuclear migration as they migrate from the apical to basal surface. (A) In the neural tube, cell bodies (blue) are located at various locations from the apical (ventricular) to the basal (pial) surface in the neural tube, appearing to form different cell layers. (B) Morphological changes account for the terminology used to describe the location of cells relative to the ventricular and pial surfaces of the neural tube. As the neural plate curves over to form the neural tube, the apical surface of the neural plate becomes the ventricular surface of the neural tube. The basal surface of the neural plate becomes the outer (pial) surface. (C) Cells along the ventricular surface extend cytoplasmic processes that contact both the pial and ventricular surface. The nucleus of the cell then migrates through the cytoplasmic process using a mechanism called interkinetic nuclear migration. The nucleus first travels to the basal (pial) surface, then returns to the apical (ventricular) surface, where it undergoes mitosis (cell division). In this example, the same cell is shown at different stages of nuclear migration. Sauer further noted that the size of the nucleus (shown in brown), an indirect measure of the amount of DNA, first increased in size as the cell approached the basal surface, then continued to increase as it approached the apical surface. [(C), Adapted from Sauer FC [1935] *J Comp Neurol* 62:377–405.]

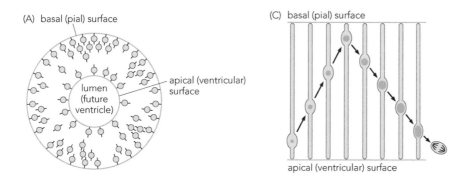

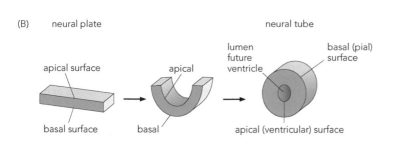

In a series of experiments conducted in multiple species, Sauer found that cells located near the ventricular surface extend thin processes that contact both the apical and basal surfaces. The nucleus of a cell then travels through these processes as the cell cycle progresses. To monitor the cell cycle at different locations from the apical to basal surfaces, Sauer measured the size of the nucleus—a method that provides an indirect measure of the amount of DNA in a cell. From these measurements, he determined that the nucleus increases in size as it approaches the pial (basal) surface and further doubles in size again as it moves back to the ventricular (apical) surface. These observations led Sauer to propose that the nucleus moves through the cytoplasm of the process to reach the basal surface, and then returns to the apical surface to divide (Figure 5.2C). The movement of the nucleus through the process was termed intermitotic migration of nuclei or, as is commonly called today, **interkinetic nuclear migration**.

Interkinetic Movements Are Linked to Stages of the Cell Cycle

F. C. Sauer's work provided the first evidence of interkinetic nuclear migration, though the interpretation of his results remained controversial for some time. After his death in 1936, his wife, Mary Elmore Sauer, followed up on his work. In the 1950s she and her colleagues used nuclear stains and the tritiated thymidine method developed in 1957 to confirm that nuclear migration was linked to phases of the cell cycle.

The proliferative **cell cycle** begins with the **G1 (gap 1)** stage, during which cells prepare for DNA synthesis. Upon exiting G1, the cells enter **S phase**, the DNA synthesis phase of the cell cycle. Once chromosomes are copied, the cells progress to the **G2 (gap 2)** stage, preparing for mitosis. The final stage of the cycle is **M phase**, when mitosis occurs, and two daughter cells are generated from the precursor cell (**Figure 5.3**A). Tritium, the radioactive label, is incorporated into DNA during the synthesis (S) phase of the cell cycle.

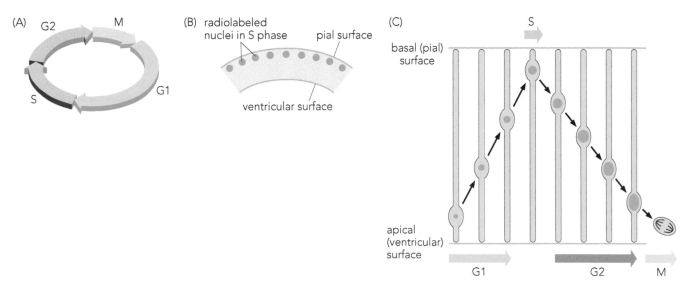

Figure 5.3 Interkinetic nuclear migration is linked to phases of the cell cycle. (A) The cell division cycle begins with gap 1 (G1) phase in which the cell prepares for DNA synthesis. The cell then enters the S phase, the phase in which DNA is synthesized. The cell progresses to the gap 2 (G2) phase, which prepares the cell for mitosis. The final stage is the M phase (mitosis) in which the cell divides to produce two daughter cells. (B) Cells that incorporated a radiolabel during the S phase of the cell cycle were located near the pial (basal) surface of the neural tube, confirming that movement of the nucleus is linked to stages of the cell cycle. (C) As a cell enters the gap 1 (G1) phase the nucleus travels slowly toward the pial (basal) surface. At the pial surface the cell enters the DNA synthesis (S) phase, then continues to the gap 2 (G2) phase when the nucleus travels quickly back to the ventricular (apical) surface, enters the mitotic (M) phase, and divides.

In 1959, M. E. Sauer reported the pattern of nuclear migration in the neural tube of the chick at the meeting of the American Association of Anatomists (now called the American Association for Anatomy). Her lab found that cells undergoing DNA synthesis were concentrated at the pial surface (Figure 5.3B), as first suggested by the nuclear measurements made by F. C. Sauer in 1935. Notably, her report was one of the first neurobiology studies to use the tritiated thymidine technique. Subsequent studies confirmed that the nuclei of neurons in the developing neural tube undergo interkinetic nuclear migration in a sequence linked to the stages of the cell cycle. Thus, nuclei of cells at the G1 phase of the cell cycle are located at the ventricular zone. When the migrating nucleus reaches the pial surface, the cell is in the S phase of the cell cycle, after which the nucleus travels back to the ventricular surface, where the cell enters the M phase and divides into two daughter cells (Figure 5.3C). The resulting daughter cells will either migrate away from the ventricular surface or extend a basal process to the pial surface and continue through another series of interkinetic movements before dividing again.

The purpose of interkinetic movements is not fully understood. It is thought that the nucleus may be exposed to growth factors or other signals in the cytoplasm as it travels through the cell process during the various stages of the cell cycle. It has also been suggested that the movement of nuclei into different layers provides more room for proliferation in the limited space of the early neural tube. Although the exact purpose is not yet known, interkinetic movements in the developing neural tube are a highly orchestrated and precise mechanism that likely provides essential cues for further development and specification of cellular characteristics.

The unique and precisely ordered patterns of interkinetic movement are seen only in the ventricular zone of the neural tube and regions of pseudostratified epithelium. At other sites of neurogenesis, including the subventricular zone of the cerebrum and external granule cell layer of the cerebellum, cells do not undergo nuclear translocation prior to cell division. Thus, the intriguing process of interkinetic movements is restricted to specific regions of the embryo.

Cell Proliferation and Migration Are Influenced by the Cell Division Plane

As more progenitor cells are generated and the wall of the neural tube thickens, layers with unique cellular characteristics emerge. Early neural tube layers include the ventricular zone (VZ) located closest to ventricular surface, the marginal zone (MZ) lying closest to the pial surface, and an intermediate zone (IZ) positioned between them. As the neural tube thickens, cell nuclei must travel increasingly greater distances to the pial membrane during interkinetic nuclear migration. The radial glia produced after initial neuroepithelial proliferation also undergo interkinetic nuclear migration, but only within the VZ. The radial glia cell bodies do not travel to the pial surface at the outer edge of the marginal zone.

Whenever the nucleus returns to the ventricular surface following interkinetic nuclear migration, the plane of cell division determines whether a new cell continues to divide or migrates away from the ventricular zone. In most species, a vertical cleavage plane (perpendicular to the ventricular surface) causes a progenitor cell to divide symmetrically. The basal process of the cell is either split between the two daughter cells, or one cell inherits the basal process while the other produces a new process. This symmetric cleavage, which predominates in the early stages, generates two identical daughter cells that remain at the ventricular surface to undergo further proliferation. This allows neuroepithelial cells to self-renew and establish a sufficient pool of progenitor cells. The neuroepithelial cells begin to

Figure 5.4 The plane of cell division determines whether a cell continues to proliferate or migrates out of the ventricular zone. (A) Symmetrical cleavage (vertical dashed line) occurs when the plane of cell division is perpendicular to the ventricular surface. In this case, the cell produces two daughter cells that remain at the ventricular (apical) surface of the neural tube. These cells continue to proliferate, producing additional daughter cells. (B) Asymmetric cleavage (oblique dashed line) occurs when the plane of cell division is shifted away from the vertical plane. With asymmetric cleavage, cell division produces a basal daughter cell that will migrate toward the pial (basal) surface as a multipolar intermediate progenitor (IP) or basal radial glial cell (bRG). The apical daughter cell (aRG) will continue to proliferate at the ventricular (apical) surface.

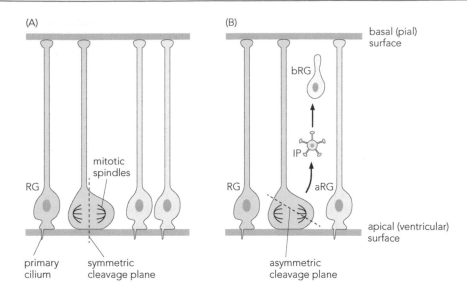

transform to RG cells at the beginning of neurogenesis (at E11 in mice, for example). These RG cells, called apical RG, either again divide symmetrically to produce additional progenitors (**Figure 5.4**A) or divide asymmetrically to both self-renew and produce cells destined to form neurons or glia.

Asymmetric division occurs when the cleavage plane is horizontal (parallel to the ventricular surface) or oblique (at an angle to the ventricular surface), producing two nonidentical daughter cells. The daughter cell closer to the ventricular surface (the new apical RG) retains the basal process and stays attached at the ventricular surface where it continues to proliferate (Figure 5.4B). In contrast, the **basal progenitor cell**, the daughter cell closer to the pial surface, does not inherit the basal process or maintains only a thin basal connection. This cell detaches from the ventricular surface and migrates away as either an **intermediate progenitor cell** (IP) or basal RG. IP cells typically divide once more outside of the VZ to further expand the number of cells generated from a single RG cell. In the spinal cord, these progenitors travel to the intermediate zone, in the cerebral cortex they travel to a layer called the subventricular zone. Some IP cells are multipotent whereas others are lineage-restricted, meaning they can only form specific types of neurons or glia.

As the basal progenitor detaches from the VZ to form an IP, it adopts a temporary multipolar morphology as it moves to a new region in the expanding neural tube. Once settled in the new area, the cell resumes its bipolar morphology, with processes extending toward the pial and ventricular surfaces. The purpose of the transient multipolar morphology is not entirely clear. The multiple processes may help the cells migrate and explore the environment for additional development cues.

Distinct Proteins Are Concentrated at the Apical and Basal Poles of Progenitor Cells

Like all epithelial cells, neuroepithelial and RG exhibit apical-basal polarity with unique proteins concentrated at each pole. These proteins serve many functions including mediating the attachment of the cellular processes to the apical and basal surfaces, directing the orientation of the mitotic spindles, and regulating the expression of genes needed for either continued proliferation or neural differentiation and migration. For example, the basal process contacts the basement membrane of the pial surface initially via a single endfoot then later by multiple endfeet. This attachment may be regulated by signals from the surrounding extracellular matrix and blood vessels.

At the ventricular surface, adherens junctions are detected in apical processes. **Adherens junctions** contain the calcium-dependent adhesion molecule cadherin and are among the cellular elements required to maintain proliferation. Adherens junctions are not found in the basal daughter cells that arise during asymmetrical cleavage because the basal progenitor cell downregulates the expression of cadherins (**Figure 5.5**A, B). To migrate out of the VZ, the basal daughter cell loses adherens junctional proteins and sheds the tip of the **primary cilium** (see **Box 4.2**) via a process called **apical abscission**. This causes the basal cell to lose its apical-basal polarity, delaminate from ventricular surface, and migrate to new location. In contrast, the proliferating, apical cell maintains the remnant of the ciliary membrane.

Whether a cell undergoes symmetrical or asymmetrical division is established by the position of the cleavage furrow which is determined by the orientation of the mitotic spindle. When the mitotic spindle is oriented for symmetrical cleavage, basal and apical polarity proteins are equally distributed in both daughter cells, and therefore proliferation continues (Figure 5.5A).

When the mitotic spindle is oriented in a horizontal or oblique plane, the daughter cells have an unequal segregation of proteins into the apical and basal poles. Initially, it appeared that horizontal cleavage was necessary for segregating apical and basal proteins, but it is now known that even a slight shift in spindle orientation away from the vertical plane provides enough of a difference in protein distribution to initiate asymmetrical division and produce two unique daughter cells (Figure 5.5B). In some species oblique divisions may predominate, while in others, such as primates, fluctuations in spindle orientation appear to influence whether cells continue to proliferate or begin to adopt a neural fate.

Much of what is known about the distribution of apical and basal proteins in vertebrate cells was first discovered in *Drosophila*, where homologous proteins regulate the formation of neuronal and

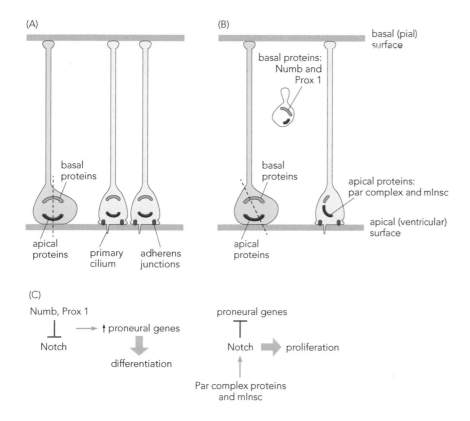

Figure 5.5 Protein distribution helps determine if a cell will continue to proliferate or migrate away. (A) In cells undergoing symmetrical (vertical) cleavage, the proteins concentrated at the basal and apical poles of the cell become equally distributed in the resulting daughter cells. Both cells are attached at the ventricular surface by adherens junctions that are needed to retain a proliferative state. (B) In cells that undergo asymmetrical (horizontal or oblique) cleavage, the basal and apical proteins become unequally distributed in the daughter cells. The migrating basal cell inherits a higher concentration of the basal proteins Numb and Prox 1, while the apical daughter cell inherits a higher concentration of apical proteins, including those of the Par complex and mInsc (mammalian Inscuteable). To migrate away from the ventricular surface, the basal cell sheds the tip of the primary cilium and loses adherens junctional proteins. (C) The differences in protein levels lead to differences in the activity of the Notch receptor that is required for continued cell proliferation. In the basal daughter cell, Numb and Prox 1 decrease the activity and expression of Notch. Decreased Notch activity in the basal daughter cell allows for the expression of proneural genes and cell differentiation. In the apical cell, the Par complex and mInsc allow the Notch receptor to have a higher level of activity. Notch receptor activity also leads to a suppression of proneural genes to help maintain the undifferentiated, proliferative state in that cell. Dashed lines indicate the plane of cleavage in A and B.

nonneuronal cells (see Chapter 6). Among the proteins involved in vertebrate neurogenesis are the Notch receptors and their corresponding membrane-bound ligands. The receptors and ligands are expressed in all precursor cells, but at differing levels. The distribution of the apical and basal proteins within a daughter cell helps to establish the level of Notch receptor activity, and therefore determines whether a cell proliferates (sustained Notch receptor activity) or migrates (reduced Notch receptor activity). Proteins unique to the apical pole of vertebrate CNS neurons include those of the Par protein complex. Par proteins recruit additional proteins, including mammalian Inscuteable (mInsc) and AGS3 (activator of G-protein signaling 3). Together these proteins appear to influence the orientation of the mitotic spindle and help keep basal proteins, such as Numb and Prox 1 (Prospero homeobox 1), concentrated at the opposite pole of the cell.

With symmetric cell division, both daughter cells maintain similar levels of Notch receptor activity, so both cells continue to proliferate. With asymmetric division, however, the higher concentration of basal proteins interferes with Notch signaling in the basal cell. Numb, for example, binds to the intracellular region of the Notch receptor to inhibit its activity and prevent proliferation of the basal daughter cell. As the basal progenitor begins to migrate out of the ventricular region, *Notch* gene expression is downregulated. Further, as cells begin to exit the cell cycle, *Prox 1* expression increases. Prox 1, like the *Drosophila* homolog Prospero, inhibits the proliferation of precursor cells by binding to the Notch promoter region, preventing *Notch* expression. Reduced Notch receptor activity or gene expression ends proliferation and allows the basal daughter cell to express **proneural genes**, the genes that direct a cell to become a neuron. Thus, the basal daughter cell is established as a neuronal progenitor cell (Figure 5.5C).

The apical daughter cell that arises following asymmetric cleavage cannot disrupt Notch receptor activity due to an insufficient level of Numb protein. Therefore, Notch activity in that cell suppresses proneural gene expression and promotes continued cell proliferation (Figure 5.5C). Because proneural genes are required to induce *Prox 1* expression, the Notch receptor activity also prevents *Prox 1* expression. Thus, Notch and Prox 1 cross-inhibit one another to carefully balance the number of proliferating and migrating progenitors so the correct number of cells is produced at each phase of neurogenesis.

Further details on Notch signaling and the regulation of gene cascades that establish cell fate are described in Chapter 6. What is significant for this stage of development is that the distribution of specific proteins into apical and basal poles during cell division establishes the first step in determining whether a cell remains in the proliferative state or migrates away from the ventricular surface to begin the differentiation process.

The Rate of Proliferation and the Length of the Cell Cycle Change over Time

The balance of symmetric and asymmetric cleavage changes over the course of development. Initially, symmetric cleavage predominates so that enough cells are produced. Over time, the need for additional new cells lessens, and the proliferation rate slows. As this happens, more and more cells reach their final cell division (terminal mitosis) and migrate out of the ventricular zone.

Proliferation rate can be estimated by monitoring the length of the cell cycle. Several methods have been developed for measuring cell cycle length, including tritiated thymidine, BrdU (bromodeoxyuridine, an analog of thymidine), and retroviral labels.

As cell cycle time lengthens, fewer cycles per day are possible, and fewer new cells are produced. In the developing mouse cortex, cells undergo about 2.5 to 3 cell cycles per day on the first day of proliferation, but by the sixth day, toward the end of this region's proliferative phase, the cells complete only one cycle per day. The length of the cell cycle, which was observed to last about 8 hours on day 1 of proliferation, increased to 18 hours by the end of the proliferative phase. Similarly, in the developing chick optic tectum, the cell cycle lengthened from about 8 to 15 hours over a 3-day period of proliferation. In most regions of the CNS, it is the length of the G1 phase that increases over time, with the length of the other stages remaining fairly constant. In some regions of the CNS, however, it appears that the time spent in both G1 and S phase is lengthened.

The purpose of the lengthening of cell cycle phases is not entirely clear. One hypothesis is that the longer G1 phase allows for exposure to growth factors that influence subsequent developmental events (**Box 5.1**). Many growth factor proteins, including members of the fibroblast growth factor (FGF), epidermal growth factor (EGF), and transforming growth factor α (TGFα) families, regulate cellular proliferation at the G1-to-S transition and therefore have the potential to impact whether cells progress through the cell cycle. Accumulating evidence suggests that exposure to such factors during G1 promotes or inhibits regulators of cell cycle progression (Box 5.1) or mediates the change from an uncommitted precursor cell to a neural progenitor.

Box 5.1 Cyclins and CDKs Regulate Progression through Cell Cycle Stages

Specific cell cycle regulators act at each stage of the cell cycle to stimulate or inhibit progression to the next stage. Much of the current understanding of cell cycle regulation originated in work with model systems such as yeast and sea urchins. Scientists soon discovered that knowledge gained from these studies is broadly applicable to most vertebrate neurons as well.

Cell cycle regulation is primarily controlled by two families of interacting proteins: cyclins and cyclin-dependent kinases. Both families contain multiple members, including some that are not associated with the cell cycle.

The **cyclins** are a group of proteins that were initially identified based on an unusual expression pattern: Levels of various cyclins oscillate throughout the cell cycle. Cyclins were first discovered in the 1980s when scientists studying extracts of fertilized sea urchin eggs noted a decrease in a particular protein band that only occurred at the time of cell division. Although many other protein bands from the analyzed extracts increased gradually over time, this protein became more abundant as the cell cycle progressed, then decreased after mitosis (**Figure 5.6**).

Cyclin-dependent kinases (CDKs) are a family of serine/threonine protein kinases that must be activated by cyclins before they can phosphorylate downstream proteins to regulate progression through the cell cycle. CDKs are related to the *Cdc* (cell division cycle) genes first identified in yeast. In contrast to the cyclical expression of the cyclins, the expression of the CDKs is stable throughout the cell cycle. It is the changes in cyclin levels that regulate the cyclical activity of these kinases.

Pairs of specific cyclins and CDKs are used at different cell cycle stages. A given cyclin binds to and activates an associated CDK. Once activated, the CDK influences subsequent substrate proteins, pushing the cell to the next phase of the cell cycle. For example, cyclin D activates CDK4 or CDK6 so that the cell enters G1, which is the phase that commits a cell to proliferation (**Figure 5.7**).

Other cyclin-CDK pairs are active at later stages of the cell cycle. The association of cyclin E with CDK2 is needed to push the cell from G1 to S phase. Cyclin A must bind to CDK2 or CDK1 for the cell to progress through S phase to G2. Cyclin B or cyclin A pairs with CDK1 to move the cell from the G2 to M phase. Note that a given cyclin such as cyclin A or specific CDK such as CDK2, may be used at different stages of the cell cycle: It is the binding of a given cyclin to a specific CDK that defines the pair's actions in the cell cycle.

The role of cyclin D in activating either CDK4 or CDK6 at the G1 phase of the cell cycle is among the best-understood cyclin/CDK interactions. As noted, some CDKs can bind to more than one cyclin.

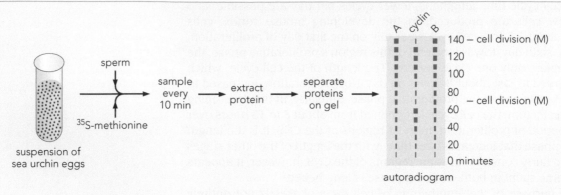

Figure 5.6 Cyclin proteins were identified in sea urchin eggs. Protein synthesis was measured from extracts of fertilized sea urchin eggs radiolabeled with ^{35}S-methionine. Samples were harvested at 10-minute intervals following fertilization and these extracts were then run on a gel. The density of the protein band represents the amount of protein synthesis. Most proteins gradually increased the amount of radiolabel, as shown by proteins A and B, indicating a linear increase in protein synthesis. However, one band showed a cyclical pattern of expression in which there was a gradual increase in radiolabel followed by a sudden decrease in signal intensity. This decrease always occurred at the mitotic phase (M) of the cell cycle. These proteins were named cyclins.

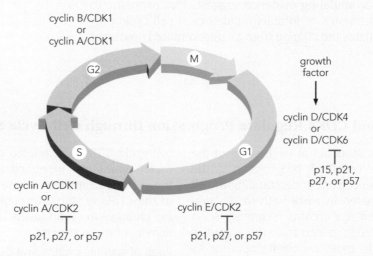

Figure 5.7 Cyclins and cyclin-dependent kinases regulate progression through the cell cycle. Cyclins are proteins that increase through the various stages of the cell cycle, then decrease after mitosis. Cyclin-dependent kinases (CDKs) are serine/threonine kinases that are activated by the cyclins to phosphorylate downstream proteins needed to drive the cell to the next phase of the cell cycle. At each cell cycle phase, specific cyclin/CDK pairs are active. Growth factors activate cyclin D and CDK4 or cyclin D and CDK6 to initiate the G1 phase of the cell cycle. The cyclin E/CDK2 pair drives the cell to the S phase. Cyclin A, coupled with CDK1 or CDK2, pushes the cell toward G2, from which cyclin B/CDK1 or cyclin A/CDK1 drives the cell to mitosis (M). The CDK inhibitory proteins p15, p21, p27, and p57 can stop the progression of the cell cycle at different phases. All four of these CDK inhibitory proteins can stop the activity of CDK4 or CDK6 associated with cyclin D in the G1 phase. In contrast, p21, p27, and p57 can halt the activity of CDK2 associated with cyclin E or cyclin A at the G1 and S phases of the cell cycle.

However, CDK4 and CDK6 only interact with cyclin D, and the pairing of cyclin D with one of these CDKs is crucial for regulating cell proliferation. Cyclin D also appears to be unique in that it can be expressed continuously, as long as a growth factor is present to stimulate cyclin D synthesis. The specific growth factor required to stimulate cyclin D expression depends on the cell type and the stage of embryonic development.

The G1 stage is also impacted by another important protein, the **retinoblastoma protein (Rb)**. Under normal conditions, Rb is unphosphorylated early in the G1 stage of the cell cycle. In the unphosphorylated state, Rb represses the EF2 transcription factor needed to progress to the S phase. However, as the cell cycle begins, Rb is phosphorylated, first by cyclin D/CDK4 or cyclin D/CDK6, and then by cyclin E/CDK2. The phosphorylated Rb no longer represses EF2, and the cell cycle continues. At the M-to-G1 transition, phosphatases return Rb to the unphosphorylated state. Thus, the repression of EF2 resumes and progression to the next S phase is halted unless Rb is again phosphorylated (**Figure 5.8**). Although Rb is often referred to as a single protein, there are actually

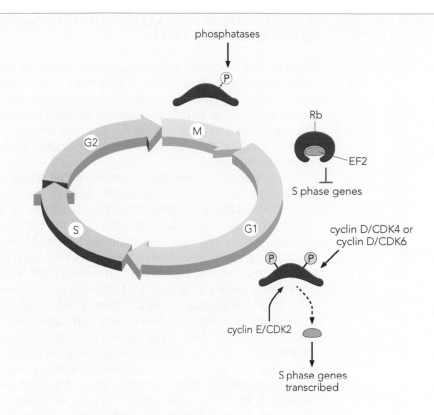

Figure 5.8 The retinoblastoma protein is used to modulate cell cycle activity. In the G1 phase of the cell cycle, retinoblastoma protein (Rb) is unphosphorylated and represses the activity of the EF2 transcription factor that is required for a cell to progress to the S phase of the cell cycle. As the cell cycle begins, Rb is phosphorylated first by cyclin D paired with either CDK4 or CDK6, then by cyclin E/CDK2. Phosphorylated Rb no longer represses EF2, so the S-phase genes are transcribed and the cell progresses to the S phase of the cell cycle. In the M phase, phosphatases remove the phosphates and Rb returns to the unphosphorylated state, thus allowing it to again repress EF2 in the G1 phase. Mutations in *Rb* lead to excess cell production and tumor formation.

several members of the Rb family, and each binds to specific transcription factors.

The importance of Rb regulation becomes apparent when the *Rb* gene is mutated. Mutations in *Rb* lead to continued cell proliferation, and thus tumor formation. *Rb* was first identified and named due to its role in a childhood retinal tumor. Children carrying two copies of the *Rb* mutation develop multiple retinal tumors in both eyes.

Additional Proteins Influence Cell Cycle Progression and Arrest

In all regions of the developing nervous system, cell proliferation must continue until the proper number of neural cells is generated. Cell cycle progression must also be inhibited at some point to prevent excess cell production, because too many neurons can be as disruptive to neural function as too few. Cell cycle progression can be halted by the action of **CDK inhibitory proteins**, a group of small proteins that block CDK activity. These CDK inhibitors include proteins such as p15, p21, p27, and p57, with the number indicating the

molecular weight of the protein. All these CDK inhibitors have the ability to inhibit the activity of CDK4 or CDK6 associated with cyclin D. In addition, p21, p27, and p57 inhibit CDK2 associated with cyclin A and cyclin E (Figure 5.7). As discussed further in Chapter 6, these CDK inhibitors not only negatively regulate cell proliferation, but often link exit from the cell cycle with the onset of cellular differentiation.

The final number of neurons is achieved not only by regulating cell proliferation, but also through the process of naturally occurring, programmed cell death that takes place in the developing nervous system, a topic covered in Chapter 8.

Although this description of cyclin/CDK pairs and substrate proteins provides only a simplified overview, it introduces the complex interactions needed to regulate the correct number of cells. The specific activities of cyclins, CDKs, Rb, CDK inhibitors, and various interacting proteins continue to be investigated, both to shed light on the basic mechanisms of cell division during development and to uncover the cellular pathways that control tumor formation.

CELLULAR MIGRATION IN THE CENTRAL NERVOUS SYSTEM

Cells do not arbitrarily settle in a nearby region of the expanding neural tube. Rather, they respond to a variety of intracellular and extracellular cues to reach specific destinations that may be temporary or permanent. In the CNS, some neural progenitors migrate relatively short distances, such as those that form cortical layers in the developing forebrain. Others migrate longer distances, such as the cells that form the granule cell layer of the cerebellum.

The migration patterns of neurons in the mammalian neocortex and cerebellum are currently among the best characterized and most fully understood in the developing CNS. Therefore, these areas are highlighted to demonstrate similarities and differences in the cell migration mechanisms utilized. The cell migration mechanisms described here and the interkinetic nuclear migrations described earlier are entirely different. This section of the chapter describes the migration of progenitors that have already undergone interkinetic nuclear migration and subsequent cell division.

Neuronal migration, like all forms of cell motility, depends on dynamic changes in the cytoskeletal elements, particularly actin and microtubules. To move forward, cells must adhere to a substrate, such as another cell or the extracellular matrix. The attachment is necessary to generate the force required for the cell to advance. Various proteins are used to link the cytoskeleton of the migrating cell to the substrate. The migration of neurons is like the advancement of the growth cone found at the tip of an extending axon or dendrite (see Chapter 7). Both neuronal migration and growth cone extension rely on extracellular signals that promote passive growth or provide instructive guidance cues. In many cases these extracellular signals initiate activity of intracellular Rho GTPase signaling pathways. The Rho family of small GTPases can activate effector proteins that influence actin and microtubule dynamics and cell adhesion. GTPases are activated by various guanine nucleotide exchange factors (GEFs) that convert GDP to GTP. GTPase activity is stopped by GTPase activating proteins (GAPS) that return Rho GTPase to the inactive GDP-bound form (**Figure 5.9**). Due to the ability of the activated Rho family of GTPases to influence cytoskeletal dynamics and cell adhesion, this intracellular pathway is often essential for neuronal migration.

In the Neocortex, Newly Generated Neurons Form Transient Layers

The adult mammalian neocortex is comprised of six cell layers (**Figure 5.10**), each of which is characterized by distinct cellular morphologies and projection patterns. Most neurons in the neocortex are excitatory projection neurons. The projection neurons in the deeper layers (layers V and VI) send axons out of the cortex to targets such as the spinal cord and

Figure 5.9 Rho family GTPases influence cell migration and adhesion. The Rho family of small GTPases are intracellular signals that are activated when guanine exchange factors (GEFs) convert GDP to GTP. The activated Rho GTPases then influence various effector proteins, such as actin and microtubules, as well as cell adhesion molecules. The activity of Rho GTPases is stopped when GTPase activating proteins (GAPs) remove phosphate (Pi) and return the GTPases to the inactive GDP-bound state. Due to the ability of Rho GTPases to influence cytoskeletal dynamics and cell adhesion, these GTPases are often activated during neuronal migration.

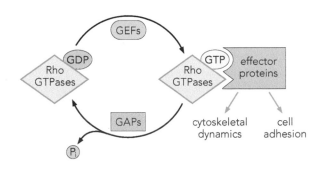

thalamus. In contrast, in the more superficial layers (layers II–IV) most projection neurons extend axons to other cortical regions, including those on the contralateral side of the brain. The projection neurons in each layer arise from the progenitor pool found at the ventricular surface.

Before these six layers arise, transient layers are formed that help direct migrating cells to the correct location and guide axons arriving from other CNS regions. In mice, these transient cortical layers form between embryonic days 11 and 18. In the developing human cerebral cortex, layering begins at about week six of gestation and continues through approximately week 20.

As early as the mid-1800s, scientists began detailing the layers of the neocortex in both adult and embryonic tissues. As shown in **Figure 5.11**, distinct embryonic layers progressively form over time. Initially, the **ventricular zone** (**VZ**) consists of a simple pseudostratified neuroepithelial region with only a single layer of cells along the lumen of the neural tube (Figure 5.11A). However, as cells in the VZ undergo asymmetric cell division and some begin to migrate away, two layers arise. The VZ persists and the **preplate** (**PP**) forms adjacent to the pial surface (Figure 5.11B). The PP therefore contains the first neurons to migrate out of the VZ. As more cells leave the VZ, the cortex begins to thicken. Cells leaving the VZ migrate through an intermediate zone (IZ) that primarily contains axons arriving from neurons located in other regions of the developing nervous system, such as the thalamus. Eventually a new cell layer called the **cortical plate** (**CP**) is formed. With the formation of the CP, the cells of the former PP separate into two layers: the **marginal zone** (**MZ**) below the pial surface and the **subplate** (**SP**) adjacent to the IZ. Some subplate neurons are also found among the axons of the IZ (Figure 5.11C). Thus, the first cells that exited the VZ are now found in the MZ and SP, while the CP becomes the destination for most subsequent migrating progenitor cells.

Each of these early formed layers of the neocortex—the MZ, CP, and SP—exhibit unique cellular characteristics. The MZ, which later forms layer I of the adult cortex, contains specialized, horizontally oriented cells named for Cajal and Gustaf Retzius, the two scientists who first described them in the 1890s. The **Cajal–Retzius** (**CR**) **cells** are thought to play a major role in neuronal patterning in the CP. The neurons migrating into the CP are primarily the pyramidal excitatory projection neurons of the neocortex.

The subplate (SP) is largely a transient layer of neurons (Figure 5.11C–E) although some SP neurons appear to persist as white matter neurons in adult rodents. During development, SP neurons in the IZ act as temporary guides and contacts for axons that arrive before the CP cells are present. This has been particularly well documented in the visual system. For example, in the cat, axons of cells arriving from the lateral geniculate nucleus (LGN) of the thalamus wait in the SP region of the IZ for about three weeks before their cortical target neurons reach the developing CP. Thus, the SP cells seem to attract and maintain the axons of LGN neurons until the correct target neurons are available.

In addition to the VZ, a second layer of proliferative cells forms between the VZ and the IZ. This **subventricular zone** (**SVZ**) lies adjacent to the VZ, just one layer further from the lumen of the neural tube (Figure 5.11D, E). Some of the cells migrating out of the VZ settle in the SVZ, where they continue to divide. The VZ eventually disappears as more and more neurons exit that layer, and the SVZ then becomes a primary source of additional neurons and most of the glial cells of the cortex. Proliferation continues in the SVZ through the middle and late stages of cortical development. In the mouse, this proliferation continues into the second postnatal week, whereas in humans the proliferation continues at least

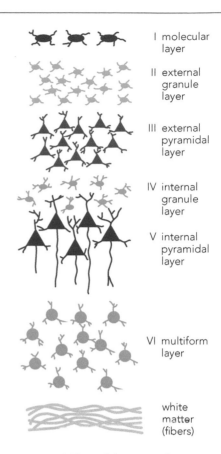

Figure 5.10 The adult mammalian neocortex is organized into six layers composed of different cell types, each with distinct projection patterns. Numbered consecutively from the pial (basal) surface to the ventricular (apical) surface are the molecular layer (layer I), the external granule layer (layer II) and external pyramidal layer (layer III), the internal granule layer (layer IV), internal pyramidal layer (layer V), and the multiform layer (layer VI). The fibers of the white matter are located beneath layer VI.

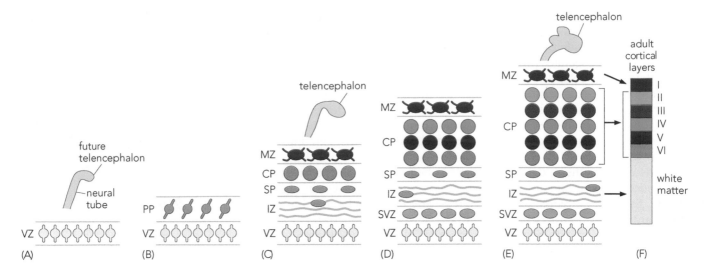

Figure 5.11 Transient embryonic layers are formed prior to the adult layers. Transient cortical layers that provide guidance cues to migrating neurons and incoming axons are illustrated in panels A–E. Above panels A, C, and E, the corresponding stage of neural tube development is shown. (A) Initially the neural tube contains only the ventricular zone (VZ), consisting of a single layer of cells along the lumen of the ventricle. (B) As these proliferative cells divide, some daughter cells begin to migrate away and form a second layer called the preplate (PP). (C) With more and more cells migrating away from the VZ, the PP expands and generates a new layer called the cortical plate (CP). The formation of the CP splits the former PP into a marginal zone (MZ) closest to the pial surface and a subplate (SP) zone adjacent to the intermediate zone (IZ). The IZ is a layer that contains the axons of neurons originating in other areas of the developing nervous system as well as some subplate neurons. (D and E) As development continues, some cells from the VZ settle in a second zone of proliferation called the subventricular zone (SVZ), which lies adjacent to the VZ. Other cells from the VZ migrate through the SVZ, IZ, and SP to reach the CP, where they form additional layers. Ultimately, five layers of cells are found in the CP region. (F) The transient layers of the developing cerebral cortex give rise to different regions of the adult cortex. The MZ will go on to form layer I of the adult cerebral cortex, while the CP will give rise to neurons of the other five layers (II–VI). The IZ will contribute to the white matter in the adult. Except for the SVZ that remains in some regions of the adult CNS, the other transient layers are lost as neuronal migration ends.

through the second year of life. Most areas of the SVZ will decrease in size as proliferation ceases. However, the SVZ persists into adulthood in the region of the lateral ventricles. In rodents, SVZ progenitor cells migrate to populate regions of the olfactory bulb and may contribute to olfactory function throughout life. The SVZ may also contribute new progenitors to the cerebral cortex after brain injury (**Box 5.2**).

In primates, the SVZ is clearly divided into an inner SVZ (iSVZ) and outer SVZ (oSVZ) separated by a thin layer of fibers. Though not as prominent, these divisions also appear in other mammals. The oSVZ zone forms during mid-stages of neurogenesis and a subtype of basal progenitors, called the outer radial glia or basal RG, are found here. Outer/basal RG divide either symmetrically to produce more outer/basal RG or asymmetrically to produce another outer/basal RG and an IP. Thus, outer/basal RG greatly increase the number of cells generated from a single RG mother cell. In addition, the basal processes of outer/basal RG serve as additional guides for neurons migrating to the CP.

The number of outer/basal RG in the oSVZ is thought to be related to the extent of gyrification (folding) of the cortex. Thus, in species such as primates and ferrets, outer/basal RG are more abundant and produce more IP cells than in species with less gyrification, such as mice. This increased capacity for proliferation is thought to be linked to the evolutionary increase in neocortical size.

Apart from the MZ that becomes layer I of the mature cerebral cortex, none of the other transient layers persist as adult cortical layers. However, as described in the next section, the cells that migrate into the CP ultimately form layers II through VI of the adult cortex (Figure 5.11F) and axons of the IZ contribute to the white matter that lies below the cortical layers.

Box 5.2 A Neurologist's Perspective: Advances in Stroke Treatment

Daniel Kinem. D.O.

Dr. Daniel Kinem majored in computer science at Edinboro University before attending medical school at the Lake Erie College of Osteopathic Medicine. He completed his residency in neurology at the University of Pittsburgh Medical Center Hamot where he now practices and trains current residents.

A cerebrovascular accident, commonly referred to as a stroke, is caused by a sudden interruption of blood supply to an area of the brain. Depending on the area of damage, the person experiencing a stroke may have sudden deficits in speech, vision, sensation, or strength. The deficits can be devastating, and permanent injury to the brain can begin after just a few minutes. However, early treatment may offer some patients a chance at more complete recovery.

Strokes can be either hemorrhagic or ischemic. Hemorrhagic strokes are caused by the *rupture* of one of the blood vessels within the brain. Often, there is immediate and permanent injury because the hemorrhaging blood damages the delicate brain tissue. Care for the patient revolves around preserving life and preventing any worsening or complications. Ischemic strokes result from *blockage* of one of the blood vessels to or within the brain. With ischemic strokes, restoring blood flow to the brain can help to prevent or limit damage to surrounding brain tissue. The area of brain tissue surrounding the infarct (the area of damaged tissue) is called the penumbra. This area may not yet be permanently damaged and therefore if we can restore blood flow in a timely manner, that area of brain can be saved.

The acute treatment of ischemic strokes largely involves attempts to "break up" the blockage in the blood vessel. The blockage is referred to as a clot or thrombus. If the patient arrives at the hospital within a specific window of time, neurologists may be able to administer medications, known as thrombolytics, which can help to dissolve the thrombus. These "clot busting drugs" became the standard of stroke care in 1995 when the results of a large, randomized trial were released from the National Institute of Neurological Disorders. This study showed that using thrombolytic medications in patients experiencing an ischemic stroke would help about 30% of the patients achieve better outcomes. The drugs must be administered within a few hours after stroke to be helpful because after about 3 hours, there is an increased risk that the medications will cause bleeding in the brain. This is called a hemorrhagic transformation. Thus, over time, the risks of administering a thrombolytic medication are greater than any potential benefit.

Some patients suffer very large, devastating strokes caused by bigger clots blocking the larger blood vessels in the brain. Because thrombolytic medications are less effective at breaking up these large clots, additional treatments were needed. From 2010 to 2014 several large clinical trials evaluated endovascular intervention for stroke. Endovascular intervention involves inserting a guide wire into a person's artery and advancing it until it reaches the clot in the brain. Once the guide wire is near the clot, an attached, very small mechanical apparatus is used to grab and pull out the clot. This procedure is known as a thrombectomy and the physician doing the procedure is known as a neurointerventionalist. These highly skilled physicians use live-action x-rays, known as fluoroscopy, to help guide clot retrieval.

The results of the thrombectomy trials were very encouraging and showed two things; the procedure was safe, and the procedure was more effective than thrombolytics alone. It also opened new possibilities for extending the treatment window, as thrombolytic medications are only beneficial within a few hours of the stroke onset.

Two trials designed to test the effectiveness of endovascular treatment with a time window up to 24 hours were reported in 2018. These landmark trials showed that certain patients with large strokes benefited from thrombectomy up to a day after they started to experience symptoms. These results have revolutionized stroke treatment. With proper patient selection, the treatment is incredibly efficacious. For a group of patients who previously had no treatment options, we can now save one out of every three from death or long-term care. With endovascular treatment, doctors are no longer bound to the 3-hour time limit for these patients.

Neurologists and interventionalists hope to extend the stroke treatment window further and are investigating new techniques and technologies. Basic scientists continue to explore the cellular changes that occur in the penumbra soon after a stroke occurs. Animal models of stroke have identified several proteins that are upregulated after stroke, including those associated with neurodegeneration and neuroprotection. For example, proteins associated with apoptosis (programmed cell death) are altered soon after a stroke, and the balance of pro-apoptotic and anti-apoptotic proteins (see Chapter 8) appears to influence the extent of neurodegeneration. Other proteins such as those involved in the Notch and Wnt/β–catenin signaling pathways are also altered in the penumbra. Several studies have focused on the Wnt/β–catenin pathway (see Figure 4.19) as its activation after an ischemic stroke, particularly in endothelial cells, appears to *decrease* the risk of thrombolytic-induced hemorrhagic transformation.

Wnt and Notch signaling are also associated with increased proliferation of neural stem cells in the subventricular zone (SVZ). Signals arising from damaged blood vessels may induce proliferation and migration of neural and glial progenitors to the penumbra soon after stroke. Although only a small percentage of these progenitors appear to survive and mature, mice with higher survival and maturation rates had better behavioral outcomes. Whether these pathways can be exploited to develop improved stroke therapies continues to be explored.

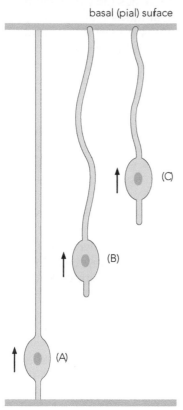

basal (pial) suface

apical (ventricular) suface

Figure 5.12 Somal translocation is one form of radial migration. Neurons generated during the earliest stages of cerebral cortical development use somal translocation to pull the cell body toward the pial surface. (A) A cell in the ventricular zone extends a process to the pial surface. (B) The attachment at the ventricular surface is lost and the cell body is pulled toward the pial surface as it migrates through the remaining process (C) to reach the pial surface.

Most Neurons Travel along Radial Glial Cells to Reach the Cortical Plate

Studies reveal two primary mechanisms by which neurons reach the PP and CP: somal translocation and migration along radial glia cells. Both forms are a type of radial migration in which neurons travel away from the ventricular surface toward the pial surface. Many neurons that are generated early in development reach the PP and MZ by **somal translocation** meaning the cell body, once detached from the ventricular surface, moves toward the pial surface in response to turning and pulling of the basal process (**Figure 5.12**).

As the cortex continues to thicken, somal translocation is no longer used. Instead, later-generated neurons reach the CP primarily by traveling along RG cell processes. Thus, RG, the first cells to differentiate in the cerebral cortex, serve as both stem cells and scaffolds for the neurons that form subsequently. The hypothesis that RG direct neural migration was first proposed by His in 1889. Cajal's work supported this hypothesis and further suggested that the radial glia later differentiate into astrocytes. Only in recent decades, however, have scientists confirmed that astrocytes arise from RG. During the neurogenesis stage, promoters of astrocyte-specific genes are hypermethylated preventing progenitors from responding to glia-inducing cues present in the extracellular environment. Later, as the neurogenesis stage ends, neural genes are repressed and the transition to gliogenesis begins allowing for the formation of not only astrocytes, but oligodendrocytes and ependymal cells.

As noted, the cell bodies of apical RG reside in the VZ whereas those of the outer/basal RG reside in the SVZ. Each RG cell extends a long process to the pial surface. A newly generated neuron attaches to the vertically oriented process of a RG cell and crawls along this process toward the CP (**Figure 5.13**). Recent studies suggest that the SP neurons provide signals for the transiently multipolar IPs to remodel their cytoskeleton and re-adopt the bipolar morphology needed to adhere to and migrate along radial glia processes. One proposed model suggests calcium influx through synapse-like connections between the processes of SP neurons and IPs directs this cytoskeletal reorganization.

As a neuron reaches the CP region, it enlarges, the soma becomes more globular in appearance, and the strength of its attachment to the RG process decreases, allowing the neuron to settle into its layer and begin to make synaptic connections.

Migrating neurons adhere to RG through protein–protein interactions. These include the adherens junctions and glycoproteins such as **astrotactin**, which localizes to the neuronal cell surface and interacts with corresponding proteins found on the RG. Astrotactin was first characterized in the cerebellum by Mary E. Hatten and colleagues in 1988. In the cerebellum, astrotactin also facilitates neuronal migration on specialized RG cells called Bergmann glia, as detailed later in this chapter.

Other molecules, particularly receptors of the integrin family, are expressed on the emerging cortical neurons to mediate their attachment and migration along the glial fibers. Integrin receptors are

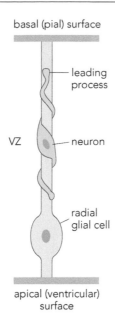

basal (pial) surface

leading process

VZ — neuron

radial glial cell

apical (ventricular) surface

Figure 5.13 Neurons also migrate out of the ventricular zone by attaching to radial glial cells. The cell bodies of radial glial (RG) cells are found in the ventricular zone (VZ). Each RG cell (yellow) extends a long process to reach the pial surface. Neurons (blue) migrating out of the VZ attach to the RG and crawl along the RG process. The neuron's leading process is used to help guide the cell toward the pial surface. Proteins present on the neurons and RG mediate the attachment, migration, and release of the neurons at the proper stage of development.

heterodimers comprised of two subunits, designated α and β. There are several different α and β subunits and the particular combination of subunits determines binding specificity. For example, the α3β1 receptor combination binds to the extracellular matrix protein laminin. During cortical development, the β1 subunit is highly expressed along with the α3, α5, or α6 subunits. These subunit combinations form receptors for the widely distributed extracellular matrix proteins laminin and fibronectin (**Table 5.1**). The expression of different integrin receptor subunits on migrating neurons allows the cells to distinguish among the available extracellular proteins. In some regions, integrin receptors are needed for the neurons to attach to RG cells. To test the necessity of integrin receptors in cortical migration, subunits of the these receptors were experimentally manipulated. For some studies, mutant mice lacking one or more subunits were generated. For others, antibodies that block the function of a specific integrin subunit were added to cultures of embryonic cortical neurons. These various studies confirmed that removing or decreasing functional integrin subunits interfered with a cortical neuron's ability to migrate normally. Thus, proper integrin receptor expression is necessary for developing cortical neurons to recognize the appropriate cues on RG and other cell surfaces.

Table 5.1 Examples of Selected Integrin Subunit Pairs That Bind to Various Extracellular Matrix Proteins Encountered by Neurons *In Vivo*.

Integrin Receptor Subunits	Extracellular Matrix Molecule
α3β1	Laminin
α5 β1	Fibronectin
α6 β1	Laminin
α7 β1	Laminin
α6 β4	Laminin
α1 β1	Laminin, Collagen
α2 β1	Laminin, Collagen

Source: Based on Reichert LF & Tomaselli KJ [1991] *Annu Rev Neurosci* 14:531–570.

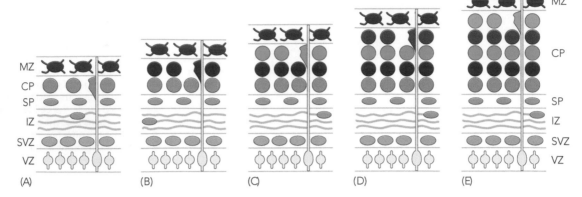

Figure 5.14 Cerebral cortical neurons rely on inside-out patterning to form layers. (A) As neurons leave the ventricular zone (VZ) by attaching to radial glial cells (RG, yellow), they begin to accumulate in the cortical plate (CP), detaching from the RG prior to reaching the marginal zone (MZ). The first cells to arrive at the CP will form the deepest, innermost layer (future layer VI). (B) The next layer of neurons (future layer V) will migrate past the first layer to form a new layer closer to the pial surface. (C) The third layer of cells (future layer IV) will migrate past the first and second layers. Similarly, the next layers (future layers III and II) will continue to migrate past existing layers until all six cortical layers have formed using this inside first, outside last pattern (D and E).

Cells in the Cortical Plate Are Layered in an Inside-Out Pattern

As cells migrate out of the VZ and SVZ to form layers in the CP, an interesting developmental pattern emerges known as "inside-out" or "inside first, outside last" patterning, because cells settle in the innermost layers of the CP prior to the outermost layers—a pattern that at first may seem counterintuitive. To form the layers of the CP, the first neurons to exit the cell cycle and leave the VZ settle in what will become the deepest layer of the CP, the future layer VI of the adult cerebral cortex (**Figure 5.14**A). As development continues, later-born cells divide and migrate past these earlier-born neurons to settle in a more superficial layer of the CP (Figure 5.14B–D). This pattern continues until the latest-born neurons originating in the SVZ reside in the outermost layer of the CP, the future layer II of the adult cerebral cortex, closest to the MZ (Figure 5.14E). As neurons arrive in the deepest layers (that is, future layers VI and V), they begin to form connections with other neurons while newly generated cells migrate to more superficial layers (for example, future layers III and II). Thus, multiple developmental events overlap each other, and the nervous system integrates these events simultaneously.

Inside-out patterning in the cerebral cortex was first reported in mice by Angelvine and Sidman. In their 1961 study they used tritiated thymidine to map cellular locations at different developmental stages in the developing rodent brain. Numerous studies throughout the 1960s and 1970s helped further define the temporal sequence of events. For example, in 1965 Berry and Rogers confirmed this patterning in rats and in 1974 Rakic reported a similar migration pattern in primates. Thus, inside-out patterning is conserved across mammalian species. However, in other vertebrates, such as reptiles, the pattern of neuronal migration is reversed and occurs in an "outside-in" manner. That is, the first cells to exit the VZ settle in the most superficial layers.

The Reeler Mutation Displays an Inverted Cell Migration Pattern

In the 1950s, mice were described with a naturally occurring mutation that results in an easily observable gait disorder. The mutant mice are called reeler mice because they move in an unsteady, staggering fashion. The reeler mutation is a recessive autosomal mutation only found in mice. Among the structural changes noted are defects in cell migration patterns in the cerebral cortex, hippocampus, and cerebellum.

In the cerebral cortex, the reeler mutation interferes with the normal association between migrating neurons and the RG. At early stages, cells in reeler mice form, undergo terminal mitosis, and begin to migrate out of the VZ—the same sequence observed in wild-type mice (**Figure 5.16**A, B). However, the cells do not migrate into the PP to form the CP and SP. Instead, cells that would normally form the SP are found in the MZ with the CR cells. Together these cells form a unique layer called the **superplate** (Figure 5.16C, D). As additional neurons migrate out of the VZ, a disorganized CP forms with these later-migrating neurons unable to move past the existing layers. Thus, the newer layers continue to stack up, one below another (Figure 5.16C, D). The result is an "outside-in" pattern of neuronal migration, in which the first cells, those that become layer VI, migrate until they are stopped in the superplate. Subsequent cells stack up below this first layer so that layers V–II are arranged in an inverted pattern.

In reeler mice, even though the cortical layering is inverted, the overall gross organization of the brain appears relatively normal, and cell survival is not impacted. Further, cells of the peripheral nervous system and other organ systems appear unaffected, indicating that the reeler mutation impacts specific cellular interactions in the developing CNS. No known human homolog of this specific genetic mutation has been identified, although the clinical syndromes described in **Box 5.3** also have altered cell migration patterns.

Box 5.3 Changes in Cortical Migration Patterns Lead to Clinical Syndromes in Humans

It is unclear why mammalian cerebral cortical neurons rely on an inside-out pattern of migration. Evolutionary change in the complexity of the cerebral cortex is one hypothesis. The early-born neurons that reach the preplate (PP) and deep layers of the cortical plate (CP) are thought to be phylogenetically older than those that migrate to the more superficial layers of the (CP). Whatever the underlying basis for this pattern of cell migration, inside-out patterning is necessary for normal development. Defects in cortical migration patterns cause congenital disorders in humans. Disruptions of the cortical migration patterns can result in mental retardation, seizures, and in some cases death. Some defects in neuronal migration result in **lissencephaly** ("smooth brain"), a condition in which the outer surface of the brain lacks the typical cortical gyri (ridges) and sulci (furrows). Rather than exhibiting an inside-out pattern of migration, the neurons in a lissencephalic brain stack one below another (**Figure 5.15A**). There are several genetic mutations that can cause lissencephaly, including mutations in the gene that codes for the protein Reelin. The clinical outcomes associated with lissencephaly are quite varied. Some children demonstrate near-typical developmental milestones and intelligence. Others have severe mental retardation, suffer frequent seizures, and die within the first decade of life.

Another human neurological disorder resulting from altered neuronal migration is **periventricular heterotopia (PH)**. In this condition, nodules of neurons form throughout the periventricular region because these cells never leave the ventricular zone to migrate to the appropriate cortical layer. The nodules form in bilaterally symmetrical patterns and appear to be highly differentiated neurons located in the wrong regions of the brain (Figure 5.15B). The unusual pattern of cell formation results in a range of deficits. Many individuals experience epileptic seizures but exhibit no other neurological symptoms. In a less common form, PH is caused by an X-linked dominant mutation. The deficits in males that inherit this mutation are much more severe because males have only one copy of the X chromosome and are therefore unable to compensate for the mutation. Many of these males die before birth due to the severity of the malformations that arise. Changes in proteins known to have roles in cell motility, migration, and adhesion are among those thought to underlie these clinical disorders.

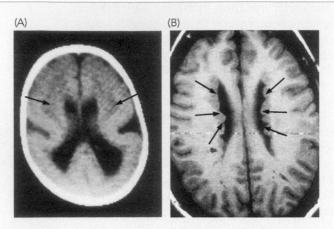

(A)   (B)

Figure 5.15 Defects in neuronal migration results in congenital disorders in humans. (A) A magnetic resonance image (MRI) illustrating the smooth brain appearance characteristic of lissencephaly. Compared to a typical brain or a brain with periventricular heterotopia (PH), the outer surface of the brain appears smooth and lacks the characteristic gyri and sulci. In addition, the ventricles are enlarged and some areas of misplaced neurons are observed in the white matter (arrows). (B) A brain scan illustrating PH. The arrows indicate nodules of neurons that cluster along the ventricles. [From Gressens P [2000] *Pediatric Res* 48:725–730.]

Cajal–Retzius Cells Release the Protein Reelin, a Stop Signal for Migrating Neurons

Because RG function primarily as a passive system for neuronal attachment and migration, additional signals must be available to direct migrating cells to detach from RG and settle in the proper cortical layer. As neurons reach the CP, the RG surface changes and the axonal fibers of neurons already in the CP release extracellular signals to promote detachment from the RG. In addition, the CR cells adjacent to the CP provide a signal to halt migration.

Studies in the reeler mutant mouse first implied that the CR cells are needed to end migration. Unlike the rest of the cells in the emerging cortex, CR cells have a horizontal orientation with cell bodies and dendritic processes parallel to the pial surface. Under normal conditions, as new neurons migrate into the CP, the MZ and its CR cells are pushed further outward and the cerebral cortex thickens. Thus, the CR cells are always in the outermost layer, in a position to provide a stop signal to migrating cells.

In 1995 two teams of researchers succeeded in identifying a gene and protein expressed in CR cells. The gene codes for an extracellular glycoprotein that was termed **Reelin**. Mice lacking *Reelin* have the same phenotype as observed with the reeler mutation. Without the Reelin signal, neurons migrate as far as they can, so that early-migrating neurons reach the outermost pial surface before they detach. Subsequent neurons migrate until they are stopped by an existing layer of neurons, resulting in an "outside-in" pattern of neuronal layers (**Figure 5.16**).

Subsequent studies suggested that the role of Reelin is more complex. For example, Reelin may provide both "stop" and "go" signals to the migrating neurons, depending on the timing and location of expression. In some instances Reelin appears more effective in mediating somal translocation than radial glia-guided migration. Reelin has also been shown to have other functions, including orienting and stabilizing the cytoskeletal elements of neural processes. Reelin appears to attract the leading process of a migrating neuron and help attach, or anchor, that process to the MZ and may also provide a signal for dendrites to anchor to the extracellular matrix of the MZ. Further, Reelin may influence the maturation of the dendritic processes and their spines, again by interacting with cytoskeletal proteins.

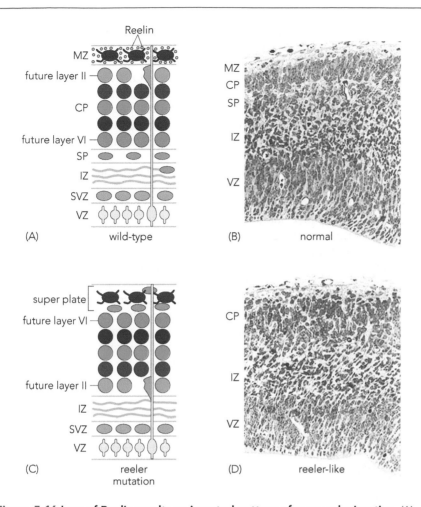

Figure 5.16 Loss of Reelin results an inverted pattern of neuronal migration. (A) A schematic diagram illustrates that in wild-type mice, the Cajal–Retzius (CR) cells of the marginal zone (MZ) produce the protein Reelin, which acts as a stop signal to prevent neurons of the cortical plate (CP) from entering the MZ. The Reelin signal instructs neurons to leave the RG (yellow) and settle in the correct layer of the expanding CP. Because the expanding cortical plate pushes the MZ, and thus the Reelin signal, outward, the newly arriving neurons are layered in the inside first, outside last pattern. (B) Photomicrograph of the cerebral cortex of a normal, wild-type mouse. (C) The normal "inside-out" pattern of neuronal migration is reversed in mice with the reeler mutation. In these mice the CR do not produce Reelin, so the cells from the ventricular zone (VZ) and subventricular zone (SVZ) continue to migrate as far as they can toward the pial surface. The cells of the subplate become mixed with the CR in the MZ, forming a layer called the superplate. Neurons that normally form the CP then migrate out of the VZ and travel as far as they can until existing layers of neurons prevent further migration. Therefore, the first cells to migrate out of the VZ (future layer VI) migrate until they are stopped at the superplate. Neurons of future layer V then migrate until they are stopped by this layer of neurons. This pattern continues, with each subsequent layer stacking up beneath the existing layers. Thus, in the absence of Reelin, CP layers are organized in an "outside first, inside last" pattern, an orientation opposite of that found in the wild-type mice. (D) A photomicrograph of the cerebral cortex reveals disorganized MZ and CP regions in the absence of Reelin signaling. [Adapted from Tissir F & Goffinet AM [2003] *Nat Rev Neurosci* 4:496–505.]

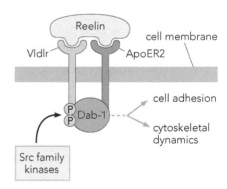

Figure 5.17 The Reelin signaling pathway influences cell adhesion and cytoskeletal dynamics. The Reelin protein binds to two low-density lipoprotein receptors, called the very-low-density lipoprotein receptor (Vldlr) and apolipoprotein E receptor 2 (ApoER2; also called low density lipoprotein 8, LRP8). To influence cell adhesion or cytoskeletal dynamics, Reelin is first internalized by the receptors. This leads to activation of the Src (sarcoma) family of kinases that in turn phosphorylate and activate the adaptor protein Disabled-1 (Dab-1). Dab-1 then activates signaling pathways that influence the expression of cell adhesion molecules that help the migrating neurons attach to the Cajal–Retzius (CR) cells. Dab-1 also influences cytoskeletal dynamics in migrating neurons to help attract the leading edge of migrating neurons to the marginal zone.

Reelin exerts its effects by binding to two low-density lipoprotein receptors: the very-low-density lipoprotein receptor (Vldlr) and apolipoprotein E receptor 2 (ApoER2; also called low-density lipoprotein 8, LRP8). These receptors are enriched in the processes of cells nearing the MZ. Reelin is internalized by the receptors, leading to the activation of Src (sarcoma) family kinases that phosphorylate, in turn, the adaptor protein Disabled-1 (Dab-1) that is attached to the intracellular domains of Vldlr and ApoER2 (**Figure 5.17**). In migrating cortical neurons, Dab-1 then activates signaling pathways that influence the expression of cell adhesion molecules such as neural cadherin (N-cadherin) to promote interactions between the migrating neurons and the CR cell.

Dab-1 activation also mediates cytoskeletal dynamics, providing a means for Reelin released by the CR cells to attract the leading edge of a migrating neuron to the MZ and position the cell properly (Figure 5.17). In the absence of Reelin, pyramidal cells in the neocortex and hippocampus lose the vertical orientation seen in wild-type mice and extend disoriented dendrites in several directions.

Mice lacking both Vldlr and ApoER2/LRP8 receptors have the same morphological and behavioral changes as those observed with *Reelin* and reeler mutations (Figure 5.17D). However, the lack of only Vldlr or ApoER2/LRP8 causes a milder phenotype. Studies suggest that the ApoER2/LRP8 receptor is more important for the layering of neurons in the neocortex, whereas Vldlr is more important for guiding neuronal migration in the cerebellum. These differences may reflect differences in how the receptors process Reelin upon internalization.

Reelin expression in vertebrates has increased over the course of evolution. Comparisons of *Reelin* expression in different vertebrate species found increased *Reelin* expression in the mammalian CNS, both in terms of the number of *Reelin*-expressing cells and the level of *Reelin* expressed in each cell. This suggests that Reelin may have an expanded role in the development of the mammalian nervous system.

Cortical Interneurons Reach Target Areas by Tangential Migration

It is now thought that radial migration patterns are primarily utilized by the glutamatergic, excitatory projection neurons (pyramidal neurons) of the cerebral cortex. These neurons arise from the VZ in the dorsal telencephalon (**pallium**) and comprise 70–80% of the neurons in the neocortex. In contrast, inhibitory interneurons, which are primarily GABAergic, contribute 20–30% of neocortical neurons. These cortical inhibitory interneurons arise from the ventrally located ganglionic eminences in the ventral telencephalon (**subpallium**). Ganglionic eminences are transient embryonic structures that give rise cortical interneurons and components of the basal ganglia and amygdala (Figure 5.18).

Whereas the radially migrating cortical projection neurons settle a short distance from their cite of origin in the VZ, the cortical interneurons from ganglionic eminences migrate tangentially over greater distances. **Tangential migration**, in which neurons follow routes parallel to the VZ, has also been observed in the diencephalon, brainstem, and spinal cord.

Figure 5.18 Cortical interneurons migrate tangentially in the neocortex. (A) Cortical interneurons originate from the medial ganglionic eminence (MGE, green) and caudal ganglionic eminence (CGE, blue) located outside the ventricular zone. These interneurons travel in a direction tangential (parallel) to the ventricular surface (green and blue arrows). (B) A coronal section through the embryonic telencephalon reveals the ventral location of the lateral and medial ganglionic eminences. Cells from the lateral ganglionic eminence (red) migrate to areas associated with the basal ganglia. Cells from the medial ganglionic eminence (green) travel along existing axons or cells to reach the neocortex. (C) The interneurons (green) from the medial ganglionic eminence migrate parallel to the ventricular zone (VZ) through the subventricular zone (SVZ), intermediate zone (IZ), or marginal zone (MZ). Tangential migration often overlaps temporally with the radial migration of other neurons (purple) along radial glia (yellow).

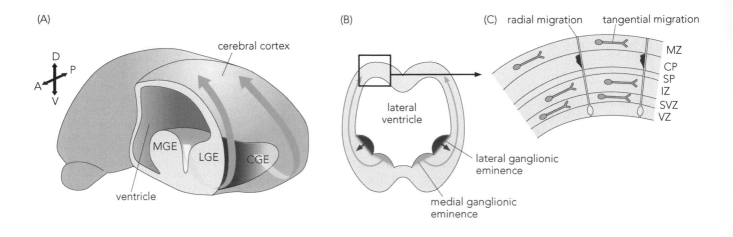

Tangential migration of cortical interneurons has been studied most extensively in mice, where approximately 70% of these neurons arise from the medial ganglionic eminence (MGE) and 30% arise from the caudal ganglionic eminence (CGE). In contrast, the lateral ganglionic eminence (LGE) appears to produce neurons that primarily migrate to the caudate and putamen (Figure 5.18A).

Originally considered a rare form of cell migration, it is now estimated that as many as 30% of the neurons in the mammalian CNS migrate tangentially. Cells migrating out of the MGE reach the neocortex then travel as two primary streams within the SVZ and MZ, with some also traveling in the IZ (Figure 5.18B, C). To reach their destinations, the interneurons respond to extracellular cues released by the cells they encounter. Early signals thought to promote migration include growth factors such as brain-derived neurotrophic factor (BDNF), hepatocyte-derived growth factor (HDF), and glial-derived neurotrophic factor (GDNF). Additional cues direct cell migration by either attracting neurons to the correct location or repelling the neurons away from an incorrect pathway. Among the signals identified as attractive signals are Neuregulin 1 (Nrg1), which binds to ErbB4 receptors on the interneurons, and the chemokine CXCL12 (C-X-C motif chemokine 12/stromal cell-derived factor 1), which binds to the CXCR4 (C-X-C chemokine receptor type 4) and CXCR7 receptors present on the interneurons. The interneurons are directed away from non-target regions by the repellent proteins of the semaphorin family. Thus, multiple attractive and repellent cues are used to guide tangentially migrating interneurons so that they enter the correct cortical layer.

In mice, most of the cortical interneurons from the MGE travel in streams along existing cells or axons to reach cortical regions around E12.5, the period of early neurogenesis and radial migration of projection neurons. Thus, the interneurons and projection neurons must coordinate their migration streams.

Several advances have been made in understanding how tangential and radial migration coordinate to form functional neural circuitry. For example, in a series of studies in the laboratory of Laurent Nguyen, it was noted that mouse cortical interneurons go through stages of steady migration and pausing. It is thought that the behaviors of tangentially migrating cells are linked to the production of neural progenitors in the VZ of the dorsal telencephalon. When the interneurons were manipulated to induce cytoskeletal changes that caused decreased pausing, more interneurons invaded the SVZ and MZ. At the same time, the cortical neural progenitors increased the number of neurons generated. Conversely, when interneurons were depleted, the number of cortical progenitors also decreased. These studies suggest that the cortical projection neurons and interneurons provide signals to ensure an appropriate complement of neurons and interneurons are generated. The importance of carefully balanced excitatory projection neurons and inhibitory interneurons is highlighted by the number of neurodevelopmental conditions, such as autistic spectrum disorders, intellectual disability, and schizophrenia, associated with altered interneuron migration.

Ongoing efforts seek to understand the specific mechanisms regulating migration patterns in various species. For example, in primates, there are conflicting reports on whether cortical inhibitory neurons arise primarily from the dorsal telencephalon or from ganglionic eminences and evidence suggests that at least some human inhibitory neurons arise in the VZ.

Cell Migration Patterns in the Cerebellum Reflect Its Distinctive Organization

Like the neocortex, the cerebellum has been used to study cell proliferation and migration for several decades. Some of the mechanisms and signaling molecules used in the cerebral cortex are also used in the cerebellar cortex.

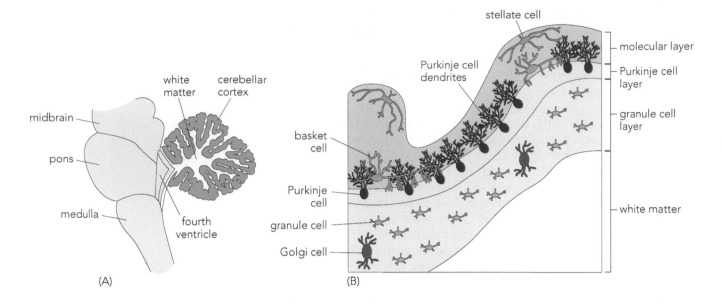

Figure 5.19 The adult cerebellar cortex is organized into the molecular, Purkinje cell, and granule cell layers. (A) The adult cerebellum is located dorsal to the pons and fourth ventricle. The cerebellar cortex is a highly convoluted structure consisting of three layers. (B) An enlarged region of the cerebellar cortex reveals the three adult layers. The molecular layer, which is the outermost of the cerebellum's three layers, primarily contains the elaborate dendritic trees of Purkinje cells. The middle, Purkinje cell layer, contains the Purkinje cell bodies. The granule cell layer is the innermost layer and contains the numerous, small granule neurons. Interneurons are also located in different layers of the adult cerebellum. These include the basket, stellate, and Golgi interneurons.

Cerebellar development also features many interesting and noteworthy differences. Part of the reason that the cerebellum has unique patterns of proliferation and migration is because the organization of the cerebellar cortex is quite different from that seen in the cerebral cortex.

The three layers of the adult cerebellar cortex include the outer **molecular layer**, the middle **Purkinje cell layer**, and the inner **granule cell layer** (**Figure 5.19**). In the mature cerebellum, the cell bodies of the Purkinje cells are found in the Purkinje cell layer, while their elaborate dendritic trees extend to form the primary component of the molecular layer. The granule neurons found in the granule cell layer are notable for their small size and incredible number. Granule cells are the most numerous neurons in the mammalian brain; in fact, they may constitute half of the total number of neurons in the brain of some species. Also dispersed within the adult cerebellar layers are various interneurons such as Golgi, stellate, and basket cells.

Cerebellar Neurons Arise from Two Zones of Proliferation

The cerebellum arises from the alar plate region in the caudal metencephalon. As described in Chapter 4, alar plates primarily give rise to sensory structures, whereas basal plates give rise to motor structures. Thus, the cerebellum is a unique derivative of the alar plate. In the caudal metencephalon, the alar plates expand in a dorsomedial direction, begin to cover the thin roof plate, and extend over the fourth ventricle. These expansions form the rhombic lips—transient structures at the edge of the fourth ventricle that extend along the posterior portion of the metencephalon (**Figure 5.20**A, B). The upper, or rostral, rhombic lip region produces cerebellar granule cells. In contrast, the lower, or caudal, rhombic lip region contributes cells of various pontine nuclei (Figure 5.20C). As the neural tube continues to develop and expand along the anterioposterior axis, the pontine flexure deepens and the rhombic lips gradually come together to form the cerebellar primordium, the precursor to the mature cerebellum (Figure 5.20D, E).

Like the cerebral cortex, at early developmental stages neuronal precursors are located in the VZ that surrounds the lumen of the neural tube—in this case the area that forms the fourth ventricle. The VZ is the site of production for Purkinje neurons, as well as the various interneurons. The Purkinje cells are the first cells produced in the cerebellum, and their progenitors migrate radially out of the VZ to establish the Purkinje cell layer (PCL; Figure 5.20D). Initially, the newly formed PCL consists of several irregular rows of Purkinje cells, but it gradually thins to the characteristic single row as the cerebellar cortex expands. The Purkinje cells are the target cells for the granule neurons that migrate to the internal granule cell layer by a different migratory pathway.

A second group of neuronal progenitors originate from the upper rhombic lips. These cells stream tangentially over the surface of the cerebellar plate to form the external granule cell layer (sometimes called the external germinal layer) that lies beneath the pial membrane. The external granule cell layer (EGL) becomes a second zone of proliferation in the cerebellum (Figure 5.20D).

The EGL is termed a "misplaced" or "displaced" germinal zone because the dividing cells originate at the cerebellar surface, rather than in the VZ. The cells of the EGL appear after mitotic activity has decreased in the VZ. In the chick embryo, for example, cells of the VZ begin to decrease mitotic activity around embryonic day 8 and stop production of new cells by embryonic day 12. In contrast, the EGL is first detectable at embryonic day 6 and new cells are produced from embryonic days 8–15. In the mouse, proliferation and migration of Purkinje cells out of the VZ occurs between E11–14, while the granule cells migrate from the rhombic lip at approximately embryonic

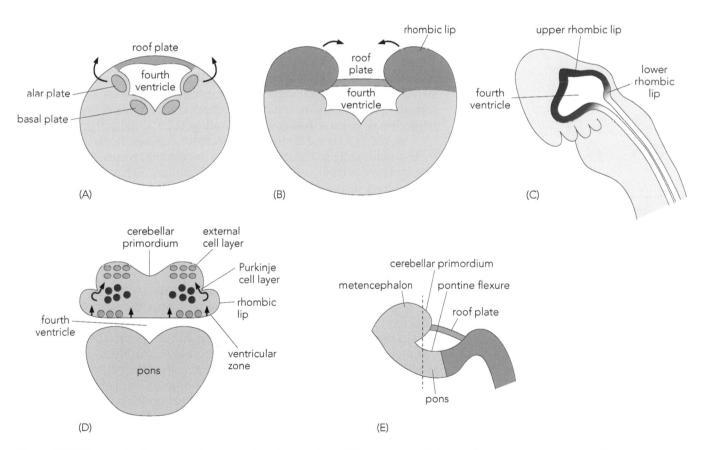

Figure 5.20 The cerebellum forms from the alar plate region of the metencephalon and generates two zones of proliferation. (A) The alar plate region of the caudal metencephalon begins to extend dorsomedially (arrows). (B) The expanding tissue extends over the thin roof plate that covers the fourth ventricle, forming the rhombic lips. (C) A dorsal view of an embryo at this stage shows the rhombic lips along the edges of the fourth ventricle. (D and E) As the rhombic lips continue to extend medially, they eventually join together to form the cerebellar primordium, the precursor of the adult cerebellum. Panel D shows a cross-section of the cerebellar primordium taken at the level indicated by the dashed line in panel E. Two areas of cell proliferation produce the neurons of the cerebellum. Cells of the ventricular zone (blue) are the first to migrate (arrowhead). These cells form the Purkinje cell layer (purple), comprised of several irregular rows of cells at this stage. A second group of cells (curved arrows) migrates from the upper rhombic lips, traveling over the surface of the cerebellar primordium until the cells lie beneath the pial membrane. These cells form the transient external granule cell layers (green).

Figure 5.21 The external granule cell layer is a secondary source of proliferating cells. The outer EGL (oEGL) consists of granule cells that continue to proliferate into postnatal life in rodents and humans. The EGL also contains several layers of premigratory granule cells in the inner EGL (iEGL). These premigratory granule cells will later migrate inward (dashed arrows) past the existing Purkinje cells to form the internal granule cell layer (IGL). The solid arrow indicates the orgin of EGL cells from the rhombic lip.

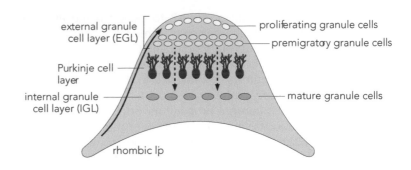

day 14 and continue to proliferate through postnatal day 15. Initially the EGL contains only one to two layers of proliferating cells, but over time the cells expand in number and several rows form, creating outer EGL (oEGL) and inner EGL (iEGL) regions. The cells in the iEGL contain premigratory granule cell neurons (**Figure 5.21**). After sufficient granule cells are generated, these cells move inward past the Purkinje cells to reach the internal granule cell layer (IGL), where they undergo terminal differentiation.

The EGL continues to produce new granule cell neurons into early postnatal life. In mice and rats, the EGL produces new neurons through the first two to three postnatal weeks, whereas in humans the EGL produces new neurons through the first two postnatal years. The EGL is ultimately lost as the cells migrate to form the IGL.

Several proteins are known to support cell cycle activity within the EGL. For example, division of granule cell precursors in the EGL is promoted by Sonic hedgehog (Shh) released by the Purkinje cells. Shh is one of the signals that induces expression of D cyclins to drive the cell to enter G1 stage of the cell cycle (Box 5.1). The transcription factor Atoh1 controls formation of the primary cilium that houses components of the Shh signaling pathway, at least during postnatal stages of granule cell proliferation.

Jagged 1, a ligand for the Notch receptor, has also been shown to promote granule cell proliferation, and as in the cerebral cortex, Notch receptor activity is associated with continued proliferation. Shh and Jagged 1 may also interact at downstream targets, such as the *Hes1* gene, to promote ongoing proliferation of cells in the EGL (see also Chapter 6, Figure 6.8).

Granule Cell Migration from External to Internal Layers of the Cerebellar Cortex Is Facilitated by Astrotactin and Neuregulin

A precise series of events must occur for the cells of the EGL to reach the IGL (**Figure 5.22**). As cells accumulate in the premigratory layers of the EGL, the cells in the deepest layer begin to extend bipolar processes parallel to the pial surface and perpendicular to the Purkinje dendrites. Each cell then extends a third process into the molecular layer—the layer that ultimately consists of Purkinje cell dendrites and interneurons. The soma of the granule cell moves along this third process through the molecular and Purkinje cell layers to reach the cell's final position in the IGL (Figure 5.22). Cells are guided through the layers along **Bergmann glia**, which are specialized RG with multiple parallel extensions. Like RG in the cerebral cortex, Bergmann glia arise from apical RG in the VZ. The cells then migrate to the Purkinje cell layer where they extend processes to the pial surface. Bergmann glial cells later form astrocytes.

In the description of cell migration in the neocortex, it was noted that the integral membrane protein astrotactin mediates adhesion between migrating neurons and RG. In the cerebellum, where astrotactin was first discovered, this interaction occurs between the astrotactin-expressing granule cells and the Bergmann glia. Hatten and colleagues first identified

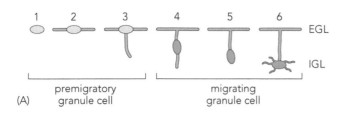

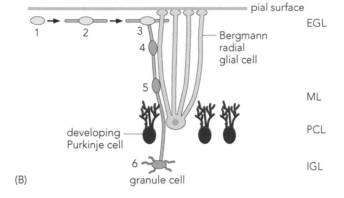

Figure 5.22 The steps necessary for migration of granule cell neurons from the external to the internal granule cell layer. Panel A illustrates the extension of processes from the granule cells in the external granule cell layer (EGL), while panel B demonstrates how the granule cells use these processes to migrate inward along Bergmann glial cells. In panel A, a premigratory granule cell in the EGL (1) first extends bipolar processes parallel to the pial surface (2). The cell then extends a third process perpendicular to the pial surface (3). The cell body of a migrating granule cell then travels through this process (4, 5) to reach the internal granule cell layer (IGL), where it will differentiate (6). Panel B illustrates how Bergmann glia (specialized radial glial cells) interact with the perpendicular process of a granule cell to guide it from the EGL through the molecular layer (ML) and the Purkinje cell layer (PCL) to the IGL. The interaction between Bergmann glia and a granule cell are illustrated beginning with step 3 in panel A. [Adapted from Govek EE, Hatten ME, & Van Aelst L [2011] *Dev Neurobiol* 71:528–553.]

astrotactin in the late 1980s by generating antibodies to granule cell proteins and testing them in cell culture assays. Their studies found that although granule cells adhere to and migrate normally along Bergmann glia in control cell cultures, granule cell adhesion and migration were blocked when anti-astrotactin antibodies were added.

Further evidence for the importance of astrotactin in cerebellar development was seen in mice lacking astrotactin. Granule cell migration in these mice is slow compared to wild-type mice due to the granule cells' decreased adhesion to Bergmann glia. The absence of astrotactin also increases granule cell death and distorts the orientation of Purkinje cell dendrites (**Figure 5.23**). These changes result in behavioral deficits in balance and coordination. Subsequently, a second member of the astrotactin family (Astn2) was identified that regulates the expression and trafficking of astrotactin (now termed Astn1) in the leading process of the migrating granule cell. Changes in *Astn2* expression have been linked to several developmental disorders, including autism. The mechanisms by which *Astn1* and *Astn2* mediate normal development are understandably important and expanding areas of research.

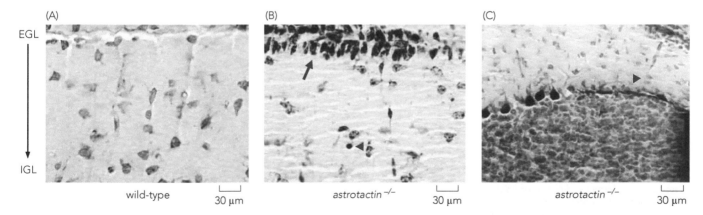

Figure 5.23 Astrotactin is necessary for normal granule cell migration in the cerebellum. (A) The cerebellum of a wild-type mouse at postnatal day 15 (P15) reveals granule cells migrating (long arrow) out of the external granule cell layer (EGL) toward the internal granule cell layer (IGL). (B) A P15 mouse lacking astrotactin (*astrotactin*⁻/⁻) shows cells remain in the EGL longer than wild-type mice (arrow) and some cells begin to die (arrowhead). (C) In the *astrotactin*⁻/⁻ mice, the Purkinje cell layer is also distorted, with some Purkinje cells oriented properly while other adjacent Purkinje cells have cell bodies and dendrites disorganized and outside the plane of the tissue section (arrowhead). [Adapted from Adams NC, Tomoda T, Cooper M et al. [2002] *Development* 129:965–972.]

Despite the identification of Astn1 as a mediator of cell adhesion over 30 years ago, only recently has a binding partner been identified. The first studies of Astn1 suggested it might function as a ligand to mediate adhesion between the granule cells and Bergmann glia. However, other studies suggested Astn1 was a receptor localized to the granule cells. Thus, in the literature, Astn1 has been described as a ligand and receptor. Yet, in all cases it was clear Astn1 was necessary for granule cell adhesion and migration.

It now appears that N-cadherin expressed on Bergmann glia interacts with a complex of N-cadherin and Astn1 expressed on granule cell neurons. Whereas homophilic binding of cadherins is common in many cellular contexts, it is insufficient to mediate granule cell migration along Bergmann glia. Therefore, the need for additional proteins had been proposed. As Hatten and colleagues examined mice in which N-cadherin was conditionally deleted from granule cell precursors and Bergmann glia, it emerged that the various domains of Astn1, including the fibronectin III and annexin-like domains, interact with cadherin domains on Bergmann glia. Thus, N-cadherin serves as a ligand for the Astn1 receptor on granule cell neurons. Moreover, Astn1 interacts with N-cadherin on the same granule cell. On granule cells, the Astn1 EGF-like (epidermal growth factor-like), MACPF (membrane attack complex/perforin), and fibronectin III domains form complexes with N-cadherin (**Figure 5.24**). Thus, Astn1 promotes and stabilizes adhesion between granule cells and Bergmann glia through both cis complexes (Astn1 and N-cadherin in the same cell) and trans complexes (Astn1 and N-cadherin in different cells).

The observation that granule cells continue to migrate, though at a slower rate, in Astn1 knockout mice suggested that additional molecules might mediate adhesion and migration of granule cells in the cerebellum. Molecules that bind integrin receptors, such as L1 cell adhesion molecule (L1CAM), may also influence granule cell migration along Bergmann glia,

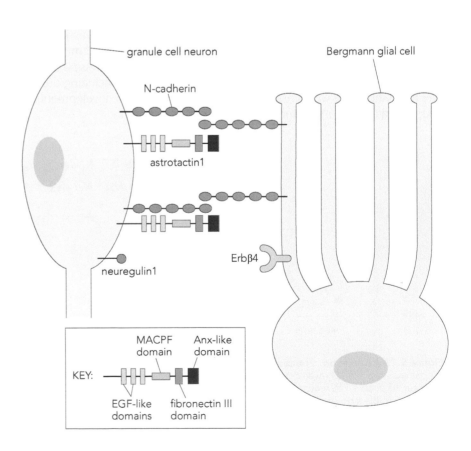

Figure 5.24 Astrotactin 1 interacts with N-cadherin to mediate granule cell adhesion and migration on Bergmann glia. Astn1 on granule cells acts as a receptor for N-cadherin on Bergmann glia because the fibronectin III and annexin-like (Anx) domains of Astrotactin 1 interact with the cadherin domains of N-cadherin. In addition, the EGF-like, MACPF, and fibronectin III domains of Astn1 interact with N-cadherin expressed on the granule cell, providing cis complexes that further stabilize adhesion. Granule cells also express neuregulin 1 (nrg1) that interacts with ErbB4 receptors on Bergmann glia during granule cell migration. Erb, erythroblastic leukemia viral oncogene; MACPF, membrane attack complex/perforin.

at least indirectly. Cerebellar granule cells also interact with the Bergmann glia through ligand-receptor binding involving the ligand neuregulin 1 (nrg1) expressed on the granule cells and the receptor ErbB4 expressed on the Bergmann glia. There are four Erb (erythroblastic leukemia viral oncogene) receptors including epidermal growth factor 1 receptor, also called Erb1. In humans, Erb1 is called Her1 (human epidermal growth factor receptor).

Nrg1 signaling appears to regulate Bergmann glial cell numbers and morphology. Several experiments found that disrupting ErbB4 receptor function or blocking Nrg1 expression led to abnormal Bergmann glia development, including a disruption in the length of the glial processes. These changes cause altered granule cell migration. Further, ErbB4 is downregulated once migration is complete, supporting a role in guiding granule cells. Soluble growth factors such as members of the FGF family appear to stimulate glial production and maintain glial morphology by binding to corresponding FGF receptors (FGFR1, FGFR2) on the Bergmann glia.

Mutant Mice Provide Clues to the Process of Neuronal Migration in the Cerebellum

As noted earlier in the chapter, spontaneously occurring mutations in mice provide insights into the mechanisms regulating cell migration and patterning. For example, reeler mice not only display altered layering in the cerebral cortex, but also show defects in Purkinje cell and granule cell migration in the cerebellar cortex. Most Purkinje cells in these mice do not migrate to form the characteristic single layer; instead, aggregates of Purkinje cells form deeper in the cerebellum (**Figure 5.25**). The granule cells of the EGL, which normally secrete Reelin protein prior to migrating to the IGL, are fewer in number, and most fail to migrate past the existing Purkinje cells. Thus, the overall cellular patterning and resulting function of the cerebellum are disrupted due to the absence of Reelin protein.

In the mouse mutant *Weaver*, named for its weaving gait with unsteady movements and tremor, the granule cell population is depleted in the cerebellum. This does not result from the decreased proliferation of granule cells, but rather from an inability of the cells, once generated, to migrate to the appropriate position. The depleted granule cell population may result from the degeneration of the Bergmann glia fibers prior to granule cell migration or from aberrant orientation of the fibers across the cerebellum. The gait deficit shows the importance of properly oriented Bergmann glia fibers for normal cerebellar development and function.

The necessity of normal granule cell and Bergmann glial morphology was further demonstrated in mice lacking the Rho GTPase Cdc42. When Cdc42 was conditionally deleted in granule cells, the cells became irregularly shaped, displayed disorganized multipolar fibers, failed to form a leading process to adhere to the Bergmann glia, and exhibited delayed

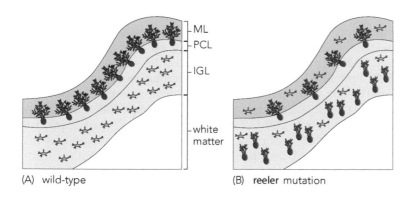

(A) wild-type   (B) **reeler** mutation

Figure 5.25 Cell layers are disorganized in reeler mice. (A) Wild-type mice form a single row of cells in the Purkinje cell layer (PCL). (B) In reeler mice, most Purkinje cells are clumped as aggregates in deeper layers of the cerebellum, closer to the ventricular surface. Granule cells, which normally secrete Reelin protein, are also fewer in number and many fail to migrate past the existing Purkinje cells (B). ML, molecular layer; IGL, internal granule cell layer.

migration. Bergmann glia displayed defects in organization and orientation secondary to the changes in granule cells. Together, these changes led to abnormal development of the cerebellar folia (folds) in the mice lacking Cdc42. How Cdc42 interacts with other GTPases to mediate granule cell polarity, fiber extension, and junctional complexes with Bergmann glia remains to be determined.

MIGRATION IN THE PERIPHERAL NERVOUS SYSTEM: EXAMPLES FROM NEURAL CREST CELLS

In 1868 Wilhelm His first described a unique population of cells in the chick embryo. These cells were named **neural crest cells** based on their site of origin—the crest of the neural folds (see Figure 4.1). Neural crest cells are a transient cell population unique to vertebrates that are found along the length of the neural tube, extending from the posterior-most region to the area of the emerging diencephalon. Neural crest cells give rise to a variety of cell types, including many neurons and supporting cells of the PNS. Specifically, neural crest cells produce ganglia of the autonomic nervous system (that is, the sympathetic chain ganglia, parasympathetic ganglia, and enteric ganglia), the dorsal root ganglia (the sensory neurons along the spinal cord), and some cranial nerve ganglia, as well as the Schwann cells and satellite cells of peripheral ganglia. Neural crest cells also give rise to melanocytes, smooth muscle cells of the aorta, chromaffin cells of the adrenal medulla, endocrine and paraendocrine cells, connective tissues, and components of the craniofacial skeleton. Thus, this single population of cells produces numerous neural and nonneural cell types distributed throughout the body.

The pathways taken and the cell fates that arise from this unique population of migrating cells are remarkably diverse. The distances traveled by neural crest cells and the mechanisms that govern their migration in the embryo differ considerably from the radial migration described in the CNS. This section provides an overview of major migratory pathways used by neural crest cells and introduces some of the extracellular cues that influence their migration patterns.

Neural Crest Cells Emerge from the Neural Plate Border

Neural crest cells arise at the lateral edges of the neural plate—at the border that lies between the emerging neural and epidermal regions. As the lateral edges of the neural plate curl over to form the neural tube, the future neural crest cells are located at the dorsal surface of the neural tube, between the neural tube and overlying epidermis (**Figure 5.26**).

The neural crest cells originate from neuroepithelial cells of the neural plate and are not considered a separate, unique population of cells until they emigrate, or **delaminate**, from the neural tube epithelium. Several growth factors and transcription factors are associated with inducing neural crest regions and specifying the neural crest cells as a separate cell population. The interactions of multiple molecules—particularly members of the BMP, Wnt, and FGF families—induce formation of the neural crest during gastrulation, thus establishing which regions are competent become neural crest cells. These molecules also regulate additional proteins that impact further specification of neural crest regions. Transcription factors such as Pax3, Pax7, Zic1, Msx1, and Msx2 are upregulated as the neural border is specified, thus delineating a region that is distinct from adjacent neural tube and nonneural ectoderm. Another group of transcription

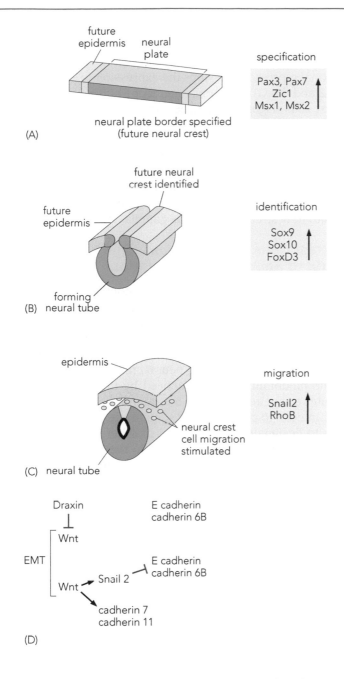

(A)

(B)

(C)

(D)

Figure 5.26 Neural crest cells arise from the dorsal surface of the neural tube. (A) Presumptive neural crest cells are specified at the neural plate border by transcription factors such as Pax3, Pax7, Zic1, Msx1, and Msx2. (B) Transcription factors such as Sox9, Sox10, and FoxD3 are upregulated in the future neural crest cells located along the dorsal region of the forming neural tube. (C) Neural crest cells are considered a separate cell population once they start to migrate from the neural tube. The transcription factor Snail2 and the GTPase RhoB stimulate the migration of neural crest cells. (D) Prior to the epithelial to mesenchymal transition (EMT) the Wnt antagonist Draxin prevents premature Wnt signaling, allowing the epithelial-associated cadherins, E cadherin and caderin6B, to be expressed. As EMT begins, Draxin is downregulated allowing Wnt signaling to promote expression of the mesenchymal-associated cadherins, cadherin-7 and cadherin-11. Wnt also activates transcription factors such as Snail2 that further repress epithelial cadherins.

factors that includes Sox9, Sox10, and FoxD3 is upregulated at the time the future neural crest region is identifiable.

Presumptive neural crest cells begin to exit the dorsal edge of the neural tube just prior to, during, or after neural tube closure, depending on the animal species and the cells' location along the anteroposterior axis. In most vertebrates, the neural crest cells undergo a cell type transition, detaching from neuroepithelium to become loosely packed, unconnected mesenchyme cells. This process, known as **epithelial-to-mesenchymal transition** (EMT), requires the cells to downregulate epithelial-related cadherins such as E-cadherin (epithelial cadherin) and cadherin6B, so cells are no longer adherent to surrounding epithelial cells. The cells also begin to express mesenchymal-related cadherins such as cadherin-7 and cadherin-11. Wnt signaling through the β-catenin pathway suppresses the epithelial cadherins while activating expression of the mesenchymal cadherins, demonstrating another important role for Wnt signaling in neural development. Wnt is regulated by a secreted antagonist called Draxin that is present in premigratory neural crest cells to prevent premature Wnt signaling. However, when EMT begins, Draxin is downregulated, allowing Wnt signaling to initiate expression of transcription factors, such as Snail2, a transcriptional repressor that stops the expression of epithelial cadherins.

Snail2 and the GTPase family member RhoB initiate migration of the neural crest cells (Figure 5.26). Whereas Snail2 influences cell adhesions mediated by cadherins, RhoB interacts with cytoskeletal elements to influence cell motility and migration.

Thus, multiple transcription factors and signaling cascades must be integrated during the early stages of neural crest specification and migration to ensure the cells exit the neural crest and enter the correct migratory stream. Subsequent signals direct migrating cells to reaggregate and coalesce into new structures, such as ganglia.

Neural Crest Cells from Different Axial Levels Contribute to Specific Cell Populations

Neural crest cells arise from four primary anatomical regions designated by their axial level—that is, their location along the anteroposterior axis (**Figure 5.27**). Cells from each region migrate along limited pathways in the embryo. For example, cells of the **cranial neural crest** originate between the midbrain and rhombomere 6 to give rise to structures associated with the head and neck. The remaining neural crest populations originate posterior to the hindbrain and are identified by their corresponding somite level. Somites are paired blocks of mesoderm that are numbered sequentially from anterior to posterior. The somites are often used as anatomical landmarks to designate the axial level of other structures. For example, cells of the **vagal neural crest** arise from the region that extends from the posterior hindbrain to somite 7. The vagal neural crest produces cells that contribute to structures in both the head and trunk, including the sensory neurons of the glossopharyngeal and vagus cranial nerves and the sympathetic, parasympathetic, and enteric ganglia of the trunk. The **trunk neural crest** extends from the somites 8–28. Trunk neural crest cells give rise to the sympathetic, parasympathetic, and dorsal root ganglia, as well as the adrenal chromaffin cells. Neural crest cells that arise posterior to somite 28 give rise to parasympathetic and enteric ganglia. These most posterior crest cells are the **sacral neural crest**, though it is now common to include the sacral neural crest as part of the trunk region.

In addition to specific cell types associated with each region, neural crest cells at all levels contribute melanocytes, glial Schwann cells, ganglion satellite cells, and endocrine cells. Many neural crest cells have the

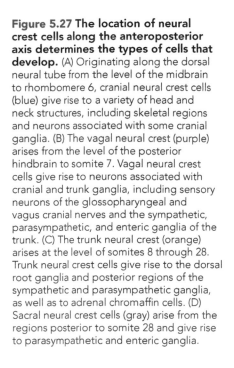

Figure 5.27 The location of neural crest cells along the anteroposterior axis determines the types of cells that develop. (A) Originating along the dorsal neural tube from the level of the midbrain to rhombomere 6, cranial neural crest cells (blue) give rise to a variety of head and neck structures, including skeletal regions and neurons associated with some cranial ganglia. (B) The vagal neural crest (purple) arises from the level of the posterior hindbrain to somite 7. Vagal neural crest cells give rise to neurons associated with cranial and trunk ganglia, including sensory neurons of the glossopharyngeal and vagus cranial nerves and the sympathetic, parasympathetic, and enteric ganglia of the trunk. (C) The trunk neural crest (orange) arises at the level of somites 8 through 28. Trunk neural crest cells give rise to the dorsal root ganglia and posterior regions of the sympathetic and parasympathetic ganglia, as well as to adrenal chromaffin cells. (D) Sacral neural crest cells (gray) arise from the regions posterior to somite 28 and give rise to parasympathetic and enteric ganglia.

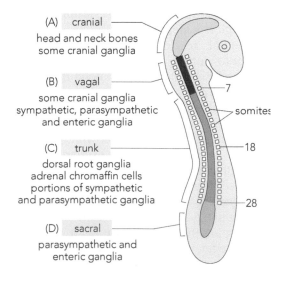

(A)   cranial
head and neck bones
some cranial ganglia

(B)   vagal
some cranial ganglia
sympathetic, parasympathetic
and enteric ganglia

(C)   trunk
dorsal root ganglia
adrenal chromaffin cells
portions of sympathetic
and parasympathetic ganglia

(D)   sacral
parasympathetic and
enteric ganglia

7

somites

18

28

potential to become any one of a diverse set of neural crest derivatives; as a result of this multipotency, some defects in neural crest development promote cancers, such as melanoma.

Cranial Neural Crest Forms Structures in the Head

The neural crest cells at the most anterior regions give rise to head structures, including cranial ganglia, endocrine cells, pigment cells, and cranioskeletal structures. The cranial neural crest cells are unique, in that they are the only crest cells to give rise to skeletal components under normal conditions. Skeletal derivatives of the cranial neural crest include the lower jaw (mandible), bones of the face, the hyoid bone of the neck, and the three bones of the middle ear: the malleus, incus, and stapes. Because so many facial structures arise from the cranial neural crest, altered neural crest development often leads to birth defects such as cleft palate.

Although neural crest cells are multipotent, some populations appear to be more restricted in their fate options. For example, cranial neural crest cells transplanted to trunk regions can form sympathetic and dorsal root ganglia as well as cells of the adrenal medulla and Schwann cells. However, neural crest cells transplanted from the trunk region to the cranial region do not readily form cartilage. As described in the following sections and in Chapter 6, neural crest cells encounter multiple extracellular cues that not only direct the migration of the cells, but also regulate cell fate options.

The cranial neural crest cells are also the source of the sensory neurons of the trigeminal, facial, glossopharyngeal, and vagus nerves (**Figure 5.28**). The only populations of cranial ganglia that are not formed from neural crest cells are those that arise from the embryonic **placodes**, thickened patches of ectoderm located in the head region (**Figure 5.29**). Lens epithelia cells, olfactory, auditory, and vestibular sensory cells, as well as various supporting cell types associated with these structures, arise from their corresponding placodes. Some placodes contribute all the neurons of a single type of ganglion. For example, statoacoustic (also called the vestibulocochlear) ganglia of the eighth cranial nerve are

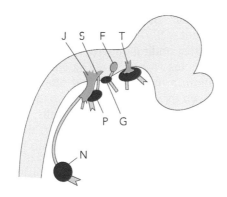

cranial nerve		proximal (neural crest)	distal (placode)
trigeminal	V	trigeminal (T)	trigeminal (T)
facial	VII	facial (F)	geniculate (G)
glossopharyngeal	IX	superior (S)	petrosal (P)
vagus	X	jugular (J)	nodose (N)

Figure 5.28 Some cranial nerve ganglia contain neurons of mixed origin. The trigeminal, facial, glossopharyngeal, and vagus nerves contain neurons derived from both the neural crest and embryonic placodes (see also Figure 5.29). The proximal regions of these ganglia are of neural crest origin (green), while the distal portions are of placode origin (purple). [Adapted from Ayer-Le Lievre CS & Le Douarin NM [1982] *Dev Biol* 94(2):291–310.]

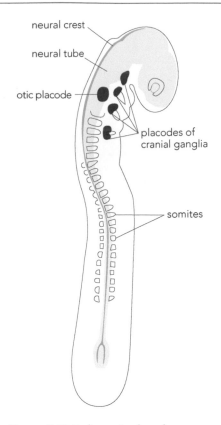

Figure 5.29 Embryonic placodes give rise to some cranial nerve ganglia. Placodes (purple) are patches of ectoderm located in the head region of the embryo. The otic placode gives rise to the embryonic inner ear (otocyst) as well as the associated statoacoustic (vestibulocochlear) ganglia of the eighth cranial nerve. Other placodes give rise to neurons found in the distal portion of cranial nerve ganglia, as shown in Figure 5.28.

derived from the otic placodes. In addition, Le Douarin and colleagues determined that some placodes generate the neurons found in the distal portions of the trigeminal, facial, glossopharyngeal, and vagus nerves (Figure 5.28). Thus, some cranial ganglia consist of both neural crest and placode-derived neurons.

Multiple Mechanisms Are Used to Direct Neural Crest Migration

The neural crest cells form as a large group along the length of the neural tube, but individual cells or small groups of cells soon begin to emigrate and migrate in restricted directions. Individual cell migration is common in mammals and birds, whereas collective migration of small groups of cells is more common in fish and amphibians. To form the many different cell types and tissues, neural crest cells often travel considerable distances throughout the embryo to reach the proper destination and aggregate into tissues.

Cranial crest cells migrate in lateral and ventral directions to form structures in the head and neck, while trunk crest cells migrate through or around somites to form melanocytes, peripheral ganglia, and related cells.

Unlike the cells of the CNS, neural crest cells continue to divide as they migrate throughout the periphery. At any level, the first cells to emigrate are generally those that travel the farthest. Thus, the neural crest cells that give rise to more ventral structures, such as sympathetic ganglia in the trunk, emigrate prior to progressively more dorsal structures, such as the dorsal root ganglia.

Once migration begins, the extracellular environment plays a large part in determining where the neural crest cells will travel. The neural crest cells produce hyaluronic acid, which is believed to alter and expand the extracellular spaces through which neural crest cells must migrate. Neural crest cells also secrete various proteases that break down cell junctions and extracellular matrix proteins. Thus, the neural crest cells themselves alter their local environment to influence migratory routes.

Additionally, permissive extracellular matrix and cell surfaces substrates located throughout the embryo permit attachment and migration of neural crest cells. Such permissive cues include the extracellular matrix (ECM) molecules fibronectin, laminin, and collagen. Neural crest cells express the corresponding integrin receptors that are needed to interact with these molecules (Table 5.1). The composition of the ECM differs based on the location in the embryo and often changes as development progresses. Thus, even though permissive substrates do not attract or specifically direct the migration of neural crest cells, the relative levels of certain ECM molecules may preferentially support the migration of different neural crest cell populations at different times to help guide them toward their destination.

Yet, neural crest cells need more than widely distributed permissive substrates to reach a target. For example, if a cell expressed the receptors for an encountered substrate, it could easily be misguided to a different region. Therefore, other cues must be present to direct the migration of the neural crest cells as they migrate along permissive pathways. Several molecules are present at the different axial levels to provide such instructive, directional cues. Some of the best characterized to date come from studies of trunk neural crest.

Trunk Neural Crest Cells Are Directed by Permissive and Inhibitory Cues

The neural crest cells that arise in the trunk region of the embryo give rise to numerous cell types, including the dorsal root ganglia and sympathetic chain ganglia. These ganglia are easily identified in the adult by

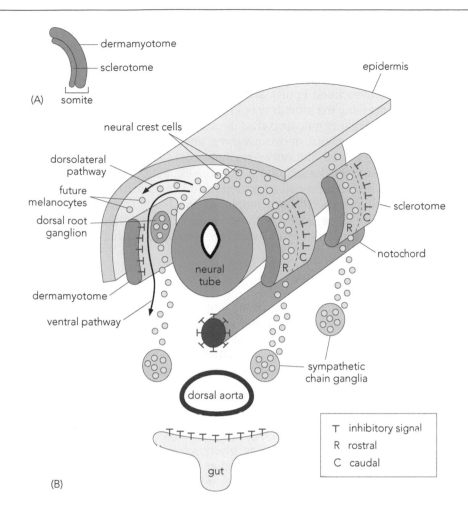

(A) somite
— dermamyotome
— sclerotome

(B)

neural crest cells
dorsolateral pathway
future melanocytes
dorsal root ganglion
dermamyotome
ventral pathway

epidermis
neural tube
sclerotome
notochord
sympathetic chain ganglia
dorsal aorta
gut

T   inhibitory signal
R   rostral
C   caudal

Figure 5.30 Multiple guidance cues direct neural crest cells to target regions in the trunk. (A) Several passive and inhibitory cues used to direct migration of neural crest cells originate in either the dermamyotome or the sclerotome of the somites, the paired blocks of mesoderm tissue along the length of the neural tube. (B) Trunk neural crest cells migrate along two main pathways: the ventral pathway and the dorsolateral pathway. Most cells enter the ventral pathway, where they coalesce into various ganglia, such as the dorsal root ganglia (DRG) and sympathetic chain ganglia. The neural crest cells are directed along this pathway by a combination of permissive and inhibitory cues. In the caudal half of each sclerotome (orange), inhibitory cues direct the neural crest cells through the rostral half of the sclerotome to produce the ladder-like patterns of dorsal root and sympathetic ganglia. Inhibitory cues also direct neural crest cells away from the dermamyotome (pink), notochord (brown), and developing gut (aqua) to ensure that the cells arrive at the correct target region. Only the cells that will form melanocytes enter the dorsolateral pathway. Melanocytes migrate later than the other neural crest populations and inhibitory cues are believed to be expressed in this region during the early stages of neural crest migration to prevent neural crest cells from accidentally entering the melanocyte pathway prematurely.

their ladder-like segregation along the body axis. This patterning arises as the neural crest cells migrate in streams through limited portions of the somites. Somites are divided into two segments, the dermamyotome and the sclerotome (**Figure 5.30**A). The **dermamyotome** (also spelled der momyotome) will later give rise to dermis, skeletal muscle, and vascular tissues, while the **sclerotome** later generates the cartilage and bone of the axial skeleton and rib cage.

In studies of mouse and chick embryos, the neural crest cells that migrate ventrally through the sclerotome go on to form the dorsal root ganglia, the sympathetic ganglia, the Schwann cells and satellite cells of these ganglia, neurons around the aorta, and chromaffin cells of the adrenal medulla. Neural crest cells that migrate through the dorsolateral pathway, located between the epidermis and dermamyotome, will form melanocytes. These general patterns of migration are largely conserved across species. However, differences have been noted in some species. For example, in *Xenopus laevis*, but not other amphibians, melanocytes migrate mainly through the ventral pathway.

Experiments in mouse and chick also noted that the neural crest cells entering the ventral pathway preferentially migrate through the rostral (anterior) portion while avoiding the caudal (posterior) region of the sclerotome (Figure 5.30B). By limiting the pathway options for migration, the segmented, ladder-like patterns of the ganglia emerge.

The extracellular surface of the rostral half of the sclerotome is favorable for neural crest cell migration because substrate molecules such as the protein tenascin are present. Conversely, the caudal segment produces inhibitory proteins that direct neural crest cells away from that region of the sclerotome. Several experiments have suggested that inhibitory cues may be the primary mechanism used to direct trunk neural crest migration. Inhibitory cues in the caudal segment are provided by the ephrin family of molecules, particularly ephrin-B1 and ephrin-B2. The neural crest cells express the associated EphB1 and EphB2 receptors and therefore avoid regions expressing the inhibitory ephrin ligands.

Other inhibitory guidance cues direct migrating neural crest cells. For example, the notochord, dermamyotome, and dorsal portion of the developing gut all produce inhibitory cues that direct sympathetic chain ganglia to coalesce near the dorsal aorta (Figure 5.30B). Inhibitory proteins known to influence neural crest cell migration include aggrecan and other chondroitin-sulfate proteoglycans, semaphorin 3A, slit, and peanut agglutinin-binding glycoproteins. Many of these same inhibitory signals also direct outgrowth of axonal processes, which are described in Chapter 7.

Melanocytes Take a Different Migratory Route Than Other Neural Crest Cells

Melanocytes, the pigment cells of the embryo, are directed along a dorsolateral pathway between the somites and overlying ectoderm (see Figure 5.30). In most vertebrates, the neural crest cells that become melanocytes enter the dorsolateral pathway after other neural crest cells have entered the ventral pathway. There is accumulating evidence to suggest that subsets of neural crest cells are prespecified to become melanocytes and enter the dorsolateral pathway. Endothelin 3, and its associated EDNRB receptor, are thought to be involved in the proliferation, migration, and delayed differentiation of melanocytes. Evidence also suggests that inhibitory cues, such as peanut agglutinin-binding glycoprotein and chondroitin-sulfate proteoglycan molecules, are expressed at the early stages of neural crest migration to prevent the other, earlier-migrating neural crest cells from entering the dorsolateral pathway. When it is time for the presumptive melanocytes to enter this pathway, the expression of the inhibitory molecules decreases to permit cell migration. It also has been proposed that presumptive melanocytes express molecules that permit their migration on otherwise inhibitory substrates.

Determining the mechanisms that specify and guide melanocyte migration remains an active area of research, both to address issues of basic science and to understand clinical conditions that arise from altered melanocyte production and migration. Alterations in melanocyte migration can contribute to tumor formation in adulthood as well as cause birth defects. Waardenburg syndrome is a well-characterized congenital disorder that results from changes in the migration patterns of the pigment-producing melanocytes. Waardenburg syndrome is a rare condition, occurring in 1:10,000–1:40,000 births. Individuals have a range of symptoms that often include changes in skin pigmentation, a white forelock of hair, and eyes that may be very pale or of two different colors. In addition, many individuals with Waardenburg syndrome have hearing loss that results because inner ear melanocytes are needed to regulate the ionic balance of inner ear fluids. Disrupted migration of these melanocytes alters inner ear function and impairs hearing. Changes in the expression of genes identified in animal models of neural crest migration, such as *Pax3* and *Sox10*, contribute to Waardenburg syndrome.

SUMMARY

This chapter highlights some of the common mechanisms used during the proliferation and migration of neurons, glia, and neural crest cells in the vertebrate nervous system. Some mechanisms are specific to a given region of the developing nervous system, such as the radial migration of neurons into the cerebral and cerebellar cortices. Others are more commonly used, such as permissive and inhibitory cues that guide the migration of numerous neuronal precursor populations in the CNS and PNS. As seen in earlier chapters, the same signaling molecules are often used throughout development, including during proliferation and migration. Extracellular matrix proteins, Rho GTPases, and ephrin–Eph interactions are just a few examples of signals used throughout different stages of neural development.

As noted at the beginning of the chapter, neuronal fate in vertebrates can be determined at the time a cell is born or by the extracellular environment that the cell encounters during migration. Some cells have the ability to take on a number of different fates given the right conditions, others are more restricted in cell fate options, and others are specified at the time of terminal mitosis. Chapter 6 provides several examples of how cell fate is determined in neuronal cell populations in vertebrates and invertebrates.

FURTHER READING

Adams NC, Tomoda T, Cooper M, et al. (2002) Mice that lack astrotactin have slowed neuronal migration. *Development* 129(4):965–972.

Albers GW, Marks MP, Kemp S, et al. (2018) Thrombectomy for stroke at 6 to 16 hours with selection by perfusion imaging. *N Engl J Med* 378(8):708–718.

Angevine JB Jr. & Sidman RL (1961) Autoradiographic study of cell migration during histogenesis of cerebral cortex in the mouse. *Nature* 192:766–768.

Anthony TE, Klein C, Fishell G & Heintz N (2004) Radial glia serve as neuronal progenitors in all regions of the central nervous system. *Neuron* 41(6):881–890.

Ayer-Le Lievre CS & Le Douarin NM (1982) The early development of cranial sensory ganglia and the potentialities of their component cells studied in quail-chick chimeras. *Dev Biol* 94(2):291–310.

Berry M & Rogers AW (1965) The migration of neuroblasts in the developing cerebral cortex. *J Anat* 99(4):691–709.

Butts T, Green MJ & Wingate RJ (2014) Development of the cerebellum: Simple steps to make a "little brain". *Development* 141(21):4031–4041.

Caviness VS Jr, Takahashi T & Nowakowski RS (1995) Numbers, time and neocortical neuronogenesis: A general developmental and evolutionary model. *Trends Neurosci* 18(9):379–383.

Consalez GG, Goldowitz D, Casoni F & Hawkes R (2021) Origins, development, and compartmentation of the granule cells of the cerebellum. *Front Neural Circuits* 14:611841.

D'Arcangelo G, Miao GG, Chen SC, et al. (1995) A protein related to extracellular matrix proteins deleted in the mouse mutant reeler. *Nature* 374(6524):719–723.

Duit S, Mayer H, Blake SM, et al. (2010) Differential functions of ApoER2 and very low density lipoprotein receptor in Reelin signaling depend on differential sorting of the receptors. *J Biol Chem* 285(7):4896–4908.

Fishell G, Mason CA & Hatten ME (1993) Dispersion of neural progenitors within the germinal zones of the forebrain. *Nature* 362(6421):636–638.

Frotscher M (1997) Dual role of Cajal–Retzius cells and reelin in cortical development. *Cell Tissue Res* 290(2):315–322.

Ghashghaei HT, Lai C & Anton ES (2007) Neuronal migration in the adult brain: Are we there yet? *Nat Rev Neurosci* 8(2):141–151.

Govek EE, Wu Z, Acehan D, et al. (2018) Cdc42 regulates neuronal polarity during cerebellar axon formation and glial-guided migration. *iScience* 1:35–48.

Goyal M, Menon BK, van Zwam WH, et al. (2016) Endovascular thrombectomy after large-vessel ischaemic stroke: A meta-analysis of individual patient data from five randomised trials. *Lancet (London, England)* 387(10029):1723–1731.

Guzelsoy G, Akkaya C, Atak D, et al. (2019) Terminal neuron localization to the upper cortical plate is controlled by the transcription factor NEUROD2. *Sci Rep* 9(1):19697.

Hack I, Hellwig S, Junghans D, et al. (2007) Divergent roles of ApoER2 and Vldlr in the migration of cortical neurons. *Development* 134(21):3883–3891.

Hatakeyama J, Wakamatsu Y, Nagafuchi A, et al. (2014) Cadherin-based adhesions in the apical endfoot are required for active Notch signaling to control neurogenesis in vertebrates. *Development* 141(8):1671–1682.

Horn Z, Behesti H & Hatten ME (2018) N-cadherin provides a *cis* and *trans* ligand for astrotactin that functions in glial-guided neuronal migration. *Proc Nat Acad Sci U S A* 115(42):10556–10563.

Kaltezioti V, Kouroupi G, Oikonomaki M, et al. (2010) Prox1 regulates the notch1-mediated inhibition of neurogenesis. *PLoS Biol* 8(12):e1000565.

Kasioulis I & Storey KG (2018) Cell biological mechanisms regulating chick neurogenesis. *Int J Dev Biol* 62(1–2–3):167–175.

Katsuyama Y & Terashima T (2009) Developmental anatomy of reeler mutant mouse. *Dev Growth Differ* 51(3):271–286.

Lee GH & D'Arcangelo G (2016) New insights into reelin-mediated signaling pathways. *Front Cell Neurosci* 10:122.

López-Bendito G, Cautinat A, Sánchez JA, et al. (2006) Tangential neuronal migration controls axon guidance: A role for neuregulin-1 in thalamocortical axon navigation. *Cell* 125(1):127–142.

Martynoga B, Drechsel D & Guillemot F (2012) Molecular control of neurogenesis: A view from the mammalian cerebral cortex. *Cold Spring Harb Perspect Biol* 4(10).

Morin X & Bellaiche Y (2011) Mitotic spindle orientation in asymmetric and symmetric cell divisions during animal development. *Dev Cell* 21(1):102–119.

National Institute of Neurological Disorders and Stroke rt-PA Stroke Study Group (1995) Tissue plasminogen activator for acute ischemic stroke. *N Engl J Med* 333(24):1581–1587.

Noden DM (1975) An analysis of migratory behavior of avian cephalic neural crest cells. *Dev Biol* 42(1):106–130.

Nogueira RG, Jadhav AP, Haussen DC, et al. (2018) Thrombectomy 6 to 24 hours after stroke with a mismatch between deficit and infarct. *N Engl J Med* 378(1):11–21.

Oberst P, Agirman G & Jabaudon D (2019) Principles of progenitor temporal patterning in the developing invertebrate and vertebrate nervous system. *Curr Opin Neurobiol* 56:185–193.

Ogawa M, Miyata T, Nakajimat K, et al. (1995) The reeler gene-associated antigen on Cajal–Retzius neurons is a crucial molecule for laminar organization of cortical neurons. *Neuron* 14(5):899–912.

Ohtaka-Maruyama C (2020) Subplate neurons as an organizer of mammalian neocortical development. *Front Neuroanat* 14:8.

Paridaen JT & Huttner WB (2014) Neurogenesis during development of the vertebrate central nervous system. *EMBO Rep* 15(4):351–364.

Pinson A & Huttner WB (2021) Neocortex expansion in development and evolution-from genes to progenitor cell biology. *Curr Opin Cell Biol* 73:9–18.

Rakic P (1974) Neurons in rhesus monkey visual cortex: Systematic relation between time of origin and eventual disposition. *Science* 183(4123):425–427.

Rakic P (2002) Neurogenesis in adult primates. *Prog Brain Res* 138:3–14.

Robinson V, Smith A, Flenniken AM & Wilkinson DG (1997) Roles of Eph receptors and ephrins in neural crest pathfinding. *Cell Tissue Res* 290(2):265–274.

Sauer FC (1935) Mitosis in the neural tube. *J Comp Neurol* 62:377–405.

Sauer ME & Chittenden AC (1959) Deoxyribonucleic acid content of cell nuclei in the neural tube of the chick embryo: Evidence for intermitotic migration of nuclei. *Exp Cell Res* 16(1):1–6.

Shimojo H, Ohtsuka T & Kageyama R (2011) Dynamic expression of notch signaling genes in neural stem/progenitor cells. *Front Neurosci* 5:78.

Sidman RL, Miale IL & Feder N (1959) Cell proliferation and migration in the primitive ependymal zone: An autoradiographic study of histogenesis in the nervous system. *Exp Neurol* 1:322–333.

Silva CG, Peyre E, Adhikari MH, et al. (2018) Cell-intrinsic control of interneuron migration drives cortical morphogenesis. *Cell* 172(5):1063–1078.e19.

Singh S & Solecki DJ (2015) Polarity transitions during neurogenesis and germinal zone exit in the developing central nervous system. *Front Cell Neurosci* 9:62.

Spear PC & Erickson CA (2012) Interkinetic nuclear migration: A mysterious process in search of a function. *Dev Growth Differ* 54(3):306–316.

Tissir F & Goffinet AM (2003) Reelin and brain development. *Nat Rev Neurosci* 4(6):496–505.

Uzdensky AB (2019) Apoptosis regulation in the penumbra after ischemic stroke: Expression of pro- and antiapoptotic proteins. *Apoptosis* 24(9–10):687–702.

Wang Y, Li G, Stanco A, et al. (2011) CXCR4 and CXCR7 have distinct functions in regulating interneuron migration. *Neuron* 69(1):61–76.

Williamson MR, Jones TA & Drew MR (2019) Functions of subventricular zone neural precursor cells in stroke recovery. *Behav Brain Res* 376:112209.

Wu CY & Taneyhill LA (2012) Annexin a6 modulates chick cranial neural crest cell emigration. *PLoS ONE* 7(9):e44903.

Xu H, Yang Y, Tang X, et al. (2013) Bergmann glia function in granule cell migration during cerebellum development. *Mol Neurobiol* 47(2):833–844.

Yang J, Yang X & Tang K (2021) Interneuron development and dysfunction. *FEBS J* 289:2318–2336.

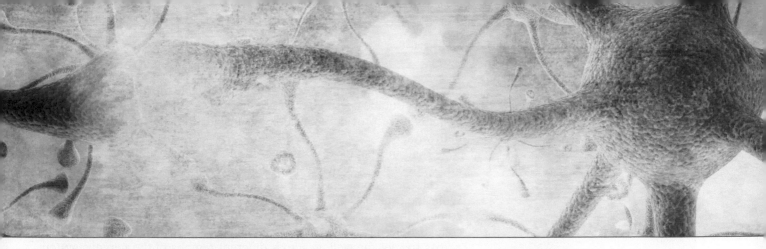

Cell Determination and Early Differentiation

6

A variety of specialized cell types is required for a functional adult nervous system and every neural, glial, sensory, and support cell must acquire highly specialized characteristics to work properly. The previous chapter discussed how vertebrate neuroepithelial cells divide, establish neural progenitors, and migrate to new locations where they ultimately differentiate into fully mature neurons. This chapter focuses on some of the common mechanisms by which cells transition from the progenitor, or precursor, stage to their ultimate cell fate. The terms "progenitor" and "precursor" are often used interchangeably, though in some organ systems or neural regions, one term is preferred over the other. Both are used in this chapter to refer to cells at the stage when they can give rise to more than one type of cell.

During early embryogenesis, neuroepithelial cells have the potential to form many cell subtypes. As development progresses, however, cells are exposed to various signals that restrict their cell fate options. Depending on the specific precursor and the signals available, a given cell may remain **multipotent**—that is, retain the ability to develop into more than one cell type—for an extended period. However, this ability only persists up until the time of **cellular determination**, the stage at which further embryonic development or experimental manipulation can no longer alter the type of cell that forms. Thus, the determined cell has acquired its **cell fate**. A determined cell will then begin to differentiate and acquire the unique characteristics associated with a given cellular subtype.

For some cell types, cell fate options become restricted early in the cell cycle in response to intrinsic cues, such as those that arise from nuclear or cytoplasmic signals inherited from a precursor cell. For other cells, fate is largely regulated by extrinsic cues encountered during migration or at the target destination. These extrinsic cues are often the same types of signals discussed in earlier chapters, such as extracellular matrix molecules and diffusible factors. A previously held view was that the fate of invertebrate precursors mostly relied on intrinsic cues, whereas vertebrate precursors primarily relied on extrinsic cues. Although these generalizations apply to some cells in these model systems, such distinctions do not apply to all cells. Further, many intrinsic and extrinsic cues overlap temporally and spatially to influence cell fate, making it difficult to establish what cues

DOI: 10.1201/9781003166078-6

predominate for any given cell population. Despite the inherent challenges of sorting out the cues that direct cell fate decisions, several animal model systems have provided considerable insight into the required signaling pathways.

Here in Chapter 6, examples from selected regions of invertebrate and vertebrate nervous systems illustrate how undifferentiated precursor cells develop as specialized neuronal, glial, sensory, or support cells. While the examples provided are by no means all-inclusive, they represent some of the most common and best-understood mechanisms underlying cellular determination. Many of these basic mechanisms are conserved across species, as well as across different regions of the nervous system in a given animal model. Common mechanisms include lateral inhibition, Notch signaling, and temporally regulated transcription factor cascades. In recent years the importance of epigenetic modifications in regulating cell fate options has also been highlighted. Epigenetic modifications that lead to changes in the accessibility of DNA binding sites provide an additional means for the nervous system to utilize the limited number of signaling pathways available to achieve the necessary range of developmental outcomes.

LATERAL INHIBITION AND NOTCH RECEPTOR SIGNALING

As introduced in Chapter 5, during early neurogenesis selected cells within the neuroepithelium begin to express **proneural genes**—the genes that provide a cell with the potential to become a neural progenitor (precursor). The expression of proneural genes leads, in turn, to the activation of transcription factors and neuron-specific genes that influence the ultimate characteristics of a neuron. Cells that do not express proneural genes later become one of the surrounding glial or other nonneuronal cell types of the nervous system. One common mechanism for specifying neuronal versus nonneuronal cells is **lateral inhibition**, a process driven by differences in Notch receptor activity in adjacent cells. Lateral inhibition is an evolutionarily conserved mechanism used by invertebrates and vertebrates to specify neural cells.

Lateral Inhibition Designates Future Neurons in *Drosophila* Neurogenic Regions

In the developing *Drosophila* nervous system, the areas of ectoderm that ultimately give rise to the neurons are called the neurogenic regions. Cells within the neurogenic region begin to express low levels of proneural genes, such as *Atonal* and members of the *Achaete-scute* complex (*Achaete, Scute, Lethal of scute,* and *Asense*). The cells that express these genes make up a **proneural cluster** (**PNC**), and at this stage of development each cell in the cluster has the potential to become a neuron. Thus, at the earliest stages the cells are equivalent, with each cell expressing low levels of proneural genes.

Through cell–cell interactions, one cell in the PNC becomes specified as a neural precursor, while the surrounding cells in the cluster become nonneuronal cells. An example of how this occurs involves the expression of the ligand Delta and the receptor Notch in cells of the PNC. In this example, the proneural genes of the *Achaete-scute* complex (*AS-C*) initiate the expression of the ligand Delta in all the cells of the proneural cluster (**Figure 6.1**A). These cells also express the receptor Notch. Thus, all cells initially express both the ligand and receptor. However, an imbalance in Delta expression begins as proneural genes lead one cell to start expressing a slightly higher level of Delta ligand (Figure 6.1B).

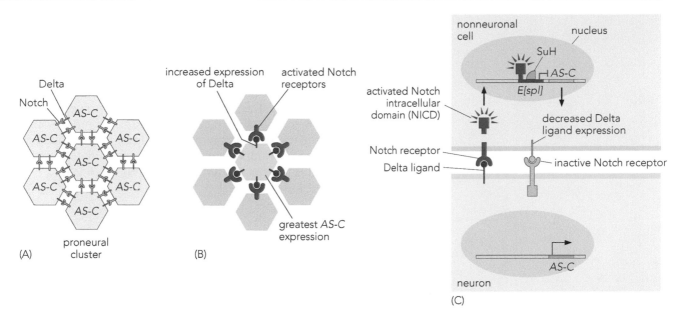

Figure 6.1 Specification of neural precursors in *Drosophila* neuroectoderm. (A) Low levels of proneural genes, such as those of the *achaete-scute* complex (*AS-C*), begin to be expressed in a subset of neuroectoderm cells called the proneural cluster (PNC). All cells in the PNC express *AS-C* genes that promote expression of Delta ligands. Notch receptors are also expressed in all cells of the PNC, so at this stage all have the potential to become neurons. (B) Some cells within the PNC begin to express higher levels of the Delta ligand. In this example, the center cell (blue) expresses a sufficient level of Delta to activate the Notch receptors in surrounding cells. (C) An enlargement of one ligand-receptor pair. When Notch is activated in an adjacent cell, the intracellular portion of the Notch receptor is cleaved and the now-activated Notch intracellular domain (NICD) travels to the nucleus, where it interacts with proteins including the DNA binding protein Suppressor of Hairless (SuH). SuH then turns on expression of *Enhancer of split* (*E[spl]*), which inhibits the expression of the proneural *AS-C* genes in the Notch-activated cell, thus leading to a nonneuronal cell fate. The inhibition of *AS-C* genes also causes a decrease in the expression of Delta ligand in that cell, thus preventing Notch activation in the adjacent (blue) cell. Because Notch is not activated in this cell, *AS-C* genes continue to be expressed at higher levels, so the cell is directed to a neuronal fate.

How the initial increase in Delta expression occurs is still unclear. What is clear is that once sufficient Delta expression is attained, Notch receptor activity on an adjacent cell initiates a signal transduction cascade that leads that cell to a nonneuronal fate. The signaling pathway is initiated when the bound Notch receptor undergoes proteolysis. The resulting Notch intracellular domain (NICD) is then transported to the nucleus, where it forms a complex with other proteins such as Mastermind and interacts with Suppressor of Hairless (SuH; Figure 6.1C, top). In the nucleus, SuH acts as a DNA-binding protein to increase the expression of *Enhancer of split* (E[spl]), which functions, in turn, as a suppressor of neural fate by inhibiting the expression of proneural *AS-C* genes. Thus, Delta binding to the Notch receptor initiates the pathway for inhibiting neural fate in the Notch-activated cell. In addition, the Notch-activated cell decreases its own expression of Delta ligand, so that cell is unable to activate the Notch receptor on a neighboring cell. Because the Notch signaling pathway is not initiated in the neighboring cell, the proneural *AS-C* genes continue to be expressed and thus direct that cell to differentiate as a neuron (Figure 6.1C, bottom).

Through this balance of Delta expression and Notch activation, the cells of the PNC are designated to adopt nonneuronal or neuronal fates. Cells that have the Notch signaling pathway activated become nonneuronal cells, whereas those that do not have the Notch signaling pathway activated become neurons. This balance must be properly maintained so that the correct number of neurons and nonneuronal cells are generated. Experimental manipulations highlight the importance of this balance. In *Drosophila* mutants that lack *AS-C* genes, the majority of neurons are absent in both the central nervous system (CNS) and peripheral nervous system (PNS). Conversely, extra copies of these genes result in extra neurons.

Lateral Inhibition Designates Stripes of Neural Precursors in the Vertebrate Spinal Cord

Lateral inhibition also impacts the development of cells within vertebrate neuroectoderm. In the *Xenopus* neural plate, for example, the region of the future spinal cord contains three longitudinal stripes of neural precursors on each side of the midline. The stripes are repositioned to ventral, intermediate, and dorsal regions as the neural tube closes. The precursors ultimately give rise to the motor neurons (medial stripes), intermediate zone neurons (center stripes), or dorsal sensory interneurons (lateral stripes). The adjacent interstripes do not produce neurons (**Figure 6.2**A).

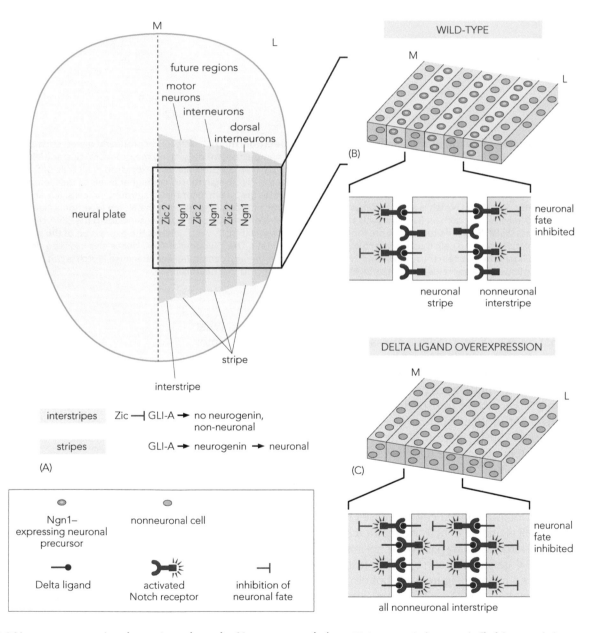

Figure 6.2 Neurons are restricted to stripes along the *Xenopus* neural plate. (A) An example from one half of the neural plate illustrates that the region of the future spinal cord is comprised of stripes of neural precursors (blue) that later give rise to motor neurons and the intermediate and dorsal interneurons. The adjacent interstripes give rise to nonneuronal cells (gray). During late gastrulation, cells in the stripe regions begin to express the proneural gene *Neurogenin 1* (*Ngn1*). *Ngn1* is activated by GLI transcription factors expressed in the midline and neural plate. The interstripes do not express *Ngn1* because Zic2 in these stripes represses GLI activator (GLI-A) forms. (B) A segment of the spinal cord with three stripes of the neural precursor cells expressing *Ngn1* (yellow), a member of the *Atonal* gene family. Normally, Delta is restricted to the stripe regions so that only Notch receptors expressed on the cells of interstripes are activated. Activation of Notch further inhibits proneural genes and represses neuronal fate. (C) Overexpression of the ligand Delta leads to production of nonneuronal cells in all stripe regions. When Delta is overexpressed, Notch is activated on cells of both stripe and interstripe regions, thus directing more cells to a nonneuronal fate.

Proneural genes are selectively expressed to establish the stripes of neural precursors. During the late stages of gastrulation, the proneural bHLH gene *Neurogenin 1* (*Ngn1*), a member of the *Atonal* gene family, is expressed in cells that will form the stripe regions. In *Xenopus*, *Ngn1* expression is activated by GLI transcription factors expressed in the midline and neural plate. However, *Ngn1* expression is prevented in interstripes by Zic2. Recall from Chapter 4 that Zic genes repress GLI activator forms. Thus, *Ngn1* expression is blocked in interstripes. These early patterning events determine where *Ngn1* can be expressed and establish which cells have the potential to develop into neurons. The importance of regulated *Ngn1* expression was seen with experimental overexpression of *Ngn1*. Overexpression led to an increase in the number of neurons in the *Xenopus* neural plate with neurons located in both stripe and interstripe regions (Figure 6.2B). The *Ngn1* gene also induces downstream expression of *NeuroD1* (also called *NeuroD*), another homolog in the *Atonal* gene family, that regulates further development of the neurons. A direct link between *Ngn1* and *NeuroD1* expression was seen in studies in which overexpression of *Ngn1* also led to overexpression of *NeuroD1*.

Once *Ngn1* designates cells in the stripe regions as the neural precursors, lateral inhibition further restricts neuronal development to the stripe regions. In the *Xenopus* spinal cord, it appears that the Delta ligand is expressed only in cells within the stripes, and this expression may be regulated by *Xenopus Achaete-scute* homolog (*Xash*) genes, such as *Xash1* or *Xash3*. In contrast to the Delta ligand, the Notch receptor is expressed in cells of both the stripe and interstripe regions, though only the Notch receptors in the interstripe region will receive Delta signals. Thus, as Notch-bearing cells in the interstripe regions are activated by Delta-expressing cells in the stripe regions (**Figure 6.2**A), neuronal fate remains suppressed. This process ensures that interstripe cells develop with a nonneuronal fate. The importance of restricted Delta expression was seen in experiments in which the overexpression of Delta activated Notch receptors in both stripe and interstripe regions, leading to an increase in the number of nonneuronal cells and the production of fewer neurons in the neural plate (Figure 6.2C).

Thus, as observed in the PNC of *Drosophila*, the *Xenopus* neural plate uses Delta and Notch signaling to pattern regions of neuronal and nonneuronal cells. The neural precursors within the stripes subsequently receive additional signals to become specific neural types, such as motor neurons and sensory interneurons. As described in Chapter 4, these signals include the ventrally derived protein Sonic hedgehog (Shh) and the dorsally derived proteins of the transforming growth factor β (TGFβ) and Wnt families.

CELLULAR DETERMINATION IN THE INVERTEBRATE NERVOUS SYSTEM

Notch signaling activity remains important after lateral inhibition. In several regions of the *Drosophila* nervous system, the uneven distribution of Notch and Numb proteins and the subsequent temporal expression of specific transcription factors further shapes the fate options available to neuronal progenitors.

Cells of the *Drosophila* PNS Arise from Epidermis and Develop in Response to Differing Levels of Notch Signaling Activity

The *Drosophila* peripheral nervous system consists of **sensory organ progenitors** (**SOPs**) that arise at various locations across the epidermis (**Figure 6.3**A). The SOPs give rise to different sensory organs, including the

Figure 6.3 Cells of the *Drosophila* PNS arise from sensory organ progenitors (SOPs). (A) A cross-section of the *Drosophila* embryo shows that SOPs originate at various locations along the epidermal ectoderm. (B) Each SOP has an unequal distribution of Numb (blue), a protein that inhibits the Notch receptor (green) and ultimately promotes a neuronal cell fate. When the SOP divides, the SOPIIb cell inherits higher levels of Numb. Because SOPIIa does not inherit sufficiently high levels of Numb, its Notch receptor can be activated by local ligands to initiate downstream signaling pathways that result in nonneuronal cell fates. Thus, the division of SOPIIa produces the nonneuronal socket and bristle cells. In contrast, when the SOPIIb divides one daughter cell expresses *Numb* at a higher level and forms a neuron. The other daughter cell does not inherit sufficient *Numb* and becomes a glial sheath cell. (C) The mature sensory bristle complex is made up of a bristle cell with a hair that extends above the cuticle, an associated socket cell, a sensory neuron, and a glial sheath cell that surrounds the neuron.

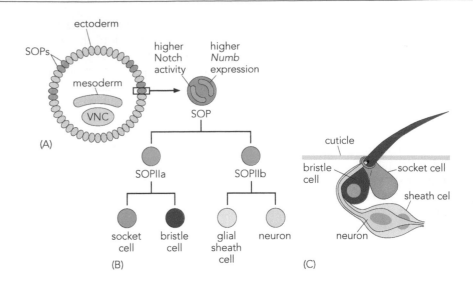

mechanosensory and chemosensory organs, as well as the chordotonal organs that contain stretch receptors. The fate of the different cell types depends, in part, on the distribution of Notch and Numb proteins in the precursor cells.

In *Drosophila*, the SOPs typically arise after the CNS cells are established. The process of lateral inhibition determines which cells from a PNC will become SOP cells. Each SOP then divides asymmetrically to produce two intrinsically different daughter cells, SOPIIa and SOPIIb (Figure 6.3B), which give rise to the four distinct cell types of the touch-sensitive sensory bristle complex. As introduced in Chapter 5, the differential distribution of Numb and Notch can influence cell development, with Numb inhibiting Notch receptor activation. The progenitor cell has an asymmetrical distribution of Numb protein so that only one daughter cell, the SOPIIb, inherits high levels of Numb. In the SOPIIa cell, which does not inherit high levels of Numb, activation of Notch signaling remains. Notch signaling initiates downstream pathways that suppress neural fate, so when the SOPIIa cell divides it produces two nonneuronal cells: a bristle and socket cell (Figure 6.3B). The Numb and Notch proteins also become distributed asymmetrically in the SOPIIb cell. When this cell divides, the daughter cell with greater Notch signaling becomes a type of glial cell called a sheath cell, but the daughter cell containing higher levels of Numb goes on to form a neuron (Figure 6.3B). Thus, levels of Numb protein influence which of the four cell types of the sensory bristle complex form (Figure 6.3C).

The importance of Notch and Numb expression levels in SOPs was seen when levels of either Notch or Numb were experimentally altered (**Figure 6.4**). When *Notch* was repressed, SOPIIa cells were unable to activate the signaling pathways that promote formation of nonneuronal socket and bristle cells. Instead, in the absence of high levels of Notch signaling activity, the SOPIIa produced only neurons (Figure 6.4B). Conversely, when *Numb* was absent, no sensory neurons formed because there was insufficient Numb present to block Notch activity in the SOPIIb cell. The *Numb* mutants were unresponsive to touch, thus behaving as if they were numb. This behavioral phenotype occurred because instead of sensory neurons, the *Numb* mutants now produced either socket and bristle cells or only socket cells (Figure 6.4C, D).

Notch signaling is also critical in sensory organs of the vertebrate nervous system. Examples describing the differentiation of sensory cells in the organ of Corti of the inner ear and the retina of the eye, provided later in this chapter, illustrate how the same general signaling pathways are used to establish structurally diverse sensory regions in different species.

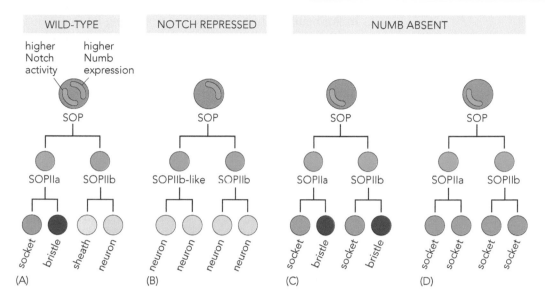

Figure 6.4 Altering Notch or Numb expression changed cell fate options of SOPIIa and SOPIIb descendants. (A) SOP cell fates in wild-type *Drosophila*. Under normal conditions SOPIIa cells have the Notch receptor activated at high levels and therefore produce the nonneuronal socket and bristle cells. In contrast, SOPIIb cells express Numb at levels sufficient to inhibit Notch signaling. These cells produce a neuron and glial sheath cell. (B) When the Notch receptor is experimentally repressed, the signaling pathways for socket and bristle cell formation cannot be activated. Thus, in the absence of Notch signaling, the SOPIIa cells function like SOPIIb cells and only produce neurons. (C, D) In the absence of *Numb*, Notch receptors are activated in both SOPIIa and SOPIIb cells. Therefore, only nonneuronal cells are produced. The absence of Numb leads to the production of socket and bristle cells (C) or only socket cells (D).

Ganglion Mother Cells Give Rise to *Drosophila* CNS Neurons

An unequal distribution of Notch and Numb proteins in progenitor cells also impacts cellular determination of neurons in the *Drosophila* brain and **ventral nerve cord** (**VNC**), a structure that is functionally analogous to the vertebrate spinal cord.

In the embryonic VNC, cells originate from a ventral stripe of ectoderm in a defined manner. The cells of the proneural cluster that were previously designated to become neurons through lateral inhibition enlarge and delaminate by moving inward to form neuroblasts (**Figure 6.5**A). Each neuroblast then divides unequally, forming one large and one small daughter cell (Figure 6.5B). The smaller cell is a ganglion mother cell (GMC). The larger cell is a neuroblast that continues to proliferate, producing another GMC and neuroblast with each cell division (Figure 6.5C). As successive divisions occur, the new GMCs are situated between the first GMC and the current neuroblast. The GMCs ultimately divide equally to produce cells with either neural or glial fates. In the VNC, the neuroblasts first produce

Figure 6.5 The *Drosophila* ventral nerve cord arises from neuroblasts that originate in the ventral ectoderm. (A) A cross-section of *Drosophila* ectoderm shows that cells of the ventral neuroectoderm enlarge and migrate inward (arrows) after invagination of the mesoderm has been completed. The areas of epidermal ectoderm located more dorsally do not give rise to CNS neurons. (B) The cells that migrate inward form neuroblasts that will later coalesce to form the ventral nerve cord. (C) Each neuroblast (Nb) divides unequally to produce two daughter cells, a new neuroblast and the first ganglion mother cell (GMC1). The resulting neuroblast (orange) again divides unevenly, forming another neuroblast (green) and a second GMC (2, orange). These asymmetric divisions continue for various lengths of time, depending on the Nb lineage. Ultimately each GMC divides equally and produces a neuron and glial cell or two neurons (not shown).

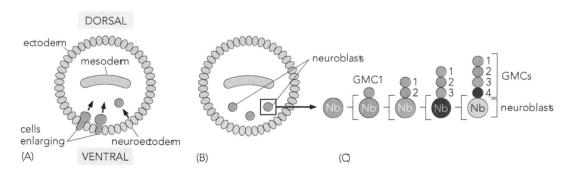

motor neurons then transition to producing interneurons. The number of GMCs generated from a neuroblast varies from a few to over 20, depending on the neuroblast lineage. Neurons in the *Drosophila* brain divide in a similar manner to those of the VNC.

Apical and Basal Polarity Proteins Are Differentially Segregated in GMCs

As described in Chapter 5, proliferating neuronal progenitors in the vertebrate CNS have specific proteins localized to the apical and basal poles of the daughter cells. When the cells divide asymmetrically, differences in the distribution of these proteins determines whether the daughter cell continues to proliferate or becomes a basal progenitor cell that migrates away from the ventricular surface (see Figure 5.5). Many of the proteins segregated to the apical or basal poles of vertebrate CNS precursors were first discovered in *Drosophila*. The homologous *Drosophila* proteins also designate which cell will continue as a proliferating neuroblast and which cell will form a GMC. As in the vertebrate CNS, the cell in which Notch activity remains high will continue to proliferate.

Drosophila apical pole proteins include the Par (Partitioning defective) complex, which consists of Par3 (Bazooka) and Par6, atypical protein kinase C (aPKC), Inscuteable (Insc), and partner of inscuteable (Pins). The basal proteins include Numb, Brat (brain tumor), Prospero, Partner of Numb (Pon), and Miranda (**Figure 6.6**). As in vertebrate neurons, the apical proteins direct the orientation of the mitotic spindles to determine the plane of cell division and direct basal proteins to the opposite pole. As a neuroblast divides, the new neuroblast inherits the apical proteins and the GMC inherits the basal proteins. The new neuroblast will divide again due to sufficient Notch signaling activity. In contrast, the GMC stops proliferating because the concentration of Numb in that cell prevents high levels of Notch signaling. Furthermore, the basal protein Prospero, now concentrated in the GMC, represses proliferation genes while activating determination genes. Thus, as in the *Drosophila* PNS, the level of Notch signaling activity first regulates the specification of neural and nonneural regions during lateral inhibition and then governs whether a cell proliferates or becomes committed to a neural fate. In the CNS, only the cells that inherit proteins that interfere with Notch signaling can commit to the GMC fate.

Cell Location and Temporal Transcription Factors Influence Cellular Determination

Intrinsic cues also help direct the fate of *Drosophila* CNS neurons. In *Drosophila*, neurons that arise from the original neuroblast do not migrate

Figure 6.6 Ganglion mother cells inherit the basal protein complex to commit to neural fate. (A) During asymmetric division of the *Drosophila* neuroblast, proteins are segregated to the apical and basal poles. Apical proteins include those of the Par complex (Par3 and Par6), atypical protein kinase C (aPKC), Inscuteable (Insc), and Partner of inscuteable (Pins). The apical proteins help orient the mitotic spindles to determine the plane of cell division. The proteins also help direct basal proteins to the opposite pole of the cell. Basal proteins include Numb, Brain tumor (BRAT), Prospero, Partner of Number (Pon), and Miranda. (B) The concentration of Numb in the GMC prevents high levels of Notch activity and therefore prevents continued proliferation. Prospero represses proliferation genes and activates determination genes so that the GMC is able to commit to the neural–glial fate. In the apical cell, Numb levels are not high enough to inhibit Notch receptor activity, so the new neuroblast continues to proliferate.

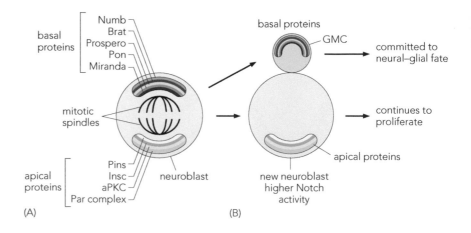

away. As a result, like most invertebrate neurons, a cell's origin is closely linked to its final position in the embryo. Thus, the first GMCs produced are found in the deeper layers of the CNS and have longer axons. In contrast, the GMCs from later cell divisions are located more superficially and have shorter axons.

A temporal sequence of transcription factor expression has been observed in neuroblasts and GMCs. These transcription factors are called **temporal transcription factors (tTFs)** or **temporal identity factors (TIFs)**. The TIFs that a cell expresses do not appear sufficient to designate its fate. Rather, cell fate is determined by a combination of transcription factor expression, cell location, and extrinsic signals. Depending on the neuroblast lineage, a given transcription factor in the VNC, for example, may designate a motor neuron or interneuron. In the VNC, five transcription factors are expressed in sequence—namely, Hunchback, Krüppel, Pdm, Castor, and Grainyhead. This same sequence is used by other neuroblast lineages in the *Drosophila* CNS, although Grainyhead may not act as a TIF in all regions.

A neuroblast first expresses Hunchback; this expression is inherited by GMC1 when the neuroblast divides (**Figure 6.7**A). The daughter neuroblast now expresses Krüppel and divides to generate the Krüppel-expressing GMC2 and a daughter neuroblast that expresses Pdm. Subsequent neuroblasts express the remaining TIFs in sequence during each subsequent division.

The timing of TIF expression is critical, as can be shown under experimental conditions. If one transcription factor is absent, only the cell type arising at that stage will be eliminated. For example, a series of experiments by Chris Doe and colleagues found that when *Hunchback* is absent, only the GMC generated during the first cell division is missing (Figure 6.7B). If a transcription factor is experimentally maintained, then those cell types will persist longer. Continued expression of *Hunchback* during the period that Krüppel-expressing cells would normally be produced, for example, led to the formation of GMCs with characteristics of the earliest cells (Figure 6.7C). In another study, one neuroblast was experimentally ablated. Although that cell never formed, the subsequent neuroblasts continued to arise in order and express the transcription factors normally present during those cell divisions (Figure 6.7D). Again, the transcription factors alone are not believed to regulate cell fate, but their presence appears to establish which cell types can form during different stages of development in the *Drosophila* CNS. As described later in this chapter, homologs of some of these TIFs have been identified in the mammalian cerebral cortex and retina, where they appear to serve similar functions.

Figure 6.7 Transcription factors are expressed in a temporal sequence in neuroblasts of the *Drosophila* ventral nerve cord. (A) The first neuroblast (Nb, blue) that arises in the ventral nerve cord expresses the transcription factor Hunchback (Hb). When this neuroblast divides, the resulting GMC (GMC1, blue) inherits Hb. The new Nb (orange) now expresses a second transcription factor called Krüppel (Kr) that is inherited by next GMC produced (GMC2). The third neuroblast (green) expresses Pdm, while the next (red) expresses Castor, and the final Nb in this lineage (yellow) expresses Grainyhead. Each of these transcription factors is also expressed in the corresponding GMC. (B–D) Experimental manipulations reveal how the timing of transcription factor gene expression impacts the cells that arise. In the first example (B), *Hunchback* was deleted from the first Nb. Only that cell failed to form (1, gray). Because the other transcription factors were not altered in dividing Nbs, the remaining GMCs (2–4) and final Nb (yellow) formed at the correct time. (C) When *Hunchback* expression was sustained and took the place of *Krüppel*, the resulting GMC now had the characteristics of GMC1 (blue cells). The other GMCs (3 and 4) and final Nb (yellow) expressed the correct transcription factor and developed as expected. (D) When a neuroblast was experimentally ablated (cell 2), only that cell failed to form. The other GMCs (1, 3, and 4) and Nb (yellow) expressed the correct transcription factors and developed at the correct time. [Adapted from Isshiki T, Pearson B, Holbrook S & Doe CQ [2001] *Cell* 106:511–521.]

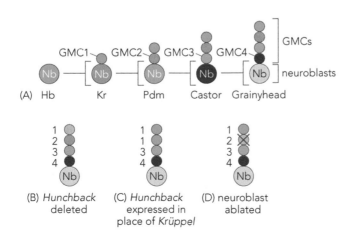

(A) Hb   Kr   Pdm   Castor   Grainyhead

GMC1   GMC2   GMC3   GMC4   GMCs
Nb   Nb   Nb   Nb   Nb   neuroblasts

(B) *Hunchback* deleted

(C) *Hunchback* expressed in place of *Krüppel*

(D) neuroblast ablated

MECHANISMS UNDERLYING FATE DETERMINATION IN VERTEBRATE CNS NEURONS

In the vertebrate nervous system, reduced Notch receptor activity, environmental cues, and the temporal expression of specific transcription factors also coordinate to influence neuronal fate options and initiate cellular differentiation. Examples of how such cues contribute to the development of cerebellar granule cells and cerebral cortical neurons are described in the following sections.

Coordinating Signals Mediate the Progressive Development of Cerebellar Granule Cells

Because the developmental events that lead to the formation and migration of cerebellar granule cells are so well documented (see Chapter 5), several studies have focused on the signals that regulate development of this highly specialized group of cells. As noted earlier, proliferation and differentiation continue through postnatal stages of cerebellar development. During these states, many signals coordinate to balance proliferation of the granule cell precursors in the outer external granule cell layer (oEGL) with the differentiation of the premigratory granule cells of the inner EGL (iEGL).

Proliferation is influenced by Shh secreted by Purkinje cells. Proliferation of granule precursor cells is further regulated by Atonal1, a unique function of a proneural transcription factor. Atonal1 (Atoh1) likely modifies Shh signaling by binding to one of several Atonal binding sites on GLI2, the activator needed to transcribe Shh-responsive genes (**Figure 6.8**A). In the cerebellum, Atoh1 also influences granule precursor cells as they transition from proliferation to differentiation. At early postnatal stages of cerebellar development, Atoh1 regulates the expression of transcription factors associated with differentiation, such as NeuroD1. One hypothesis is that over time, the differentiation-specific, Atoh1-regulated transcription factors accumulate to surpass the proliferative effects of Atoh1, shifting the precursor pool from one that is primarily proliferating to one that is mainly differentiating (Figure 6.8B).

Granule cell differentiation is further influenced by the level of Notch receptor activity. Notch signaling regulates whether precursors in the EGL continue to proliferate or commit to the granule cell fate. Manipulations of Notch activity *in vivo* revealed that if Notch activity is experimentally increased, granule cells proliferate longer. Conversely, if Notch receptor activity is inhibited, cells stop proliferating early and begin to express transcription factors characteristic of committed granule cell neurons.

Mikio Hashino and colleagues described two populations of granule cell precursors. Notch2-expressing granule precursor cells—the signal receiving "Notch on" precursors and Jag1-expressing cells—the signal sending "Notch off" precursors. When Jag1, a Notch activating ligand, binds to the Notch2 receptor, the receptor bearing cell cleaves NICD which in turn upregulates *Hes1* (*Hairy/Enhancer of Split*) expression. *Hes1*, a mammalian downstream target of Notch, inhibits proneural *NeuroD1* gene expression to keep the cell in a proliferative state (Figure 6.8A). In contrast, the Jag1-expressing, Notch off granule cell precursor stops proliferating and begins to express proneural genes such as *NeuroD1* to differentiate as a granule cell (Figure 6.8B).

A combination of extrinsic signals appears to regulate subsequent development of cerebellar granule cells. For example, Pax6 activates BMP to influence the differentiation and migration of the committed granule cells of the iEGL (Figure 6.8) and brain-derived neurotrophic factor

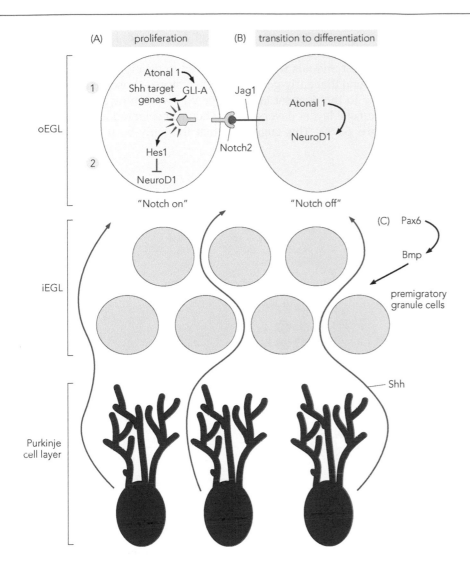

Figure 6.8 Multiple signals influence proliferation and differentiation of cerebellar granule cells. Sonic hedgehog (Shh) secreted by the Purkinje cells (purple) promotes proliferation of granule cell precursors in the outer region of the external granule cell layer (oEGL). (A) Atonal1 in the oEGL precursor cells has a unique function, helping to promote proliferation, likely by binding to a GLI activator (GLI-A) to influence Shh target genes (1). (B) As granule cells transition to the differentiation stage, Atonal1 promotes greater NeuroD1 expression causing a shift in Atonal1 function that allows cells to enter the differentiation stage. Differentiation is further regulated by the Notch signaling pathway. The Notch ligand Jagged 1 (Jag1), expressed on a subset of granule cell precursors, actives the Notch receptor on adjacent precursors. The activated Notch intracellular domain (NICD) in the receptor-bearing "Notch-on" cell upregulates the expression of *Hes1* that inhibits the proneural gene *NeuroD1*, thereby keeping the cell in the proliferative state (A, 2). In contrast, the Jag1 expressing "Notch-off" cell expressing *NeuroD1* begins to differentiate as a granule cell neuron. (C) Pax 6 activates BMP to influence differentiation of the premigratory cells in the inner EGL (iEGL).

(BDNF) supports later developmental events, including granule cell survival, the differentiation of granule cell processes, and the migration of the cells to the internal granule cell layer (IGL). Cerebellar granule cells must therefore integrate multiple signals to progress from a granule cell precursor to a fully differentiated granule cell neuron.

Temporally Regulated Transcription Factor Networks Help Mediate the Fate of Cerebral Cortical Neurons

In the vertebrate cerebral cortex, the time of neurogenesis is linked to the migratory destination and cell fate (see Chapter 5). Those cells born early migrate to the deepest layers of the emerging cortical plate (CP), while later-born neurons migrate to more superficial layers, thereby creating the "inside first, outside last" pattern of cortical development. The link between migration time and cortical layer destination suggested that there are both temporal and environmental cues to direct the neurons to the correct cortical layer. *In vivo* and *in vitro* studies have confirmed that both types of cues are important for cell fate determination of cortical neurons.

For example, in the 1990s a series of transplantation studies in the ferret demonstrated that cortical neural progenitors become progressively restricted in their cell fate options. The ferret is a popular animal model for studies of CNS development and the time of cortical layer formation is well documented. To monitor the fate of cortical neurons, the researchers harvested neurons from the ventricular zone of a donor cortex at one stage of development. The cells were then dissociated, labeled with tritiated thymidine, and injected into a host cortex at a different stage of development. The host animals were then allowed to develop for several days. The studies found that early-generated progenitors—those that would normally migrate to the deepest layer (layer VI)—were now able to migrate to more superficial layers (layers II/III) when transplanted to an older host cortex (**Figure 6.9**A). This suggested that the early-born neurons could

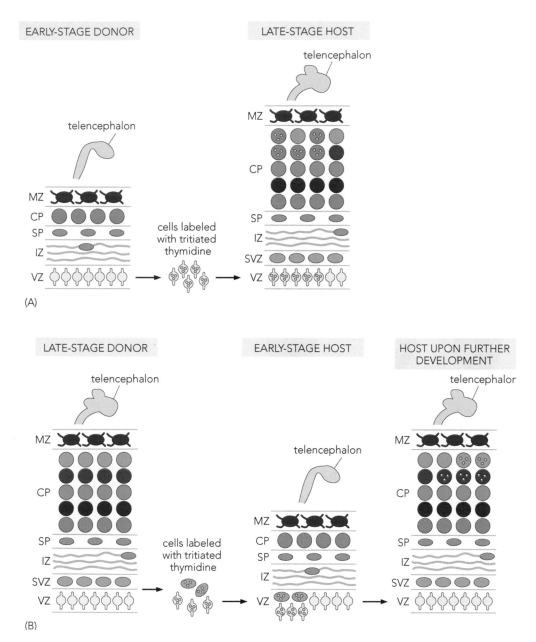

Figure 6.9 Cell fate options of cortical neurons become restricted as development progresses. (A) Neuronal progenitors were harvested from the ventricular zone (VZ) at an early stage of cortical development. The cells were dissociated, labeled with tritiated thymidine (yellow), and injected into the VZ of an older host embryo. In the older host embryo, the donor cells migrated past the deep layer (layer VI), where they would normally settle, and instead migrated to superficial layers (layers II/III) consistent with the age of the host embryo. Thus, early-stage cortical progenitors could alter their fate and migrate to a new layer. However, this effect was only seen if the cells were harvested early in the cell cycle, prior to the final mitotic division; cell fate could not be altered in post-mitotic cortical neurons. (B) Cortical progenitors were harvested from the VZ and subventricular zone (SVZ) of a late-stage donor embryo at a time the progenitors would migrate to superficial layers II/III. The labeled cells from the older donor were injected into an early-stage host embryo at a stage when the progenitors would migrate to deep layer VI. The injected donor neurons continued to migrate to the superficial layers (layers II/III; pink, green), indicating that older cortical neurons could not switch fate even in a new environment. MZ, marginal zone; CP, cortical plate; SP, subplate; IZ, intermediate zone.

respond to environmental cues in the host environment and migrate to a new destination. This effect was only seen when the cells were harvested early in the cell cycle, prior to the final mitotic division. Those cells harvested in late stages of the cell cycle still migrated to their original, deeper layer (layer VI). This finding suggested there were also intrinsic temporal cues present to influence cell fate options. Thus, by the time a cortical cell is post-mitotic, cell fate is established and unable to be altered, even when placed in a new host environment.

In contrast to early-stage neurons injected into older hosts, late-stage progenitors that normally migrate to layer II/III did not migrate to the deeper layer VI when transplanted to younger hosts, even when the cells were harvested early in the cell cycle (Figure 6.9B).

A third set of experiments further supported the hypothesis that the fate potential of the cortical neurons becomes gradually restricted during normal development. Mid-stage progenitors, those that would migrate to layer IV, were able to migrate to a new location (layer II/III) if transplanted to an older host. However, these progenitors were unable to migrate to the deeper layer VI when transplanted to a younger host. Together, the transplantation experiments suggested that early-mid-stage progenitors are multipotent early in the cell cycle and can adopt new cortical fates when placed in an older host environment. However, the progenitors gradually lose the ability to change fate as they reach mid-late stages of development. At these stages, the cells remain committed to the fate associated with their time of migration.

More recently, transcriptomics has added further insight into how fate is determined from a seemingly uniform group of cortical progenitors. Different transcription factor networks appear to subdivide progenitors into two broad subclasses: those that will form the deeper layers (layer V and VI) and those that form the more superficial (upper) layers (layers II–IV). It is thought that the transcriptional networks gradually change as cells switch from a progenitor cell to a post-mitotic neuron responsive to environment cues. These changes could explain, in part, why early-born neurons remained multipotent when transplanted, but later born progenitors did not.

Transcription factors specific to neurons located in different cortical layers at different stages of development have been identified. For example, Tbrl (T-box brain 1), FoxP2 (Forkhead box protein P2), and Fezf2 (Forebrain embryonic zinc finger-like protein 2) are associated with early-born cells found in deeper layers. In contrast, Cuxl (cut like homeobox 1), Cux2, Satb2 (SATB homeobox 2), and Brn2 (brain-2; also called POU class 3 homeobox 2, POU3f2) are specific for later-born cells that settle in more superficial layers (layers II–IV).

At some developmental stages, some of these transcription factors are found in cells of adjacent deep and superficial layers, suggesting daughter cells maintain both progenitor and neuron-specific transcriptional networks until they settle into their final layer and begin to express layer-specific markers. For example, Satb2 is found in both early-born cells that settle in deep layers and later-born cells that settle in superficial layers. Some combinations of transcription factors including Fezf2, Tbr1, and SatB2 appear to be mutually repressive and may therefore coordinate to prevent later born cells from adopting the fate of cells in an adjacent deeper layer. For example, SatB2 in superficial layers and Fezf2 in deeper layers cross-repress one another.

Increasing evidence suggests some progenitors are specified from early stages. For example, a subpopulation of radial glial cells destined for superficial layers expresses the Sox9 transcription factor. Sox9 regulates the cell cycle to keep these cells in proliferative state longer, until it is time for them to migrate. Ongoing studies continue to sort out how various transcription factor networks integrate to establish cortical layer cell fate. In general, transcription factors important for cell cycle regulation are highly expressed early, prior to migration while those important for the differentiation of axons, dendrites, and synapses are expressed later in postmitotic cells.

Temporal identity factors homologous to those found in *Drosophila* also play a role in cerebral cortical fate potential. For example, *Ikaros*, the mammalian ortholog of *Hunchback*, is expressed in cortical progenitors. In mice, *Ikaros* is detected in early-stage progenitors of the ventricular zone, but is decreased at later stages. When *Ikaros* was overexpressed in mice, the number of progenitor neurons was increased. If *Ikaros* expression was sustained, more early-born, Tbr1-positive neurons were generated and fewer late-born neurons were present (**Figure 6.10**). Thus, cells expressing markers for layer VI were increased, while those for layers III and IV were decreased. If *Ikaros* was misexpressed in later-born progenitors, early-born fates could not be generated, consistent with the idea that *Ikaros* encodes a temporal factor utilized only by early-generated progenitor cells, similar to the function of *Hunchback* in *Drosophila* neuroblasts. *Ikaros* appears to provide the early-generated neurons with the ability to adopt deep-layer cortical fates. The expression of other transcription factors is then needed for the cells to differentiate into mature cortical neurons with the characteristics typical of cells in that layer. Currently, it appears that stage-specific transcriptional networks, rather than single temporal identity factors, are critical for mammalian neocortical development.

Epigenetic Factors Influence Determination and Differentiation in Vertebrate Neurons

In recent decades studies have also begun to focus on how **epigenetic** factors influence cell fate options in the developing nervous system. Epigenetic mechanisms play a very important role in regulating gene activation and repression by controlling the accessibility of DNA binding sites to transcription factors. Common epigenetic modifications include DNA methylation—the process by which methyl groups are added to DNA at the promoter region, often to repress gene transcription; noncoding RNAs—RNAs that are not translated into protein but instead influence gene expression at transcriptional or post-transcriptional stages; and histone

Figure 6.10 The temporal identity factor Ikaros influences the fate of early generated cortical neurons. *Ikaros*, the mammalian ortholog of the *Drosophila Hunchback*, is expressed in early-stage progenitors in the VZ, but not later-stage progenitor neurons. (A) Early-born neurons that settle in future layer VI are positive for the transcription factor Tbr1 (Tbr1+). (B) In experimental mice expressing *Ikaros* through the stages that later-born neurons are generated, the number of early-born Tbr1+ cells (yellow outline) increased and migrated throughout more superficial layers. However, if *Ikaros* was expressed in later-born progenitors, cell fate was not altered (not shown). These results support the hypothesis that the temporal identity factor Ikaros, like Hunchback, only influences the fate of early-generated progenitor cells.

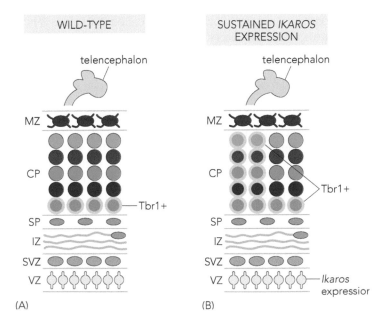

modifications—post-translational modifications to the histone proteins that wrap around the DNA strand. Histone modifications include the recruitment of histone modifiers or alterations in chromatin structure. Chromatin is comprised of the DNA strand and the histone proteins. Changes in chromatin structure influence the accessibility of target genes. For example, when chromatin is in a lightly packed (euchromatin) state, the corresponding promoter region of a target gene becomes accessible to transcription factors.

Epigenetic mechanisms are widely used throughout the developing embryo. Research in the developing vertebrate CNS has revealed several important epigenetic modifications that determine whether a cell remains in the proliferative state or begins the determination process. One example comes from genes related to the *Drosophila* gene *Brahma* (*Brm*). The group of related factors in vertebrates includes ATP-dependent chromatin-remodeling enzymes of the BAF complex. BAF stands for Brg1 (brahma-related gene 1) and Brm- (brahma-) associated factors. This group of proteins determines the accessibility of DNA binding sites.

Brg1 appears to be particularly important during neural proliferation, whereas Brm is required for cell fate determination of progenitors and differentiation of post-mitotic neurons. Brg1 and other subunits are needed to maintain Notch signaling and repress proneural genes to keep cells in a proliferative state. In contrast, Brm and other subunits activate the transcription of neuron-specific genes such as *Neurogenin 1* and *NeuroD1*.

The BAF complex is comprised of at least 15 subunits whose composition changes as cells progress from a proliferative to post-mitotic state. In the vertebrate CNS, the neural embryonic stem (ES) cells express subunits that comprise the esBAF complex (**Figure 6.11**), whereas neural progenitor (np) cells express slightly different subunits in the npBAF complex. Post-mitotic neurons (n) express a third group of subunits to make up the nBAF complex. The changes in subunit composition correlate with the transition

Figure 6.11 Developmental changes in the subunit composition of the BAF complex influence proliferation and fate options of vertebrate neurons. Subunits of the BAF (brahma-related gene 1 and brahma-associated factors) complex change as neurons progress from proliferative to post-mitotic stages. The subunit composition influences the accessibility of DNA binding sites for transcription factors and therefore whether target genes are expressed. The BAF complex is comprised of multiple subunits. (A) In embryonic stem (ES) cells, the BAF complex (esBAF) includes two 155 subunits, as well as a 45a and a 53a subunit that are important during neural development. (B) In neural progenitor (np) cells, the BAF complex (npBAF) continues to express the 45a and 53a subunits, but exchanges one of the 155 subunits with 170. (C) In post-mitotic neurons (n), the BAF complex (nBAF) no longer expresses the 43a and 53a subunits, but instead expresses the 45b and 53b subunits. nBAF continues to express the 170 subunit that alters the chromatin state so that transcription factors such as Pax6 can access their target genes and induce cellular characteristics of nonproliferating and post-mitotic neurons. In esBAF, the lack of the BAF 170 subunit means the chromatin is tightly packed, in which case Pax6 and other binding sites are less accessible in ES cells, thus preventing them from adopting a neural fate prematurely. [Adapted from Yoo AS & Crabtree GR [2009] *Curr Opin Neurobiol* 19:120–126.]

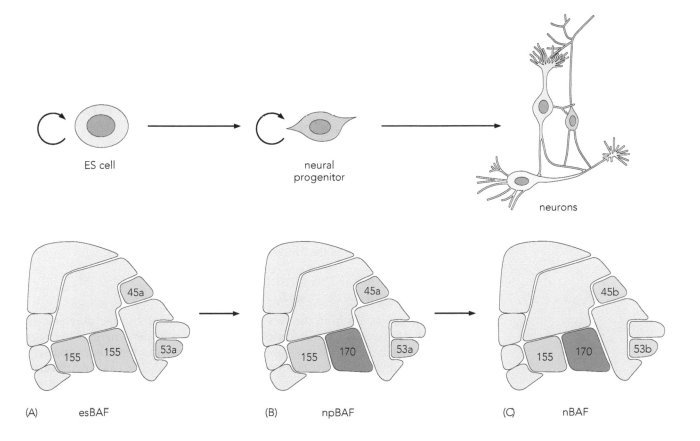

(A)   esBAF   (B)   npBAF   (C)   nBAF

to each stage of neuronal development. Thus, the subunit composition of the BAF complex, as well as the dosage of individual subunits, influences whether the cells continue to proliferate or progress to the post-mitotic state. For example, in the developing cerebral cortex and cerebellum, the BAF subunits BAF45a and BAF53a are required for the continued proliferation of neural progenitor cells. In post-mitotic neurons, these subunits are exchanged with BAF45b and BAF53b (Figure 6.11). The changes in subunit expression are correlated with the transition from proliferating progenitors to committed cortical and granule cell fates.

Further, changes in BAF subunit composition have been linked to changes in the activity of the transcription factor Pax6. In the developing CNS, Pax6 plays a wide range of roles throughout the developing forebrain (see Chapter 3). In the cerebral cortex, Pax6 activates target genes such as *Tbr2* (T-box brain 2), which is detected in nonproliferating progenitors (for example, the basal progenitor cells). Pax6 then activates *Cux1*, which is detected in post-mitotic neurons, and *Tle1* (transducin-like enhancer protein-1), which is needed for the survival of post-mitotic neurons. During early stages of neurogenesis (E12.5–E14.5 in mouse), two BAF155 subunits are highly expressed (the esBAF complex). BAF155 inhibits the euchromatin state of Pax6 target genes. This means the DNA promoter regions for *Tbr2*, *Cux1*, and *Tle1* are not easily accessible. Thus, Pax6 cannot readily bind to the target genes and initiate their expression in proliferating cells. As subunits of the npBAF complex begin to be expressed, one of the BAF155 subunits is replaced with BAF170 (Figure 6.11). The decreased expression of BAF155 and concurrent increase in BAF170 expression leads to greater accessibility to DNA promoter regions so that Pax6 can initiate the expression of the target genes in the neural progenitor and post-mitotic neurons at the times they are needed.

These examples indicate one way epigenetic regulation of transcription factor binding sites can influence whether genes necessary for determination and subsequent differentiation are expressed. In addition to the role of the BAF complex in CNS neurons, other BAF subunit complexes are associated with the differentiation of Schwann cells and oligodendrocytes. Thus, epigenetic modifications provide another means by which the limited number of available transcription factors can exert a wide range of effects in the developing nervous system.

DETERMINATION AND DIFFERENTIATION OF NEURAL CREST-DERIVED NEURONS

The experimental accessibility of the neural crest has allowed investigators to study the fate of different neural crest-derived cell populations. As discussed in Chapter 5, the fate options available to neural crest cells are probably the most varied in the nervous system, and each neural crest cell population relies on specific signals for determination and differentiation. Most neural crest cells appear to be particularly influenced by extrinsic signals encountered as they migrate toward their destinations.

Environmental Cues Influence the Fate of Parasympathetic and Sympathetic Neurons

Neural crest cells from the caudal hindbrain through the sacral region are divided into vagal and trunk populations. Among the derivatives of the vagal and trunk neural crest cells are neurons in the parasympathetic and sympathetic divisions of the autonomic nervous system. The vagal neural crest gives rise mainly to parasympathetic neurons that innervate the gut and utilize the neurotransmitter acetylcholine. In contrast, sympathetic neurons that innervate smooth muscle cells and utilize the neurotransmitter norepinephrine (also called noradrenaline) are derived from the trunk neural crest. The vagal and trunk populations of neural crest cells have been utilized extensively to evaluate whether neural crest cell fate is predetermined or regulated by cues from the extracellular environment.

The influence of the environment on the fate options of parasympathetic and sympathetic neurons from the vagal and trunk regions of the neural tube was first described in the now-classic studies of Nicole Le Douarin and colleagues in the 1970s. These studies relied on transplantation techniques pioneered by Le Douarin in which the neural crest cells of quail were transplanted to a chick embryo at a similar stage of development. In such cases, the quail cells integrate into the chick host and differentiate as if they were chick cells. Because the quail cells can be identified histologically by the increased heterochromatin in the nucleus (see Figure 3.12), investigators are able to determine the fate of the transplanted quail cells.

Using this chick–quail chimera method, Le Douarin and colleagues transplanted neural crest cells from one region of the neural tube of a quail embryo to a different region of a chick neural tube. Because neural crest development progresses from rostral-to-caudal (anterior-to-posterior), cells from the vagal region develop prior to those of the trunk region. Vagal-to-trunk and trunk-to-vagal transplantations were performed with these progressive developmental differences accounted for so that neighboring donor and host cells were at similar stages of development (**Figure 6.12**).

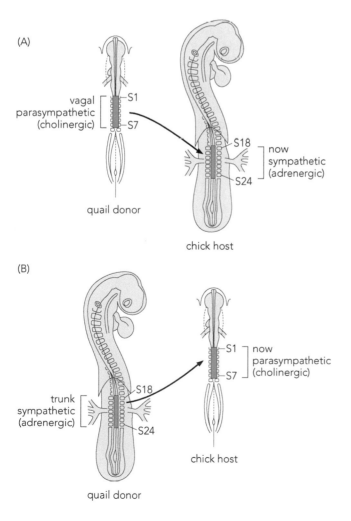

Figure 6.12 Transplanted quail neural crest cells adopted the fate of the chick host tissue. (A) When neural crest cells from the vagal region of a quail donor—cells that normally differentiate as parasympathetic, cholinergic neurons—were grafted to the trunk region of the chick host, the transplanted cells migrated along the typical route of trunk cells and adopted a sympathetic, adrenergic fate. (B) Conversely, when quail trunk neural crest cells—cells that normally become sympathetic, adrenergic neurons—were grafted to the vagal region of the chick embryo, the cells migrated along routes typical of vagal neural crest cells and adopted a parasympathetic, cholinergic fate. The size difference between the quail donors and chick hosts reflects differences in the developmental stage of the embryos. Development in the vagal region precedes trunk development, so to ensure that the transplanted donor cells are at the same developmental stage as their neighboring host cells, cells from the vagal region of a younger donor were transplanted into the trunk region of an older host (A), whereas cells from the trunk region of an older donor were transplanted into the vagal region of a younger host (B). [Adapted from Le Douarin NM [1980] *Nature* 286:663–669.]

When cells from the vagal neural crest were transplanted to trunk regions, the majority of the vagal cells now migrated along the route of trunk-derived neural crest cells and developed into sympathetic neurons. Similarly, when trunk neural crest cells were transplanted to vagal regions, the trunk cells took the expected migratory route for vagal crest cells and became parasympathetic neurons. Thus, the fate of parasympathetic and sympathetic neurons was not predetermined but appeared to depend on the axial level and associated environmental cues encountered along a particular migratory route. More recent studies suggest environmental cues alter neural crest fate by regulating the expression of specific transcription factors, at least during defined stages of development. For example, BMPs activate transcription factors to specify subtypes of sympathetic neurons while Wnt signaling activates the expression of neurogenins (Ngns) that are important for development of dorsal root ganglion (DRG) neurons.

Wnt influences *Ngn2* expression in early-migrating DRG neurons and *Ngn1* expression in late-migrating DRG neurons. These Ngns then activate the expression of the neuronal differentiation marker, NeuroD1. Cells expressing Ngns also increase their expression of Delta-like ligand 1 (Dll1) that binds to Notch receptors expressed on surrounding cells. The cells with activated Notch signaling are prevented from developing as neurons so only those cells expressing Ngns adopt a neural fate.

Sympathetic Neurons Can Change Neurotransmitter Production Later in Development

During normal development, all sympathetic neurons originate as adrenergic neurons that produce norepinephrine (noradrenaline). Most sympathetic neurons will innervate tissues such as skin and smooth muscle cells, and these neurons remain adrenergic (**Figure 6.13**A). BMPs from the dorsal aorta appear to provide the local environmental signal that induces the production of the enzymes needed to synthesize norepinephrine. BMPs also activate various combinations of transcription factors such as members of the Phox (paired-like homeobox) family and GATA3 (GATA binding protein 3). The particular combinations of transcription factors expressed in a given cell influence the differentiation of sympathetic neuron subtypes.

Figure 6.13 Neurotransmitter production can change during postnatal development. (A) All sympathetic neurons are initially adrenergic, producing the neurotransmitter norepinephrine (also called noradrenaline). At the time of innervation, most sympathetic neurons, such as those that innervate smooth muscle, continue to produce norepinephrine (upper panel, wild-type). These sympathetic neurons remain adrenergic into adulthood. However, some sympathetic neurons switch neurotransmitter fate to become cholinergic. For example, at the time that sympathetic nerve fibers innervate sweat gland tissues, their neurons begin to produce acetylcholine (lower panel, wild-type). (B) Transplantation studies revealed that changing the target tissue altered sympathetic neuron neurotransmitter production. When sweat gland tissue (a cholinergic target) replaced smooth muscle tissue (an adrenergic target), the sympathetic neurons switched from their normal adrenergic fate and became cholinergic. (C) Conversely, when the sweat gland was replaced with the parotid gland (another adrenergic target), the sympathetic neurons did not switch to cholinergic as usual, but remained adrenergic.

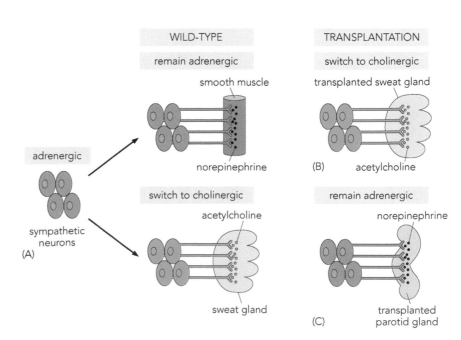

A smaller population of sympathetic neurons innervates other target cells such as sweat glands. Although these sympathetic neurons initially produce norepinephrine, once they begin to innervate their target tissues, they stop expressing the enzymes needed for norepinephrine synthesis and begin to express the enzymes needed to make acetylcholine. Thus, these sympathetic neurons switch from adrenergic to cholinergic (Figure 6.13A). In rats, this takes place during the second postnatal week at the time of sweat gland innervation—a relatively late stage to switch an aspect of cell fate. Transplantation, lesion, and cell culture experiments demonstrated that the cells that innervate sweat glands arise from the same population of sympathetic neurons. Thus, the change in neurotransmitter is not a result of differential survival of a subset of neurons.

The signal that instructs the neurons to stop producing norepinephrine and start producing acetylcholine appears to come from the target tissue itself. Evidence for the role of the target tissue again comes from multiple experimental approaches. For example, if a tissue that contains sweat glands, such as the footpad of rat, is transplanted to a region that does not have many sweat glands, such as the skin of the thoracic region, the arriving sympathetic neurons innervate the transplanted footpad and over a period of three to six weeks switch neurotransmitters, becoming cholinergic sympathetic neurons (Figure 6.13B). Conversely, when an adrenergic target, the parotid gland, is transplanted to the footpad region, the innervating sympathetic neurons remain adrenergic and do not switch to cholinergic neurons (Figure 6.13C). Similar results were found in tissue culture studies. Co-culturing adrenergic sympathetic neurons with sweat gland tissue caused the neurons to switch to cholinergic sympathetic neurons. In contrast, sympathetic neurons remained adrenergic when cultured with an adrenergic target.

The identity of the diffusible sweat gland-derived signal remains uncertain. Several candidate molecules have been identified, including cytokines such as ciliary neurotrophic factor (CNTF) and leukemia inhibitory factor (LIF). Scientists continue to study the growth factors and signal transduction cascades that regulate this aspect of sympathetic neuron differentiation, both during development and in response to injury (**Box 6.1**).

DETERMINATION OF MYELINATING GLIA IN THE PERIPHERAL AND CENTRAL NERVOUS SYSTEM

In addition to generating multiple subtypes of neurons, the nervous system also produces numerous subtypes of glia. Many of the glial subtypes adopt their fates after neuronal fates are specified. Among the glia produced in the vertebrate peripheral and central nervous systems are the myelinating glia that wrap around axons to speed the conduction of action potentials.

Neuregulin Influences Determination of Myelinating Schwann Cells in the PNS

The myelinating glia of the peripheral nervous system are the Schwann cells. Schwann cell determination and differentiation typically occur after neural fate specification has begun. Signals from the associated neural-crest-derived neurons induce the formation of Schwann cells and initiate the myelination of peripheral axons. These signals include members of the Neuregulin (Nrg) family of proteins. There are four members

Box 6.1 Developing Neuroscientists: Gp130 Cytokines Play Key Roles in Regulating Transmitter Phenotype during Development and in Response to Injury

Richard Zigmond, Ph.D.

Richard Zigmond received his undergraduate degree from Harvard University in 1966 and his Ph.D. from Rockefeller University in 1971. Since 1989 he has been a faculty member in the Neurosciences Department at Case Western Reserve University, where his research focuses on the changes in gene expression in sympathetic and sensory neurons after axotomy. He is currently investigating interactions between the immune and nervous systems and macrophages and neutrophils in the degeneration and regeneration of axotomized sensory neurons.

The vast majority of sympathetic neurons use norepinephrine as their primary neurotransmitter. In the 1970s and 1980s, a group of researchers that included Edwin Furshpan, Story Landis, Paul Patterson, David Potter, and colleagues discovered that when neonatal rat sympathetic neurons from the superior cervical ganglion (SCG) were cultured with certain nonneuronal cells—for example, heart cells—they underwent a switch in the transmitter they synthesized and released. The neurons switched from producing norepinephrine to producing acetylcholine. Patterson's laboratory went on to determine that the factor released by cultured heart cells is leukemia inhibitory factor (LIF), a protein previously known primarily in the immune system. Landis examined whether a similar "cholinergic switch" ever occurs in sympathetic neurons *in vivo*. Basing her studies on previous observations that sympathetic innervation of adult sweat glands uses acetylcholine as its transmitter, her lab discovered that, *prior* to innervating their targets, these neurons synthesize norepinephrine, but that *after* contact with sweat glands, the neurons synthesize acetylcholine—remarkably analogous to the situation that had been found in cell culture. Elegant tissue transplant studies followed and firmly established that it was the target sweat glands that acted on the sympathetic neurons to trigger the cholinergic switch.

Whether the sweat glands acted on the neurons through LIF was not determined immediately. A major advance came from the availability of animals in which the gene for LIF had been knocked out (LIF$^{-/-}$). Studies on these animals produced some rather surprising results. Landis's laboratory found that, in LIF$^{-/-}$ mice, the cholinergic switch occurred just as in wild-type animals. How do we account for the fact that while LIF can trigger the cholinergic switch in culture, it is not required *in vivo*?

It is now recognized that LIF belongs to a family of peptides that does not have a lot of amino acid sequence homology, but does have a common three-dimensional structure and acts through a common receptor system that includes the signaling subunit gp130 (see Chapter 8). These LIF-related cytokines are often referred to as **gp130 cytokines**. Further studies in neonatal sweat glands and in adult SCGs established that in fact other members of the gp130 family, in addition to LIF, were present and almost certainly are involved in switches in transmitter expression.

In adult animals, changes in the neurotransmitters that sympathetic neurons synthesize and release can also be dramatically altered in response to severing the cells' axons (axotomy). For example, SCG neurons begin to express several additional neuropeptides after axotomy, including vasoactive intestinal peptide (VIP) and galanin. Our lab found that the increases in VIP and galanin were significantly reduced, though not totally abolished in LIF$^{-/-}$ mice. This led us to question why this may be so.

Here again the development of mutant mice allowed the research to move forward. The laboratory of Hermann Rohrer made a conditional knockout of the gp130 receptor subunit in neurons synthesizing norepinephrine. The researchers found that these mice did not undergo a cholinergic switch. Instead, the neurons innervating sweat glands remained adrenergic. These studies demonstrated that these neurons required the binding of LIF-related cytokines to gp130 in order to induce changes in neurotransmitter synthesis.

Using gp130-knockout mice, our laboratory then found that the changes in SCG neurons that occur in response to neuronal injury also depend on gp130 signaling. For example, the increases in expression of VIP and galanin that are normally observed after axotomy were completely abolished in the absence of gp130. Further, increases in nerve fiber outgrowth that are typically seen *in vitro* following nerve injury were absent in neurons harvested from gp130-knockout mice. Together, these studies support the hypothesis that gp130 cytokines are necessary for the inducing characteristics of sympathetic neurons during development and in response to injury.

of the Nrg family (Nrg1–Nrg4) as well as several isoforms of Nrg1 that are either tethered to the cell membrane or released from the cell surface following proteolytic cleavage. In the PNS, Nrg1 stimulates Schwann cell proliferation, survival, migration, and myelination. As discussed in Chapter 5, Nrg1 also plays a role in the development of Bergmann glia in the cerebellum.

Since the 1990s, numerous studies have demonstrated the importance of Nrgs in Schwann cell determination and myelin formation. For example, in cell cultures of neural crest precursor cells from the dorsal root ganglia (DRG), the addition of Nrg1 led to a decrease in the number of neurons and an increase in the number of Schwann cells. Because the total number of cells remained the same, the results suggested that Nrg1 signaling suppresses neural fate while promoting Schwann cell fate (**Figure 6.14**). Nrg1 was initially called glial growth factor (GGF) due to this ability to stimulate the production of glial Schwann cells *in vitro*.

The Nrgs signal through ErbB receptors that are found on neural and Schwann cell precursors. Expression studies demonstrated that the Nrg1 ligands and ErbB receptors are distributed on neuronal and Schwann cell precursors at the correct developmental stages to initiate Schwann cell fate and stimulate myelination. Schwann cells were not produced in mice lacking *Nrg1*, *ErbB2*, or *ErbB3*, confirming the importance of this signaling pathway in Schwann cell development. One model proposes that once developing neurons begin to express sufficient Nrg1, they are able to activate ErbB receptors on adjacent neural crest cells, initiating the MAPK/Erk Kinase (MEK) pathway that drives them to become Schwann cells. Axon-derived Nrg1 then stimulates the myelination process, inducing the Schwann cells to extend cytoplasmic processes to wrap around the axon (**Figure 6.14**).

Studies continue to explore the mechanisms that govern the determination and differentiation of peripheral glial cells. The signaling pathways that drive proliferation and tumor formation in adults are also active areas of research (**Box 6.2**). Other efforts focus on understanding the similarities and differences that underlie myelination in the PNS and CNS.

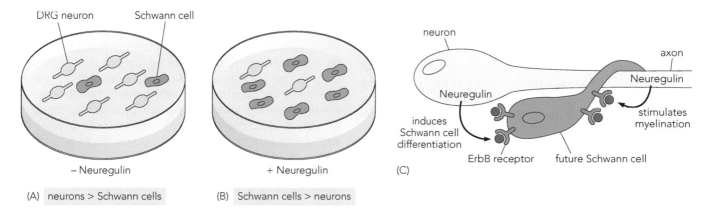

(A) neurons > Schwann cells   (B) Schwann cells > neurons

Figure 6.14 Neuregulin suppresses neuronal fate and stimulates Schwann cell determination and myelination. (A) When precursor cells from dorsal root ganglia (DRG) were placed in a cell culture dish lacking Neuregulin 1 (Nrg1), most of the precursor cells differentiated into DRG neurons. (B) When Neuregulin 1 was added to these cell cultures, the number of Schwann cells increased, although the total number of cells remained the same. These results suggested that Neuregulins stimulate differentiation of Schwann cells while suppressing differentiation of neurons. (C) Peripheral neurons produce Neuregulin 1, which binds to ErbB receptors on adjacent neural crest precursor cells, signaling those precursors to differentiate into Schwann cells. Additionally, Neuregulin from the axon initiates the myelination process so that extensions from the Schwann cell begin to wrap around the peripheral axon.

Box 6.2 Surgeons' Perspective: Diagnosis, Treatment, and Pathogenesis of Vestibular Schwannomas

Sean Carroll, D.O. and Robert Archbold, D.O.

Dr. Carroll is the program director for the Otolaryngology-Head and Neck Surgery Residency Program at UPMC Hamot. He completed his pre-med major at Gannon University, graduated from the Midwestern University-Chicago College of Osteopathic Medicine and completed his residency training with Ohio University.

Dr. Archbold graduated from the University of California at Santa Barbara and Lake Erie College of Osteopathic Medicine. He is a fourth-year resident in Otolaryngology Head and Neck Surgery at UPMC Hamot.

Schwann cells sometimes form tumors called schwannomas. These tumors can develop along any peripheral or cranial nerve but often arise from the vestibular division of the cochleovestibular nerve (cranial nerve VIII, see Figure 6.23A). Although frequently called an acoustic neuroma, the more accurate term is vestibular schwannoma. **Vestibular schwannomas** are benign tumors (neoplasms) that account for roughly 90% of the intracranial tumors found in the cranial fossa (the portion of the skull that houses the cerebellum and brainstem). Specifically, vestibular schwannomas are found in the cerebellopontine angle (CPA), a site within the cranial fossa filled with cerebrospinal fluid and bounded by the brainstem, cerebellum, and temporal bone of the skull.

On average, 2,000–3,000 new cases of vestibular schwannoma are diagnosed annually in the United States. Most cases are sporadic and occur in patients between 40 and 60 years old, with males and females equally affected. Vestibular schwannomas typically develop in the internal auditory canal (IAC) that extends from the cranial fossa to the inner ear. Occasionally a schwannoma forms in the CPA medial to the porus acusticus, the medial opening of the IAC.

As otolaryngologists and head and neck surgeons, we focus on diagnosis, monitoring, and surgical treatments of vestibular schwannomas. Fortunately, these tumors usually develop slowly, with average annual growth rates of 0.2 cm. However, in rare cases, vestibular schwannomas exhibit a rapid growth rate. If Schwann cell proliferation continues unabated, these neoplasms can be fatal due to compression of the brainstem, leading to increased intracranial pressure and death over the course of five to 15 years.

For most patients, symptoms arise gradually as the tumor grows and exerts pressure on surrounding neurovascular structures. With a vestibular schwannoma, approximately 90% of patients will have some degree of asymmetrical sensorineural hearing loss, with worse hearing in the ear on the side of the schwannoma. It is, therefore, important to evaluate any patient with asymmetric hearing loss for this type of tumor. About 70% of these patients will also have tinnitus (ringing in the ear), commonly, or worse, in the ear with hearing loss. Despite these tumors arising from Schwann cells of the vestibular nerve, only half of patients will experience any dizziness, with most reporting a mild sense of being off balance, and only 10% being truly vertiginous (experiencing the sensation of room-spinning).

Patients may experience cranial nerve palsies when the tumor exceeds 2–3 cm in greatest dimension, though these are rare. If the trigeminal nerve (CN V) is affected, the patient may complain of facial numbness or neuralgia (severe nerve pain). If the facial nerve (CN VII) is compressed, a patient will present with some level of facial nerve paralysis.

An asymmetric sensorineural hearing loss may be the first indication of a vestibular schwannoma. With additional hearing tests, patients usually exhibit decreased speech-discrimination (ability to understand words presented at a comfortable listening level), exceeding what would be expected given the amount of hearing loss. Many also have a paradoxical worsening of word-recognition scores when words are presented at a louder level, a phenomenon termed rollover.

Magnetic resonance imaging (MRI) can detect tumors as small as 1.5 mm, whereas computerized tomography (CT) is unable to detect vestibular schwannomas smaller than 15 mm. Therefore, CT imaging is reserved for those patients who are unable to undergo MRI due to the presence of an implantable device such as a cardiac pacemaker or cochlear implants, or for those patients with claustrophobia.

The treatment of vestibular schwannomas depends on several factors, chief among them being tumor size. Other patient characteristics pertinent in the treatment decision include age, general health, pre-existing hearing loss, and patient preference. There are three different treatment options: observation with serial imaging, radiosurgery, and microsurgical resection.

Small tumors, defined as less than 1.5 cm in maximum dimension, may be treated with any of the three approaches. Due to the slow growth of vestibular schwannomas, observation with serial imaging is often chosen for small tumors. This is also indicated in elderly patients or those too frail to undergo a more extensive intervention.

Medium-sized tumors are those measuring 1.5–3.0 cm, and for these, observation is no longer feasible. As such, patients may elect radiosurgery or microsurgery.

Radiosurgery involves treatment with a single fraction of radiation. The goal of this type of therapy is to prevent new tumor growth, with tumor control reported in 90–95% of patients ten years after treatment. Currently, the Gamma Knife is the most used radiosurgery platform. Gamma Knife is not a knife and there is no incision. Instead, it is a form of stereotactic radiosurgery that delivers about 200 tiny beams of radiation directly to the tumor. The benefits of choosing radiosurgery include that it is an outpatient treatment, there are no posttreatment restrictions, and essentially no recovery time. Side effects are few; however, radiosurgery will accelerate the natural course of hearing loss, though this is gradual over time.

Once tumors grow beyond 3.0 cm, they are classified as large tumors and are only amenable to surgical intervention. There are several surgical approaches, and the one chosen is tailored to the patient's specific pathology, tumor size, functional status, and surgeon's expertise. One common approach is called the translabyrinthine approach. This approach provides wide, direct access to the CPA with the best visualization of the facial nerve so the surgeon can avoid injuring that nerve. The main drawback is that hearing will be sacrificed on the operated side as CN VIII is damaged. Regardless of which surgical approach is used, gross tumor resection (removal of the tumor in its entirety) is the goal. In some cases, subtotal resection (some tumor remaining) is acceptable to limit cranial nerve damage or the risk of stroke. Tumor remnants would then be treated with radiosurgery.

As clinicians, we are also interested in the underlying causes of schwannomas. Although most vestibular schwannomas arise sporadically, two genetic diseases are commonly associated with these tumors. **Neurofibromatosis type 1** (NF-1, or von Recklinghausen disease) is inherited in an autosomal dominate fashion with variable penetrance. The defect occurs in the NF-1 gene on chromosome 17, resulting in the development of neuromas from Schwann cells of any nerve in the body. Over 5% of those with an NF-1 mutation develop a unilateral vestibular schwannoma.

The second genetic disease is **neurofibromatosis type 2** (NF-2), which is characterized by bilateral vestibular schwannomas in up to 96% of affected patients. These tumors develop earlier in life than the unilateral version, often before the age of 21. Because of this, when a vestibular schwannoma is diagnosed before the age of 30, the contralateral ear is closely followed for the development of a second schwannoma. Patients with NF-2 also commonly have other cranial nerve schwannomas, meningiomas (tumors arising in the meninges surrounding the brain and spinal cord), and ependymomas (tumors arising from ependymal cells). The defective gene in NF-2 is found on chromosome 22.

The normal NF-2 gene produces a protein product called **Merlin** (moesin-ezrin-radixin-like protein). Merlin is a scaffold protein structurally similar to the ezrin, radixin, moesin (ERM) family of cytoskeletal linker proteins. Merlin normally interacts with integrins and proteins in adherens junctions. A loss of Merlin is thought to interfere with attachment of Schwann cells to peripheral axons. Under normal conditions, Merlin acts as a tumor suppressor, inhibiting the expression and activity of tyrosine kinase receptors, including epidermal growth factor receptors (EGFRs). Merlin can also directly inhibit the mitogen-activated protein kinase (MAP kinase) pathway used in cell proliferation. When mutations in the NF-2 gene occur, Merlin can no longer inhibit these pathways and tumors, including schwannomas, develop.

Although benign and slow growing, vestibular schwannomas can impact the quality of life for those experiencing them. Understanding what drives Schwann cells to proliferate in adulthood could lead to ways to prevent or better treat at least some forms of schwannoma.

Precursor Cells in the Optic Nerve Are Used to Study Oligodendrocyte Development

The myelinating glia of the CNS are the oligodendrocytes. To investigate the mechanisms regulating determination and differentiation of oligodendrocytes, many studies have utilized preparations of the optic nerve. The optic nerve became a useful experimental system due to its relatively simple composition. The optic nerve contains the axons of the retinal ganglion cells (RGCs) and two types of glial cells: type 1 astrocytes, which contact blood vessels that run through the optic nerve, and oligodendrocytes. The type 1 astrocytes arise from the epithelial cells in the optic stalk, whereas the oligodendrocytes arise from **oligodendrocyte precursor** cells (OPCs) that originate in the ventricular zone near the third ventricle and migrate into the optic nerve.

Because the optic nerve contains no neuronal cell bodies, experiments that focus specifically on glial cell differentiation can be designed. Since the 1970s, numerous discoveries have revealed that the glial cells in the optic nerve first proliferate then form different glial types in response to a combination of extrinsic cues and intrinsic timing mechanisms. *In vitro* analysis of optic nerve cells has proved particularly useful for studying the development of oligodendrocytes.

In contrast to the cellular composition of the optic nerve *in vivo*, early studies of rat optic nerve cell cultures revealed that three different glial types were present (**Figure 6.15**). The cultures not only included type 1 astrocytes and oligodendrocytes, but also type 2 astrocytes. A single precursor population of cells gave rise to both the oligodendrocytes and type 2 astrocytes, whereas the type 1 astrocytes were derived from a separate progenitor pool. Because of these initial observations, OPCs were originally called O2A cells—that is, these precursors gave rise to either oligodendrocytes (O) or type 2 astrocytes (2A). However, later studies revealed that type 2 astrocytes were generated only under certain cell culture conditions and did not normally contribute to the optic nerve *in vivo*. The observation that OPCs can give rise to type 2 astrocytes in cell culture demonstrated that OPCs have a degree of plasticity that allows them to differentiate into type 2 astrocytes when provided with the proper signals. However, *in vivo*, developing OPCs produce signals to suppress the production of type 2 astrocytes.

Deciphering the signaling mechanisms that regulate glial fate in the optic nerve was aided by the identification of proteins and other antigens that are selectively present on the surfaces of specific cell types. For example, antibodies against rat neural antigen-2 (RAN-2) selectively bind to type 1, but not type 2, astrocytes. Anti-galactocerebroside (GC) binds only to differentiated oligodendrocytes, while antibodies against A2B5 bind OPCs and type 2 astrocytes. The restricted production of these antigens led to the development of a cell culture method to selectively harvest a given cell population (**Figure 6.16**). Cell culture dishes are coated with an antibody to one of the cell-type-specific antigens. When dissociated optic nerve cells are added to the dish, the cells that produce that antigen bind to the antibody and adhere to the culture surface. The adherent cells can then be studied and the loose cells can be transferred to another dish coated with a different antibody. When the final dish is coated with A2B5, a purified population of OPCs results.

By successfully isolating OPCs *in vitro*, researchers determined that OPCs rely on platelet-derived growth factor (PDGF) or neurotrophin-3 (NT-3) during the proliferative phase of development. Once the proliferation phase ends, a second signal is required to promote either the oligodendrocyte or type 2 astrocyte fate. Oligodendrocytes require signals such as thyroid hormone (TH) or retinoic acid (RA), whereas the type 2 astrocytes require signals such as BMPs.

In vivo studies confirmed the necessity of these factors for oligodendrocyte development. For example, PDGF and NT-3 are produced by type 1 astrocytes in the optic nerve to promote OPC proliferation. TH is present throughout the developing nervous system and the receptors for TH are localized to OPCs and oligodendrocytes. The necessity of TH in oligodendrocyte development was seen in hypothyroid rats and mice. The optic nerves of these animals had fewer oligodendrocytes and exhibited delayed myelination throughout the CNS. Conversely, in hyperthyroid mice myelination was accelerated, further demonstrating the importance of this hormone in regulating oligodendrocyte development in the CNS.

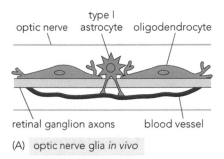

(A) optic nerve glia *in vivo*

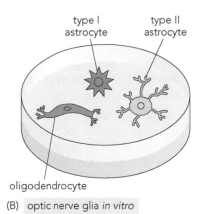

(B) optic nerve glia *in vitro*

Figure 6.15 Identification of optic nerve glial types *in vivo* and *in vitro*.
(A) *In vivo*, the adult optic nerve contains two glial cell types: oligodendrocytes, which myelinate axons of the retinal ganglion cells, and type 1 astrocytes, which contact blood vessels. (B) *In vitro*, optic nerve progenitor cells generate three types of glial cells: type 1 astrocytes, oligodendrocytes, and type 2 astrocytes. The type 1 astrocytes are derived from one progenitor pool, while oligodendrocyte precursor cells (OPCs) give rise to the oligodendrocytes and type 2 astrocytes found in cell culture. Type 2 astrocytes are not found *in vivo* due to the presence of molecules that suppress the formation of type 2 astrocytes in the optic nerve.

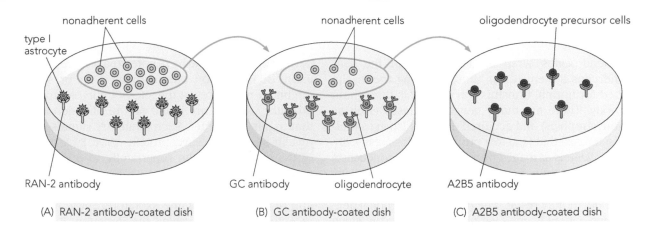

(A) RAN-2 antibody-coated dish   (B) GC antibody-coated dish   (C) A2B5 antibody-coated dish

Figure 6.16 Cell cultures are used to harvest specific optic nerve glia. Cell culture dishes are coated with an antibody that selectively binds an antigen on one of the optic nerve glial types. When dissociated optic nerve cells are added to the dish, cells that produce the corresponding antigen bind to the culture dish. Scientists then study the cells that adhere to the dish, or remove the loose cells and transfer them to a culture dish coated with a different antibody. In this example, the first dish (A) is coated with RAN-2 antibody, which selectively binds type 1 astrocytes. The loose cells are transferred to a second dish (B) coated with GC antibody to bind any differentiated oligodendrocytes. The final dish is coated with A2B5 antibody to bind OPCs (C). Thus, at the end of the three culture preparations, only OPCs are present in the culture dish. These cells can then be used to identify molecules that regulate differentiation of oligodendrocytes or type 2 astrocytes.

Internal Clocks Establish When Oligodendrocytes Will Start to Form

Studies also revealed that OPCs proliferate for a specific number of cycles before they are directed to an oligodendrocyte fate. Under normal conditions, the OPCs typically divide a maximum of eight times before they stop proliferating. This observation suggested that the cells use an internal clock to monitor the length of the proliferation phase. Intracellular mechanisms that help regulate this clock include the cyclin-dependent kinase (CDK) inhibitors that signal when cells should exit the cell cycle and initiate differentiation. The expression of two CDK inhibitors, p27 and p57, increases in OPCs during the proliferation phase and reaches a plateau at the time cells commit to the oligodendrocyte fate. These CDK inhibitors, part of the cip/kip (CDK interacting protein/kinase inhibitory protein) family normally inhibit the cyclinE/CDK2 pair that drives the G1–S transition of the cell cycle (see **Box 5.1**). Thus, OPCs continue to proliferate until their intracellular levels of CDK inhibitor are sufficient to end the cell cycle. If the expression of these CDK inhibitors is kept below a certain threshold, proliferation continues.

Signals that regulate levels of p27 and p57 in OPCs have also been identified. For example, a protein called Inhibitor of differentiation 4 (Id4) is expressed at high levels in proliferating OPCs but diminishes as the cells switch from the proliferation to determination phase (**Figure 6.17**). If Id4 is overexpressed in OPCs, proliferation is extended and determination does not occur. Id4 also interacts with p57. When Id4 is expressed at a sufficient level, it suppresses p57. As Id4 expression decreases over time, p57 levels rise. Thus, proliferation ceases, and cells begin to respond to fate determination signals such as TH.

Identifying the signals that regulate the development of these myelinating glia is important not only in terms of understanding normal development, but also for potential therapeutic treatments in demyelinating diseases such as multiple sclerosis (**Box 6.3**). Therefore, research continues

Figure 6.17 Levels of Id4 and p57 interact to regulate differentiation of oligodendrocytes. In OPCs, expression of the inhibitor of differentiation 4 (Id4) protein is highest during the proliferation phase, decreases gradually, and reaches its lowest level when the cells stop proliferating. In contrast, the expression of the CDK inhibitor p57 is initially low, increases gradually during the proliferation phase, and peaks just prior to differentiation. Id4 normally suppresses p57 expression. Thus, when Id4 expression falls below a critical level, p57 is no longer suppressed, and cells switch from the proliferative phase to the nonproliferative phase. The nonproliferating OPC is then able to respond to differentiation factors, such as thyroid hormone (TH), and become a mature oligodendrocyte.

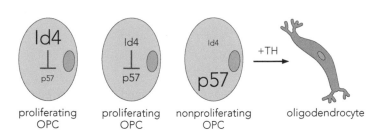

proliferating OPC   proliferating OPC   nonproliferating OPC   oligodendrocyte

Box 6.3 The Clinical Significance of Oligodendrocytes

Oligodendrocytes wrap around axons in the central nervous system (CNS), thus providing the myelin needed for normal transmission of signals throughout the body. There are many genetic diseases (such as leukodystrophies) and acquired diseases (such as multiple sclerosis) that lead to a progressive loss of myelin. When myelin is damaged or missing, nerve conduction is impaired, resulting in a variety of functional deficits. In severe cases, particularly in some of the genetic forms, death may result. The degree of functional deficit in any patient is quite variable, depending on which areas of the CNS have lost myelin. An active area of research involves studying ways to produce new oligodendrocytes or activate oligodendrocyte precursor cells that remain in the adult central nervous system. By studying the signals that regulate formation of oligodendrocyte precursor cells in the embryo and adult, scientists hope to one day develop targeted treatments to repair areas of damage and halt the progression of these currently incurable demyelinating diseases.

In the case of demyelinating diseases, having more healthy oligodendrocytes could lead to improved function of the nervous system. However, in other cases, oligodendrocytes impede repair of the CNS damage. Unlike axons of the peripheral nervous system (PNS), the axons of CNS neurons cannot regenerate after damage. Thus, any injury to CNS axons, such as occurs in spinal cord and traumatic brain injuries, is permanent. Scientists have noted that one reason CNS axons do not regrow after damage is the presence of inhibitory molecules at the site of injury. Many of the inhibitory signals are found on the oligodendrocytes that axons encounter. Such inhibitory signals include myelin-associated glycoprotein (MAG), a member of the immunoglobulin superfamily, and isoforms of neurite outgrowth inhibitor (NOGO), a member of the reticulon family of membrane proteins. Both MAG and NOGO bind the same axonal receptor: NOGO-66 receptor (NgR), also called the Reticulon 4 receptor (RTN4R). Such inhibitory proteins are not found on PNS Schwann cells. In fact, studies have shown that CNS axons can regenerate and grow across Schwann cells but not oligodendrocytes.

It is not clear why oligodendrocytes in the CNS would produce inhibitory proteins. One idea is that due to the complexity of synaptic connections in the CNS, any attempt to regenerate axonal connections could result in faulty innervation patterns that might lead to undesirable behavioral consequences. Scientists are working to develop ways to overcome innate inhibitory signals so that damaged areas of CNS can be selectively treated to regrow new, functional axonal connections. These examples show how understanding the biology of just one cell type in the CNS, the oligodendrocyte, has the potential to impact numerous disease processes.

to explore the various signals and intracellular pathways that regulate the survival, proliferation, and differentiation of oligodendrocytes in the optic nerve and other regions of the CNS.

DEVELOPMENT OF SPECIALIZED SENSORY CELLS

The nervous system is comprised of not only neurons and glia but also specialized sensory cells such as those used for vision and hearing. Many of the same signaling pathways that regulate determination and initial differentiation of neurons and glia in other regions of the nervous system also influence development of these sensory cell populations. This section provides examples of how the fates of cells are determined in the *Drosophila* eye, vertebrate ear, and vertebrate retina. Each example begins with an overview of the anatomical organization of the sensory system, then explains how processes such as lateral inhibition, transcription factor cascades, or internal timing mechanisms direct development of specialized sensory cell types.

Cell–Cell Contact Regulates Cell Fate in the Compound Eye of *Drosophila*

The compound eyes of *Drosophila* are derived from two eye-antennal imaginal discs that give rise to the head structures of the adult fly. Within the eye disc portion, located most anteriorly, transcription factors including

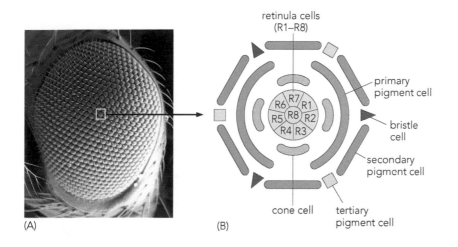

Figure 6.18 Cells types of the *Drosophila* ommatidium. (A) The compound eye of the adult *Drosophila* is seen in this scanning electron micrograph. Each eye consists of several hundred hexagonally shaped ommatidia. A single ommatidium is outlined. (B) The cells of each ommatidium are arranged in a precise order with eight photoreceptor cells, called retinula cells (R1–R8), surrounded by four cone cells. Two primary pigment cells lie adjacent to the cone cells and are surrounded by secondary pigment cells, tertiary pigment cells, and bristle cells. [(A), From Jackson GR [2008] *PLoS Biol* 6:e53. Published under Creative Commons Attribution License.]

homothorax (*Hth*), *teashirt* (*Tsh*), and the Pax6 homologues *Eyeless* (*Ey*) and *Twin of eyeless* (*Toy*) specify the eye region, promote proliferation, and suppress differentiation. Development of the eye begins at the first larval instar stage (see Chapter 1) and proliferation continues until the late second instar or early third instar stage.

Each compound eye is made up of about 750–800 hexagonal units called **ommatidia** organized into vertical columns. Each ommatidium contains several cell types organized in a precise pattern, including eight photoreceptor cells located at the center (**Figure 6.18**). These special sensory neurons are **retinula cells**, commonly called **R cells**. Each of the eight R cells is characterized by its spectral sensitivities and connections within the brain. Each ommatidium also includes the cone cells that cover the R cells and two primary pigment cells located just outside the photoreceptor cluster. There are also secondary and tertiary pigment cells and bristle cells located at the edges of the ommatidium. These cells are shared with adjacent ommatidia. The cells of developing ommatidia are initially equivalent and have the potential to become any of these cell types (**Box 6.4**).

Box 6.4 Developing Neuroscientists: Mosaic Analysis of Cell Fate Specification

Adam Haberman, Ph.D.

Adam Haberman is an associate professor of biology at the University of San Diego. He received his bachelors degree in biochemistry from the University of Texas at Austin and his Ph.D. in cell biology from the Johns Hopkins University School of Medicine. Using Drosophila, Dr. Haberman's research focuses on the cellular mechanisms that differentially promote neuronal survival or neuronal degeneration.

Concepts we now think of as biological principles were once hotly debated questions. In the 1970s, many scientists were trying to determine if cell fates were determined by a cell's lineage or its environment. If lineage determined cell fate, then the two daughter cells produced by a particular mitosis would always have the same two cell fates, no matter what cells were neighboring them. If environment determined cell fate, then a cell could only adopt a certain fate if it received specific signals from neighboring cells. The development of the eyes of the fruit fly *Drosophila melanogaster* turned out to be a useful system for addressing this question.

Fly eyes contain about 800 identical units, called ommatidia, that each have more than 20 specific cells. All these cells come from the eye imaginal disc, a small patch of cells that are specified early in development. By just watching the patterning of the imaginal disc, it was not possible to determine if cell fates were determined by lineage or by environment. However, a genetic technique called mitotic recombination made it possible to map cell fate. Cells were irradiated to cause rearrangement of chromosome arms during mitosis. While the parental cell was heterozygous for a mutation, its two daughter cells were each different.

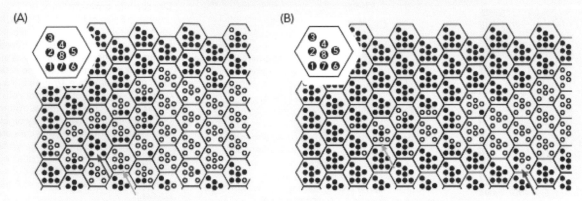

Figure 6.19 Mosaic analysis comparing the effects of mutations in Sevenless and Boss led to an understanding of how these genes functioned in R7 formation. (A) The diagram represents a cross-section through many ommatidia, with each hexagon representing an ommatidium. Within each hexagon, the circles represent the R1–R8 photoreceptors. Open circles represent mutant cells that lack the *Sevenless* (*Sev*) gene. Mutant cells are detected by their lack of the cell marker, the white gene (W^-). Filled circles represent wild-type cells (W^+) that carry both the white marker and the *Sev* gene. By analyzing hundreds of ommatidia, scientists established that the R7 cell must express the *Sev* gene for an ommatidium to have an R7 cell (the middle cell in the bottom row within each ommatidium in this example). This gene was named *Sevenless* because cells lacking the gene did not form an R7 photoreceptor. The arrows indicate the different cell patterns that were observed in these experiments. The red arrow points at an ommatidium in which R1–R5 and R8 are all wild-type, yet R7 is missing. The blue arrow points to an ommatidium in which all cells except R7 are mutant, yet R7 is present. Collectively, these data indicated that *Sev* is needed only in the presumptive R7 for R7 cell fate determination. (B) Similar methods were used to determine the role of *Bride of sevenless* (*Boss*) in R7 formation. The open dots indicate cells that lack the wild-type gene *Boss*, while the filled circles represent those that carry the *Boss* gene. In this panel, the blue arrow points to an ommatidium in which only R2 and R8 are wild- type, yet R7 is present. The red arrow points to an ommatidium in which R2 is wild-type, but R8 is mutant for *Boss*. In this case, R7 is absent. Together, these data indicated that *Boss* is specifically required in R8 for the R7 cell to form. We now know that *Boss* encodes a membrane-bound signaling protein and that *Sev* encodes its receptor on the surface of the presumptive R7 cell. However, the basic principles of how these genes worked were determined without knowing what kinds of proteins they encoded. The ability to decipher so much information through manipulation of fly genetics demonstrates the power of mosaic analyses. [(A), Adapted from Tomlinson A & Ready DF [1987] *Dev Biol* 123:264–275. With permission from Elsevier Inc.; (B), adapted from Reinke R & Zipursky SL [1988] Cell 55:321–330. With permission from Elsevier Inc.]

One daughter carried two wild-type copies of the gene, and the other was homozygous for the mutation. As those two daughter cells underwent mitosis, they created copies of themselves with their unique genetics. Since little cell migration occurs during eye development, these groups of cells stayed near each other in patches called clones. The resulting eyes were a mosaic of clones, and every cell in a clone shared a lineage, since they all derived from a single cell.

These clones were only useful if they could be identified. Therefore, rearranged chromosomes carried mutations called markers, which resulted in cellular changes that were easy to see under a microscope. The most common marker was a mutation in the *White* (*W*) gene. Eye cells with a wild-type *W* gene (W^+ cells) created pigment granules filled with easily seen pigments that give the eye its red color. Cells with two mutant copies of the *W* gene (W^- cells) made no pigment granules, so the eyes appeared white. Mosaic eyes were mostly red but contained white patches created by W^- clones.

When X-rays induced mitotic recombination in random parts of the eye, each resulting mosaic eye was unique. Researchers looked a hundreds of mosaic eyes and mapped which cell fates could come from the same clone. Under the microscope, it was easy to see which cells in each ommatidium were W^-. The researchers made diagrams showing the location of W^- cells in each ommatidium and looked for patterns.

What they discovered was that there was no relationship between cell fate and lineage in the fly eye. There were no rules stating, for instance, that if an R5 cell came from a clone, the neighboring R4 cell had to come from the same clone. The fate of each cell was independent of whether it shared a lineage with another cell. Therefore, cell fate could not be determined by lineage, but had to be determined by environment. Today we know that cell fate decisions are determined by signaling between cells, but proof that environmental signaling occurred was a significant result at the time.

Fly biologists also used mosaic analysis to understand how these environmental signals worked. Researchers had identified mutations in two genes, named *Sevenless* (*Sev*) and *Bride of sevenless* (*Boss*), in which eyes had no R7 cells. However, it was unclear how these genes worked. To determine the signals encoded by these genes, scientists placed a marker mutation on the same chromosome arm as the mutation they wanted to follow. Then they could make clones that were all mutant for *Sev* or *Boss* and which could be easily distinguished from the rest of the eye. After analyzing hundreds of mutant ommatidia, some patterns became clear. There were no ommatidia containing R7 cells that lacked *Sev* (Figure 6.19A). This meant that *Sev* was only required in the R7 cell and must be a signal-receiving gene. Additional studies found that no ommatidia that had an R8 cell lacking *Boss* had an R7 cell. Therefore, *Boss* had to be a signal-sending gene, and it could only send the signal from the R8 cell (Figure 6.19B).

The precise location and order of photoreceptor cell determination has been recognized for over 35 years. However, because the retinula cells were identified and numbered prior to knowing the order they are determined, the numbering does not reflect the order of retinula production. In fact, the R8 cell forms first and is required for the other cells to develop. Following R8, the R2 and R5 cells are the next to be determined and are formed at the same time. The next pair to form is R3 and R4, followed by the R1 and R6 pair, and finally R7 (**Figure 6.20**).

Signaling through epidermal growth factor (EGF) receptors is required for the cells of the imaginal disc to become photoreceptor cells. If EGF signaling is blocked, the cells become one of the nonphotoreceptor cell types. However, the EGF pathway does not determine which type of photoreceptor cell (R1–R8) a cell will become. Additional local signals establish final photoreceptor cell fates. These signals begin soon after the eye disc is specified.

The eye disc region is divided, or compartmentalized, into dorsal and ventral halves and the boundary formed between the two establishes the **morphogenetic furrow** (MF), the site where the cells of the *Drosophila* compound eye begin to form. During the third instar stage, Hedgehog (Hh) signaling induces a group of cells at the posterior margin of the eye disc to express *Dpp* (*Decapentaplegic*). *Dpp* expressing cells influence cells located anteriorly to establish a "pre-proneural zone." The morphogenetic furrow then moves as a "wave" from posterior to anterior, inducing new pre-proneural zones. Hh also works with *Eyes absent* (*Eya*), *Sine Oculis* (*So*), and *Dachshund* (*Dac*) to activate basal expression levels of the proneural gene *Atonal* (*Ato*) in a stripe of cells along the morphogenetic furrow. *Ato* expression is gradually limited to stripes of 12–15 cells in the epithelial sheet of the eye imaginal disc (**Figure 6.21**). Ultimately *Ato* is expressed in only one cell, the cell that differentiates as R8.

The R8 cell fate arises largely as the result of lateral inhibition. Proneural *Ato* activates expression of *Senseless*, a gene coding for a zinc-finger transcription factor often co-expressed with *Ato*. *Ato* and *Senseless* then positively regulate one another so their expression gradually becomes restricted to about ten cells arranged in a rosette pattern (Figure 6.21A, C).

Figure 6.20 The eight photoreceptor cells are determined in a precise order. R8 is the first photoreceptor cell fate to be determined and is required for the formation of the other retinula cells. Following determination of R8, the R2 and R5 cells develop at the same time. The next pair to be determined is R3 and R4, followed by R1 and R6. R7 is the last photoreceptor cell fate to be determined.

Figure 6.21 Transcription factor levels and *Atonal* expression identify cells that have the potential to form the R8 photoreceptor cell. (A) Atonal and the transcription factor Senseless positively regulate the expression of one another while the transcription factors Senseless and Rough repress one another. When levels of Senseless are higher than those of Rough, Senseless inhibits Rough. When Rough levels are higher, Senseless becomes repressed, directing a cell to a non-R8 fate. (B) *Atonal* and *Senseless* are initially expressed in stripes of 12–15 cells in the epithelial sheet that gives rise to the *Drosophila* eye (green cells). However, as development continues, *Atonal* and *Senseless* expression become progressively more restricted. (C) Expression of *Atonal* and *Senseless* are first limited to a rosette containing about ten cells. In a subset of cells, Notch receptor activation leads to increased expression of *Enhancer of split* (*E[spl]*), which inhibits, in turn, the expression of *Atonal*, thus leading to a decrease in *Senseless* expression as well (tan cells). These cells are prevented from becoming an R8 cell. (D) Three cells remain in the R8 equivalence group. Two cells express the transcription factor Rough (Ro, red cells) at levels sufficient to inhibit Senseless and prevent formation of an R8 cell. In the cell lacking high levels of Ro, Senseless levels are sufficient to promote expression of *Atonal* and those cells adopt the R8 fate (green cell in each equivalence group). [(B–D), Adapted from Tsachaki M & Sprecher SG [2012] *Dev Dyn* 241:40–56.]

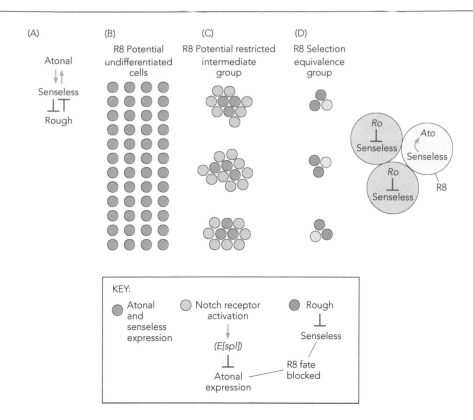

Delta ligand is also expressed on the cell that becomes R8. Lateral inhibition through the Notch-Delta signaling pathway leads to activation of *(E[spl])* in a subset of the remaining cells to inhibit further *Ato* expression and prevent those cells in the rosette from adopting the R8 fate.

Three cells remain in what is called the R8 equivalence group (Figure 6.21D). Two of the three cells will produce the transcription factors Rough and Senseless. Rough and Senseless can repress one another, with the transcription factor expressed at the higher level prevailing. In two of the cells, the level of Rough becomes sufficient to inhibit Senseless. In contrast, the remaining cell in the equivalence group only expresses Senseless and is thus able to increase the expression of *Ato* and become the R8 photoreceptor (Figure 6.21D). R8 will then provide additional signals that help direct the fates of other R cells.

Cell–Cell Contacts and Gene Expression Patterns Establish R1–R7 Photoreceptor Cell Types

The signaling pathways that regulate the determination of the R7 photoreceptor were among the first to be discovered. In the 1970s, Seymour Benzer and colleagues identified mutants that lacked the R7 photoreceptor cell. These mutants developed a cone cell in the usual location of the R7 photoreceptor. Benzer's lab, as well as Gerald Rubin's lab, then discovered that the mutants lack a tyrosine kinase receptor found in wild-type R7 cells. This receptor, named Sevenless (Sev) after the mutant phenotype, activates a Ras signal transduction cascade that leads to the activation of transcription factors that induce the expression of genes necessary for the R7 fate (**Figure 6.22**A). If the Ras pathway is experimentally inhibited, then the prospective R7 cell becomes a cone cell instead, just as when the Sev receptor itself is absent. The Sev receptor is activated by a membrane-bound ligand found on the surface of the centrally located R8 cell. This ligand, identified by Lawrence Zipursky and colleagues, was named Boss (Bride of Sevenless). Boss activates the Sev

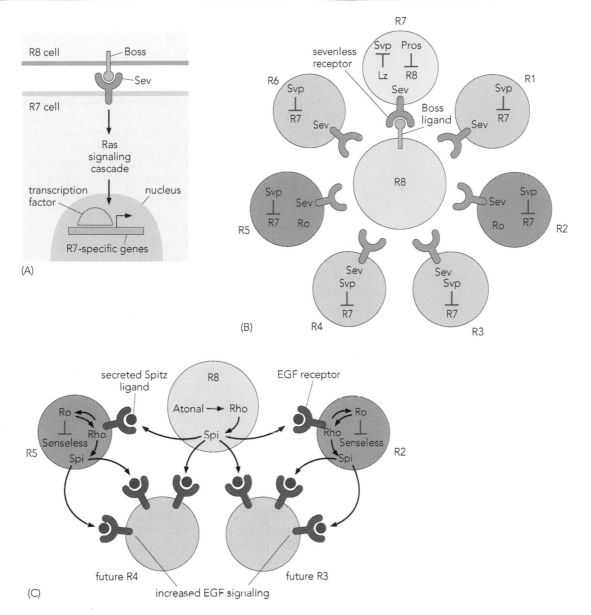

Figure 6.22 Signaling pathways that lead to the formation of R7 and the remaining photoreceptors. (A) The discovery of a *Drosophila* mutant that lacked the R7 cell led to the identification of a tyrosine kinase receptor called Sevenless (Sev). The Sev receptor is normally expressed by the R7 photoreceptor cell and activated by the Bride of Sevenless (Boss) ligand expressed on R8. When Boss binds to Sev, the Ras signal transduction cascade (Ras/Raf/MEK/Erk) is activated, leading to the formation of R7. (B) R1–R6 cells also contact R8 and transiently express the Sev receptor. However, the expression of other genes directs these cells to different photoreceptor fates. For example, the R1, R3, R4, and R6 cells (orange) express *Seven up* (*Svp*), which prevents continued expression of the Sev receptor and therefore those cells do not adopt an R7 fate. In R7 (yellow), *Lz* inhibits *Svp* so that Sev expression continues. R7 also expresses Prospero (Pros) to block R8 formation. R2 and R5 (pink) require the expression of other transcription factor encoding genes, such as *Rough* (*Ro*), to differentiate as non-R7 cells. (C) *Atonal* expression in R8 activates Rhomboid (Rho) that in turn cleaves the EGF ligand Spitz (Spi) so that it can be secreted. Spi binds to EGF receptors on adjacent cells and initiates expression of *Ro* in two of the cells, directing them away from the R8 fate by suppressing Senseless. These cells ultimately adopt R2 and R5 fates. *Ro* expression in these two cells then induces expression of Rho, cleavage of Spi, and activation of EGF receptors on adjacent cells leading them to adopt other fates.

receptor on R7 to initiate the Ras signaling pathway (see Figure 1.21). In the absence of Boss, the R7 cell fails to form.

Perhaps surprisingly, even though only one cell adopts the R7 fate, the Sev receptor is expressed transiently in other cells, including the R1–R6 photoreceptors and the cone cells. Because the cone cells are not normally in contact with R8 and Boss is not diffusible, it makes sense that these cells do not normally adopt an R7 fate despite expressing the Sev receptor. However, the R1–R6 photoreceptors that express the Sev receptor are in contact with R8. Yet, these photoreceptors still do not adopt an R7 fate. One reason

R1–R6 cells do not acquire the R7 fate is because they also express genes for transcription factors, such as *Seven up* (*Svp*), which blocks cells from adopting the R7 fate. For example, R1, R3, R4, and R6 express the Sev receptor but also express *Svp* and therefore do not form R7 cells even though they are in contact with R8 (the orange cells in Figure 6.22B). The importance of *Svp* in normal photoreceptor formation is seen in mutants lacking the gene. When *Svp* is absent, additional photoreceptors adopt the R7 fate.

The importance of *Svp* in preventing R7 formation is further noted by the expression of the gene *Lozenge* (*Lz*). During normal R7 development, *Lz* represses the expression of *Svp* in R7 cells, thus preventing them from adopting any other photoreceptor fate (the yellow cell in Figure 6.22B). R7 also expresses Prospero (Pros), which suppresses the R8 fate and confirms the R7 identity.

Additionally, R8 provides signals to direct the fate of the other surrounding cells. For example, the expression of *Ato* in R8 activates Rhomboid (Rho) that then cleaves an EGF ligand called Spitz (Spi). Once cleaved the ligand is secreted to activate EGF receptors located on adjacent cells. This EGF signaling leads to the expression of Rough (Ro) in two yet unspecified cells. Rough suppresses R8 fate in these cells allowing them to develop as R2 and R5 (Figure 6.22C). The expression of Rough in R2 and R5 also induces expression of *Rho* leading to an increase in EGF signaling in cells adjacent to R2 and R5, directing them to R1 and R6 fates.

These are just a few examples of the transcriptional networks that are utilized during differentiation of the fly ommatidium. Additional networks have been identified, and more continue to be discovered, highlighting the complexity of the signaling cascades used to determine even a single cell type in the developing nervous system.

Cells of the Vertebrate Inner Ear Arise from the Otic Vesicle

The mature vertebrate inner ear consists of a ventral auditory region that senses sound and a dorsal vestibular region that processes information related to balance (**Figure 6.23**A). Both regions convey sensory information to the brain via the corresponding branch of the eighth cranial nerve.

The sensory epithelium in both the auditory and vestibular regions of the inner ear consists of sensory hair cells surrounded by various supporting cells. The hair cells have rows of stereocilia, or "hairs," that project from the apical surface. Each stereocilium has a dense core of actin and contains ion channels necessary for signal transduction. The inner ears of species as diverse as zebrafish, chicks, and mice all feature precisely organized patterns of hair cells and supporting cells. Although the anatomical organization and patterning of these cells differs somewhat across these animal models, the signaling pathways that determine hair cell or supporting cell fate are largely conserved.

In the mature mammalian auditory system, the sensory and supporting cells are located in the organ of Corti that extends the length of the cochlear spiral (Figure 6.23B). The organ of Corti lies on the basilar membrane and is surrounded by the scala media, a fluid-filled chamber that lies between two other scalae: the scala vestibuli and scala tympani. The hair cells are aligned in distinct rows along this epithelium (Figure 6.23C). The cochleae of mammals typically contain one row of inner hair cells and three rows of outer hair cells, although there can be as many as four or five rows of outer hair cells in some regions. The tectorial membrane lies above the stereocilia of the hair cells. The inner hair cells are innervated by most of the afferent nerve fibers of the spiral ganglion neurons. One branch of the spiral ganglion neurons extends to the inner hair cells and the other to the brainstem. The outer hair cells, in contrast, receive fewer afferent fibers, but are innervated by most of the efferent auditory nerve

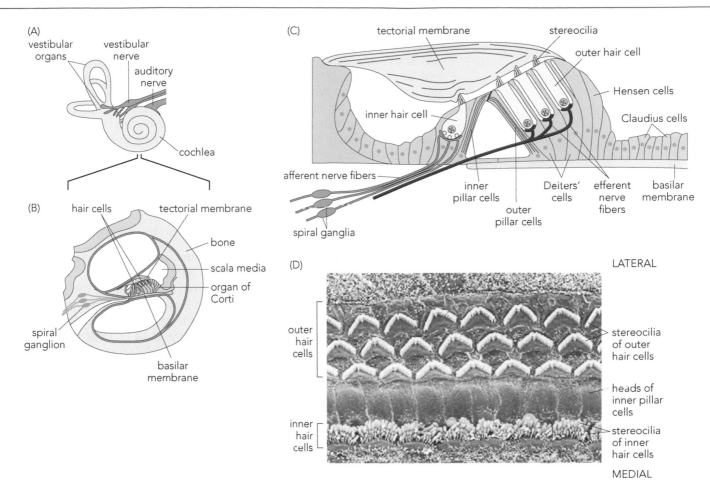

Figure 6.23 The mature inner ear is comprised of auditory and vestibular regions that have precisely organized cellular arrangements. (A) The inner ear is comprised of an auditory portion that processes sounds and a vestibular portion that processes information related to balance. Each region sends information from sensory hair cells through the corresponding segment of the eighth cranial nerve to the brain. The entire inner ear is encased in a bony capsule (blue). (B) A cross-section through the cochlea illustrates the location of the auditory sensory epithelium, the organ of Corti. The organ of Corti runs the length of the cochlear spiral and lies on the basilar membrane. Much of the organ of Corti is covered by the tectorial membrane. The organ of Corti is in the fluid-filled scala media. The spiral ganglion neurons that relay afferent information are found in the central bony core of the cochlea. (C) The cells of the organ of Corti are precisely organized. The inner hair cells are arranged in a single row located more medially, closest to the spiral ganglion neurons. The outer hair cells are organized in three rows closer to the lateral edge of the organ of Corti. The tectorial membrane lies above the stereocilia of both types of hair cells. The inner and outer hair cells are surrounded by various supporting cells, including the inner and outer pillar cells, respectively. Outer hair cells are also surrounded by the cells of Deiters. Additional supporting cells are found lateral to the outer hair cells, including Hensen and Claudius cells. (D) A scanning electron micrograph of the surface of the organ of Corti shows the organized patterning of the inner hair cells, outer hair cells, and inner pillar cells. This surface view also illustrates the organization of the stereocilia. The outer hair cell stereocilia are arranged in a "W"-like pattern, whereas the inner hair cell stereocilia have a shallower "U"-like pattern. [(D), From Hudspeth AJ [2013] *Neuron* 80:536–537.]

fibers (Figure 6.23C). The cell bodies of the efferent nerve fibers originate in brainstem nuclei. This efferent innervation influences motility of the outer hair cells and helps modify acoustic vibrations along the organ of Corti.

The various supporting cells, which include the pillar cells and the cells of Deiters, Henson, and Claudius, are located below and adjacent to hair cells and may extend cellular processes to surround the hair cells (Figure 6.23C). These supporting cells have various functions, such as providing structural support to the organ of Corti, maintaining ionic homeostasis, and clearing excess neurotransmitters from around hair cells. One of the most striking observations about the organ of Corti is the very precise and consistent organization of the hair and supporting cell populations (Figure 6.23D).

Most of the sensory cells, supporting cells, and neurons of the inner ear derive from the **otic vesicles**, or **otocysts**. As described in Chapter 5, each otocyst arises from the otic placode, a thickened patch of ectoderm adjacent to rhombomere 5 on each side of the hindbrain (see Figure 5.29). The otic placode invaginates to produce the otic pit, which then closes and sinks below the surface ectoderm to form the otocyst. During embryogenesis, signals from surrounding tissues direct the spherical otocyst to gradually transform along its A/P, D/V, and M/L axes. For example, Shh from the floorplate and notochord and Wnt from the hindbrain regulate formation of the D/V axis, similar to their actions in other parts of the nervous system. Retinoic acid (RA) influences development along the A/P axis, while FGF3 originating in the hindbrain may help influence development along the M/L axis.

The cochlear sensory epithelium arises from a prosensory region located in the central portion of the emerging cochlear duct. Initially all areas of the cochlear duct have the potential to develop as prosensory cells, but over time nonsensory regions are prevented from adopting sensory cell fates.

Notch Signaling Specifies Hair Cells in the Organ of Corti

The patterning of cell types within the mammalian organ of Corti is established using many of the same signaling mechanisms that specify cells in the proneural regions of *Drosophila*. The cells that develop along the cochlear duct are initially homogeneous in appearance. However, the transcription factor Sox2 (Sry-related HMG box 2) soon begins to be expressed in a restricted area, designating the prosensory region (**Figure 6.24**A). Cells within this prosensory region first proliferate to ensure that a sufficient pool of precursor cells is obtained. Once sufficient precursors are available, the cells require additional signals to differentiate into sensory hair cells or supporting cells.

Like the PNC in *Drosophila*, the prosensory region determines which cells become hair cells and which become supporting cells. Hair cells are the first cell type to differentiate in the developing organ of Corti and rely on the proneural bHLH transcription factor Atonal homolog 1 (Atoh1; also called Math1 in mice). *Atoh1/Math1* is first expressed at the time of terminal mitosis in the subset of cells that will become hair cells. Mice lacking *Atoh1/Math1* have a loss of hair cells, as well as a secondary, indirect loss of supporting cells. Further evidence for the importance of *Atoh1/Math1* in hair cell development was seen when the gene was misexpressed in regions of the organ of Corti that do not normally produce hair cells. Misexpression of *Atoh1/Math1* led to hair cell formation in these regions. Thus, *Atoh1/Math1* specifies hair cell fate.

Like the inititation of neuronal cell fate in *Drosophila* and *Xenopus*, *Atonal*-related genes initiate sensory hair cell fate by regulating the expression of cell surface ligands that activate the Notch receptor. In the inner ear *Atoh1/Math1* induces expression of the ligands Jagged2 (Jag2) and Delta-like1 (Dll1), both of which bind and activate the Notch receptor. Notch receptors are initially expressed throughout the developing sensory epithelium. However, Jag2 and Dll1 ligands are expressed only in those cells that become hair cells. Ligand expression in the future hair cells appears to result from a decrease in the expression of the Inhibitors of differentiation

Figure 6.24 Lateral inhibition establishes hair cell and supporting cell fates. (A) The transcription factor Sox2 is expressed in a subset of cells in the developing cochlear epithelium. The Sox2-expressing cells (blue) form the prosensory region. (B) Lateral inhibition determines which cells in the prosensory region will become supporting cells (tan) and which will become hair cells (green). Hair cells require the activation of the *Atonal homolog-1* (*Atoh1*) gene that promotes expression of the ligands Delta-like1 (Dll1) and Jagged2 (Jag2), which bind and activate the Notch receptor on an adjacent cell. The activated Notch intracellular domain (NICD) travels to the nucleus, where it activates *Hairy/ Enhancer of Split* (*Hes*) genes that inhibit *Atoh1*, thus suppressing hair cell fate and leading to a supporting cell fate. In addition, elevated levels of Inhibitors of differentiation (Id) proteins inhibit *Atoh1*, leading to decreased Dll1 and Jag2 expression and thus preventing supporting cells from adopting the hair cell fate. In contrast, cells with low levels of Id proteins maintain expression of *Atoh1* and differentiate as hair cells. [Adapted from Puligilla C & Kelley MW [2009] in *Encyclopedia of Neuroscience* [LR Squire ed], pp. 999–1004.]

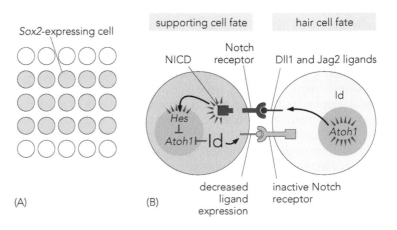

(A)   (B)

(Id) proteins that would otherwise block *Atoh1/Math1* expression. Thus, in the future hair cells, *Atoh1/Math1* levels are sufficient to induce ligand expression. In contrast, cells with elevated Id levels cannot produce Jag2 or Dll1 ligands. The Jag2 and Dll1 ligands on future hair cells bind to the Notch receptors on adjacent cells activating signaling pathways that lead to formation of one of the supporting cell types (Figure 6.24B).

Under normal conditions, Notch signaling promotes supporting cell differentiation by activating *Hairy/Enhancer of split-1* (*Hes1*) and *Hes5*, two downstream targets of mammalian Notch that function like the *(E[spl])* gene in *Drosophila*. *Hes1* and *Hes5* also inhibit *Atoh1/Math1* to prevent hair cell fate in Notch-activated cells. Evidence for the importance of this pathway was seen when the *Notch* gene was disrupted in mice. Decreased Notch signaling led to increased production of hair cells—particularly inner hair cells. Conversely, overexpression of *Hes1* led to a decrease in hair cells and an increase in supporting cells. Thus, much like the patterning of neuronal and nonneuronal cells in *Drosophila* neuroepithelium (see Figure 6.1), the sensory hair cells and nonsensory supporting cells of the organ of Corti rely on the Notch signaling pathway and associated *Hes* genes to establish cell types in a precise and orderly array.

The signals needed to establish inner versus outer hair cells and the different subtypes of supporting cells are only beginning to be defined. For example, FGF8 induces pillar cell fate. FGF8 is secreted by the inner hair cells at the same time prosensory cells in the cochlear duct express the FGF3 receptor and the FGF antagonist Sprouty 2 (see Figure 3.14). At some point the future outer hair cells downregulate their FGF3 receptor expression so they are not influenced by the FGF8 released from the inner hair cells. However, the concentration of FGF8 that reaches the FGF3 receptor-expressing cells in the presumptive pillar cell region is sufficiently high to overcome the inhibitory effects of Sprouty 2. FGF signaling also activates Hey2 (Hes-related family bHLH transcription factor 2) to suppress *Atoh1/Math1* required for hair cell fate. Thus, these cells are prevented from adopting a hair cell fate while being directed to a pillar cell fate. In contrast, presumptive supporting cells located further from the inner hair cells, such as the Deiters' cells that surround outer hair cells, do not receive a sufficient concentration of FGF8 to overcome Sprouty 2 activity and therefore are unable to adopt a pillar cell fate (**Figure 6.25**).

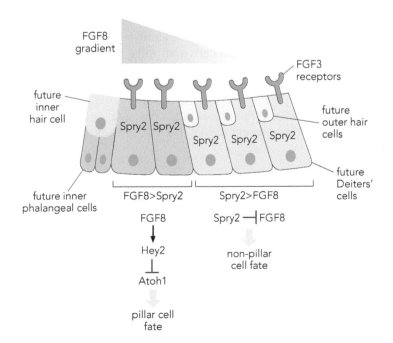

Figure 6.25 FGF signaling induces pillar cell fate. FGF8 is secreted by future inner hair cells at the same time that prosensory cells express FGF3 receptors and the FGF antagonist Sprouty 2 (Spry2). The concentration of FGF8 near future pillar cells is high enough to overcome the inhibitory effects of Spry2 and drive the cells to a pillar cell fate. At the same time, FGF8 signaling activates Hey2 that suppresses *Atonal1* and prevents the cells from adopting a hair cell fate. Future outer hair cells begin to downregulate the FGF3 receptors, so they are not influenced by FGF8. Although the surrounding future supporting cells continue to express FGF receptors they do not receive enough FGF8 signal to overcome the inhibitory effects of Spry2 and therefore do not adopt the pillar cell fate.

Researchers continue to investigate the signals involved in hair cell and supporting cell differentiation in various animal models. In mammals, cochlear hair cells are unable to regenerate after terminal mitosis. However, in other species, such as birds, hair cells can regenerate following trauma induced by excessive noise or exposure to certain ototoxic drugs. By investigating the mechanisms that regulate hair cell differentiation during development, scientists hope to one day stimulate similar pathways in the adult to produce new hair cells and alleviate hearing loss in humans. This is one more example of how discoveries stemming from basic experimental research can shape potential therapeutic treatments.

Cells of the Vertebrate Retina Are Derived from the Optic Cup

The sensory epithelium of the vertebrate eye processes visual stimuli. The components of the mature eye derive largely from extensions of the embryonic forebrain and surrounding tissues. The sensory epithelium of the eye first begins to form in the anterior region of the neural plate. The neural plate forms bilateral optic grooves as the neural folds begin to curve upward. Once the neural tube begins to close, these grooves evaginate, extending outward. With the closure of the neural tube, these extensions form the optic vesicles (**Figure 6.26**). As the diencephalon continues to expand, the optic vesicles also extend outward, eventually contacting the surface ectoderm. This contact induces the surface ectoderm to form the lens placode. The lens placode invaginates to form the lens pit then ultimately pinches off to form the lens vesicle that is the precursor to the adult lens. At the same time, the optic vesicles invaginate to form the optic cups. The anterior portion of the optic cup gives rise to the iris, the colored portion of the adult eye. In the posterior region of the optic cup, two distinct layers form—namely, the neural retina at the inner layer and the future pigment epithelium at the outer layer. The optic stalk represents the remaining connection between the forebrain and eye and later forms the optic nerve. The optic nerve contains the axons of the retinal ganglion cells that transmit visual information to the CNS. Structures such as the cornea and ciliary muscles are not derived from the optic cups, but instead originate from multiple tissues, including the surrounding surface ectoderm and mesoderm, as well as neural crest and mesenchymal cells.

Visual stimuli in the form of light enters the front of the eye and passes through the cornea and lens to reach the retina at the back of the eye (**Figure 6.27**A). The retinal sensory epithelium is organized into a laminar structure consisting of six neural cell types and a specialized population of glial cells called the Müller glia. In the mature retina, retinal ganglion cells are located in the layer closest to the lens and furthest from the retinal pigment epithelium (Figure 6.27B). The rod and cone photoreceptors are found in the outermost retinal layer, adjacent to the pigment epithelium. Situated between these layers are interneurons. Light enters the eye and

Figure 6.26 The optic vesicles give rise to the structures of the vertebrate eye. (A) The optic vesicles form as extensions of each side of the diencephalon. B–D illustrate developmental changes of a single eye. (B) The optic vesicle continues to expand outward until it reaches the surface ectoderm where it induces the ectoderm to form the lens placode. The optic vesicle also begins to invaginate, forming the optic cup. (C) The lens placode subsequently invaginates to form a lens pit (not shown), which then pinches off to form the lens vesicle. The optic cups continue to expand and the optic stalk, the remaining connection between the forebrain and the developing eye, is established. The optic stalk later forms the optic nerve. (D) The inside layer of the optic cup gives rise to the neural retina, while the outer layer forms the pigment epithelium. The lens vesicle gives rise to the lens of the adult eye. In the human embryo, these events take place at approximately gestation days 24–35. [Adapted from Larsen WJ [1993] *Human Embryology*. With permission from Churchill Livingstone.]

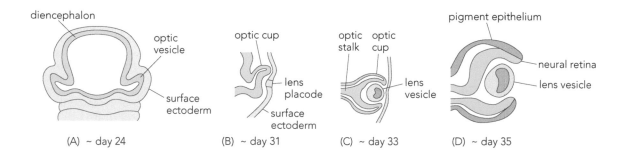

(A) ~ day 24   (B) ~ day 31   (C) ~ day 33   (D) ~ day 35

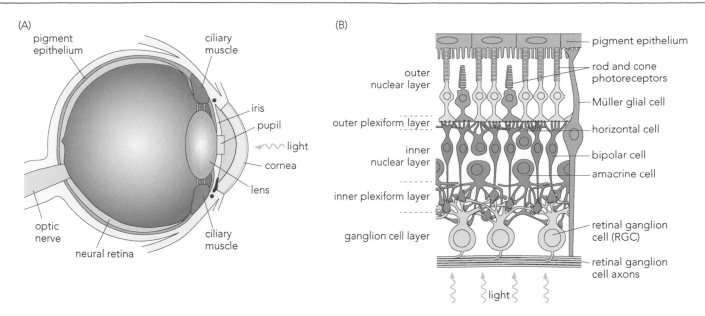

Figure 6.27 The eye is designed to process light stimuli. (A) The adult eye processes visual information in response to light that enters through the cornea and lens. The iris regulates the diameter of the pupil based on the intensity of light, and the ciliary muscles modify the shape of the lens for near and far vision. The light reaches the neural retina, the visual sensory epithelium that lies just in front of the pigment epithelium at the back of the eye. Axons of the retinal ganglion cells form the optic nerve. (B) The sensory epithelium of the neural retina consists of several precisely organized cellular layers: the outer nuclear layer, the inner nuclear layer, and the retinal ganglion cell layer. Between these cellular layers are areas of synaptic contacts called plexiform layers. The outermost cellular layer, closest to the pigment epithelium, is the outer nuclear layer. The outer nuclear layer contains the cell bodies of the rod and cone photoreceptor cells that detect light stimuli. The pigment epithelium is closest to these cells because they help remove and recycle molecules necessary for photoreceptor function. The inner nuclear layer contains cell bodies of the bipolar, horizontal, and amacrine interneurons and the Müller glia. The processes of the Müller glia span the retinal layers. The retinal ganglion cells are found in the innermost layer, closest to the lens. This means light must pass through the ganglion cell layer to reach the rods and cones. The rods and cones then relay information back through the interneurons of the inner nuclear layer to stimulate the retinal ganglion cells. The axons of the retinal ganglion cells travel in the optic nerve to the brain.

passes through the layers of ganglion cells and interneurons before reaching the rod and cone photoreceptors. The photoreceptors then relay signals through the interneurons to reach the ganglion cell layer. The retinal ganglion cells in turn project nerve fibers to the brain via the optic nerve. The organization of the retinal layers can seem counterintuitive, since the light must travel through the retina to reach the photoreceptors that then signal back to ganglion cells at the innermost layer. This seemingly flipped arrangement appears to reflect the importance of having the photoreceptors adjacent to the retinal pigment epithelium that helps remove and recycle molecules needed for photoreceptor function.

The names of the retinal layers are based on their histological appearance. Cell bodies are found in nuclear layers, and synaptic contacts are found in plexiform layers (Figure 6.27B). The outer nuclear layer contains the cell bodies of the rods and cones, whereas the inner nuclear layer contains cell bodies of the various interneurons and the Müller glia. The interneurons of the retina include horizontal, bipolar, and amacrine cells. These cells relay or modify information transmitted from photoreceptors to retinal ganglion cells. Spanning the layers of the retina are the processes of the Müller glia cells. Like other glia, Müller glia have diverse roles, such as serving as a source of stem cells, providing nutritional support to retinal cells, and contributing to signal transduction cascades.

The Vertebrate Retina Cells Are Generated in a Specific Order and Organized in a Precise Pattern

All retinal cell types arise from a single population of epithelial cells, the **retinal progenitor cells** (**RPCs**). In most vertebrates, the order in which the retinal cells differentiate is similar. Typically, the first cells generated are

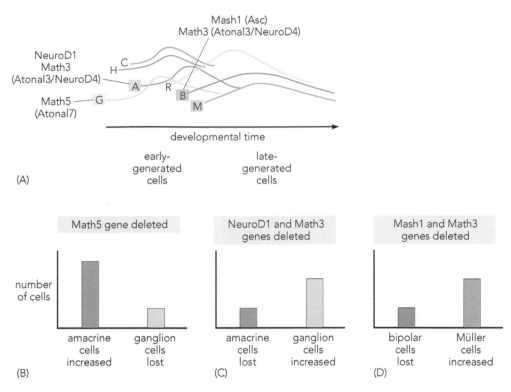

Figure 6.28 Developmentally regulated transcription factors determine the fate of retinal cell types. (A) Different retinal cell types arise at different stages of development and express specific transcription factors. Early-generated cells, produced during embryonic development, include the retinal ganglion (G), horizontal (H), cone (C), and amacrine (A) cells. Later-generated cells include the rod (R), bipolar (B) and Müller (M) cells. The production of late-generated cells peaks during late embryogenesis and in some species continues through early postnatal life. Retinal ganglion cells express the transcription factor Math5 (Atonal 7); amacrine cells express NeuroD1 and Math3 (Atonal3/NeuroD4); bipolar cells express Mash1 (Asc1) and Math3. (B) If Math5 is deleted, the number of retinal ganglion cells decreases while the number of amacrine cells, another early-generated cell type, increases. Conversely, if NeuroD1 and Math3 are deleted, amacrine cell numbers decrease while ganglion cells increase (C). (D) If Mash1 and Math3 are deleted fewer bipolar cells are produced while the number of Müller cells increases. [(A), Adapted from Napier HRL & Link BA [2009] in *Dev Neurobiol* [G Lemke ed], pp. 251–258. With permission from Elsevier.]

the retinal ganglion cells followed in sequence by the horizontal cells, cone cells, and amacrine cells. These four cell types are the early-generated cell types. The late-generated cell types include the rod cells, bipolar cells, and lastly the Müller glia (**Figure 6.28**A). Though the RPCs are categorized as "early-generated" or "late-generated," the timing of production for the different cell types often overlaps. Thus, early-generated cells, such as ganglion cells and amacrine cells, are still being produced when production of late-generated cells, such as the rod photoreceptors and bipolar cells, begins.

RPC proliferation is regulated in part by Delta ligands activating Notch receptors and their downstream signaling molecules, Hes1 and Hes5. Notch activation keeps the RPCs in a proliferative state until a sufficient progenitor pool is available. Once Notch signaling decreases in a group of RPCs, those cells begin to express various transcription factors that direct the cells to a particular retinal cell fate.

The sequential production of the different types of retinal cells appears to be influenced by the cellular environment into which a cell is born, and the combinations of transcription factors activated in each cell type. Neurogenin 2 (Ngn2), for example, is associated with early RPCs that give rise to retinal ganglion cells. In addition to Ngn2, Math5 (Atonal7), is important for the differentiation of retinal ganglion cells. Mice lacking the *Math5* gene had a decrease in the number of retinal ganglion cells but an increase in amacrine cells (Figure 6.28B). Amacrine cell differentiation requires *NeuroD* and *Math3(Atoh3/NeuroD4)*. When both *NeuroD* and *Math3* genes were mutated, the number of amacrine cells decreased and the number of retinal ganglion cells increased (Figure 6.28C). The development of later-generated bipolar cells also requires expression of bHLH transcription factors, but in this case the cells require *Asc1/Mash1* (achaet-scute homolog1/mammalian achaet-scute homologue 1) and *Math3*. Deletion of *Asc1/Mash1* and *Math3* genes in mice results in the loss of bipolar cells and an increase in the latest born cells, the Müller glial cells (Figure 6.28D). Together, these observations demonstrate that RPCs have the capacity to become multiple cell types, with specific transcription factors expressed at each stage of development. Any alteration in transcription factor expression can cause a dramatic shift in the types of retinal cells produced.

While studies in various species have established that most RPCs are multipotent and give rise to numerous retinal cell types, recent studies suggest that the early-born cone and horizontal cells arise from a separate, fate-restricted subpopulation of RPCs. These fate-restricted cells express transcription factors such as Otx2 and Onecut1 that are not expressed in the multipotent RPCs. Whether other fate restricted subpopulations are present in the RPCs is not yet clear.

Temporal Identity Factors Play a Role in Vertebrate Retinal Development

Studies of the mouse retina detected genes related to the *Drosophila* temporal identity transcription factors *Hunchback* and *Castor* (see Figure 6.7). As noted earlier, the ortholog of *Hunchback* is *Ikaros*, also called *Ikaros/Znfn1a1* (*Ikzf1*). The ortholog of *Castor* is *Casz1*. In mice, *Casz1* is downstream of *Ikzf1*, just as *Castor* is downstream of *Hunchback*.

In mice, *Ikzf1* is expressed in early RPCs. Overexpression of *Ikzf1* leads to production of more retinal cells with early fates, whereas loss of *Ikzf1* results in fewer cells with early fates.

Casz1 begins to be expressed after E14.5, in mid-late-stage RPCs. *Casz1* appears to regulate the progression to later cell fates while also preventing continued generation of early cell fates. Prior to E14.5, *Ikzf1* represses the expression of *Casz1* so early cell fates are established. However, at later stages, *Ikzf1* is no longer expressed, so *Casz1* expression directs cells to later retinal cell fates. Mice with a conditional deletion of *Casz1* in RPCs produced fewer late-born retinal cell types but more early-born cell types. The early-born horizontal, cone, and amacrine cells increased in number, while the number of later-born rod cells decreased. Together, these studies suggest that temporal identity factors may be present in vertebrate RPCs. How these interact with other transcription factor networks identified in RPCs continues to be studied.

SUMMARY

This chapter introduces some of the many ways in which cell fates are determined in the nervous system. Many of these cellular mechanisms are conserved across invertebrate and vertebrate species. The determination of cell fate typically begins with the differential expression of proneural genes, directing some cells toward neural or sensory cell fates and others toward nonneuronal, supporting cell fates. Subsequent development often results from a combination of environmental signals and temporally regulated transcription factor cascades. In both invertebrates and vertebrates, the timing of transcription factor expression is often critical in directing a cell toward a particular fate. While illustrating fate determination of several diverse cell types in various regions of the invertebrate and vertebrate nervous systems, this discussion is incomplete both in scope and detail, representing only a small selection of the ways in which nervous system cell fates are determined. Nevertheless, these selected examples provide insight into the numerous steps required for a neural precursor to establish a particular cell fate.

FURTHER READING

Adachi T, Miyashita S, Yamashita M, et al. (2021) Notch signaling between cerebellar granule cell progenitors. *eNeuro* 8(3):ENEURO.0468-20.2021.

Alsiö JM, Tarchini B, Cayouette M & Livesey FJ (2013) Ikaros promotes early-born neuronal fates in the cerebral cortex. *Proc Natl Acad Sci USA* 110(8):E716–725.

Banerjee U, Renfranz PJ, Hinton DR, et al. (1987) The Sevenless protein is expressed apically in cell membranes of developing *Drosophila* retina; it is not restricted to cell R7. *Cell* 51(1):151–158.

Banerjee U, Renfranz PJ, Pollock JA & Benzer S (1987) Molecular characterization and expression of sevenless, a gene involved in neuronal pattern formation in the *Drosophila* eye. *Cell* 49(2):281–291.

Bhatt S, Diaz R & Trainor PA (2013) Signals and switches in Mammalian neural crest cell differentiation. *Cold Spring Harb Perspect Biol* 5(2).

Billon N, Jolicoeur C, Tokumoto Y, et al. (2002) Normal timing of oligodendrocyte development depends on thyroid hormone receptor alpha 1 (TRα1). *EMBO J* 21(23):6452–6460.

Brown NL, Patel S, Brzezinski J & Glaser T (2001) Math5 is required for retinal ganglion cell and optic nerve formation. *Development* 128(13):2497–2508.

Chitnis AB (1995) The role of Notch in lateral inhibition and cell fate specification. *Mol Cell Neurosci* 6(4):311–321.

Desai AR & McConnell SK (2000) Progressive restriction in fate potential by neural progenitors during cerebral cortical development. *Development* 127(13):2863–2872.

Driver EC & Kelley MW (2020) Development of the cochlea. *Development* 147(12):dev162263.

Dugas JC, Ibrahim A & Barres BA (2007) A crucial role for p57(Kip2) in the intracellular timer that controls oligodendrocyte differentiation. *J Neurosci* 27(23):6185–6196.

Fabra-Beser J, Alves Medeiros de Araujo J, Marques-Coelho D, et al. (2021) Differential expression levels of Sox9 in early neocortical radial glial cells regulate the decision between stem cell maintenance and differentiation. *J Neurosci* 41(33):6969–6986.

Francis NJ & Landis SC (1999) Cellular and molecular determinants of sympathetic neuron development. *Annu Rev Neurosci* 22:541–566.

Frantz GD & McConnell SK (1996) Restriction of late cerebral cortical progenitors to an upper-layer fate. *Neuron* 17(1):55–61.

Hafen E, Basler K, Edstroem JE & Rubin GM (1987) Sevenless, a cell-specific homeotic gene of Drosophila, encodes a putative transmembrane receptor with a tyrosine kinase domain. *Science* 236(4797):55–63.

Hart AC, Krämer H, Van Vactor DL Jr, et al. (1990) Induction of cell fate in the Drosophila retina: The bride of sevenless protein is predicted to contain a large extracellular domain and seven transmembrane segments. *Genes Dev* 4(11):1835–1847.

Hilton DA & Hanemann CO (2014) Schwannomas and their pathogenesis. *Brain Pathol (Zurich, Switzerland)* 24(3):205–220.

Hirono K, Kohwi M, Clark MQ, et al. (2017) The Hunchback temporal transcription factor establishes, but is not required to maintain, early-born neuronal identity. *Neural Dev* 12(1):1.

Isshiki T, Pearson B, Holbrook B & Doe CQ (2001) *Drosophila* neuroblasts sequentially express transcription factors which specify the temporal identity of their neuronal progeny. *Cell* 106(4):511–521.

Kelley MW (2006) Hair cell development: Commitment through differentiation. *Brain Res* 1091(1):172–185.

Krämer H, Cagan RL & Zipursky SL (1991) Interaction of bride of sevenless membrane-bound ligand and the sevenless tyrosine-kinase receptor. *Nature* 352(6332):207–212.

Le Douarin NM (1980) The ontogeny of the neural crest in avian embryo chimaeras. *Nature* 286(5774):663–669.

Lemke GE & Brockes JP (1984) Identification and purification of glial growth factor. *J Neurosci* 4(1):75–83.

Leone DP, Srinivasan K, Chen B, et al. (2008) The determination of projection neuron identity in the developing cerebral cortex. *Curr Opin Neurobiol* 18(1):28–35.

Lessard J, Wu JI, Ranish JA, et al. (2007) An essential switch in subunit composition of a chromatin remodeling complex during neural development. *Neuron* 55(2):201–215.

Livesey FJ & Cepko CL (2001) Vertebrate neural cell-fate determination: Lessons from the retina. *Nat Rev Neurosci* 2(2):109–118.

Lütolf S, Radtke F, Aguet M, et al. (2002) Notch1 is required for neuronal and glial differentiation in the cerebellum. *Development* 129(2):373–385.

Manuel MN, Mi D, Mason JO & Price DJ (2015) Regulation of cerebral cortical neurogenesis by the Pax6 transcription factor. *Front Cell Neurosci* 9:70.

Mattar P, Ericson J, Blackshaw S & Cayouette M (2015) A conserved regulatory logic controls temporal identity in mouse neural progenitors. *Neuron* 85(3):497–504.

Narayanan R & Tuoc TC (2014) Roles of chromatin remodeling BAF complex in neural differentiation and reprogramming. *Cell Tissue Res* 356(3):575–584.

Nguyen-Ba-Charvet KT & Rebsam A (2020) Neurogenesis and specification of retinal ganglion cells. *Int J Mol Sci* 21(2):451.

Raff M (2011) Looking back. *Annu Rev Cell Dev Biol* 27:1–23.

Reinke R & Zipursky SL (1988) Cell–cell interaction in the *Drosophila* retina: The bride of sevenless gene is required in photoreceptor cell R8 for R7 cell development. *Cell* 55(2):321–330.

Solecki DJ, Liu XL, Tomoda T, et al. (2001) Activated Notch2 signaling inhibits differentiation of cerebellar granule neuron precursors by maintaining proliferation. *Neuron* 31(4):557–568.

Srivastava R, Kumar M, Peineau S, et al. (2013) Conditional induction of Math1 specifies embryonic stem cells to cerebellar granule neuron lineage and promotes differentiation into mature granule neurons. *Stem Cells* 31(4):652–665.

Stanke M, Duong CV, Pape M, et al. (2006) Target-dependent specification of the neurotransmitter phenotype: Cholinergic differentiation of sympathetic neurons is mediated in vivo by gp130 signaling. *Development* 133(1):141–150.

Tsachaki M & Sprecher SG (2012) Genetic and developmental mechanisms underlying the formation of the *Drosophila* compound eye. *Dev Dyn* 241(1):40–56.

Tuoc TC, Narayanan R & Stoykova A (2013) BAF chromatin remodeling complex: Cortical size regulation and beyond. *Cell Cycle* 12(18):2953–2959.

Willardsen MI & Link BA (2011) Cell biological regulation of division fate in vertebrate neuroepithelial cells. *Dev Dyn* 240(8):1865–1879.

Yang Y, Lewis R & Miller RH (2011) Interactions between oligodendrocyte precursors control the onset of CNS myelination. *Dev Biol* 350(1):127–138.

Yiu G & He Z (2006) Glial inhibition of CNS axon regeneration. *Nat Rev Neurosci* 7(8):617–627.

Yoo AS & Crabtree GR (2009) ATP-dependent chromatin remodeling in neural development. *Curr Opin Neurobiol* 19(2):120–126.

Zhang Y, Long J, Ren J, et al. (2021) Potential molecular biomarkers of vestibular schwannoma growth: Progress and prospects. *Front Oncol* 11:731441.

Zigmond RE (2011) gp130 cytokines are positive signals triggering changes in gene expression and axon outgrowth in peripheral neurons following injury. *Front Mol Neurosci* 4:62.

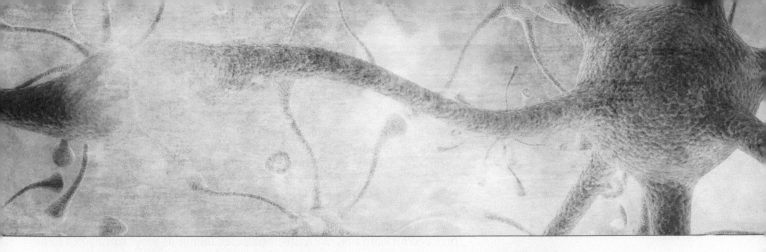

Neurite Outgrowth, Axonal Pathfinding, and Initial Target Selection

7

Unlike other embryonic cells, neurons extend long processes—axons and dendrites—from their cell bodies. The formation of axons and dendrites, collectively called nerve fibers or neurites, is an aspect of cellular differentiation unique to neurons. During embryogenesis, extending nerve fibers are first guided to the correct target tissue. Once at the target region, the fibers seek and contact appropriate partner cells then begin to differentiate the cellular machinery needed to form a presynaptic nerve terminal. Signals that regulate the initial outgrowth and guidance of nerve fibers in different invertebrate and vertebrate animal models are described in this chapter. Examples of guidance cues that influence axonal growth from spinal motor neurons to limb skeletal muscle, commissural interneurons to the spinal cord midline, and retinal ganglion cells to the optic tectum are provided to illustrate how the local environment, intermediate targets, and final target cells direct neurite extension. Chapters 9 and 10 then describe mechanisms used to form mature synaptic connections between presynaptic and postsynaptic partner cells. Because all aspects of nervous system function rely on synaptic communication, proper neural connectivity, or wiring, is crucial. As seen in this chapter, developmental neurobiologists study every aspect of this wiring from the initial extension of a neurite along an embryonic path to the final selection of an appropriate target cell.

GROWTH CONE MOTILITY AND PATHFINDING

As discussed in Chapter 5, scientists in the nineteenth century debated whether neurons were connected through a syncytial network of cytoplasmically interconnected cells or by contacts established between individual cells. As these two possibilities were debated, several prominent scientists, such as Wilhelm His, Albert von Koelliker, and Santiago Ramón y Cajal, reported observations of individual neurons extending nerve fibers directly from their cell bodies to contact other cells.

DOI: 10.1201/9781003166078-7

Early Neurobiologists Identify the Growth Cone as the Motile End of a Nerve Fiber

Cajal and others in the late nineteenth century recognized that the tip of the nerve fiber was a unique, motile structure necessary for nerve fiber extension. Cajal is credited with naming this structure the **growth cone**. In 1890, he proposed how a neuron might extend a process through embryonic tissues to reach a target cell. Cajal's detailed description of a growth cone as "endowed with exquisite chemical sensitivity, rapid amoeboid movements, and a certain motive force, thanks to which it is able to proceed forward and overcome obstacles met in its way" turned out to be remarkably accurate, despite being based entirely on his observations of sections of fixed tissue (**Figure 7.1**). Today, time-lapse video recordings detail growth cone motility and extension under different conditions and reveal that a growth cone is in constant motion as it extends and samples the environment to identify growth-promoting and growth-inhibiting regions. Growth cone motility is much like that of other highly motile cells such as fibroblasts, leading some to refer to the growth cone as a "fibroblast on a leash."

In Vitro and *In Vivo* Experiments Confirm Neurite Outgrowth from Neuronal Cell Bodies

Cajal's explanations of growth cone movements and nerve fiber outgrowth still did not convince everyone that nerve fibers extended directly from the cell body. However, 20 years later, in 1910, Ross G. Harrison published a paper describing the extension of an axon directly from the cell body and the formation and branching of a growth cone (**Figure 7.2**). Harrison's observations stemmed from a tissue culture method he developed to grow embryonic frog spinal cords in a hanging droplet of clotted frog lymph. The lymph provided nutrients and structural support for the neurons so they could survive and extend nerve fibers. Although some scientists at the time criticized the work and felt tissue culture did not replicate what was happening *in vivo*, this finding is now seen as a pivotal moment in developmental neurobiology, demonstrating axonal outgrowth from individual cells while pioneering the use of tissue culture for experimental analysis—a method that continues to be of great value today.

In vivo evidence of nerve fiber outgrowth was then reported by Carl C. Speidel in the 1930s. He first described axonal outgrowth in the tadpole where neurite outgrowth was visible in the transparent tail of the intact animal. Thus, results from both *in vitro* and *in vivo* preparations confirmed that neural processes extend from the neuronal cell body.

Figure 7.1 Observations of growth cones in fixed tissue predicted their movements. (A) Cajal's drawings of growth cones from an embryonic day 4 chick spinal cord. (B) A photomicrograph from one of Cajal's original slides shows a single axon and growth cone extending in the embryonic chick spinal cord. In 1890, Cajal proposed that growth cones were highly motile and actively sampled the local environment to reach a target cell. [From De Carlos JA & Borrell J [2007] *Brain Res Revs* 55:8–16.]

(A)

growth cone

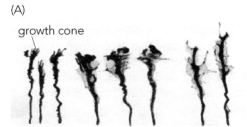

(B)

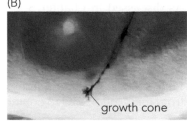

growth cone

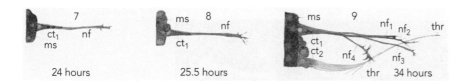

Figure 7.2 Tissue culture preparations confirmed the outgrowth of nerve fibers from neuronal cell bodies. In 1910 Ross Harrison described a tissue culture method that revealed nerve fibers extended directly from neuronal cell bodies. Here some of his original drawings from experiments conducted in 1908 document the increasing length and branching of the nerve fibers at 24 hours (his panel 7), 25.5 hours (panel 8), and 34 hours (panel 9). During this period, the length of the nerve fiber (nf) increased (panel 7 and 8), then formed four distinct branches (nf1–nf4) by 34 hours. The labels he used in his drawings refer to cells (ct1, ct2), masses of cells (ms), nerve fibers (nf), and fimbrin filaments (thr). [From Harrison RG [1959] J Experimtl Zool 142:5–73.]

Substrate Binding Influences Cytoskeletal Structures to Promote Growth Cone Motility

As noted in Harrison's hanging droplet cultures, the nerve fibers needed to attach to the fibrillar components of the lymph to extend. Other scientists soon confirmed the necessity of attachment, noting that growth cones could not extend in a fluid medium or on substrates that did not support attachment. It was soon understood that attachment to a substrate provided the tension needed for neurite extension.

The structural components of the growth cone play a critical role in the attachment and extension of nerve fibers. Like other highly motile cells, growth cones contain abundant actin filaments and microtubules organized in a manner that permits cellular movements. Microtubules are concentrated in the axon, whereas actin is concentrated in the distal portion of the growth cone (**Figure 7.3**A). However, actin and microtubules overlap in the cell body and proximal portion of the growth cone. The growth cone itself contains the cellular machinery required for motility, as demonstrated in cell culture experiments in which the fibroblast-like movements of growth cones temporarily continued after the growth cone was severed from the axon. Thus, growth cone motility does not depend entirely on the cellular components localized to the cell body.

Distinct growth cone regions are designated based on morphologic appearance and the cytoskeletal elements present. The region that lies closest to the neurite shaft is the **central mound** or **central domain** (**C-domain**). This central region contains many organelles as well as bundles of stable microtubules that originate in the neurite shaft (Figure 7.3B). Some microtubules, called **dynamic microtubules**, extend into the distal region of the growth cone, where the individual motile polymers help establish the direction of growth cone extension (Figure 7.3B, C).

The leading edge of the growth cone, a region also called the **peripheral domain** (**P-domain**), consists of sheetlike projections. These flattened projections are the **lamellipodia**—motile "veils" that contain a fanlike meshwork of filamentous (fibrillar) actin (F-actin) and subunits of globular actin (G-actin), but few organelles. The lamellipodia are found between thin projections containing rodlike bundles of F-actin. These projections, called **filopodia** or microspikes, have receptor proteins that sample the surrounding environment. If even a single filopodium touches an attractive cue, the entire growth cone will turn toward that cue. In contrast, filopodial contact with an inhibitory cue causes the growth cone to turn away or temporarily collapse back toward the axon shaft.

When the neuronal cell body and emerging nerve fibers are attached to a substrate the microtubules in the central mound region become stabilized. This stability, coupled with the temporary movement of dynamic microtubules into the peripheral region of the growth cone, helps advance the growth cone forward. However, turning of the growth cone requires polymerization of actin subunits located at the ends of the actin bundles in the lamellipodia and filopodia.

Actin subunits are characterized by a barbed (plus end) and a pointed (minus end). During polymerization, actin subunits are added to the barbed

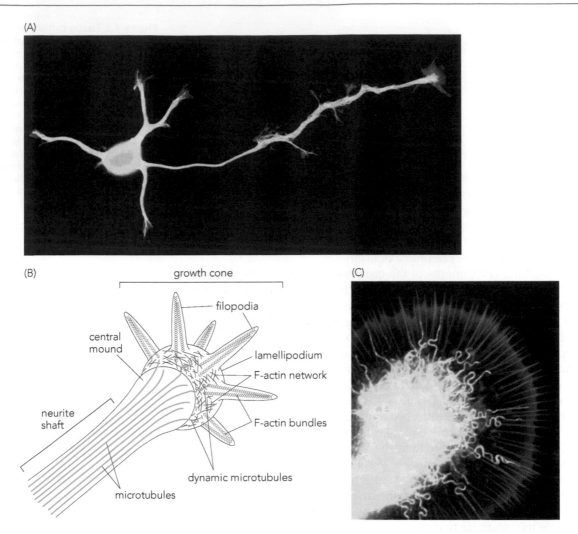

Figure 7.3 Cytoskeletal elements regulate growth cone motility and axon extension. Distinct cytoskeletal elements are concentrated in the neuronal cell body, axon, and growth cones. (A) The axon shaft (green) is characterized by microtubules. The distal region of the growth cone and emerging axonal branches are characterized by actin (red). Microtubules and actin overlap in the cell body and proximal growth cone (yellow). (B) A schematic representation of the distribution of cytoskeletal elements reveals that the neurite shaft contains microtubules that extend into the central mound region at the proximal end of the growth cone. A smaller number of dynamic microtubules extend past the central mound and help direct growth cone advancement. In the distal region of the growth cone, mesh-like F-actin networks fill the flattened lamellipodium. Thin projections called filopodia extend outward from the lamellipodium. Filopodia, which contain bundled F-actin, sample substrates and determine which pathway a growth cone will follow. (C) Photomicrograph of a growth cone showing microtubules labeled green and actin labeled red. Note the dynamic microtubules extending into the distal region of the growth cone and the actin bundles concentrated in the filopodia. [(A), From Kalil K, Szebenyi G & Dent EW [2000] *J Neurobiol* 44:145–158; (B), Adapted from Lowery LA & Van Vactor D [2009] *Nat Rev Mol Cell Biol* 10:332–343; (C), Courtesy of Elizabeth Hogan.]

ends of actin filaments. Actin polymerization begins with **nucleation**—the formation of a small complex of approximately three G-actin monomers. Continued addition of these actin monomers leads to the elongation of the actin filament. In the growth cone, this occurs at the leading (distal) edge of the lamellipodia and filopodia where the addition of the actin subunits pushes the cell membranes of the leading edge forward (**Figure 7.4**). Depolymerization occurs in the proximal region of the growth cone when actin subunits are removed from the pointed ends of filaments, those located closest to the central mound.

The movement of actin subunits is further facilitated by myosin II, which acts as a motor protein to move actin rearward, toward the central mound where the subunits are removed. The released subunits are recycled for later addition to the leading edge, creating the cyclical or treadmilling action of actin subunits. During neurite outgrowth, the addition of actin

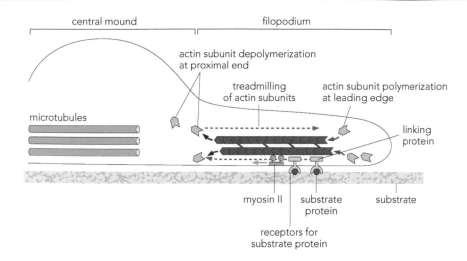

central mound   filopodium

actin subunit depolymerization
at proximal end

treadmilling   actin subunit polymerization
of actin subunits   at leading edge

microtubules

linking
protein

myosin II   substrate   substrate
protein

receptors for
substrate protein

Figure 7.4 Recycling of actin subunits and substrate attachment lead to forward movement of growth cones. In a filopodium, actin subunits polymerize and depolymerize in a cycle that results in net forward movement of the growth cone. Actin subunits are added to the barbed, distal end of the actin filaments (red) at the leading edge of the growth cone (polymerization of subunits, pink arrows), while other subunits are released at the pointed, proximal ends of the filaments (depolymerization of subunits, black arrows). The released actin subunits are recycled and used again (dashed arrows). Myosin II acts as a protein motor to help move actin in a retrograde direction (blue arrow), away from the leading edge of the growth cone. Receptors for substrate proteins attach the growth cone to the substrate. These receptors interact with linking proteins to stabilize the actin filament bundles and temporarily slow the retrograde flow of actin. This provides the tension that allows the growth cone to move forward. In such cases, depolymerization proceeds at a slower rate than polymerization, so the recycling, or treadmilling, of subunits contributes to the net forward movement of the growth cone. [Adapted from Letourneau PC [2009] In *Encyclopedia of Neuroscience* [LR Squire ed], pp. 1139–1145.]

subunits at the leading edge occurs at a rate faster than the depolymerization near the central mound, causing forward extension of the growth cone.

Actin-Binding Proteins Regulate Actin Polymerization and Depolymerization

Several actin-binding proteins interact to regulate the polymerization and depolymerization of the actin filaments. These proteins may promote the addition of new actin subunits at a barbed end, sever existing actin filaments, or cap the barbed end of a filament. In many cases, actin-binding proteins induce the formation of new barbed ends that promote branching of existing actin filaments. Among the actin-binding proteins shown to influence the creation of new barbed ends are the Arp2/3 complex (actin-related proteins 2 and 3) and formins. Arp2/3 binds to the side of an existing actin filament. Such binding initiates nucleation of G-actin subunits to promote branching of the filament. This is often seen in the lamellipodia (**Figure 7.5**A). In contrast, formins function primarily in filopodia, where they bind and add new actin subunits to the barbed ends of the existing actin filaments (Figure 7.5B). New barbed ends can also be exposed by severing an existing actin filament. The actin-binding proteins ADF (actin depolymerizing factor) and cofilin are two examples of such severing proteins that create new sites for the addition of G-actin subunits (Figure 7.5C). Another group of actin-binding proteins are the capping proteins such as CapZ that bind to the barbed end of the actin filaments to prevent the loss of existing actin subunits and stabilize the growth cone (Figure 7.5D). Capping proteins can be inhibited by other actin-binding proteins to allow polymerization of the actin filaments during periods of growth cone extension.

Whether a growth cone stabilizes, branches, or extends appears to depend on the balance of actin-binding protein activity and the concentration of available G-actin subunits. For example, in some contexts, when the levels of ADF/cofilin are higher, increased severing occurs causing actin filaments to break down, thus preventing growth cone extension. However, if the available levels of G-actin are high enough, the additional barbed ends created by ADF/cofilin provide, instead, multiple new sites for polymerization and therefore increased growth cone movement and extension.

Rho Family GTPases Influence Cytoskeletal Dynamics

The actin polymerization needed for growth cone motility and neurite extension is also influenced by the Rho family of small GTPases. As noted in Chapter 5, Rho family GTPases activate effector proteins that influence cytoskeletal dynamics (see Figure 5.9). In some cases, specific Rho family GTPases enhance actin polymerization and growth cone advancement, whereas other

Figure 7.5 Actin-binding proteins help regulate actin polymerization and depolymerization. Different actin-binding proteins attach to or sever existing actin filaments to create new sites for actin polymerization. (A) New barbed ends are formed along existing actin filaments when the actin-binding protein complex Arp2/3 attaches to an existing actin filament and initiates nucleation of G-actin subunits at that site. This branching is often seen in the lamellipodia of the growth cone. (B) The actin-binding protein formin typically attaches to the barbed ends of actin filaments in the filopodia to help recruit additional actin subunits to the elongating filopodia. (C) New barbed ends can be created by the actin-severing proteins ADF or cofilin. The exposed ends create new sites for the addition of G-actin subunits to promote growth cone extension. In some cases, the levels of ADF and cofilin exceed the levels of available G-actin, so the severing causes a breakdown of the actin filaments and growth cone collapse. (D) Capping proteins such as CapZ attach to the barbed end of actin filaments to prevent the loss of the existing subunits and stabilize the growth cone.

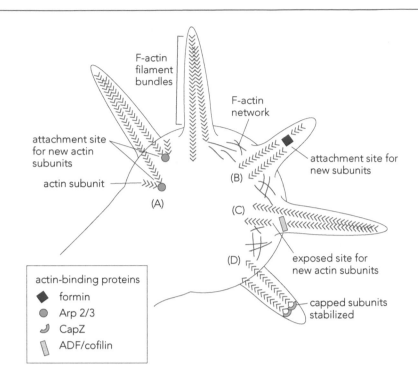

Rho family GTPases increase depolymerization and growth cone retraction (**Table 7.1**). For example, the Rho family GTPase Cdc42 (cell division control protein 42) activates N-WASP (Neural-Wiskott-Aldrich syndrome), whereas Rac1 (Ras-related C3 botulinum toxin substrate 1) activates WAVE (WASP-family verprolin-homologous protein). Either of these proteins can then activate the Arp2/3-dependent polymerization of actin filaments to promote growth cone advancement. In contrast, RhoA (Ras homolog family member A) activity is most often associated with the retraction of growth cones by activating ROCK (Rho-associated coiled-coil-containing protein kinase). ROCK then increases the activity of the myosin II motor protein that regulates retrograde flow of actin. With an increase in retrograde flow, growth cone advancement is prevented and the growth cone retracts.

Intracellular signals have also been implicated in regulating cytoskeletal organization and growth cone motility. For example, the level of intracellular calcium can determine whether a growth cone is active or inactive. For many growth cones, motility requires an optimal window of calcium concentration. If the cytosolic calcium level is too low or too high, the growth cone is inactive. The mechanism by which this occurs is unclear, but it appears that calcium helps regulate processes such as actin flow and the addition of plasma membrane to the extending growth cone. In at least

Table 7.1 Rho Family Small GTPases Influence Cytoskeletal Proteins and Growth Cone Behavior.

Rho Family Small GTPase	Protein Activated	Effect on Actin	Growth Cone Response
Cdc42	N-WASP	↑ Arp2/3-mediated actin polymerization	Extension
Rac1	WAVE	↑ Arp2/3-mediated actin polymerization	Extension
RhoA	ROCK	↑ myosin II-mediated retrograde flow of actin	Retraction

Note: Arp2/3, actin-related proteins 2 and 3; cdc42, cell division control protein 42; N-WASP, Neural-Wiskott-Aldrich syndrome; Rac1, Ras-related C3 botulinum toxin substrate 1; RhoA, Ras homolog family member A; ROCK, Rho-associated coiled-coil-containing protein kinase; WAVE, WASP-family verprolin-homologous protein.

some contexts, specific levels of calcium are associated with the activation of different actin-binding proteins or Rho GTPases.

GROWTH CONE SUBSTRATE PREFERENCES *IN VITRO* AND *IN VIVO*

Because the growth cone cytoskeleton reorganizes in response to environmental cues, the resulting morphology reflects different growth conditions. For example, a complex morphology with a broad lamellipodium and multiple filopodia is characteristic of a growth cone exploring the surface on which it is growing, actively seeking the most favorable substrate on which to advance. A smaller, less elaborate growth cone is typical of a neurite extending on a favorable substrate where cue-seeking behavior is unnecessary. A growth cone that has stopped extending and is about to collapse or change direction has few or no filopodia. When a neurite reaches its target cell, the growth cone adopts a simple morphology as it prepares to transform into a presynaptic connection.

In Vitro Studies Confirm That Growth Cones Actively Select a Favorable Substrate for Extension

In vitro studies reveal how growth cones distinguish among different substrates and actively select a particular pathway. In one classic study from 1975, Paul Letourneau demonstrated that embryonic chick dorsal root ganglion (DRG) neurons were able to distinguish and preferentially grow along lanes of specific substrates. In these experiments, a pattern was made on cell culture dishes by placing a small grid in a dish before coating the dish with the metal palladium (Pd). When the grid was removed, the dish contained squares of Pd and lanes without Pd (the areas once covered by the grid). Depending on the type of petri dish used, the Pd-free lanes consisted of either untreated petri dish plastic or tissue culture plastic that is positively charged to attract the net negative charge of neuronal membranes. In other cases, the lanes were coated with polyornithine (a poly-cationic amino acid) or collagen, two substrates shown to be favorable for the adhesion of other cell types.

When DRG neurons were cultured in the various dishes, the neurons did not adhere to the untreated petri dish plastic, but were able to adhere to and extended neurites along areas of tissue culture plastic or Pd. However, when the tissue culture lanes were coated with the other substrates, the neurites avoided the areas of Pd and instead grew along the polyornithine or collagen pathways (**Figure 7.6**A). The neurites also grew on top of any glia encountered in the dish. Thus, the neurites extended along whichever substrate was most favorable for neuronal attachment and neurite extension.

These studies also demonstrated the importance of the growth cone filopodia in pathway selection. As a growth cone was advancing along one substrate, the filopodia were observed to extend in multiple directions. As the filopodia sampled the environment, a single filopodium would periodically contact an adjacent substrate (Figure 7.6B). If a filopodium contacted a surface that was less adherent, it rapidly withdrew. However, if even one filopodium contacted the substrate that was more adherent, the entire growth cone would turn and extend in that direction. Thus, the growth cone filopodia sampled available substrates and guided the growth cone along the most favorable of the choices.

Subsequent investigations determined that the growth cones from different neuronal populations often have specific substrate preferences. It has also been noted that the most adherent substrate is not always best for optimal growth; if a substrate is too sticky, neurite extension is not possible because the growth cone cytoskeletal elements are unable to move freely.

(A)
neural cell body   neurites

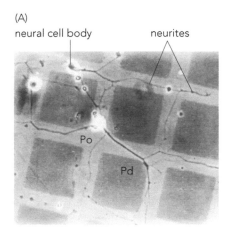

(B)

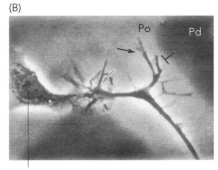

glial cell

Figure 7.6 Dorsal root ganglion (DRG) neurons detect differences in substrates and preferentially grow along those that are more adherent. When DRG neurons were placed in cell culture with squares and lanes of different substrates, the DRG axons identified the most favorable substrates for growth. (A) When presented with a choice of palladium (squares, Pd) or polyornithine (lanes, PO), the neurites extended along the more adherent polyornithine-coated lanes. (B) Growth cones were observed to extend filopodia to each surface and sample the choices available before choosing the more adherent substrate. Here, one branch of the growth cone extends across the surface of a glial cell (green arrow), while the other continues along the lane of polyornithine (PO, pink arrow). The growth cone avoids the square covered with Pd (pink bar). [From Letourneau PC [1975] *Dev Biol* 44(1):92–101.]

Extracellular Matrix Molecules and Growth Cone Receptors Interact to Direct Neurite Extension

Within the embryo, pathway selection is mediated by proteins present on the embryonic cell surfaces or in the surrounding extracellular matrix. Specific proteins expressed in the cellular environment provide guidance cues that lead an extending nerve fiber to its correct target cell. Among the pathway cues encountered *in vivo* are extracellular matrix (ECM) molecules, such as the laminins, collagens, and fibronectins. These large extracellular molecules are not diffusible and remain near the cell that secretes them, thereby providing local cues to extending neurites.

Growth cones express corresponding integrin receptor proteins that allow them to bind to specific ECM substrates. Each integrin receptor is a heterodimer of two different integrin subunits. As discussed in Chapter 5, the pairing of specific subunits produces the binding specificity for a particular ECM molecule. For example, in vertebrates, the α5β1 subunits form a receptor to bind fibronectin, the α6β1 subunits bind laminins, and the α1β1 subunits bind both collagens and laminins (see Table 5.1). Growth cones extending from different neuronal populations express distinct integrin receptor combinations at specific developmental stages to respond to spatially and temporally regulated ECM guidance cues.

Net forward movement of the growth cone can only occur if the integrin receptors that bind to a substrate also interact with the actin filaments. Such an interaction is possible because the integrin receptors are also associated with various linkage, or clutch, proteins. Linkage proteins include vinculin and talin, which cluster near the site where the growth cone contacts the substrate. Because these proteins link the integrin receptor to the actin filaments, the actin filaments are stabilized at the site of substrate contact and actin polymerization at the leading edge of the growth cone occurs (**Figure 7.7**). In this way a favorable substrate promotes actin polymerization and growth cone advancement.

In some regions of the embryo, a growth cone may encounter an unfavorable or inhibitory substrate. Such substrate cues can block integrin-mediated adhesion, causing the growth cone to turn away from the inhibitory cue or to collapse, rather than extend forward. Growth cone turning and collapse result from a combination of changes in cytoskeletal dynamics, including cessation of actin polymerization at the barbed ends of actin filaments; an increase in the retrograde flow of actin; greater severing of actin filaments, leading to actin breakdown; and the removal of capping proteins that would otherwise stabilize the barbed ends of the actin filaments.

Pioneer Axons and Axonal Fasciculation Aid Pathway Selection

While the spatially and temporally restricted expression of ECM proteins throughout the embryo can guide growth cones along *in vivo* pathways

Figure 7.7 Integrin receptors interact with clutch proteins to link actin filaments to substrate. Growth cones must adhere to a substrate to advance forward. Integrin receptors in the cell membrane of the growth cone bind to extracellular matrix (ECM) proteins on the substrate. The receptors also interact with various clutch (linking) proteins, such as talin and vinculin, which contact actin filaments to stabilize F-actin bundles so that additional actin subunits can be added at the growth cone's leading edge. [Adapted from Alberts B, Johnson A, Lewis J et al. [2015] *Molecular Biology of the Cell*, 6th ed. Garland Science.]

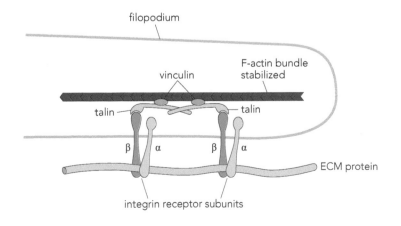

their broad expression does not offer the degree of specificity needed for each neuronal population to reach the correct target. Thus, additional cues are required to ensure neurites from a particular neuronal population travel toward the correct target tissue. For many extending axons, early guidance cues are found on adjacent nerve fibers.

In any given neuronal population, a single axon, or a small set of axons, must be the first to extend in the direction of a target tissue. These first nerve fibers, typically called **pioneer axons**, often respond to local ECM guidance cues. Pioneer axons then express other guidance cues along their axon shafts for subsequent nerve fibers to follow, thereby directing axons from the same neuronal population toward the same target tissue. Additionally, during the initial stages of outgrowth, axons from the same neuronal population tend to preferentially bundle together (fasciculate) to ensure a group of nerve fibers extends to the same target.

Fasciculation is promoted by adhesion molecules, such as neural cell adhesion molecule (NCAM) and neural cadherin (N-cadherin), which are expressed on adjacent axons (**Figure 7.8**). In 1987, NCAM was the first neural adhesion molecule identified. Additional adhesion molecules have since been discovered, including L1, neuron–glia cell adhesion molecule (NgCAM), and transient axonal glycoprotein-1 (TAG-1). Members of the IgG superfamily, these molecules contain IgG domains and fibronectin type III repeats. The adhesion molecules often use homophilic binding in which they recognize a binding partner of the same structure on an adjacent cell. There are also some instances of heterophilic binding with other molecules in the IgG superfamily, other ECM molecules, or members of the semaphorin family (**Box 7.1**). Unlike the IgG superfamily, members of the cadherin family, including N-cadherin, are calcium-dependent. Calcium helps stabilize the ectodomains of these adhesion molecules. Thus, adhesion between axons expressing N-cadherin takes place only in the presence of calcium (Figure 7.8).

In addition to promoting fasciculation of axons bundles, in some instances neural adhesion molecules decrease axonal bundling. For example, a post-translational modification of NCAM occurs in which polysialic acid (PSA) is added to the NCAM molecule. As described later in this chapter, PSA NCAM is expressed on motor neuron axons as they reach their target muscle. This highly sialylated form of NCAM interferes with the bundling of axons, allowing the axons to branch away from each other, spread across the muscle surface, and contact individual target cells.

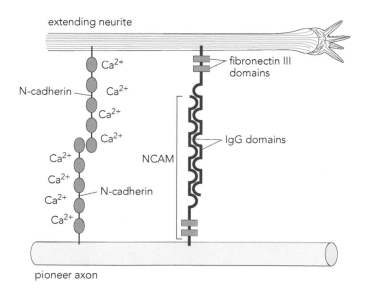

extending neurite

Ca^{2+}

N-cadherin

Ca^{2+}

Ca^{2+}

Ca^{2+}

Ca^{2+}

Ca^{2+}

Ca^{2+}

Ca^{2+}

N-cadherin

fibronectin III domains

IgG domains

NCAM

pioneer axon

Figure 7.8 Pioneer axons use cell adhesion molecules to guide other axons. Cell adhesion molecules, such as neural cadherin (N-cadherin) and neural cell adhesion molecule (NCAM), are located along the axon shafts of pioneer axons to promote fasciculation and direct related axons to the same target region. Homophilic binding between adjacent, extending neurites allows the axons to recognize each other. N-cadherin is a member of the cadherin family and requires calcium (Ca^{2+}) for binding to take place. NCAM is a member of the IgG superfamily and consists of IgG domains and fibronectin III domains.

Box 7.1 Semaphorins Provide Repulsive Cues to Direct Dorsal Root Ganglion Axons Entering the Spinal Cord

Throughout the developing nervous system, axons rely on repulsive guidance cues to avoid inappropriate axonal pathways and target regions. Among the most widely expressed and best-characterized repulsive guidance cues are those of the semaphorin family. The first semaphorin (Sema) protein was identified as an inhibitory cue for grasshopper sensory axons in 1992. The name derives from its function in directing axons to a particular pathway, much like the role of a semaphore. In the 1980s, a similar molecule was purified from chick brain. This protein, called collapsin, caused the stalling and retraction of dorsal root ganglion (DRG) axons *in vitro*. As additional members of the semaphorin family were identified, it became apparent that collapsin was homolgous to Sema3A (originally called SemaIII). *In vivo*, Sema3A is expressed in the ventral spinal cord at the time DRG axons innervate target laminae in the dorsal horn of the spinal cord, suggesting a role in repelling axons away from the ventral region (**Figure 7.9**A). *In vitro*, COS cells expressing Sema3A also repelled DRG axons (Figure 7.9B). Thus, ventrally expressed Sema3A provides a repulsive guidance cue for sensory axons to guide them to correct targets in the dorsal region.

There are currently 21 known semaphorins in vertebrates, five in invertebrates, and two in viruses, some of which, such as Sema3A, are secreted proteins. Others are membrane-associated through either a transmembrane domain or GPI linkage. Semaphorins are found in numerous tissues, but some of the best-characterized functions have been identified in the nervous system, where they are found in both neuronal and nonneuronal cells. Semaphorins have multiple roles in neural development and may also be active during neural injuries and pathologies.

To identify receptors for Sema3A, scientists made fusion proteins that contained a segment of Sema3A fused to alkaline phosphatase (AP), an enzyme that can be visualized using a colorimetric reaction. With this method DRG neurons that bound the Sema3A-AP fusion protein could be identified (**Figure 7.10**). This method ultimately led to the identification of the receptor neuropilin-1. **Neuropilins** are now known to be a receptor component for many semaphorins.

Around the same time, other scientists studying viruses identified a different receptor called plexin. **Plexins** were also subsequently detected in neurons. Both neuropilin and plexin components are required to transduce the Sema3A signal that directs axons away from inappropriate target regions. The activation of the receptors alters cytoskeletal organization, leading to growth cone retraction and collapse. The large size of the semaphorin family and the associated receptor families, together with the distribution of these molecules in numerous tissues, suggests they have multiple, diverse functions. Scientists continue to sort out intracellular mechanisms that allow these molecules to regulate different cellular events, including axon retraction, synapse formation, cell migration, and vascular development.

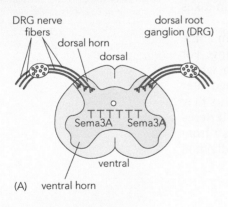

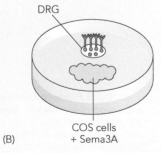

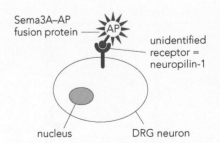

Figure 7.9 Semaphorin 3A provides a repulsive cue to direct growth of dorsal root ganglion (DRG) axons. (A) The axons of DRG neurons that enter the central nervous system project to cells within the dorsal horn of the spinal cord. The ventral spinal cord expresses semaphorin 3A (Sema3A) to prevent DRG axons from entering the ventral spinal cord. (B) *In vitro*, DRG neurites are repelled by Sema3A-expressing COS cells.

Figure 7.10 Neuropilin is identified as a receptor for semaphorin 3A. To identify receptors for semaphorins, fusion proteins consisting of Sema3A and alkaline phosphatase (AP) were incubated with DRG neurons. Using a colorimetric assay to detect the AP, scientists identified neurons that bound the Sema3A and ultimately isolated the previously unidentified receptor as neuropilin-1.

Research in Invertebrate Models Leads to the Labeled Pathway Hypothesis

Much of what is known about how pioneer axons function comes from studies in three invertebrate nervous systems: grasshopper, *Drosophila*, and *C. elegans*. Compared to tracking nerve fibers in vertebrate systems, following individual neurites in an invertebrate system is easier. Invertebrate neurons are larger, fewer in number, and readily identified by their location in the nervous system. Together these properties allow scientists to observe the initial outgrowth and turning of subsets of neurons as they travel through different embryonic regions.

Studies of pioneer axons by Corey Goodman and colleagues tracked pathways taken by specific neurons in developing grasshoppers. The grasshopper embryo consists of 17 segments divided into half segments on each side of the body. Each half segment contains ganglia and within these ganglia are several identified neurons. For example, in each half of a thoracic segment identified neurons include Q1, Q2, Q5, Q6, G, and C. The scientists first determined that the axons of the G and C neurons connect to the opposite side of the body by traveling along a transverse-oriented bundle of nerve fibers called a **commissure**. Once on the opposite side, the G and C axons join the A/P (anterior/posterior) fascicle, but none of the other longitudinally oriented axon bundles in the area.

To determine how these axons reach the opposite side and choose the correct fascicle, studies using both electron microscopic reconstructions of developing growth cones and *in vivo* manipulations were completed. It was found that the growth cones of G neurons detect differences between the axons of the A and P neurons traveling within the A/P fascicle. The growth cones of G neurons preferred to extend along the P axons, to which they adhered more tightly (**Figure 7.11**). Thus, the filopodia of the G neuron growth cones could distinguish the A/P bundle from other axon bundles in the region, then further differentiate P axons from A axons. Together, the observations suggested that axons express molecules that guide growth cones to join and travel in the correct fascicle. These observations led to the **labeled pathway hypothesis** described by Goodman and colleagues in 1982. This hypothesis states that the axons of pioneer neurons differentially express molecules, or labels, on their surfaces. Later-extending growth cones then choose which axon pathway to follow based on the expression patterns of these labels.

To test the labeled pathway hypothesis directly, finely sharpened microelectrodes were used to selectively ablate specific A and P axons within the A/P fascicle, then the growth of individual G axons was documented. The axons of G neurons crossed the commissure as expected, but upon reaching the opposite side, the axons grew in a random manner, as if seeking the axonal guidance cues normally present in the intact A/P fascicle. The ablation of either A or P axons disrupted G axon outgrowth, with the loss of P axons being particularly disruptive. Even though there were over 20 other axon bundles in the area, the G axons did not show a preference for any of these longitudinal fascicles, further demonstrating that the G axons recognized a cue specific to axons located in the A/P fascicle.

Fasciclins Are Expressed on Axonal Surfaces

Axons travel with a particular bundle until they reach a **choice point**, where they select the next axon bundle to join. For example, the axons of G neurons in the grasshopper first choose to grow along axons in the transverse commissure, then they choose to grow along axons in the A/P fascicle. For the neurites to switch to another axonal bundle, there must

Figure 7.11 The growth patterns of grasshopper axons led to the labeled pathway hypothesis. In the grasshopper embryo, G neurons extend axons across a transverse commissure to reach the contralateral longitudinal anterior–posterior (A/P) fascicle. Other neurons, including A1, A2 and P1, P2, extend axons within the A/P fascicle. Studies revealed that the axons of G neurons preferentially grow along A and P axons, showing the greatest adherence to P axons. The axons of G neurons ignored axons in other fascicles located in the area (not shown), even when A and P axons were missing. *Source:* Adapted from Bastiani MJ, Raper JA & Goodman CS [1984] *J Neurosci* 4(9):2311–2328.

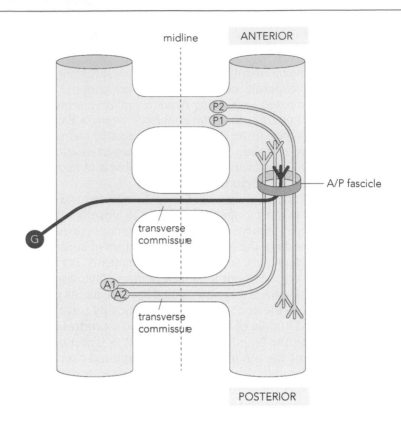

be changes in the expression of, and the response to, axonal cues encountered at the choice point.

Studies in invertebrates identified a family of molecules that helps early axons associate with different fascicles as they travel throughout the embryo. These molecules, called **fasciclins**, are expressed on the surfaces of early axons in animal models such as the grasshopper and *Drosophila*. In *Drosophila*, fasciclin II (Fas II) was later determined to be an ortholog of NCAM. Like NCAM, the fasciclins are members of the IgG superfamily and use homophilic binding to mediate adhesion.

Fasciclins are often expressed on discrete regions of axons to direct the growth cones of other axons at choice points. For example, in grasshopper and *Drosophila*, axons traveling along a longitudinal pathway express Fas II, allowing them to interact with other Fas II-expressing axons. However, axons preparing to cross a transverse commissure downregulate Fas II expression and begin to express fasciclin I (Fas I) so they can interact with Fas I-expressing axons in the transverse commissure. Once the axons exit the transverse commissure on the opposite side, they again express Fas II in order to adhere to and grow along the axons present in the longitudinal pathway (**Figure 7.12**). Axonal surface guidance cues are used throughout the developing nervous system and appear particularly important for directing early-extending axons toward a target tissue.

Vertebrate Motor Neurons Rely on Local Guidance Cues

As described in Chapter 4, motor neurons in the ventral half of the spinal cord extend axons to innervate the skeletal muscles that control movement. As the spinal motor axons exit the spinal cord, they form discrete axon bundles at intervals along the length of the spinal cord (**Figure 7.13**). The motor axons in the ventral root join the sensory axons of the dorsal root ganglia to form the spinal nerves. Several studies over

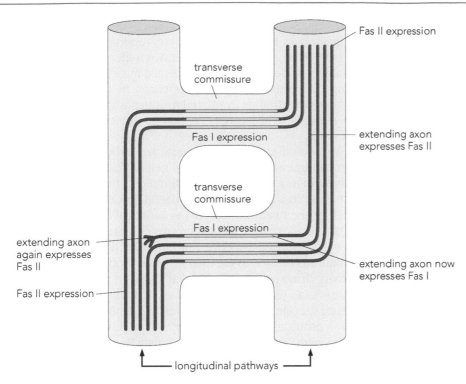

transverse commissure

Fas I expression

transverse commissure

Fas I expression

extending axon again expresses Fas II

Fas II expression

Fas II expression

extending axon expresses Fas II

extending axon now expresses Fas I

longitudinal pathways

Figure 7.12 Axons adjust fasciclin expression patterns to cross the midline. Fasciclins bind to one other via homophilic binding mechanisms. In this example from invertebrates, fasciclin I (Fas I) is expressed in the transverse commissures, whereas fasciclin II (Fas II) is expressed in longitudinal pathways. To cross the midline, axons decrease expression of Fas II and increase expression of Fas I, allowing homophilic binding with other Fas I-expressing axons in the commissures. Once the axons have reached the contralateral side, they again express Fas II to bind to other Fas II-expressing axons projecting longitudinally. [Adapted from Bastiani MJ, Harrelson AL, Snow PM & Goodman CS [1987] *Cell* 48(5):745–755.]

the past century have shed light on how the stereotyped, segmental patterns of axons emerge as the motor neurons project fibers toward different limb muscles.

During the first half of the twentieth century, scientists studying the role of limb muscles in guiding motor axons formulated seemingly

(A)

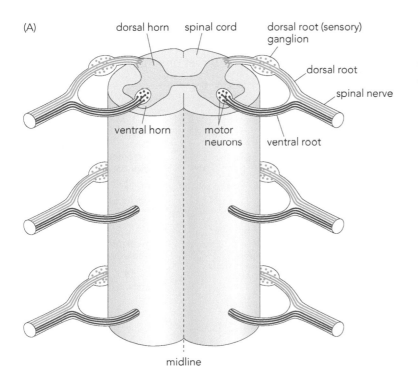

dorsal horn   spinal cord   dorsal root (sensory) ganglion

dorsal root

spinal nerve

ventral horn   motor neurons   ventral root

midline

(B)

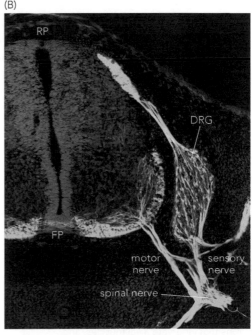

RP

DRG

FP

motor nerve   sensory nerve

spinal nerve

Figure 7.13 Spinal nerves form discrete axonal bundles along the length of the spinal cord. (A) Spinal nerves are comprised of motor axons from the ventral root (red) and sensory axons (green) from the dorsal root. These axon bundles form at intervals along the length of the spinal cord, creating a ladder-like pattern. (B) A section of an embryonic chick spinal cord reveals the origins of the ventral motor nerve and dorsal sensory nerve roots that comprise the spinal nerves (green). Pink region, ventricular zone; DRG, dorsal root ganglion; RP, roof plate; FP, floor plate. [(B), Courtesy of Jason Newbern.]

conflicting hypotheses. Some concluded that muscle-derived cues guided motor axons to the correct target muscle. Others reported that local environmental cues guided the axons to the target where they then innervated the closest available muscle, rather than a specific muscle. Some of these conflicting findings were likely due to differences in the experimental approaches used, as some studies evaluated re-innervation to regenerated amphibian limb muscles, while others looked at innervation to grafted embryonic chick limbs.

In the late 1970s and early 1980s, several key experiments in chick embryos by Lynn Landmesser and colleagues helped resolve the differing hypotheses for how motor neuron axons are guided to a target muscle region. The chick embryo was a particularly useful model system because by the second half of the twentieth century anatomical and electrophysiological studies had revealed that specific groups of motor neurons within the chick spinal cord innervated specific limb muscles (**Figure 7.14**). For example, motor neurons located in the medial, or ventromedial, motor column innervate axial (trunk) muscles, whereas those in the lateral motor column innervate distal (limb) muscles. Thus, it was possible to map axonal growth from specific spinal cord regions to specific muscles.

Landmesser's group also took advantage of useful anatomical features of the lumbosacral region of the chick spinal cord. Normally, lumbosacral segment 1 (LS1) is located rostral (anterior) to LS2 and LS3, and the size of the motor neuron pools in the lateral motor column (LMC) increases from rostral to caudal. Thus, LS3 has a larger pool of motor neurons in the LMC than does LS1 (Figure 7.14).

These features were used to test what happened to motor neurons and their axons when a portion of chick spinal cord was rotated 180 degrees, inverting the rostral–caudal orientation of that spinal cord segment (**Figure 7.15**A). When a spinal cord segment from thoracic segment 7 (T7) through LS3 was rotated, the size of the motor neuron pools did not change. The LS3 motor neuron pool remained larger than the LS1 pool, even when rotated to a more rostral position. Anatomical tracing methods were then used to test whether axons from the inverted motor neuron pools were still guided to the correct target or simply innervated the closest muscle.

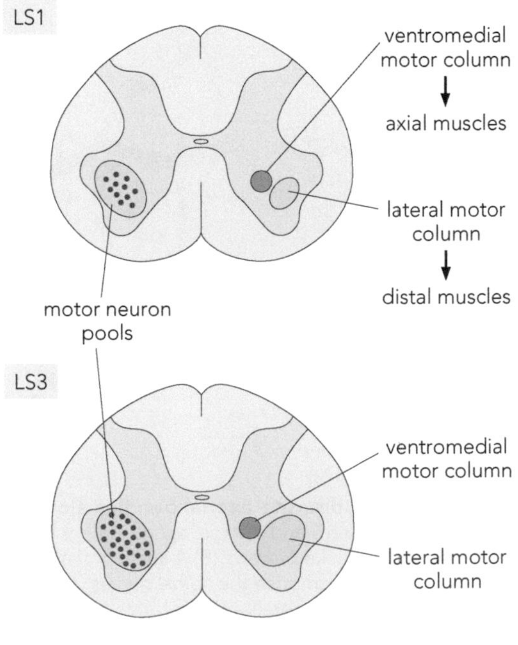

Figure 7.14 Specific pools of motor neurons vary in size and innervate different muscles. The motor neuron pools in the ventral horn are divided into a ventromedial motor column with motor neurons that innervate axial muscles, and a lateral motor column with motor neurons that innervate distal muscles. The size of the lateral motor column increases from lumbosacral segment 1 (LS1) to lumbosacral segment 3 (LS3), providing an anatomical marker for scientists investigating axonal guidance from different spinal cord levels.

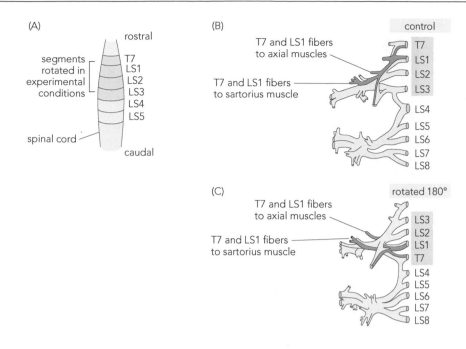

(A)

rostral

segments rotated in experimental conditions

T7
LS1
LS2
LS3
LS4
LS5

spinal cord

caudal

(B)   control

T7 and LS1 fibers to axial muscles

T7 and LS1 fibers to sartorius muscle

T7
LS1
LS2
LS3
LS4
LS5
LS6
LS7
LS8

(C)   rotated 180°

T7 and LS1 fibers to axial muscles

T7 and LS1 fibers to sartorius muscle

LS3
LS2
LS1
T7
LS4
LS5
LS6
LS7
LS8

Figure 7.15 Motor axons grow in specific patterns to reach target muscles. (A) A segment of the chick spinal cord from thoracic level 7 (T7) to lumbosacral level 3 (LS3) was used to assess the growth of motor axons to axial and sartorius muscles in control and experimental (rotated) preparations. In the experimental group, the segment was rotated 180 degrees along the rostral–caudal (anterior–posterior) axis. (B) A schematic outline of projections (dark pink) in normal, control conditions. Axons from T7 and lumbosacral segment 1 (LS1) normally project to axial muscles and the sartorius muscle located in the hindlimb of the chick embryo. (C) The correct motor axon projections persisted following spinal cord rotation. When the motor axon pathways were visualized approximately 6 days after rotation, the motor axons (dark pink) were found to innervate the correct muscles. That is, in both control and rotated spinal cord segments, motor axons from T7 and LS1 grew to the axial and sartorius muscles. Motor axons from the rotated spinal cord had to cross over inappropriate muscles to reach the correct target, suggesting that nerve fibers did not simply grow to the closest muscle, but used guidance cues to locate the correct muscle. [(B and C), Adapted from Lance-Jones C & Landmesser L [1980] *J Physiol* 302:581–602.]

The researchers found that axons extending from the rotated spinal segments still projected to the correct muscle with very few axons projecting to a non-target muscle. For example, axons from T7 and LS1 grew along new paths to reach the correct axial and sartorius (distal) muscles (Figure 7.15B, C). Electrophysiological recordings confirmed that the muscles were innervated by axons originating in the correct spinal cord segment. Thus, the axons did not grow passively to the closest muscle group as some earlier studies suggested, but specifically sought the correct target muscle and grew along novel routes when necessary.

Subsequent studies noted that if spinal cord segments were displaced too far from their original site, the motor axons were unable to contact the correct muscle. This observation helped clarify some of the contradictory findings of earlier studies, in which both correct and random innervation patterns were reported following surgical manipulations. Thus, if spinal cord segments are placed too far away from local guidance cues, motor axons cannot find their correct target.

Several Molecules Help Direct Motor Axons to Muscles

Motor axons are directed to the correct muscle using many different cues. Like neural crest cells (see Chapter 5), axons of motor neurons rely on guidance cues found in the somites. These cues lead the axons to grow along the rostral (anterior) segment of the somites—the region also favored by neural crest cells. This preferential growth along the rostral segments leads to the ladder-like segmentation of motor axons seen in Figure 7.13. Repulsive cues produced by members of the ephrin and semaphorin families, as well as peanut-agglutinin-binding proteins, direct axons away from the caudal (posterior) segments of the somites (**Figure 7.16**). Other cues such as fibroblast growth factor (FGF) and hepatocyte growth factor (HGF) may promote the growth of axons along the rostral segment.

As in other systems, adhesion molecules are also present at key locations to direct the growth of motor neuron axons. For example, the axons of the LMC that innervate limb muscles are divided into different branches, or plexuses. The cervical and brachial plexuses innervate forelimb muscles, whereas the lumbar and sacral plexuses innervate hindlimb muscles. The

Figure 7.16 Multiple cues direct the motor axons through the rostral (anterior) segments of somites. (A) A ladder-like pattern of motor axons arises as motor axons (white) project through the rostral, but not the caudal (C), segment of the somite. (B) Several studies have detected inhibitory cues in the caudal (posterior) segments of somites, including peanut-agglutinin-binding protein (PNA-binding protein) and members of the ephrin and semaphorin families. These proteins repel motor axons from the caudal regions of somites. Fibroblast growth factor (FGF) and hepatocyte growth factor (HGF) may promote motor axon growth through the rostral (anterior) portions of the somites. NT, neural tube. [From Cook GMW, Sousa C, Schaeffer J, et al. [2020] *eLife* 9:e54612.]

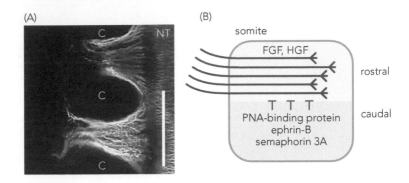

fibers of each plexus first travel together until they reach the proximal portion of the limb bud. At this choice point, the fibers separate. Fibers from the lateral portion of the LMC travel together through dorsal segment of the limb bud to reach specific extensor muscles. Fibers from the medial portion of the LMC travel to the ventral segment to contact flexor muscles (**Figure 7.17**A). Various cues present at the proximal limb choice point regulate the binding and branching of motor axons into the dorsal or ventral pathway. For example, L1 and NCAM adhesion molecules are important in the bundling and guidance of motor axons in the dorsal pathway (Figure 7.17B). L1 is associated with the axons themselves; in the absence of L1, there is a decrease in fasciculation and an increase in nerve branching. NCAM, found both on axons and on developing muscle cells (myotubes), influences axon bundling as well as axonal branching across the surfaces of muscles. In the absence of NCAM, there is some decrease in adhesion between adjacent axons, but the main effect is a decrease in branching along muscle surfaces. The level of PSA associated with NCAM determines how much branching occurs. For example, NCAM without PSA, or with low levels of PSA, promotes fasciculation and keeps axons bundled at the developmental stages when the axons need to travel together. In contrast, NCAM with higher levels of PSA (PSA-NCAM) interferes with axon–axon binding and thus increases the branching of nerve fibers at the time of target innervation. It is hypothesized that this decrease in fasciculation helps individual nerve fibers to respond better to local guidance cues near the target muscle (Figure 7.17). In recent years, scientists have also noted that the level of spontaneous electrical activity present in developing muscles regulates levels of PSA-NCAM. If electrical activity is blocked *in ovo* at the time when motor axons choose the dorsal or ventral pathway, PSA levels decrease, causing the axons to remain bundled together and dorsal–ventral pathfinding errors to occur.

A decrease in electrical activity also decreased the expression of EphA4 receptors normally found on axons projecting along the dorsal pathway. Under normal conditions, ephrin A ligands are expressed in the mesenchyme of the ventral portion of the limb, where they provide inhibitory cues to direct EphA4-expressing dorsal axons away from the ventral pathway (Figure 7.17B). If spontaneous electrical activity is blocked at the time the axons reach the choice point, however, then the levels of EphA4 decrease, so the axons are no longer inhibited by ephrin A ligands in the ventral pathway, contributing to the pathfinding errors observed in these experiments. Thus, spontaneous electrical activity influences the expression of several proteins necessary for normal pathfinding.

Other factors such as members of the glial-cell-derived neurotrophic factor (GDNF) and semaphorin families (see **Box 7.1**) interact to direct the pathfinding choices of motor neuron axons as they travel to the correct limb muscle. Once at the correct target muscle, the growth cone transforms into a presynaptic nerve terminal. The molecular mechanisms regulating synaptic connections between motor neurons and muscles are discussed in Chapter 9.

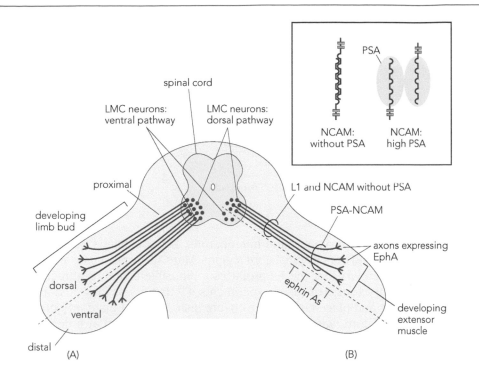

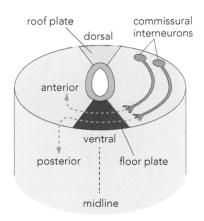

Figure 7.17 Multiple cues guide motor axons along the dorsal pathway. (A) Axons from motor neurons in the lateral motor column (LMC) initially extend together as a single bundle until they reach the proximal region of the developing limb bud. At this choice point the axons separate and travel along two distinct pathways: the dorsal pathway, which directs motor axons from the lateral region of the LMC, and the ventral pathway, which directs motor axons from the medial region of the LMC. (B) Several guidance cues interact to direct motor axons along the dorsal pathway. Adhesion molecules such as L1 and NCAM keep these axons grouped together as they approach the limb bud. Ephrin A molecules in the ventral region of the limb bud then provide inhibitory cues to prevent EphA-expressing motor axons from entering the ventral pathway. As axons approach the muscle, NCAM expresses high levels of polysialic acid (PSA). PSA interferes with homophilic binding (inset) to decrease fasciculation and promote axonal branching across the muscle. [Adapted from Bonanomi D & Pfaff SL [2010] *Cold Spring Harb Perspect Biol* 2(3):a001735.]

INTERMEDIATE, MIDLINE TARGETS FOR SPINAL COMMISSURAL AXONS

As noted in Chapter 4, the dorsal half of the vertebrate neural tube gives rise to multiple cell types, including spinal sensory interneurons located on each side of the midline. Axons from some of these interneurons remain on the **ipsilateral** side of the spinal cord, projecting to targets on the side where they originated. Another subset of interneuron axons crosses the midline, forming a commissure, to reach the **contralateral** side. In vertebrates, these spinal commissural interneurons first grow ventrally toward the midline floor plate (FP) then cross the midline and travel to their target tissues, with the majority projecting anteriorly (**Figure 7.18**). These contralateral axons never cross back to the side of the spinal cord from which they originate. Ipsilateral and contralateral projections are needed to integrate information from the two sides of the body and in some species, to coordinate the left–right patterning of limb movements. Several integrated guidance cues are needed to ensure proper ipsilateral and contralateral wiring is achieved.

The following section examines midline guidance cues that regulate growth of vertebrate and invertebrate commissural interneurons. As will be seen, homologous molecules have similar, as well as unique, functions in guiding axons toward and across the midline in these different animal models.

The Axons of Vertebrate Commissural Interneurons Are Attracted to the Floor Plate

In the 1890s, Cajal published intricate drawings depicting the growth of axons toward the midline. Based on his examination of fixed tissue sections (**Figure 7.19**A), he suggested an active mechanism by which the FP provided signals to attract the commissural axons. Thus, the axons of commissural interneurons would first extend to the FP, an **intermediate target** for these neurons, before crossing the midline and responding to guidance cues provided by their target cells.

Nearly a century later, scientists in the 1980s developed a standard *in vitro* assay to test whether the vertebrate FP was really a source of attractive

Figure 7.18 Spinal commissural interneurons are attracted to the ventral midline. Spinal commissural interneurons in the dorsolateral region of the developing vertebrate spinal cord project axons toward the floor plate at the ventral midline. Commissural axons cross the midline and project anteriorly or posteriorly (dashed arrows) to reach targets on the contralateral side.

Figure 7.19 Floor plate tissue secretes an attractive guidance cue for commissural interneurons. (A) In the 1890s, Cajal proposed that the axons of dorsal commissural interneurons are attracted to the floor plate. In this image from one of his original slides from a 3-day-old chick embryo, the dorsal commissural axons are seen approaching the floor plate at the ventral midline. (B) A century later, *in vitro* studies found that segments of the floor plate provided a diffusible signal to attract commissural interneuron axons. Conversely, roof plate tissue failed to attract the commissural interneuron axons. (C) Multiple segments of floor plate tissue attracted axons of spinal commissural interneurons, further indicating the axons did not simply grow in a pre-programmed direction, but responded to the floor-plate-derived signal. [(A), From de Castro F [2007] *Brain Res Rev* 55:481–489.]

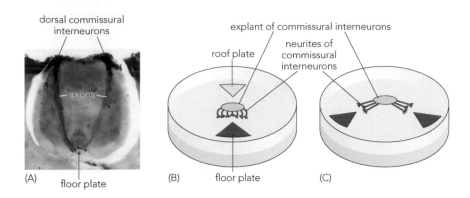

guidance cues for commissural interneurons. Using explants from rat or chick ventral and dorsal spinal cord regions, Marc Tessier-Lavigne and colleagues tested whether the FP produces a diffusible signal to attract the commissural axons. For these experiments, FP tissue was dissected from spinal cords and placed in a tissue culture dish. The roof plate (RP) of the dorsal spinal cord was also surgically removed and placed at a distance from the FP region. In these preparations, all tissues were placed in a collagen gel matrix that provided support for the tissues while limiting the diffusion of any signals from the FP or RP, thus allowing for assessment of directed axonal outgrowth.

These assays showed that the midline target FP, but not the non-target RP, attracted axons of commissural neurons (Figure 7.19B). When additional FP regions were added at different locations in the dish, the commissural axons were directed toward those FP tissues as well (Figure 7.17C). These results indicated that the axons were not just growing in a pre-programmed direction, but that some sort of FP-derived factor directed the growth of the axons. This provided an example of **chemotropism**—the directed growth of neuronal processes. This research also established the developmental stages of the FP secreted the signal. These criteria proved very helpful for later experiments testing whether putative factors were likely to be active *in vivo*.

Laminin-Like Midline Guidance Cues Are Found in Invertebrate and Vertebrate Animal Models

In the late 1980s through the mid-1990s, labs studying invertebrate and vertebrate animal models identified various midline guidance cues. Although it was not known at the time of the initial studies, these different labs were all studying homologs of the same guidance signal. In 1990, the first description of a specific midline guidance cue was published by Edward Hedgecock and colleagues. Experiments in *C. elegans* identified gene mutations that caused uncoordinated movements. The sixth uncoordinated (Unc) gene discovered, *Unc-6*, coded for a laminin-like protein. Based on its location in the ventral nerve cord and the timing of its expression, *Unc-6* was proposed to function in circumferential axon guidance, attracting axons from the dorsal commissural interneurons toward the ventral region along the circumference of the body (**Figure 7.20**A). It is now known that Unc-6 is a secreted protein that associates with basement membrane and cell surfaces, creating a ventral–dorsal gradient to which axons of the dorsal interneurons respond.

In 1994, Tessier-Lavigne and colleagues identified proteins from chick brain extracts that directed outgrowth of vertebrate commissural interneurons. They called the isolated proteins Netrin-1 and Netrin-2, from the Sanskrit word *netr*, meaning to guide. Netrin-1 was found to best mimic the FP-derived signal. For example, when *Netrin-1* was transfected into the

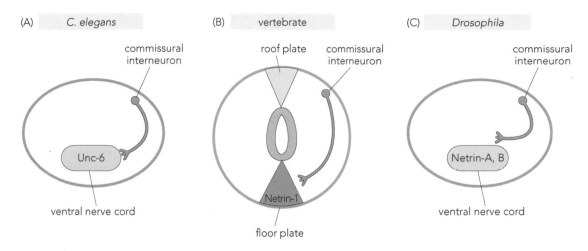

Figure 7.20 The ventral midline releases a laminin-like signal that attracts commissural interneurons. Cross-sections of different animal models reveal the origin of midline signals in *C. elegans*, vertebrates, and *Drosophila*. (A) The first midline attractive cue was identified in the ventral nerve cord (VNC) of *C. elegans*. The VNC releases the laminin-like protein, Unc-6, to attract dorsal commissural interneurons toward the midline. (B) A family of related proteins, called Netrins, was subsequently discovered in vertebrates. Netrin-1 mimics the diffusible floor-plate-derived signal previously identified *in vitro*. (C) In *Drosophila*, the related proteins, Netrin A and Netrin B, are found in the VNC. These Netrins guide commissural interneurons toward the VNC, where they join an adjacent commissure.

COS cell line, the Netrin-1-expressing COS cells acted like FP cells in the collagen gel assay, attracting commissural axons. Importantly, at the time chick commissural axons grow toward the floor plate *in vivo*, Netrin-1 is expressed in the FP in a ventral–dorsal gradient (Figure 7.20B). Thus, the timing and location of Netrin-1 production are consistent with its role as a midline guidance cue. Soon after they were described in chick embryos, *Netrin* genes were recognized as homologs of *Unc-6*.

Two labs then published studies in 1996 that revealed that homologous midline attractive cues were also active in *Drosophila*. The midline of the *Drosophila* ventral nerve cord contains neurons and glia that function much like the FP in vertebrates. The axons of dorsal commissural axons grow toward the ventral nerve cord to join an adjacent commissure. *Drosophila* has two genes homologous to *Unc-6*, called *Netrin-A* and *Netrin-B*, which appear to have overlapping functions at the midline (Figure 7.20C). In flies lacking only one of these Netrins, most commissural axons continue to cross the midline. Fly Netrins appear to function as short-range guidance cues closely associated with the cells that produce them, so they only attract nearby axons that reached the midline using other cues.

Since the original characterization of vertebrate Netrin-1 and Netrin-2 in chick, studies have found that vertebrates also express Netrin-1 in the ventricular zone (VZ) where it may interact with FP-derived Netrin-1 to guide commissural axons to the ventral midline. Other studies suggest Netrin-1 in the VZ, rather than the FP, is the primary source of the midline guidance cue.

Two additional secreted vertebrate netrins, called Netrin-3 and Netrin-4, and two membrane-bound Netrin proteins, called Netrin-G1 and Netrin-G2, which are coupled to the cell membrane by a glycosylphosphatidylinositol (GPI) linkage, have been identified. Structurally, three of the secreted vertebrate Netrins (Netrin-1, Netrin-2, and Netrin-3) are the most similar to each other and can function to some extent as midline attractants. These are the Netrins that are homologous to *C. elegans* Unc-6 and *Drosophila* Netrins A and B. To date, only chick and zebrafish have been found to express Netrin-2, the function of which is unclear. Netrin-4 and the membrane-bound G Netrins are found only in vertebrates and appear to have different functions than Netrins 1–3. Recent studies suggest Netrin-4 and the G Netrins may refine or maintain synaptic connections.

Homologous Receptors Mediate Midline Attractive and Repulsive Guidance Cues

The commissural interneurons in invertebrates and vertebrates express homologous receptors that mediate the Unc-6/Netrin guidance cues. In *C. elegans*, the axons of the dorsal interneurons express two receptors, Unc-40 and Unc-5. The growth response of the axons differs depending on which receptor is activated by Unc-6. For axons expressing the Unc-40 receptor, Unc-6 acts as an attractive guidance cue, directing axons toward the midline. However, for axons expressing the Unc-5 receptor, Unc-6 functions as a repulsive guidance cue, directing the axons away from the midline (**Figure 7.21**A).

Like *C. elegans*, *Drosophila* and vertebrate interneurons express Unc-5 receptors that cause axons to be repelled by, rather than attracted to, Netrins. There is only one form of this receptor in *C. elegans* and *Drosophila*, but four isoforms of the Unc-5 receptor are found in mammals (UNC5A–UNC5D). The function of these various isoforms is not yet clear.

In *Drosophila*, receptors homologous to Unc-40 are called the Frazzled receptors (Fra; also abbreviated Frl in the literature). Like the Unc-40-expressing neurons in *C. elegans*, interneurons expressing Fra receptors are attracted to the *Drosophila* Netrins (Figure 7.21B).

In vertebrate commissural interneurons, the homologous receptor is called DCC (deleted in colorectal cancer). When DCC is activated by FP-derived Netrin-1, axons are attracted toward the midline (Figure 7.21C). Yet, in mice lacking the *Dcc* gene the decrease in commissural axons crossing the midline was not as great as the decrease seen in mice lacking *Netrin-1*. This suggested that additional receptors are needed to transduce the Netrin-1 signal. One such receptor is Neogenin-1 (Neo1). When both *Dcc* and *Neo1* are absent in mice, there is a greater loss of commissural axons compared to the loss observed with deletion of either receptor gene alone. Recent studies suggest that Netrin-1 binds to different regions of DCC and Neo1 to cross-link the receptors and optimize the cellular response.

How the Netrins regulate both attraction and repulsion by activating different receptors is not fully understood. In some cases expression and activation of Unc-5 alone is sufficient to direct axons away from the Unc-6/Netrin. However, in other cases both Unc-5 and Unc-40/Fra/DCC receptor types appear necessary to initiate the repulsive guidance cue. Several lines of evidence suggest that when the two receptor types are expressed at the same time in a growth cone, the cytoplasmic regions of the Unc-5 and Unc-40/Fra/DCC receptors interact, causing the Netrin signal to change from an attractive cue to a repulsive cue.

Figure 7.21 Differential expression of receptors determines if the midline signal is attractive or repulsive. (A) In *C. elegans*, two receptors that bind Unc-6 were identified in commissural interneurons. Axons expressing the Unc-40 receptor are attracted to the midline Unc-6. In contrast, axons of commissural interneurons expressing the Unc-5 receptor, alone or in combination with Unc-40, are repelled by Unc-6 at the midline. (B) In *Drosophila*, commissural interneurons expressing Frazzled (Fra) receptors, homologs of Unc-40, are attracted to Netrins A and B in the ventral nerve cord, while those expressing Unc-5 are repelled by the midline. As in *C. elegans*, the Unc-5 receptor alone, or in combination with the Unc-40 receptor homolog, mediates the repulsive effect of the midline signal. (C) Similarly, vertebrate commissural interneurons expressing the DCC (deleted in colorectal cancer) receptor, a homolog of Unc-40, are attracted to the midline, whereas those expressing Unc-5, alone or in combination with DCC, are repelled by floor-plate-derived Netrin-1.

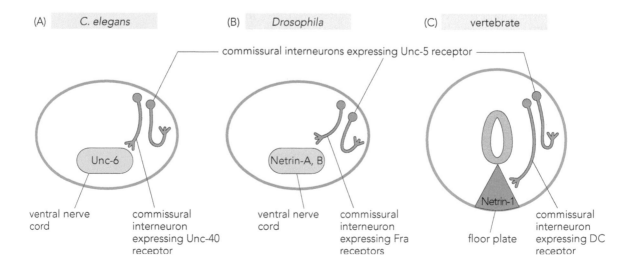

(A)   *C. elegans*   (B)   *Drosophila*   (C)   vertebrate

commissural interneurons expressing Unc-5 receptor

Unc-6

Netrin-A, B

Netrin-1

ventral nerve cord

commissural interneuron expressing Unc-40 receptor

ventral nerve cord

commissural interneuron expressing Fra receptors

floor plate

commissural interneuron expressing DC receptor

Slit Proteins Provide Additional Axonal Guidance Cues at the Midline

While Netrin-related molecules play a major role in midline guidance, they are not the only signals used by commissural axons. In the 1990s, other midline guidance cues were discovered including a group of large, secreted proteins called Slits. Slit proteins are expressed in *Drosophila*, *C. elegans*, and vertebrates. *Drosophila* and *C. elegans* each express one Slit protein that is localized at the ventral midline. In contrast, vertebrates express three Slit proteins (Slit1, Slit2, and Slit3), all of which are expressed in the FP.

The normal expression of Slit at the midline regulates growth of commissural axons in two important ways: repelling axons of ipsilateral pathways to prevent those fibers from ever crossing to the contralateral side and repelling the crossed axons of the contralateral pathway to prevent them from re-crossing back to the original side (**Figure 7.22**A, C).

The importance of the repulsive Slit cue was first seen in *Drosophila* mutants lacking the *Slit* gene. In these mutants, all axons grew toward the midline, including those that normally remain on the ipsilateral side. The ipsilateral axons never crossed to the contralateral side, however, but collapsed at the midline instead. The contralateral fibers crossed to the opposite side as in normal conditions, but also re-crossed back to the original side in the absence of *Slit* (Figure 7.22B, D).

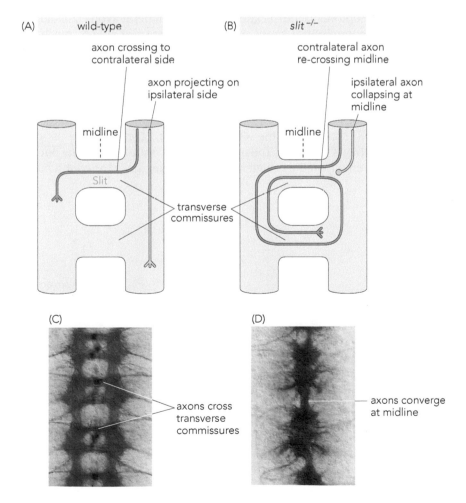

Figure 7.22 Slit proteins inhibit midline crossing of commissural axons. (A) In wild-type *Drosophila*, commissural interneurons that contact targets on the contralateral side must cross a transverse commissure. Axons of interneurons that project to ipsilateral targets must remain on the original side. Slit protein is expressed in the midline to prevent ipsilateral axons from crossing to the contralateral side as well as prevent contralateral axons from returning to the original side. (B) The importance of midline Slit proteins is observed in *Drosophila* lacking *Slit* (*Slit⁻/⁻*). In these mutants, rather than traveling to their normal targets, ipsilateral axons approach the midline, where they collapse. The axons of commissural interneurons, which should project to contralateral targets, freely cross the transverse commissures multiple times. (C, D) Comparison of axonal growth patterns at the *Drosophila* midline in wild-type (C) and *Slit⁻/⁻* mutants (D) showing that in the absence of *Slit*, axons converge at the midline. Axons are labeled brown and the midline glia are blue. [(C and D), From Seeger M, Tear G, Ferres-Marco D & Goodman CS [1993] *Neuron* 10:409–426.]

Slit Proteins Repel Commissural Axons away from the Midline by Activating Robo Receptors

It was another *Drosophila* mutant that led to the identity of the receptors for Slit. The receptors were named Robo (Roundabout) based on the phenotype observed in flies lacking the gene. Unlike the patterns observed in normal conditions, the contralaterally projecting commissural axons in *Robo* mutants continued to cross and re-cross at the midline multiple times, traveling in a roundabout fashion (**Figure 7.23**A).

There are three identified Robo receptors in *Drosophila* (Robo1, Robo2, and Robo3). Robo1 and Robo2 are important for directing the growth of ipsilaterally and contralaterally projecting axons of *Drosophila* commissural neurons. The response of these axons to midline Slits depends on which of these Robo receptors is expressed. In *Drosophila* lacking only *Robo1*, the contralaterally projecting commissural axons freely crossed the midline multiple times. However, if both *Robo1* and *Robo2* were absent, all axons grew toward the midline, including those that should remain on the ipsilateral side. Thus, as with the absence of *Slit*, the ipsilateral axons were no longer repelled from the midline.

Further investigations of Robo1 and Robo2 suggested that under normal growth conditions both Robo1 and Robo2 are expressed in ipsilateral axons where the receptors work together to respond to the repulsive signal provided the midline Slit protein. This midline repulsion keeps the ipsilateral projections on the correct side. In contrast, the contralaterally projecting axons express only Robo1, so for these axons Slit functions to prevent the re-crossing of axons that have already crossed to the opposite side (Figure 7.23B). However, given that Slit is expressed at the midline at the time of commissural axon crossing, the differential expression of Robo1 and Robo2 receptors on the ipsilateral and contralateral axons still did not explain how commissural axons expressing Robo1 are able to reach the contralateral side in the first place. The discovery of another mutation in *Drosophila* called *Commissureless* provided insight into how this occurs.

Robo Signaling Is Regulated by Additional Proteins Expressed on Commissural Axons

Under normal conditions, the *Commissureless* (*Comm*) gene is only expressed in axons that cross the midline to the contralateral side, but not in axons that remain on the ipsilateral side. However, once the contralateral

Figure 7.23 Slit proteins repel axons by activating Robo receptors. (A) The first receptor for Slit was identified in a *Drosophila* mutation. The newly identified receptor was named Roundabout (Robo) because in the absence of the gene, the contralaterally projecting axons grew across the midline multiple times. (B) Additional Robo receptors have been identified. Robo1 (Robo) is expressed on axons that cross the midline, whereas axons projecting on the ipsilateral side express both Robo1 and Robo2. Both *Robo1* and *Robo2* appear necessary for ipsilateral projections to respond to the repulsive midline Slit protein.

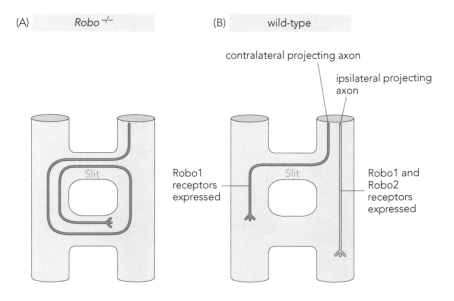

axons cross the midline, *Comm* expression is downregulated so the axons are repelled by midline Slit, thus preventing re-crossing to the original side. The necessity of *Comm* was seen in the mutants that lacked this gene. In the absence of *Comm*, the majority of the contralaterally projecting commissural axons were prevented from crossing the midline (**Figure 7.24**A).

Subsequent studies in *Drosophila* indicated that *Comm* encodes a small transmembrane protein that interacts with Robo1, preventing it from reaching the growth cone surface, perhaps by directing Robo1 to endosomes for degradation. Thus, when *Comm* is present in commissural axons, Robo1 is unavailable, and the repulsive activity of Slit is not transduced. Thus, the axons only respond to the attractive Netrin signal at the midline and are guided to the contralateral side. Once on the contralateral side, *Comm* activity is downregulated so Robo1 expression increases. These Robo1-expressing axons are repelled by the midline Slit and therefore prevented from re-crossing (Figure 7.24B).

It first appeared that there was no vertebrate homologue of the *Drosophila Comm* gene. However, a variant of Robo3, called Robo3.1 (also called Rig-1), appeared to function much like *Comm* as it stopped Robo1 signaling. Early studies found that Robo3.1 interfered with the ability of Robo1 to transduce the repulsive signal produced by the midline Slits. However, it was unclear how this occurred. As details of vertebrate isoforms of Robo3 emerged, it was suggested that Robo3.1 is present in commissural neuron cell bodies when fibers reach the midline where it acts to suppress the repulsive midline Slit signal by targeting Robo1 for degradation. Without Robo1 receptor activity, the axons cross to the contralateral side. However, once crossed, *Robo3.2* expression transiently increases and the Robo3.2 receptor is transported to the axon fibers where it transduces the repulsive Slit cue to prevent crossed axons from re-crossing.

Unique roles for Robo3 isoforms in different vertebrate species continue to be discovered. Mammalian isoforms of Robo3, for example, do not appear to directly bind to Slits, whereas non-mammalian isoforms do. In 2017, a functional *Comm* homologue, *PRRG4* (proline rich and Gla domain 4) was described. In cell culture experiments, *PRRG4*, like *Comm*, prevented Robo1 from reaching the cell surface. How Robo and Robo-like receptors interact with other signaling molecules, such as Netrins and their DCC receptors, continues to be investigated.

Shh Phosphorylates Zip Code Binding Proteins to Increase Local Translation of Actin and Direct Growth of Vertebrate Commissural Axons

As critical as Netrins and Slits are for directing axonal growth at the midline, they are not the only cues available to commissural interneurons. Sonic hedgehog (Shh) provides a secreted, attractive guidance cue that

Figure 7.24 *Comm* mutants reveal the mechanisms by which commissural axons cross the midline. (A) In *Drosophila Comm*$^{-/-}$ mutants, the commissures fail to form, revealing the importance of this gene in guiding commissural interneurons to the contralateral side. (B) Netrins attract commissural interneurons to the midline, while Slit proteins repel Robo1-expressing axons. When commissural axons expressing *Comm* (*Comm*$^{+/+}$) reach the midline, Robo1 signaling is inactivated so that the axons will not respond to the inhibitory Slit signal. These axons will respond, however, to the attractive signal provided by Netrins. Once the axons reach the contralateral side, *Comm* is downregulated and *Robo1* expression is upregulated so the axons can respond to the inhibitory midline Slit and are thus prevented from re-crossing back to the original side. [(A), From Seeger M, Tear G, Ferres-Marco D & Goodman CS [1993] *Neuron* 10:409–426.]

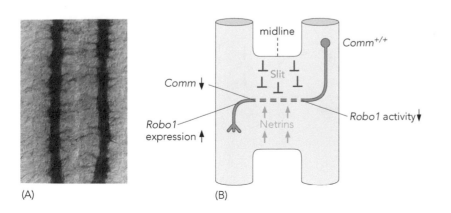

(A)　　　　　　　　(B)

causes commissural axons to turn toward the FP as they approach the midline. For a growth cone to turn toward an attractive midline guidance cue, there must be changes in the organization of the cytoskeletal elements. When a growth cone turns toward an attractive cue, actin polymerization becomes concentrated on the side of the growth cone closest to that cue. In at least some contexts, new actin protein is translated in the growth cone itself rather than in the cell body. Shh can increase this local translation of actin by inducing phosphorylation of zipcode binding protein 1 (ZBP1).

In the cell body, ZBP1 attaches to a specific nucleotide sequence called the zipcode that is located at the 3′ untranslated region of the beta actin mRNA that produces both G-actin and F-actin. Once ZBP1 attaches to this nucleotide sequence of beta actin mRNA, the mRNA is transported along microtubules to reach the growth cone. As long as ZBP1 remains unphosphorylated, the associated beta actin mRNA remains untranslated. However, in the growth cone, extrinsic signals such as Shh can lead to phosphorylation of ZBP1, causing the release of the mRNA strand and local translation of the beta actin mRNA.

The Shh signaling pathway used in commissural axons is slightly different from the canonical Shh pathways described in Chapter 4. In commissural axons, Shh binds to the receptors Patched (Ptc) and BOC (brother of CDO; CAM-related/downregulated by oncogenes), a receptor related to Robo. Shh binding to Ptc and BOC receptors releases the inhibition of Smoothened (Smo) so that it can travel to the cell membrane (see Figure 4.8). Smo then activates Src family kinases that in turn phosphorylate ZBP1, thus leading to the release of the beta actin mRNA and the local translation of actin in the growth cone (**Figure 7.25**).

A recent series of experiments in rats, mice, and chicks by Frédéric Charron and colleagues demonstrated that a gradient of Shh is responsible for the local accumulation of beta actin and the turning of commissural axon growth cones at the midline. For example, *in vitro* studies revealed that in cultures lacking Shh, rat commissural neurons extend without turning (**Figure 7.26**A). However, when added to cell culture chambers that included a gradient of Shh, the growth cones turned and continued to extend toward the increasing concentration of Shh (Figure 7.26B). When

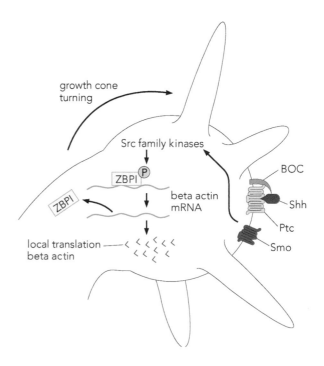

Figure 7.25 Sonic hedgehog (Shh) initiates local translation of beta actin in the growth cone. Binding of Shh to the BOC and Patched (Ptc) receptors located on commissural axons allows Smoothened (Smo) to move to the cell membrane (see Figure 4.8). Smo then activates the Src family kinases that phosphorylate the zipcode binding protein 1 (ZBP1) attached to a segment of beta actin mRNA. The phosphorylated ZBP1 detaches from the mRNA strand. Once the ZBP1 is released, local translation of actin occurs in the growth cone, ultimately causing the growth cone to turn toward the source of Shh. BOC (brother of CDO; CAM-related/downregulated by oncogenes).

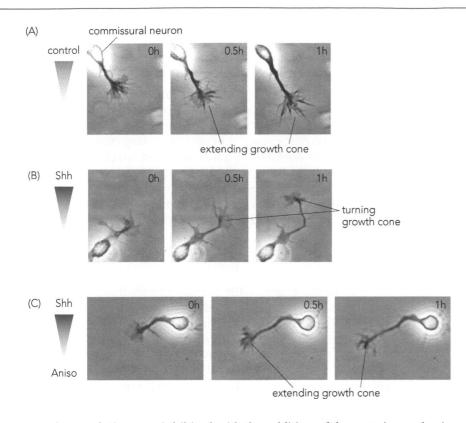

Figure 7.26 Protein translation induced by a gradient of Shh mediates growth cone turning in rat commissural axons *in vitro*. The trajectory of axonal growth from embryonic rat commissural neurons was observed over a 1-hour period in different cell culture conditions. (A) When embryonic rat commissural neurons were grown in cultures lacking Shh, the axons extended across the substrate without turning. (B) When the commissural neurons were grown in culture conditions in which a gradient of Shh was provided, the axons turned toward the higher concentrations of Shh. (C) To test whether protein translation was necessary for the Shh-induced growth cone turning, the protein synthesis inhibitor anisomycin (aniso) was added to the cell cultures. When protein synthesis was inhibited, the Shh gradient was no longer able to induce growth cone turning. These experiments support the hypothesis that guidance cues mediate local translation of actin during periods of growth cone turning. [Adapted from Lepelletier L, Langlois SD, Kent CB et al. [2017] *J Neurosci* 37:1685–1695.]

protein translation was inhibited with the addition of the protein synthesis inhibitor anisomycin, the growth cones no longer turned (Figure 7.26C). This suggested that protein translation was needed for growth cones to turn toward the Shh gradient. The studies also noted that Shh induced the phosphorylation of ZBP1 and increased the amount of beta actin in the growth cones. These results supported the hypothesis that gradients of Shh were responsible for the changes in actin translation associated with the turning response.

In mice lacking the *Zbp1* gene the commissural interneurons displayed a disorganized trajectory to the floor plate. Similar results were seen in chick embryos when a mutant form of *Zbp1* was electroporated into chick spinal cord. In these experiments, the mutated ZBP1 could still bind to beta mRNA, but could not release it for translation.

These studies also support previous findings that local translation of beta actin is mediated by ZBP1 to induce the turning of growth cones in response to extrinsic guidance cues. For example, Netrin-1 is associated with local translation of beta actin in *Xenopus* retinal ganglion axons, and brain-derived neurotrophic factor (BDNF) was shown to increase local beta actin translation in *Xenopus* spinal commissural axons. However, other studies suggest that local protein translation is not necessary to induce changes in growth cone cytoskeletal organization or alter the trajectory of growth. Research investigating whether these differences are related to the animal model used, the age or neuronal subtype investigated, the guidance cues tested, or metabolic differences associated with various cell culture conditions are being explored.

THE RETINOTECTAL SYSTEM AND THE CHEMOAFFINITY HYPOTHESIS

The preceding sections described examples of the many cues that guide axon bundles and individual axons toward, or away from, a particular target tissue. This section explores how axons are selectively matched to an individual target cell using examples from the vertebrate retino-tectal system. The retinotectal system has been a popular model system for investigating axonal pathfinding and target cell recognition since the 1920s. There are many advantages to using this system as an experimental model. The retinal ganglion cells and their target tissue, the optic tectum, are easily identified and accessible for experimental manipulations in many vertebrate species. Additionally, because the retinotectal system has been studied for so many years, it is very well characterized and thus provides scientists with a wealth of information on which to draw. The following sections describe the findings that first led scientists to investigate axon-target recognition in the retinotectal system and the subsequent experiments that identified specific cues that regulate the correct mapping of retinal ganglion cell axons across the optic tectum.

Early Studies of Axon-Target Recognition Focused on Physical Cues and Neural Activity

In the 1920s and 1930s, many of the prevailing hypotheses about how axons locate specific target cells focused on the roles of physical guidance cues, such as blood vessels and other tissues within the embryo, or the necessity of matching neural activity patterns between a target cell and an innervating nerve fiber. Many favored the idea that once axons reached their target tissue, a given target cell formed a synaptic connection only with the axon that provided a matching pattern of neural activity. The notion of a target cell responding to the neural activity of an axon was known as the **resonance hypothesis**—that is, the target cells would resonate only with axons providing matching electrical activity. This hypothesis was developed over several years largely through the work of Paul Weiss and colleagues.

While the physical environment and electrical activity are important for some aspects of axonal guidance and target recognition, they cannot fully explain how connectivity is established in the nervous system. Among the most pivotal experiments that reshaped how scientists think about axonal guidance mechanisms were those conducted by Roger Sperry from the 1940s to the 1960s. Although Sperry worked with Weiss, he saw limits to the resonance hypothesis and so began a series of experiments using the retinotectal system of amphibians to test how axons recognize and make proper connections with target cells.

Amphibian Retinal Ganglion Cell Axons Regenerate to Reestablish Neural Connections

In amphibians, the axons of retinal ganglion cells (RGCs) travel in the optic nerve to reach their midbrain target, the optic tectum—a structure analogous to the superior colliculus in mammals. Studies in the early 1900s revealed that if the optic nerve was crushed or severed, the retinal axons would regenerate and reestablish connections within the optic tectum. In the 1920s, Robert Matthey demonstrated that after the retinal ganglion cell axons of the adult newt were severed, the animal's vision was restored within a few weeks' time. The novelty and significance of Matthey's findings were later summarized by Sperry (1956):

> He had severed the optic nerve in adult newts, or salamanders, and they later recovered their vision! New nerve fibers had sprouted from the cut stump and had managed to grow back to the visual centers of the brain. That an adult animal could regenerate the optic nerve . . . was surprising enough, but that that it could also re-establish the complex network of nerve-fiber connections between the eye and a multitude of precisely

located points in the brain seemed to border on the incredible. And yet, this was the only possible explanation, for without question the newts had regained normal vision.

In the 1940s, Leon Stone's lab expanded on Matthey's work and found that the optic nerve of salamanders could be sectioned multiple times, and each time the retinal axons regenerated and vision was restored. This group also found that if an eye from one salamander was transplanted to another salamander, the retinal axons of the transplanted eye regenerated and restored vision. Because vision was restored even after multiple surgeries, Sperry and others recognized that the retinotectal system would provide a means of testing how axonal connections are established with specific target cells.

At the time of Sperry's initial investigations, some scientists thought the regenerating fibers grew to the tectum in a more or less random manner, then through visual experience those connections that proved useful remained, whereas those that were nonfunctional were lost. Others hypothesized that the axons regrew in a systematic manner to reestablish their original connections with specific target cells. This latter hypothesis was consistent with the idea that a chemical cue directed the axons to a particular target cell. Ideas about "chemiotropism" had been first proposed in the 1890s by both Langley and Cajal (see Chapter 9) but had not yet been formerly tested.

To test whether retinal axons relied on target-derived, chemical cues to contact a specific region of the tectum, Sperry modified the eye surgery technique used by Matthey and Stone. For his experiments, Sperry cut the muscles around the newt eye so he could rotate the eye 180 degrees (**Figure 7.27**A). After the eye was rotated, Sperry severed the optic nerve. In his 1943 report, Sperry completed this surgery on 58 adult newts and noted they recovered visual function over a period ranging from 28–95 days. Prior to this recovery, the animals behaved as if blind. However, unlike the animals in the studies by Matthey and Stone, which did not have eyes rotated, these animals did not recover normal visual function. The newts

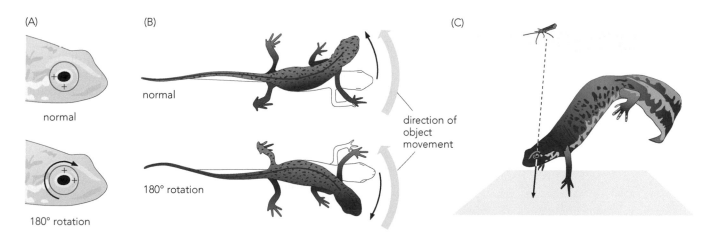

Figure 7.27 Surgical rotation of the newt eye causes the visual world to be reversed. Studies by Roger Sperry from the 1940s through the 1960s indicated that retinotectal projections returned to the original target cells after experimental manipulations. (A) In these drawings from Sperry's original work, the muscles around the eye were severed and the eye was rotated 180 degrees. The + signs mark the original and rotated locations of the eye. After the eye was rotated, the optic nerve was cut and the animal was allowed to recover. (B) The behavior of the newt indicated that the visual world was also reversed. The large arrows indicate the direction of movement of an object, while the small arrows indicate the direction of head movement. The experimental animals always responded as if the object were moving in the opposite direction. In this example, the head always turned away from the direction of movement. (C) When bait was presented above the head of an experimental animal, the newt reached for the bottom of the tank. In all animals with rotated eyes, no amount of practice or recovery time improved the animals' performance. [Adapted from Sperry R [1956] *Sci Amer* 194:48–52.]

with rotated eyes could see, but they behaved as if the world were reversed (Figure 7.27B, C). Thus, the behavior was consistent with the new orientation of the eye. These findings suggested that the axons of retinal ganglion cells did not travel to the nearest cells in the optic tectum, but instead established connections with their original target cells.

As Sperry (1956) later summarized:

> When a piece of bait was held above the newt's head, it would begin digging into the pebbles and sand on the bottom of the aquarium. When the lure was presented in front of its head, it would turn around and start searching in the rear; when the bait was behind it, the animal would lunge forward.

In the newts with rotated eyes, vision never corrected to the preoperative state and practice did not improve the animals' performance. Even those animals that survived for two years continued to behave as if the visual world were rotated 180 degrees. Experiments in frogs, toads, and fish revealed similar results following eye rotations. As in the newts, the visual world was inverted and no amount of practice compensated for the altered visual field. However, if the eye was later returned to the original position, the animals' vision returned to normal.

Sperry proposed a mechanism by which retinal axons put out numerous branches and tested different cells until "eventually the growing tip encounters a cell surface for which it has a specific chemical affinity and to which it adheres." This formed the basis of what came to be known as the **chemoaffinity hypothesis**. In the strictest sense, chemoaffinity would require matching "chemical tags," as Sperry called them, between each axon and each target cell. Although other scientists questioned whether there could be a strict one-to-one matching of chemical tags between individual axons and target cells, Sperry's experiments clearly demonstrated that there was some form of specific recognition between axons and their target cells.

Retinotectal Maps Are Found in Normal and Experimental Conditions

To better understand the organization of retinal axons within the tectum, axonal connections were mapped using histological preparations and electrophysiological recordings. With these methods, scientists determined that the axons of RGCs located on the temporal side of the eye innervate the anterior portion of the tectum, whereas the axons of RGCs on the nasal side of the eye innervate the posterior portion of the tectum (**Figure 7.28**). This is one example of a **topographic map**—the consistent, primarily invariant projection of axons from one region of the nervous system onto another. The behavioral results obtained after the eye rotation

Figure 7.28 Normal patterns of retinotectal projections. Axons of retinal ganglion cells (RGCs) from the nasal half of the retina normally project to the posterior portion of the tectum, whereas RGC axons from the temporal half of the retina project to the anterior region of the tectum.

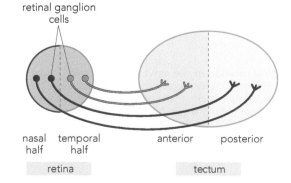

retinal ganglion cells

nasal half   temporal half   anterior   posterior

retina   tectum

surgeries indicated that the temporal axons must still grow to the anterior tectum and the nasal axons to the posterior tectum, even after experimental manipulations.

Throughout the 1950s, 1960s, and 1970s, scientists used a variety of techniques to further evaluate the mapping of retinotectal projections under different experimental conditions. Many of the surgical manipulations were done using frogs, salamanders, or fish, though some were also conducted in chick embryos. One of the methods used to investigate retinotectal mapping was to create a "compound eye" in which, for example, two nasal sections of the retina were grafted into a single eye so that nasal RGCs were also on the temporal side of the eye. These extra nasal axons still mapped to the posterior region of the tectum indicating that the location of the retinal ganglion fibers in the eye (temporal or nasal side) did not establish the location of axonal projections in the tectum. These studies supported Sperry's hypothesis that "chemical tags" were used to direct retinal ganglion axons to the correct region of the tectum.

Further support for the importance of chemical cues came from studies in which an eye was treated with tetrodotoxin (TTX), a chemical that blocks action potentials. Even with RGC electrical activity blocked, mapping to the correct tectal location occurred. These experiments again suggested that topographic mapping in the tectum was not due to retinal axons firing in resonance with target cells, but was more likely due to chemical cues present on the axons and target cells.

Some Experimental Evidence Contradicts the Chemoaffinity Hypothesis

The chemoaffinity hypothesis remained popular for quite some time, but scientists also observed situations in which chemoaffinity did not appear to account for the mapping of axons in the optic tectum. For example, studies in the 1970s revealed that if half of the retina were surgically removed and the animals were given sufficient time to recover, the remaining axons would ultimately grow over the entire tectum. In these types of experiments it was often noted that while the retinal axons initially grew to the original anterior or posterior tectal area, over time the axons expanded to contact non-target regions as well (**Figure 7.29**). This suggested that there was not a strict one-to-one matching of retinal axons and tectal cells or any rigid boundaries to limit where axons could grow.

In other experiments, half the tectum was surgically removed. In these preparations, the RGC axons from the temporal and nasal sides of the retina would all converge on the remaining tectal space. Axons that would normally innervate the missing portion of the tectum would now make connections with cells in the remaining half of the tectum, forming a "compressed" retinotectal map. For example, temporal axons would converge on the remaining posterior tectum if the anterior portion were missing.

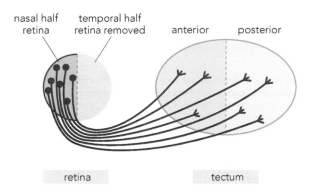

nasal half retina   temporal half retina removed   anterior   posterior

retina   tectum

Figure 7.29 Retinal projections grow to the original tectal site initially but spread out given sufficient recovery time. When the temporal portion of the retina was surgically removed (gray area), the remaining RGCs from the nasal portion initially projected to the posterior region of the tectum as usual. However, when the animals were given a sufficient recovery time, the axons from nasal retinal ganglion cells expanded to contact cells in the available anterior portion of the tectum.

Additional experiments conducted throughout the 1970s and 1980s noted species and age differences that influenced the ability of retinal axons to map onto new tectal regions. However, the general consensus that emerged during this period was that regenerating RGC axons prefer to map to the original target areas if possible, but are able to make connections in new target regions if necessary. These findings suggested there is not a specific matching of axons and target cells, so many scientists began to dismiss the chemoaffinity hypothesis and look for other mechanisms that could direct retinal axons to specific areas of the tectum. Yet, by the 1990s, the chemoaffinity hypothesis re-emerged as a viable mechanism for retinotectal mapping.

A "Stripe Assay" Reveals Growth Preferences for Temporal Retinal Axons

A major advance in understanding how retinal axons selectively grow to a given region of the tectum came from the lab of Friedrich Bonhoeffer in the 1980s. The lab developed a unique cell culture method called a "stripe assay" to analyze the growth of chick retinal axons on cell membranes extracted from tectal cells. The researchers first removed the tectum, separated the anterior and posterior regions, homogenized each region, then treated the homogenates so that only the cell membranes remained.

The cell membranes from the anterior or posterior tectal cells were then added to a small silicon device with channels 90 microns in diameter and 90 microns apart. Anterior or posterior cell membranes were suctioned through the channels onto a filter membrane that permitted growth of retinal axons but did not promote axonal growth on its own (**Figure 7.30**). Once one of the tectal cell membrane samples (for example, anterior or posterior) was firmly adhered to the filter membrane, a second tectal membrane sample would be suctioned onto the remaining stripes, or the stripes would be left untreated.

The filter membranes were then transferred to a cell culture dish and RGCs from either the temporal or nasal side of the retina were placed at the edge of the membrane and covered with a culture media that promoted survival and growth of RGCs. The extension of the temporal and nasal axons was then compared on the untreated lanes and lanes of anterior or posterior tectal cell membranes (**Figure 7.31**). When untreated lanes

(A)

(B)   channels

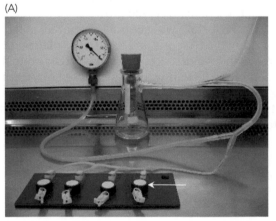

Figure 7.30 An assay is developed to evaluate differential growth of retinal axons on anterior and posterior tectal cell membranes. Scientists developed a cell culture method to produce precise lanes of anterior and posterior tectal cell membranes. Lanes of tectal cell membranes were coated onto a filter membrane that supported growth of RGC axons (arrow, A) by suctioning the cell membranes through channels of a silicon device (B) using the vacuum system shown in panel A. Each filter membrane would contain lanes of cell membranes from the anterior or posterior region of the tectum or would be left untreated as shown in Figure 7.31. [From Knoll B, Weinl C, Nordheim A & Bonhoeffer F [2007] *Nat Protoc* 2:1216–1224.]

and lanes with cell membranes from a single tectal region were compared, both the temporal and nasal axons grew freely along the stripes of tectal cell membranes. This indicated that temporal and nasal axons could grow along membranes from either anterior or posterior regions of the tectum if no other tectal membrane choice were available (Figure 7.31A, B). This observation was reminiscent of the surgical manipulations that revealed temporal and nasal axons had the ability to grow on either portion of the tectum if the normal target region was removed.

When the RGCs were presented with stripes of alternating anterior and posterior tectal membranes, differences were seen. The temporal axons showed a clear preference for stripes containing anterior tectal cell membranes. Thus, temporal axons grew along the lanes of anterior tectal membranes, but not the lanes with posterior tectal cell membranes (Figure 7.31C). Notably, the temporal axons only grew along anterior stripes at the developmental stages that retinotectal maps were formed *in vivo*. In contrast to the temporal axons, the nasal axons had no preference and grew on stripes of anterior or posterior cell membranes (Figure 7.31D).

The stripe assay was then used to evaluate the nature of the chemical signals regulating temporal axon growth. To test whether membrane-bound guidance cues were located on the tectal cells, proteins on the tectal cell membranes were denatured by exposing them to heat or fixation. When denatured anterior cell membranes were included in the stripe

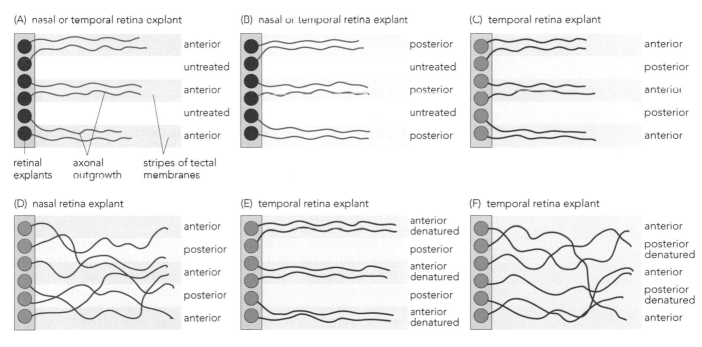

Figure 7.31 Stripe assays reveal how membranes from tectal cells direct growth of retinal ganglion cell axons. Schematic representations of the results obtained in different stripe assays. For all assays, explants of RGCs from the nasal or temporal half of the retina were placed at the edge of the filter membrane and cultured under conditions that allowed for growth. (A) When nasal or temporal retinal ganglion neurons were given the choice between anterior tectal cell membranes or untreated lanes, axons from both retinal regions preferred to grow along the lanes coated with anterior membranes. (B) Both nasal and temporal axons also preferred to grow along lanes coated with posterior tectal cell membranes rather than untreated lanes. These experiments revealed that if no other tectal cells were available, nasal and temporal axons could grow on either tectal cell type. (C) When given a choice between anterior and posterior tectal cell membranes, temporal RGC axons showed a distinct preference. In these assays, temporal axons only grew in the anterior lanes. (D) In contrast, axons from the nasal side of the retina showed no preference and readily grew along both lanes. (E) The specific growth preference of temporal retinal axons led the scientists to test whether the anterior tectal cells provided an attractive guidance cue for temporal axons, or the posterior membranes provided an inhibitory cue. When anterior tectal cells were denatured to inactivate cell surface proteins, temporal axons continued to prefer those lanes, indicating there was not an attractive protein on the anterior cell membranes. (F) When posterior cell membranes were denatured, temporal axons grew across all lanes. This indicated that the posterior tectal cell membranes normally expressed an inhibitory protein that directed temporal axons away from posterior tectum. [Adapted from Walter J, Henke-Fahle S & Bonhoeffer F [1987] *Development* 101(4):909–913.]

assay, the temporal axons still grew on those lanes, suggesting a membrane-bound attractive protein was not normally present on anterior tectal cells (Figure 7.31E). Surprisingly, the temporal retinal axons also grew on the denatured posterior membranes—the same membranes that temporal axons normally avoided if given a choice (Figure 7.31F). This finding suggested that under normal *in vivo* conditions, posterior tectal cells produced an inhibitory protein. Thus, temporal axons would grow into the anterior region of the tectum but stop at the posterior region in response to the inhibitory cell surface cue present on those tectal cells.

While examples of inhibitory and repellent cues have already been described in this chapter, it is important to emphasize that the discovery of inhibitory guidance cues was a novel and largely unexpected observation in the 1980s.

The results of the stripe assay experiments provided a renewed interest in the chemoaffinity hypothesis and the mechanisms underlying retinotectal mapping. Soon many other labs used the stripe assay, as well as other cell culture preparations, to investigate the biochemical nature of the inhibitory cues in the posterior tectum. For example, labs using co-cultures of RGCs and dissociated tectal cells separated into different compartments of a cell culture dish found that growth cones of temporal axons collapsed and retracted when they touched cells from the posterior tectum. This effect was not observed when temporal axons encountered anterior cells or when nasal axons encountered anterior or posterior tectal cells. As other experiments began to focus on the biochemical nature of the inhibitory cue, it was noted that when posterior tectal cells were treated with the enzyme phospholipase C, the inhibitory cue was lost. This suggested the protein was tethered to the cell membrane by glycophosphatidylinositol (GPI).

Retinotectal Chemoaffinity Cues Are Finally Identified in the 1990s

In 1995 two papers were published that identified the long sought after "chemical tags" that guide retinal axons within the tectum. Work from Bonhoeffer's lab identified a molecule initially called RAGS (for repulsive axon guidance signal). The protein was expressed at the time that retinotectal mapping occurred. Further, when overexpressed in COS cells, retinal axons from both temporal and nasal regions retracted upon contact with the RAGS-expressing cells. RAGS was later found to be one of the GPI-linked ligands of the ephrin family, ephrin A5.

The second paper published at the same time by John Flanagan and colleagues identified another protein of the ephrin family. The ligand was called Elf-1 (Eph ligand family-1), but is now called ephrin A2. Ephrin A2 is expressed in a gradient across the tectum, with the lower concentrations in the more anterior regions and the higher concentrations in the posterior regions. Further, a receptor for this ligand, now called EphA3, was expressed in a gradient across the retina, with the highest concentrations of receptors on the temporal axons and the lowest concentrations on the nasal axons (**Figure 7.32**).

The model that emerged from these studies suggests that the gradient of ephrin A ligands in the tectum directs the mapping of axons arriving from different regions of the retina. Thus, temporal axons that express the highest concentration of EphA receptors are restricted to the anterior-most regions of the tectum, because the higher concentration of receptors makes these axons most sensitive to the inhibitory cue.

In contrast, the nasal axons with fewer receptors grow across the anterior and mid-regions of the tectum and into the posterior tectum, until a sufficient threshold of the inhibitory signal is reached. Axons from neurons located in intermediate regions of the retina express an intermediate

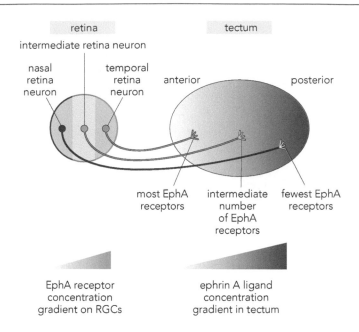

Figure 7.32 Gradients of receptor and ligand expression provide a model for retinotectal mapping. Retinal ganglion cells (RGCs) on the nasal side of the retinal express low levels of EphA receptors, whereas increasingly higher levels of EphA receptor expression are found at progressively more temporal regions. Thus, RGCs at the temporal-most region of the retina will have the highest concentration of EphA receptors. In the tectum, the inhibitory ephrin A ligand is also expressed in a gradient with low concentrations at the anterior region and progressively higher concentrations at more-posterior regions. Thus, the temporal axons with the highest levels of EphA receptors will be inhibited by the low concentrations of ephrin A in the anterior-most region of the tectum and will not grow past that region. Axons from RGCs located at intermediate regions of the retina with a lower concentration of EphA receptors will contact regions of the tectum with slightly more ephrin A ligand. In contrast, axons at the nasal region of the retina will have axons with the lowest concentrations of EphA receptors and therefore will be able to grow past low and intermediate levels of ephrin A in the tectum and will be inhibited only by the higher levels of ephrin A found in the most posterior regions of the tectum. Thus, the graded distributions of the receptors and the inhibitory ligand allow for graded mapping of temporal and nasal RGC axons along the anterior–posterior axis of the tectum.

number of EphA receptors and therefore travel past the anterior-most regions of the tectum before encountering sufficient inhibitory signal in the mid-regions of the tectum (**Figure 7.32**).

After 50 years, the matching chemical tags envisioned by Sperry were found to be Eph receptors distributed in a gradient on the retinal axons that were activated by ephrin ligands found in a gradient across the tectum. These studies also quickly changed the opinions of many who had previously dismissed the chemoaffinity hypothesis, demonstrating how ideas find favor at different periods of history based on the available data.

Ephrin A5 (RAGS) has now been shown to inhibit both temporal and nasal axons, and it is thought to prevent axons from leaving the posterior tectum. Consistent with this idea, mice lacking *ephrin A5* have retinal axons misprojecting within the posterior regions of the superior colliculus (tectum) while other retinal axons invade the inferior colliculus—a non-target region that is part of the central auditory system.

Gradients of ephrin/Eph molecules have since been identified in other regions of the nervous system, including the olfactory bulb, central auditory system, hippocampus, and thalamocortical projections, suggesting such chemoaffinity cues help establish topographic maps in these regions as well. These molecules do not account for all mapping in the nervous system, and ephrins may be just one of many signals used during neural development. For example, **Box 7.2** describes how odorant receptors and axonal "presorting" contribute to initial mapping in the olfactory bulb.

Box 7.2 Developing Neuroscientists: Wiring the Nose

Roman Corfas, Ph.D.

Roman Corfas graduated from Oberlin College in 2008 with a major in neuroscience. He completed his Ph.D. in neurogenetics and behavior with Dr. Leslie B. Vosshall at Rockefeller University in 2015. After a postdoctoral research fellowship at the California Institute of Technology, he moved to the University of Texas at Austin where he now teaches.

Sniff a container of leftover food and you will quickly know if it is tasty and nourishing or rotten and dangerous. Olfaction, the perception of smell, begins with a molecular event: A volatile odor chemical binds to a sensory receptor. So, when you inhale the smell of vanilla, vanillin molecules are entering your nose, binding odorant receptor proteins, and exciting neurons. These neurons, called olfactory sensory neurons (OSNs), or olfactory receptor neurons, sample the air for chemicals and relay this information to the central nervous system. The ability of this system to

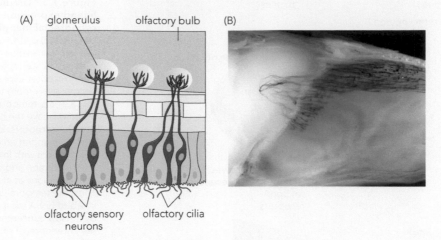

(A) glomerulus   olfactory bulb   (B)

olfactory sensory   olfactory cilia
neurons

Figure 7.33 Each odorant can be recognized by its unique pattern of OSN activation. (A) Schematic representation of the projections from olfactory sensory neurons (OSNs) in the olfactory epithelium of the nasal cavity to the glomeruli in the olfactory bulb in the central nervous system. OSNs dispersed throughout the olfactory epithelium express olfactory receptors (ORs) for distinct odorants. The ORs are located on the olfactory cilia, the dendrites of the OSNs. Those OSNs expressing a particular OR then converge on the same bilaterally symmetrical glomeruli on the medial and lateral side of each olfactory bulb. Thus, each odorant activates a specific set of glomeruli. (B) Mouse brain whole mount showing OSNs expressing a single OR and labeled with a blue axonal protein marker converging on a single glomerulus (left) in the olfactory bulb. [(B), Courtesy of Masayo Omura and Peter Mombaerts.]

distinguish so many stimuli (theoretically over 1 trillion) depends on a highly organized genetic and anatomical developmental process.

Olfaction has been well characterized in mice, whose genome encodes ~1000 distinct olfactory receptors (ORs). This represents the largest known mammalian gene family, composing 3–5% of all genes. Each gene encodes a different G-protein-coupled receptor that detects a specific set of odorants, and when activated, initiates a signal transduction cascade that results in an action potential. This OR activation is the first step of smell perception, but how is this information organized so that a mouse recognizes an odorant? Experiments have shown that mature OSNs express a single OR gene at high levels, meaning that each of these neurons will be tuned to respond to a specific set of odorants. A given odorant may bind to multiple ORs, thus leading to the activation of a particular suite of OSNs. In this manner, each odorant can be recognized by its unique pattern of OSN activation—a combinatorial code. Our ability to detect such a vast array of odors is due to the enormous number of possible combinations in this system.

Once odorants bind to ORs located on the olfactory cilia, the dendrites of the OSNs, olfactory information is transmitted to the central nervous system to elicit a smell "percept." So, how is the combinatorial code conveyed from the nose to the brain? OSNs residing in the nose project their axons to a part of the brain called the olfactory bulb, which is organized into several thousand synaptic specializations called glomeruli (**Figure 7.33**A). In the nose, OSNs of different types are dispersed throughout the olfactory epithelium, but their axons become spatially arranged according to the OR they express as they travel to the brain. Axons of OSNs expressing a particular OR converge and terminate on the same bilaterally symmetric glomeruli in both the medial and lateral side of each olfactory bulb (Figure 7.33B). In this manner, the combinatorial code is preserved and spatially organized in the central nervous system—each odorant will activate a specific set of olfactory glomeruli.

Guidance cues, such as the semaphorins and their neuropilin receptors, are distributed in a graded and complementary manner in the OSNs, where they appear to help guide the axons toward the correct target in the olfactory bulb. In contrast to other regions of the nervous system, these cues seem to primarily direct "presorting" of the axons prior to arriving at the target so that they are on a path to the correct region of the olfactory bulb. Additional olfactory-bulb-derived cues are presumed to then direct precise topographic mapping of the OSN axons within the olfactory bulb.

This elaborate orchestration of gene expression and neuronal wiring during the development of the olfactory system allows us to recognize and respond to an extraordinary number of odorants in our environment. One of the most remarkable features of the olfactory system is that OSNs are continually replaced by populations of newborn precursor cells. This means that all the developmental processes, such as restricted OR expression and subsequent axon guidance, are occurring throughout our lifetime.

Eph/Ephrin Signaling Proves to Be More Complex Than Originally Thought

The renewed interest in chemoaffinity and the role of Eph/ephrin signaling led to a remarkable number of discoveries over the past 25 years. Yet, precisely how ephrin–Eph interactions regulate formation of topographic maps in the nervous system is still not fully understood. It is clear that the signaling mechanisms are much more elaborate than originally modeled in the retinotectal system. For example, as noted in earlier chapters, Eph/ephrin signaling includes forward signaling (the ligand initiating signaling in the receptor-bearing cell) and reverse signaling (the receptor initiating signaling in the ligand-bearing cell; **Figure 7.34**A). This indicates that any topographic map has the potential to be shaped by signals initiated in either partner cell.

Another surprising finding is that ephrin ligands and Eph receptors participate in cis interactions. When ephrin ligands and Eph receptors are expressed in the same cell, the ligands attenuate the response of the receptors to extrinsic sources of ephrins (Figure 7.34B). Thus, the ligands acting in cis do not activate the receptors, but rather interfere with receptor activation by any ephrins expressed on adjacent cells (trans interactions). In retinal axons, ephrin A5 ligand is expressed from a high nasal to low temporal gradient, a gradient opposite to that of the EphA3 receptor expressed from a low nasal to high temporal gradient. The ephrin A5 ligand on the nasal axons can bind the EphA3 receptors on those axons and attenuate their response to tectal-derived ephrin A ligands. It is thought that the gradient of the cis-acting ligand further sharpens the targeting of the retinal axons to the correct location along mid and posterior areas of the tectum.

Axonal pathfinding in other regions of the nervous system also appears to be influenced by the balance of ephrin A and EphA family members signaling in cis and trans. In studies of motor neurons, for example, experimentally increased levels of ephrin A ligands acting in cis attenuated normal receptor signaling and interfered with transduction of the inhibitory signal. However, when levels of ephrin A ligands acting in cis were lowered, trans interactions dominated and the inhibitory cue needed to direct the motor neuron axons away from an incorrect target region prevailed. Subtle changes in the expression levels of ephrin A ligands acting in cis may be regulated at various stages of development to carefully control

(A) trans interactions

(B) cis interactions

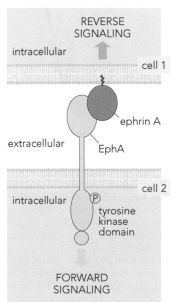

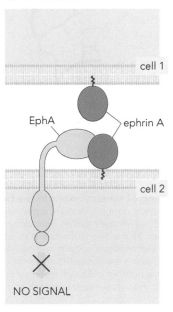

Figure 7.34 Multiple mechanisms for Eph/ephrin signaling. (A) Eph-ephrin signaling occurs by trans interactions when a receptor and ligand located on adjacent cells interact. Trans interactions can result in forward signaling when the ligand initiates signaling in the receptor-bearing cell (cell 2) or reverse signaling when the receptor initiates signaling in the ligand-bearing cell (cell 1). (B) When an EphA receptor and ephrin A ligand are located in the same cell (cell 2), the ligand can bind to the receptor. These cis interactions do not initiate signaling in the receptor-bearing cell (cell 2), but instead prevent the EphA receptor from interacting with ephrin A ligands on an adjacent cell (cell 1) so that no signal is transduced. [Adapted from Egea J & Klein R [2007] *Trends Cell Biol* 17:230–238.]

axonal pathfinding and mapping to target cells. This is another example of how the developing nervous system uses the limited number of signaling molecules available to full advantage to ensure precise wiring of the billions of neural connections required in the vertebrate nervous system.

Axonal Self-Avoidance as a Mechanism for Chemoaffinity

Other forms of chemoaffinity have been proposed to contribute to the complex yet highly ordered arrangements of connections in the nervous system. In the 1980s, for example, scientists studying leech neurons described the process of self-avoidance in which nerve fibers from the same neuron were repelled from one another. In contrast, nerve fibers from different neurons were able to intermingle. Self-avoidance ensured that axons and dendrites from the same neuron were uniformly spread across a target field while also allowing axons and dendrites from different neurons to intermingle and overlap within the target field. Subsequent studies in *Drosophila* found that Dscams (Down syndrome cell adhesion molecules), members of the IgG superfamily, were important for regulating this self-avoidance (**Figure 7.35**A). The most remarkable aspect of Dscams is the vast number of isoforms that exist. To date, in *Drosophila*, over 19,000 isoforms of Dscams that differ in the structure of their ectodomains have been identified. The numerous Dscam isoforms available suggest a means for a remarkably large number of molecular cues to be generated from a single gene.

Because axons and dendrites from the same neuron express the same isoform, they are repelled from one another. In contrast, the axons and dendrites from other neurons express different isoforms, allowing nerve fibers from the different cells to overlap. The importance of Dscams in mediating axonal patterning across a target tissue was seen when Dscam levels were altered in *Drosophila*. When two different sensory neurons were forced to express the same Dscam isoform, the axons from the different neurons could no longer intermingle and cross, but instead avoided one another (Figure 7.35B). Conversely, if Dscam was absent in neurons, the repulsive cue was lost and the axons and dendrites from a single neuron

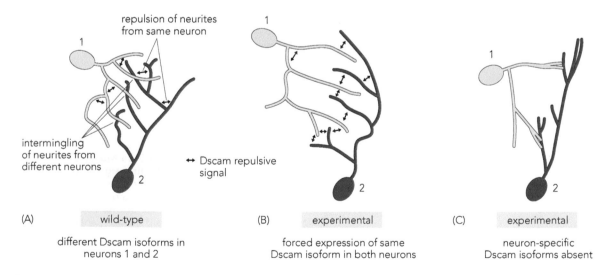

Figure 7.35 Self-avoidance of axons results in patterning of sensory axons in the body wall of *Drosophila*. (A) Different Dscam isoforms are expressed in different sensory neurons. The Dscams normally provide repulsive cues (double-headed arrows) that prevent neurites from the same neuron (1 or 2) from intermingling. However, the neurites from different neurons (1 and 2) overlap and cross one another because each expresses a different isoform of Dscam. This allows two neurons to innervate a shared target region. (B) In experimental conditions in which two neurons were forced to express the same Dscam isoform, the neurites from both cells now avoided one another, as well as their own neurites (double-headed arrows). (C) If Dscams are absent from neurons, neurites from a single neuron are no longer repelled from each other and therefore intermingle. The large number of Dscam isoforms present in the nervous system suggests these molecules may provide a type of chemoaffinity signal to help individual neurites branch and map to precise target locations. *Source:* Adapted from Zipursky SL & Sanes JR [2010] *Cell* 143(3):343–353.

intermingled and overlapped (Figure 7.35C). The normally repulsive interactions between neuronal processes expressing different Dscam isoforms led to the suggestion that these molecules act as a form of chemoaffinity, directing individual axons and dendrites to specific target cells (Figure 7.35C).

Although fewer Dscam isoforms have been identified in vertebrates, it is thought that the Dscams and related proteins function in a similar manner to those found in *Drosophila*. In mammals, protocadherins, a large subgroup in the cadherin family expressed in the nervous system, repel neurites from a single neuron. Like the Dscams, protocadherins have multiple isoforms and may help regulate the specificity of axonal and dendritic branching at different targets.

SUMMARY

Chapter 7 describes just some of the many signals that a nerve fiber must integrate as it extends from the neuronal cell body to reach a target cell. Growth cones grow along different embryonic surfaces and actively select pathways and target cells using available permissive, attractive, and repulsive cues.

Many of the discoveries made in the past few decades have significantly changed how neurobiologists view guidance cues. For example, as revealed by the studies described in this chapter, a single protein can have different functions in different regions of the developing embryo, such as being attractive to one set of axons at one stage of development and repulsive to another group of axons at a different age. Additionally, a single protein family can have produce both secreted and cell surface-bound ligands. Such findings address the long-perplexing question of how the nervous system can generate all the proteins necessary to establish the billions of neural connections required for proper function. As some scientists have long suspected, there does not need to be a single protein for each axon, or even for a group of axons. Instead, the nervous system effectively uses the limited number of proteins available to produce subtle, cell-specific responses in different contexts.

FURTHER READING

Axel R (2005) Scents and sensibility: A molecular logic of olfactory perception (Nobel lecture). *Angew Chem Int Ed* 44:6110–6127.

Bastiani MJ, Harrelson AL, Snow PM & Goodman CS (1987) Expression of fasciclin I and II glycoproteins on subsets of axon pathways during neuronal development in the grasshopper. *Cell* 48(5):745–755.

Bastiani MJ, Raper JA & Goodman CS (1984) Pathfinding by neuronal growth cones in grasshopper embryos. III. Selective affinity of the G growth cone for the P cells within the A/P fascicle. *J Neurosci* 4(9):2311–2328.

Bonanomi D & Pfaff SL (2010) Motor axon pathfinding. *Cold Spring Harb Perspect Biol* 2(3):a001735.

Buck L & Axel R (1991) A novel multigene family may encode odorant receptors: A molecular basis for odor recognition. *Cell* 65:175–187.

Bushdid C, Magnasco MO, Vosshall LB & Keller A (2014) Humans can discriminate more than 1 trillion olfactory stimuli. *Science* 343:1370–1372.

Cook GM, Sousa C, Schaeffer J, et al. (2020) Regulation of nerve growth and patterning by cell surface protein disulphide isomerase. *eLife* 9:e54612.

Cypher C & Letourneau PC (1992) Growth cone motility. *Curr Opin Cell Biol* 4(1):4–7.

de Castro F, Lopez-Mascaraque L & De Carlos JA (2007) Cajal: Lessons on brain development. *Brain Res Rev* 55(2):481–489.

Drescher U, Kremoser C, Handwerker C, et al. (1995) In vitro guidance of retinal ganglion cell axons by RAGS, a 25 kDa tectal protein related to ligands for Eph receptor tyrosine kinases. *Cell* 82(3):359–370.

Gomez TM & Letourneau PC (2014) Actin dynamics in growth cone motility and navigation. *J Neurochem* 129(2):221–234.

Goodman CS (1996) Mechanisms and molecules that control growth cone guidance. *Annu Rev Neurosci* 19:341–377.

Gorla M & Bashaw GJ (2020) Molecular mechanisms regulating axon responsiveness at the midline. *Dev Biol* 466(1–2):12–21.

Hanson MG & Landmesser LT (2004) Normal patterns of spontaneous activity are required for correct motor axon guidance and the expression of specific guidance molecules. *Neuron* 43(5):687–701.

Imai F & Yoshida Y (2015) Axon guidance in the spinal cord. In *Semaphorins* (Kumanogoh A ed), pp. 39–63. Springer.

Kao TJ & Kania A (2011) Ephrin-mediated cis-attenuation of Eph receptor signaling is essential for spinal motor axon guidance. *Neuron* 71(1):76–91.

Kennedy TE, Serafini T, de la Torre JR & Tessier-Lavigne M (1994) Netrins are diffusible chemotropic factors for commissural axons in the embryonic spinal cord. *Cell* 78(3):425–435.

Klein R (2012) Eph/ephrin signalling during development. *Development* 139(22):4105–4109.

Kolodkin AL, Levengood DV, Rowe EG, et al. (1997) Neuropilin is a semaphorin III receptor. *Cell* 22(90):753–762.

Komiyama T & Luo L (2006) Development of wiring specificity in the olfactory system. *Curr Opin Neurobiol* 16:67–73.

Lance-Jones C & Landmesser L (1980) Motoneurone projection patterns in the chick hind limb following early partial reversals of the spinal cord. *J Physiol* 302:581–602.

Lepelletier L, Langlois SD, Kent CB, et al. (2017) Sonic Hedgehog guides axons via zipcode binding protein 1-mediated local translation. *J Neurosci* 37(7):1685–1695.

Letourneau PC (1975) Cell-to-substratum adhesion and guidance of axonal elongation. *Dev Biol* 44(1):92–101.

Letourneau PC (2009) Axonal pathfinding: Extracellular matrix role. In *Encyclopedia of Neuroscience* (Squire LR ed), pp. 1139–1145. Academic Press.

Lin CH, Thompson CA & Forscher P (1994) Cytoskeletal reorganization underlying growth cone motility. *Curr Opin Neurobiol* 4(5):640–647.

Lowery LA & Van Vactor D (2009) The trip of the tip: Understanding the growth cone machinery. *Nat Rev Mol Cell Biol* 10(5):332–343.

Messersmith EK, Leonardo ED, Shatz CJ, et al. (1995) Semaphorin III can function as a selective chemorepellent to pattern sensory projections in the spinal cord. *Neuron* 14(5):949–959.

Mombaerts P (2006) Axonal wiring in the mouse olfactory system. *Annu Rev Cell Dev Biol* 22:713–737.

Mueller BK (1999) Growth cone guidance: First steps towards a deeper understanding. *Annu Rev Neurosci* 22:351–388.

Nakamoto M, Cheng HJ, Friedman GC, et al. (1996) Topographically specific effects of ELF-1 on retinal axon guidance in vitro and retinal axon mapping in vivo. *Cell* 86(5):755–766.

Pasquale EB (2016) Exosomes expand the sphere of influence of Eph receptors and ephrins. *J Cell Biol* 214(1):5–7.

Placzek M, Tessier-Lavigne M, Yamada T, et al. (1990) Guidance of developing axons by diffusible chemoattractants. *Cold Spring Harb Symp Quant Biol* 55:279–289.

Roche FK, Marsick BM & Letourneau PC (2009) Protein synthesis in distal axons is not required for growth cone responses to guidance cues. *J Neurosci* 29(3):638–652.

Rutishauser U & Landmesser L (1991) Polysialic acid on the surface of axons regulates patterns of normal and activity-dependent innervation. *Trends Neurosci* 14(12):528–532.

Serafini T, Kennedy TE, Galko MJ, et al. (1994) The netrins define a family of axon outgrowth-promoting proteins homologous to C. elegans UNC-6. *Cell* 78(3):409–424.

Sperry RW (1947) Nature of functional recovery following regeneration of the oculomotor nerve in amphibians. *Anat Rec* 97(3):293–316.

Tessier-Lavigne M & Kolodkin AL (eds) (2011) *Neuronal Guidance: The Biology of Brain Wiring.* Cold Spring Harbor Laboratory Press.

Udin SB & Fawcett JW (1988) Formation of topographic maps. *Annu Rev Neurosci* 11:289–327.

Vitriol EA & Zheng JQ (2012) Growth cone travel in space and time: The cellular ensemble of cytoskeleton, adhesion, and membrane. *Neuron* 73(6):1068–1081.

Walter J, Henke-Fahle S & Bonhoeffer F (1987) Avoidance of posterior tectal membranes by temporal retinal axons. *Development* 101(4):909–913.

Walter J, Kern-Veits B, Huf J, et al. (1987) Recognition of position-specific properties of tectal cell membranes by retinal axons in vitro. *Development* 101(4):685–696.

Yazdani U & Terman JR (2006) The semaphorins. *Genome Biol* 7(3):211.

Zipursky SL & Sanes JR (2010) Chemoaffinity revisited: Dscams, protocadherins, and neural circuit assembly. *Cell* 143(3):343–353.

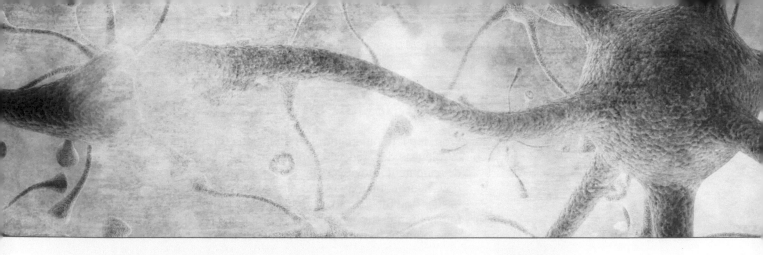

Neuronal Survival and Programmed Cell Death

8

C hapter 7 described mechanisms by which neurons extend nerve fibers along distinct pathways to reach target tissues. Once a nerve fiber reaches its final target region, the target tissue plays an important role in determining whether that neuron will survive. Many target tissues, particularly those in the peripheral nervous system (PNS), secrete survival-promoting (trophic) proteins to regulate the final number of innervating neurons. As proposed by the **neurotrophic hypothesis**, the release of limited amounts of these proteins at the time of neural innervation helps ensure that only the correct number of neurons survive in a given neuronal population. The neurons that do not obtain the target-derived proteins die off, thus preventing survival of any excess neurons (**Figure 8.1**).

Another means by which neuronal numbers are regulated in the developing nervous system is through **programmed cell death (PCD)**, a process by which subsets of neurons produced during neurogenesis are later eliminated. Programmed cell death is now understood to be a developmentally necessary event that further shapes the final, correct number of neurons in the nervous system. This chapter focuses on the discovery and role of target-derived proteins, their receptors, and the intracellular signaling pathways that intersect to regulate neuronal survival and programmed cell death during neural development.

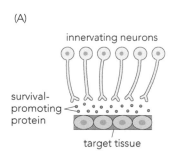
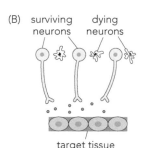

Figure 8.1 Target tissues support the required number of innervating neurons. (A) The cells in a target tissue secrete limited amounts of survival-promoting proteins at the time of neuronal innervation. (B) The neurotrophic hypothesis proposes that only those neurons that receive sufficient quantities of the target-derived protein survive. Those neurons that do not receive enough of the protein die. This process helps ensure that only the required number of neurons survive.

DOI: 10.1201/9781003166078-8

GROWTH FACTORS REGULATE NEURONAL SURVIVAL

From the late nineteenth through to the mid-twentieth century, neuroembryologists debated what role, if any, target tissues served during neural development. However, by the 1950s, several studies had documented how neuronal survival was altered by the presence or absence of target tissues and the idea that target-derived proteins are important in promoting neuronal survival and preventing neuronal death during embryonic development became firmly established.

The Death of Nerve Cells Was Not Initially Recognized as a Normal Developmental Event

Most scientists in the late 1800s and early 1900s did not suspect that cell death is an essential part of neural development. There were many reasons for this, both theoretical and technical. Although Wilhelm Roux, writing in 1881, suggested that neuronal death was a normal part of embryogenesis, Cajal and others felt that, even if neuronal death occurred for some developing neurons, "the immense majority of the neuroblasts survive to term and succeed in collaborating with the normal structures of the adult nervous system." This view was widely held into the mid-twentieth century, as scientists seemed reluctant to consider the possibility that healthy neurons were initially over-produced, with a percentage being eliminated later in development. It seemed much more plausible that the nervous system would produce only the number of neurons needed.

The failure to consider cell death as a normal part of neural development was further influenced by technical considerations. On the slides of fixed tissue sections examined by early neuroembryologists, it was difficult to detect cell death at different developmental stages because any dying cells were quickly removed by phagocytosis. Therefore, cell death might not be evident in any given tissue section viewed. Cell death was later documented in fixed tissues by examining serial tissue sections and comparing detailed cell counts at different developmental stages. However, because most of the early researchers did not expect cell death, they were not comparing neuronal numbers in the tissue sections they collected. In the 1940s, as scientists began to compare neuronal numbers at different developmental stages, more and more investigators reported neuronal death during embryogenesis. From these observations it was gradually accepted that cell death was a required aspect of neural development.

Studies Reveal That Target Tissue Size Affects the Number of Neurons That Survive

Changes in cell number were first reported following various surgical manipulations of target tissues. In 1909, for example, Márian Shorey found that removing the limb buds from chick or amphibian embryos resulted in **hypoplasia** of the motor and dorsal root ganglion (DRG) neurons innervating the limb bud. The hypoplasia reflected a decrease in the size of the motor neuron pools and dorsal root ganglia, as well as the individual neurons within these populations. Others noted that grafting extra limb tissue resulted in **hyperplasia**, an increase in the size of the innervating neuronal populations and individual neurons. At the time, it was thought that target tissues somehow regulated neuronal proliferation, migration, or differentiation to establish the size of the innervating neuronal populations. However, neuronal survival was not yet suspected to be influenced by target tissues.

In the late 1930s and early 1940s, Viktor Hamburger, a pioneer in the study of neuron-target interactions, found a direct correlation between the size of the target tissue and the extent of hypoplasia observed in the innervating neural populations. Determining whether the cue originating in the target tissue controlled neuronal proliferation, migration, or differentiation became the focus of subsequent studies conducted by Hamburger and others. However, the scientists found that none of the proposed mechanisms accounted for the changes in cell number. In the early 1940s, Italian researchers Rita Levi-Montalcini and Giuseppe Levi published papers indicating that target tissues might instead control neuronal survival. While they too reported hypoplasia of the innervating motor and DRG neurons in the absence of target limb bud tissue, they also noted a new key detail: When target tissues were removed, the nuclear material in many of the innervating neurons became **pyknotic** (irreversibly condensed), indicating that the cells were dying in the absence of target tissue. After World War II ended, Hamburger invited Levi-Montalcini to work with him. Their subsequent studies confirmed her earlier observations and further determined that extra limb bud tissue increased the number of surviving neurons (**Figure 8.2**). Thus, through careful analysis of neuronal cell numbers, the idea emerged that target tissues regulated neuronal survival, rather than the proliferation, migration, or differentiation of innervating neurons, as originally hypothesized.

Another important observation was that a certain percentage of cells died even under normal conditions. The number of neurons exhibiting this type of cell death varied with the cell population examined, but several neuronal types—including spinal motor neurons and neurons from dorsal root and sympathetic ganglia—revealed a decrease in neuronal number at the time of innervation, even in the presence of target tissues. As described later in the chapter, the significance of this observation became clear over subsequent decades as studies of neuron-target interactions continued.

Some Tumor Tissues Mimic the Effect of Extra Limb Buds on Nerve Fiber Growth

Studies by Levi-Montalcini and Hamburger were among the most influential in explaining neuron-target interactions, and the 1940s and 1950s proved to be pivotal decades for revealing how target tissues regulate the innervation and survival of associated neurons. It was during this period that the neurotrophic hypothesis emerged (see Figure 8.1) and the first

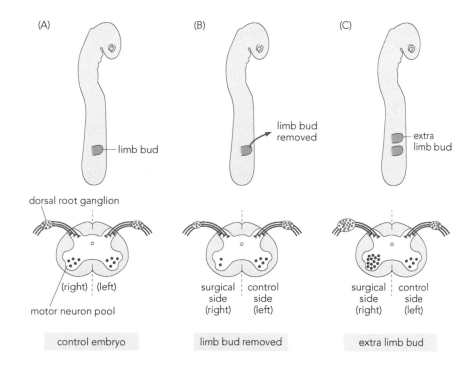

Figure 8.2 Target tissue size regulates the number of surviving neurons. In chick embryos, the amount of limb bud tissue ultimately determines how many motor and dorsal root ganglion neurons survive. (A) In the control embryo, the motor neuron pools and the dorsal root ganglia (DRG) are of equal size on either side of the embryo. (B) Removing the limb bud on one side, in this example the right (surgical) side of the embryo, causes a decrease in the number of motor and dorsal root ganglion neurons only on that side. (C) Grafting an extra limb bud to the right (surgical) side of the embryo increases the number of motor and dorsal root ganglion neurons on that side compared to the contralateral, control side. Together, these experiments demonstrate that the size of the target tissue directly influences the number of neurons that survive.

target-derived, survival-promoting nerve growth factor was identified. The discovery of this first nerve growth factor is one of the best documented in the history of developmental neurobiology. There are many reasons for this, including the importance of the finding in understanding how neuronal numbers are established, the fortunate events that led to its purification, and the impact this discovery had on subsequent studies, including those that identified other nerve growth factors. The discovery of the first nerve growth factor also illustrates how the techniques available at any given period shape the types of discoveries that can be made and demonstrates how detailed and careful studies from one decade can be of ongoing benefit to scientists in future decades.

The first major finding that altered how scientists approached studies of target-derived factors was reported by Elmer Bueker, a student in Hamburger's lab during the 1940s. Because embryonic limb buds contain dividing cells and gradually increase in size during development, it was thought that other tissues containing rapidly dividing cells, such as tumors, might prove to be a better experimental tool than grafted limb buds. Therefore, Bueker grafted different tumors onto chick embryos to see if neuronal populations were influenced by the size of the tumor graft.

Bueker tested three tumors: an adenocarcinoma, the Rous sarcoma, and sarcoma 180. He implanted each into the body wall of early-stage chick embryos, prior to the period of neural innervation. Only the sarcoma 180 tumor survived and integrated into the body wall of the embryo. After five days *in ovo*, nerve fibers from DRG neurons extended into the growing tumor. In addition, the ganglia appeared larger on the side of the tumor graft compared to ganglia on the opposite side. Bueker proposed that chemical properties of the tumor promoted the growth of nerve fibers and increased neuronal numbers.

These findings, published in 1948, influenced subsequent studies by Levi-Montalcini, who followed up on this effect. She first repeated Bueker's studies, extending the survival times of the embryos. She also tested additional mouse sarcoma tumors (sarcoma 37 and sarcoma 1). She confirmed that the tumors did in fact lead to extensive nerve fiber growth into the tumor and an increase in the size of dorsal root ganglia on the side of the sarcoma graft. She also noted that sympathetic ganglia on that side showed an even greater increase in volume than the dorsal root ganglia (**Figure 8.3**). In contrast to what was observed with limb bud grafts, however, motor neuron number was not affected by the presence of the tumor.

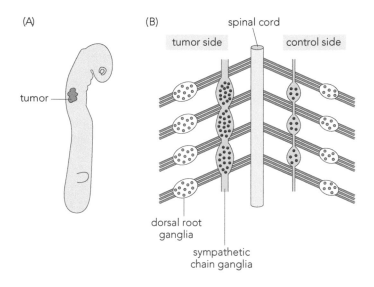

Figure 8.3 Tumor tissue promotes survival of dorsal root and sympathetic neurons. (A) To test whether a rapidly dividing tissue produced a factor to promote neuronal survival or outgrowth, a sarcoma 180 tumor was implanted onto the body wall of a chick embryo. (B) Studies by both Bueker and Levi-Montalcini revealed that the DRG, as well as the sympathetic chain ganglia, increased in size on the side of the tumor graft. The size of the ganglia on the contralateral side did not change. Unlike grafts of limb bud tissue, tumor tissue did not influence the size of the motor neuron pools. [Adapted from Levi-Montalcini R & Hamburger V [1951] *J Experimtl Zool* 116:321–361.]

In Vitro Studies Led to a Bioassay Method to Study Nerve Growth Factors

As Levi-Montalcini and Hamburger continued to investigate the developmental effects of the sarcoma tumors, they discovered that sympathetic nerve fibers not only invaded the tumor, but also grew into veins and other tissues located away from the tumor. This led to the idea that the tumors released a factor into the embryonic circulatory system that could influence nerve fiber growth. To test this hypothesis, fragments of tumor were implanted on the chorioallantoic membrane surrounding early-stage embryos. This method prevented direct contact between the tumor and embryo, but allowed any factors released by the tumor to enter the circulatory system. In these preparations, sympathetic ganglia were again increased in size, and nerve fibers invaded blood vessels and other tissues, confirming that the tumor released a diffusible factor that promoted neuronal survival and nerve fiber outgrowth.

Realizing it would be difficult to investigate the nature of the tumor-derived factor using only *in ovo* preparations, Levi-Montalcini traveled to Brazil in the early 1950s to learn neuronal tissue culture methods, a technique that was not yet commonly used by neuroembryologists. In Brazil, Levi-Montalcini learned how to successfully culture dorsal root and sympathetic ganglia as **explants**—individual intact tissues (for example, a ganglion) grown in a cell culture dish. This allowed her to test whether the tumor tissues placed at a distance away from a ganglion directly influenced nerve fiber growth *in vitro*. She found that when a sarcoma tumor segment and a ganglion were cultured in the same dish, nerve fibers grew outward from the ganglion explant, forming what was described as a "halo effect." In control cultures of ganglia grown in dishes with other tumors or various mouse tissues, neurite outgrowth was limited or did not occur at all. Thus, there appeared to be something released by the sarcoma tumors that promoted extensive neurite outgrowth from chick dorsal root and sympathetic ganglion explants.

Levi-Montalcini sent detailed drawings of her findings to Hamburger as she collected data on the outgrowth from ganglia at different developmental stages (**Figure 8.4**). In 1954 they published their first paper using this cell culture method to demonstrate the outgrowth-promoting properties of substances released by the sarcoma tumor cells. Levi-Montalcini's bioassay method soon became a primary means of testing whether different substances promoted outgrowth from ganglia. Such assays remain a popular tissue culture method today.

The Factor Released by Sarcoma 180 Is Found to Be a Protein

Upon returning to Hamburger's lab, Levi-Montalcini began collaborating with a postdoctoral fellow, the biochemist Stanley Cohen, who had joined Hamburger's lab to identify the factor that promotes nerve fiber growth. He

Figure 8.4 First images to reveal the effects of tumor tissue on the outgrowth of dorsal root ganglia *in vitro*. In 1952, Rita Levi-Montalcini developed a tissue culture method to assay outgrowth of dorsal root and sympathetic ganglia. A ganglion was co-cultured with fragments of sarcoma tissue (S) to test for the presence of diffusible, growth-promoting factors. This image is one of Levi-Montalcini's original India ink drawings, illustrating a dense halo of nerve fiber outgrowth extending from a dorsal root ganglion explant (right) after 24 hours in culture with a piece of sarcoma 180 tumor (left, S). [From Levi-Montalcini R [1987] *Science* 237:1154–1162.]

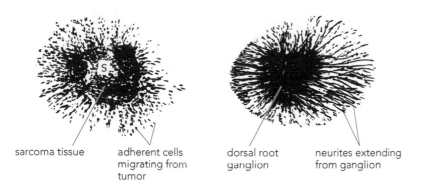

sarcoma tissue   adherent cells migrating from tumor   dorsal root ganglion   neurites extending from ganglion

Figure 8.5 Snake venom contains a nerve growth factor. In studies designed to test whether extracts of tumor tissue contained nucleic acids, snake venom was added to cultures of chick ganglia. Snake venom contains phosphodiesterases—enzymes known to break down nucleic acids. (A) In the presence of tumor extract only, chick dorsal root or sympathetic ganglia produced a dense halo of neurite outgrowth. (B) When snake venom was added to the tumor extracts, a dense halo of outgrowth was again noted and appeared greater than that observed in cultures treated with tumor extract only. (C) Surprisingly, the control cultures that contained only snake venom produced greater neurite outgrowth than the cultures treated with tumor extract. (D–G) Images from Cohen and Levi-Montalcini's 1956 paper revealing that the nerve-growth-promoting activity from snake venom extracts was greater than that from control medium or tumor extract. (D) A dorsal root ganglion grown in a standard cell culture medium without additional extracts (control) does not produce neurite outgrowth. (E) Neurite outgrowth occurred in the presence of an unprocessed snake venom sample. (F) Denser and more extensive growth was observed when the snake venom was partially purified. (G) Neurite outgrowth was less dense when grown in the presence of a partially purified sample of sarcoma 180 tumor extract. [(D–G), From Cohen S & Levi-Montalcini R [1956] *Proc Natl Acad Sci USA* 42:571–574.]

subjected extracts of sarcoma 180 tissues to various treatments to characterize the type of factor involved. The first studies, published in 1954, revealed that the substance released by the sarcoma 180 tumor was a nucleoprotein, though at this stage it was unclear if the active substance was a nucleic acid or a protein.

Cohen's studies suggested that the factor might be a protein because the nerve-growth-promoting activity was destroyed by proteases—enzymes that break down proteins. However, there was also concern that viruses or other contaminating nucleic acids could be in the extracts. A colleague suggested he treat the tumor extracts to remove any DNA or RNA in the samples. To eliminate these potential contaminants, the tumor extracts were treated with snake venom containing phosphodiesterases—enzymes that break down nucleic acids. The snake venom was milked from pit vipers (either rattlesnakes or moccasins) living at the Ross Allen Reptile Institute, a former theme park and educational center located in Silver Springs, Florida. That the scientists had to order snake venom from a reptile institute to remove nucleic acids from the tumor extract samples is just one reminder that the reagents and techniques available to biochemists in the 1950s were quite different from those available today.

Explants of chick dorsal root and sympathetic ganglia were cultured with the venom-treated sarcoma extracts and the extent of neurite outgrowth was compared to ganglia cultured with untreated sarcoma extracts. In addition, ganglia were treated with snake venom alone to confirm that the venom was not toxic to the ganglia. Ganglia grown in the presence of venom-treated sarcoma extracts extended nerve fibers, consistent with the hypothesis that the active substance was a protein and not a nucleic acid (**Figure 8.5**A, B). They also observed that ganglia grown in the presence

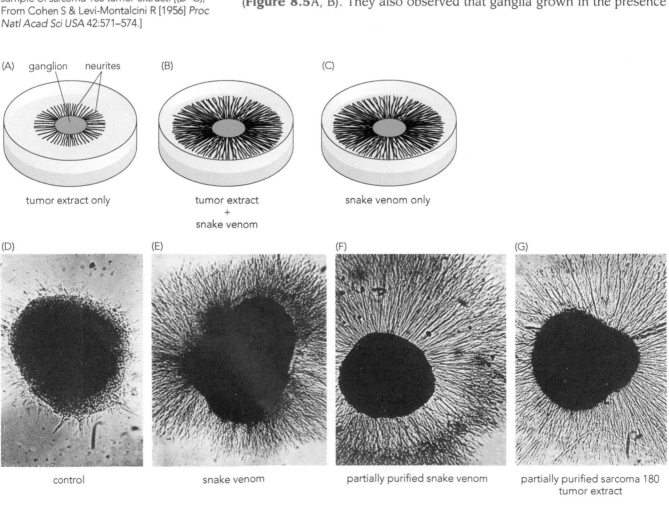

(A) ganglion neurites

tumor extract only

(B)

tumor extract
+
snake venom

(C)

snake venom only

(D)

control

(E)

snake venom

(F)

partially purified snake venom

(G)

partially purified sarcoma 180
tumor extract

of the venom-treated sarcoma extracts appeared to produce more neurite outgrowth than ganglia treated with tumor extract alone (Figure 8.5C). In addition, a surprising and ultimately more influential discovery was made: The ganglia treated solely with snake venom exhibited extensive neurite outgrowth, exceeding the outgrowth produced by tumor extracts alone. The scientists subsequently found that snake venom was the most potent promoter of neurite outgrowth they had found to date (Figure 8.5D–G).

Nerve Growth Factor Is Identified in Salivary Glands

Cohen then sought to purify the active protein. However, protein purification requires large quantities of starting material, and because snake venom glands are very small and not readily available, they were not a good source of material. Cohen recalled that snake venom glands are modified salivary glands. Salivary glands are not only larger but are more easily obtained from common animal models such as mice. He therefore tested salivary glands from male mice and found that the growth-promoting factor was present there as well. It was fortunate that Cohen started with male mice, because salivary glands from female mice do not produce the same level of the factor. Remarkably, for reasons that remain unknown, the male mouse salivary gland is the most abundant known source of the nerve-growth-promoting factor.

Using these salivary glands as raw material, by 1956 Cohen was able to purify a protein that they called the nerve-growth-stimulating factor, now called nerve growth factor (NGF). Like the tumor extracts and snake venom, purified NGF promoted the outgrowth of dorsal root and sympathetic ganglia *in vitro* (**Figure 8.6**A). To confirm that NGF is needed for normal development, the scientists developed an antiserum that blocked the activity of NGF. When this antiserum was added to cultures of NGF-treated DRG, neurite outgrowth was inhibited, demonstrating that the antiserum blocked NGF activity (Figure 8.6B). Further, when this antiserum was injected into newborn mice at the time when sympathetic neurons innervate target tissues, the sympathetic ganglia were greatly reduced in size, with very few neurons remaining in any single ganglion (Figure 8.6C). This "immunosympathectomy" was further evidence that NGF normally acts *in vivo* to promote the outgrowth and survival of sympathetic neurons. Additional studies revealed that injections of NGF rescued neurons

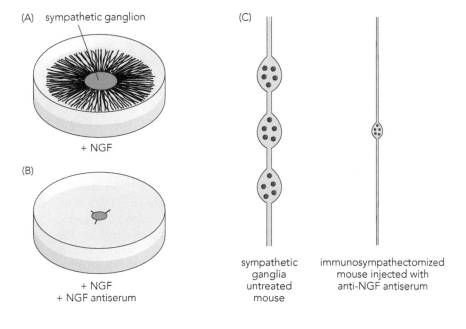

(A) sympathetic ganglion

+ NGF

(B)

+ NGF
+ NGF antiserum

(C)

sympathetic
ganglia
untreated
mouse

immunosympathectomized
mouse injected with
anti-NGF antiserum

Figure 8.6 Developing sympathetic neurons require Nerve growth factor (NGF). (A) When the purified NGF protein was added to cultures of sympathetic ganglia, a dense halo of neurite outgrowth was observed. (B) NGF-induced outgrowth was blocked by the addition of antiserum that inhibits NGF activity. (C) The NGF antiserum also blocked the effects of NGF *in vivo*. When NGF antiserum was injected into newborn mice at the time of target innervation, sympathetic ganglia were greatly reduced in size, with few neurons remaining. These studies demonstrated that NGF is required *in vivo* at the time of target innervation to promote survival of innervating sympathetic neurons. [(C), Adapted from Levi-Montalcini R & Booker B [1960] *Proc Natl Acad Sci USA* 46:384–391.]

that would normally die during development. Thus, *in vitro* as well as *in vivo* manipulations confirmed that NGF was important for normal development of sympathetic and dorsal root ganglion neurons at the time of target innervation. The significance of the discovery of NGF was highlighted in 1986 when Rita Levi-Montalcini and Stanley Cohen received the Nobel Prize in Physiology or Medicine for their work (**Box 8.1**).

Further characterization of NGF became possible as laboratory techniques evolved. In the 1970s, the amino acid sequence of NGF was determined, and in the 1980s, NGF was cloned. In the 1990s, when techniques for producing gene knockout mice were first developed, mice lacking the *Ngf* gene were generated. These mice had drastically decreased numbers of neurons in dorsal root and sympathetic ganglia, consistent with the various *in vivo* and *in vitro* studies conducted over the years.

Studies of NGF Lead to the Discovery of Brain-Derived Neurotrophic Factor

Once purified NGF was available, researchers from other labs began to investigate the growth factor's functions in other aspects of neural development. NGF was found to affect not only the neurons in dorsal root and sympathetic ganglia, but also the neurons in most neural-crest-derived ganglia in the PNS and the basal forebrain cholinergic neurons in the CNS. However, NGF did not influence all neuronal populations studied. Motor neurons in the spinal cord and neurons in most placode-derived PNS ganglia failed to respond to NGF. Such findings suggested that additional growth factors were present in other target tissues to regulate survival and outgrowth of their corresponding neuronal populations.

Box 8.1 NGF and the Intersection of Two Extraordinary Careers

Rita Levi-Montalcini and Stanley Cohen shared the Nobel Prize in Physiology or Medicine in 1986 for their discovery of NGF and its role in neural development. Levi-Montalcini (**Figure 8.7**) was born in Turin, Italy, in 1909. She received a degree in medicine and surgery from the University of Turin in 1936, where she

Figure 8.7 Rita Levi-Montalcini shown in her office in 1990. [From Aloe L [2004] *Trends Cell Biol* 14:395–399.]

continued to work as a research scientist until 1938, when laws forbade individuals of Jewish descent from working in universities or the medical profession. In autobiographical writings she describes how she and her family had to relocate to the countryside during World War II and how she managed to continue her studies in a makeshift lab set up in a bedroom. This research, when later published, motivated Viktor Hamburger to invite her to work in his lab. Originally anticipating a one or two year stay, she ultimately spent nearly 30 years at Washington University in St. Louis working first in Hamburger's lab, and later her own. Levi-Montalcini lived to be 103 years old, and throughout her life she continued to study NGF, first in the United States and then in her native Italy. Over the years, she also wrote extensively about her experiences working with Hamburger, Cohen, and others, providing a personal perspective on the process of discovering the first known nerve growth factor.

Stanley Cohen (**Figure 8.8**) was born in 1922 in Brooklyn, New York. He received a bachelor's degree in biology and chemistry from Brooklyn College in 1943, a master's degree in zoology from Oberlin College in 1945, and a Ph.D. in biochemistry from the University of Michigan in 1948. Cohen made remarkable and

Figure 8.8 Stanley Cohen shown talking with students on a visit to Oberlin College. [Courtesy of Oberlin College Archives.]

diverse contributions throughout his career. In the process of purifying NGF, Cohen injected early postnatal mice daily with the crude extract from adult male mouse salivary glands. He observed that not only did the treated mice develop an enlarged sympathetic nervous system, but that their eyes opened and teeth erupted earlier than untreated controls. In contrast, injections of purified NGF enlarged the sympathetic nervous system, but did not alter eye or tooth development. Recognizing that the crude extract must contain another active component, he went on to identify epidermal growth factor (EGF) and, later, its receptor (EGFR), which was the first ligand-activated tyrosine kinase to be identified. These discoveries ultimately led to the use of anti-receptor antibodies and tyrosine kinase inhibitors as therapies for cancers that overexpress proteins in the EGF family. Cohen died in 2020 at the age of 97. Throughout his career he devoted time to educating and encouraging young scientists.

In the late 1970s, Yves-Alain Barde, Hans Thoenen, and colleagues described a novel factor detected in a glioma-cell-conditioned medium. When glioma cells were grown in cell culture, they secreted proteins into the surrounding cell culture medium. This "conditioned medium" contained a factor that shared some, but not all, properties of NGF. Like NGF, the glioma-cell-conditioned medium promoted growth of DRG explants. In contrast to NGF, however, the biological activity was not blocked by antibodies to NGF (**Figure 8.9**).

As the scientists searched for other, more abundant sources of this growth-promoting factor, they found a similar biological activity present in extracts of whole pig brain. This tissue was then used to purify a growth-promoting substance that was distinct from NGF. In 1982, more than 25 years after the purification of NGF, Barde, Edgar, and Thoenen reported the discovery of brain-derived neurotrophic factor (BDNF). As noted previously, it requires large quantities of starting material to purify a protein. This was clearly illustrated in the discovery of BDNF, during which 3 kilograms of pig brain were processed to generate just 2 micrograms of protein.

BDNF is distinct from NGF, not only because antibodies to NGF failed to inhibit BDNF activity, but also because BDNF promoted the survival and neurite outgrowth of neuronal populations that were not influenced by NGF. For example, the placode-derived neurons of the nodose ganglia were responsive to BDNF, but not NGF (**Figure 8.10**). Subsequent studies

(A) dorsal root ganglion

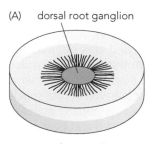

+ glioma cell-conditioned medium

(B) dorsal root ganglion

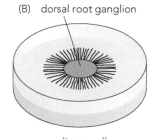

+ glioma cell-conditioned medium + NGF antibodies

Figure 8.9 Glioma-cell-conditioned medium contains proteins distinct from NGF. (A) A dorsal root ganglion cultured in the presence of growth medium conditioned by glioma cells produces a dense halo of neurite outgrowth. (B) Addition of antibodies that block NGF activity does not limit this neurite outgrowth, suggesting that the glioma cell-conditioned medium contains a nerve-growth-promoting factor distinct from NGF.

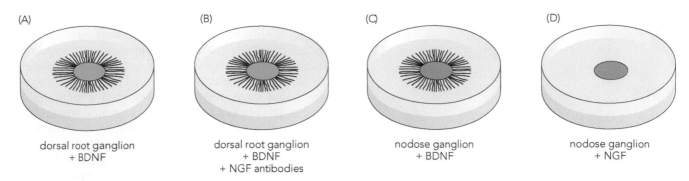

Figure 8.10 The biological activity of brain-derived neurotrophic factor (BDNF) differs from that of NGF. (A) Like glioma-cell-conditioned medium, BDNF promotes neurite outgrowth from a dorsal root ganglion explant. (B) BDNF-induced outgrowth is not inhibited when antibodies specific to NGF are added, indicating that the neurite-outgrowth-promoting properties of BDNF are distinct from those of NGF. (C) BDNF also promotes outgrowth of nodose ganglia, a neuronal population that is unresponsive to NGF (D).

demonstrated that BDNF and NGF have overlapping but distinct spatial and temporal patterns of expression, further emphasizing that they are two different molecules.

Discoveries of Other NGF-Related Growth Factors Rapidly Followed

In 1989, Barde and colleagues determined the amino acid sequence of BDNF noting it was very similar to NGF, sharing about 50% sequence identity. Thus, NGF and BDNF are part of the same gene family. This finding prompted several labs to look for additional NGF family members.

Whereas it took decades to purify and sequence NGF and several years to purify and sequence BDNF, four labs were able to identify a third member of the NGF family of growth factors within months of the publication of the BDNF amino acid sequence using molecular techniques rather than protein purification methods. The new factor was called various names by the different groups who discovered it, including NGF-2 and hippocampal-derived neurotrophic factor, but is now called neurotrophin-3 (NT-3). The entire family of NGF-related proteins is now known as the **neurotrophins**.

NT-3, like NGF and BDNF, promotes survival and outgrowth of dorsal root ganglia, though different subpopulations of neurons within the ganglia are influenced by the different neurotrophins. The developmental pattern of NT-3 expression also differs from the expression patterns of NGF and BDNF. For example, NT-3 is generally expressed at higher levels during development than in adulthood, whereas BDNF levels are often lower in developing tissues but higher in adult tissues, where it appears to play roles in neuronal maintenance and synaptogenesis.

A fourth member of the neurotrophin family was found in *Xenopus* in 1991 and was named neurotrophin-4 (NT-4). A mammalian homolog of this protein was identified shortly after and was originally called mammalian NT-4 by one lab and NT-5 by another. Now often designated as NT-4/5, these proteins differ from other neurotrophin family members in that NT-4/5 sequences are not as highly conserved across species, explaining why the *Xenopus* and mammalian proteins initially appeared to represent two distinct neurotrophins. Neurotrophin-6 and neurotrophin-7 were discovered in 1994 and 1998, respectively, but have only been detected in fish. To date their functions are not as well characterized as those of the first four neurotrophins.

NGF SIGNALING MECHANISMS AND NEUROTROPHIN RECEPTORS

In addition to characterizing the biological activity and tissue distribution of NGF and the other neurotrophins, scientists also investigated the ways such growth factors exerted their survival-promoting effects. Because neuronal cell bodies are often located at a distance from target tissues, the mechanism by which growth factors reach the cell body to promote cell survival was initially unclear. For many years, scientists suspected that the nerve fibers must express receptors that take up the growth factor and transport it to the cell body. However, this hypothesis could not be tested until appropriate techniques and assays were available.

NGF Is Transported from the Nerve Terminal to the Cell Body

By the early 1970s scientists had developed a method to evaluate whether NGF was taken up by nerve terminals and transported **retrogradely** to the cell body. In the first study, published in 1974, Ian Hendry and colleagues injected radiolabeled NGF (^{125}I-NGF) into the mouse iris, a tissue that is innervated by sympathetic neurons of the superior cervical ganglia (SCG). If NGF were transported along the nerve fibers to the cell bodies as hypothesized, the ^{125}I-NGF signal would be detected in neurons of the SCG. Using this method, the scientists confirmed that NGF was transported back to the cell bodies of the sympathetic neurons; the radiolabeled NGF was detected in the neurons of the SCG (**Figure 8.11**). Proteins similar in size to NGF were not transported to the cell bodies, nor were NGF proteins whose structure was modified, indicating that the observed retrograde transport was specific for NGF. Later studies found that radiolabeled NGF was also transported retrogradely when injected near sciatic or spinal nerve terminals of adult mice. Thus, both developing and mature nerve fibers expressed unidentified receptors that could specifically bind NGF and initiate transport of the growth factor to the cell body.

The importance of NGF retrograde transport for neuronal survival was demonstrated *in vitro*. In the mid-1970s, a cell culture technique was developed called the Campenot chamber, named for its inventor, Robert Campenot. With this method, dissociated neurons are placed in a central chamber, and the neurites extend and grow beneath a silicon barrier to reach one of two adjacent chambers (**Figure 8.12**A, B). The central (proximal) or adjacent (distal) chambers can be individually treated with NGF to limit growth factor exposure to the cell body or nerve terminals. When NGF was added to the cell bodies of sympathetic neurons located in the proximal chamber, the neurons survived and extended neurites into the distal chambers (Figure 8.12C). Further, the addition of NGF to one or both distal chambers still promoted the survival of sympathetic neurons (Figure 8.12D, E). However, if NGF was withdrawn from one of the distal chambers, the neurites in that chamber retracted and their associated cell bodies degenerated (Figure 8.12F). Together, these latter studies provided

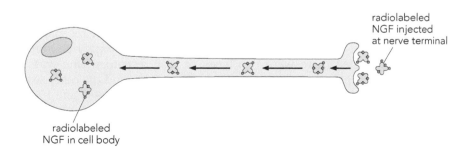

radiolabeled NGF injected at nerve terminal

radiolabeled NGF in cell body

Figure 8.11 NGF is retrogradely transported to the neuronal cell body. When radiolabeled NGF was injected in the region of nerve terminals, the NGF was transported to the cell body (arrows). These studies demonstrated that NGF did not need to be expressed near the cell body to exert its effects. Instead, NGF produced at sites distant from the cell body could travel via retrograde transport to influence signal transduction pathways and gene expression.

Figure 8.12 *In vitro* studies demonstrate NGF is retrogradely transported to the neuronal cell body and required for neuronal survival. A cell culture method was developed that allowed scientists to selectively add NGF to cell bodies or nerve terminals. (A) A chamber is added to a petri dish. This chamber consists of a single central chamber (the proximal chamber) and two distal chambers. Silicon barriers separate the proximal and distal chambers. (B) When the cell bodies of sympathetic neurons are placed in the proximal chamber (P) under culture conditions that support neuronal survival and allow neurites to grow, the neurites extend beneath the silicon barriers and enter the distal chambers (D1 and D2). The lower panel is an enlarged view of the boxed area in the upper panel. (C) When NGF was added to the proximal chamber, neurites extended into both distal chambers (D1 and D2). (D) NGF, added to one distal chamber only (D1), promoted neurite outgrowth to the treated side. However, neurites did not grow to the untreated chamber (D2). (E) When NGF was added to both distal chambers, neurites extended to both D1 and D2. (F) If NGF was subsequently withdrawn from one chamber (D2), the neurites that extended to that chamber retracted and their associated cell bodies degenerated. Together these experiments demonstrated that NGF is retrogradely transported to the neuronal cell body to promote neuronal survival. [Adapted from Campenot RB, Lund K & Mok S-A [2009] *Nat Protoc* 4:1869–1887.]

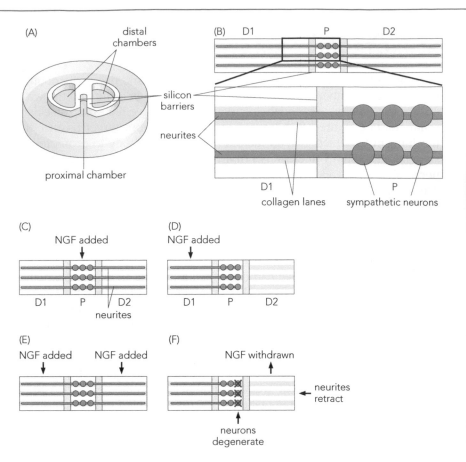

additional evidence that NGF from distant targets must bind to receptors on the nerve terminals and be transported back to the cell body to influence cell survival.

NGF Receptors Are First Identified in the PC12 Cell Line

Primary cell cultures consist of cells taken directly from an animal model and grown *in vitro*. Examples of primary cell cultures include explants of whole ganglia, such as those used by Levi-Montalcini, and the dissociated cell preparations used by Campenot and others. **Cell line cultures** are also important experimental tools. A cell line consists of immortalized cells derived from a single cell and therefore contains a homogeneous cell population. This characteristic allows researchers to assess the effects of growth factors on a uniform cell population, rather than the mixed cell populations found in ganglia. Cell lines also have the advantage of providing large numbers of cells to analyze without the need to dissect cells from animals for each experiment. A cell line was identified in the 1970s that proved very useful for studying NGF. PC12 cells, an adrenal chromaffin cell line derived from a rat pheochromocytoma, develop characteristics of sympathetic neurons when treated with NGF (**Figure 8.13**). Importantly, like primary sympathetic neuronal cultures, PC12 cells bind radiolabeled NGF. PC12 cells became an important tool for testing the mechanisms by which NGF exerted its effects and for identifying the structure of NGF receptors.

By the mid-1980s, evidence from PC12 cells and primary neuronal cultures suggested there were two types of NGF receptors. One was a low-affinity receptor that released bound NGF quickly, and the other was a high-affinity receptor that released bound NGF slowly. In 1986 and 1987, the laboratories of Moses Chao and Eric Shooter identified a glycoprotein that binds NGF with low affinity. Structural analysis of what was then called the LNGFR (low-affinity NGF receptor) revealed that the receptor lacks an

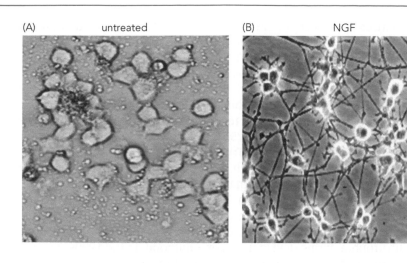

(A) untreated   (B) NGF

Figure 8.13 The PC12 cell line is used to study NGF signaling mechanisms. The PC12 cell line is an adrenal chromaffin cell line derived from a rat pheochromocytoma that normally lacks characteristics of neurons (A). When treated with NGF, these cells develop characteristics of sympathetic neurons (B). PC12 cells first became an important experimental tool in the 1970s and continue to be used by researchers today. [From Rocco ML, Balzamino BO, Passeri PP et al. [2015] *PLoS One* 10(4): e0124810.]

intracellular kinase domain. Thus, the kinetics and structure of the LNGFR suggested that this receptor alone could not be the sole signaling receptor for NGF, even though it clearly bound NGF and was expressed on many NGF-dependent neurons, as well as on Schwann cells and other nonneuronal cells.

Later, following the discoveries of BDNF, NT-3, and NT-4/5, those neurotrophins were also found to bind to the LNGFR with affinity and kinetics similar to NGF. Thus, the LNGFR was not NGF-specific (**Figure 8.14**A). The name of the receptor was subsequently changed to reflect its size; it is now known as the p75^{LNGFR} or p75NTR (p75 neurotrophin receptor). Scientists knew that the p75NTR alone could not mediate the biological effects of NGF, and several studies suggested the presence of an additional high-affinity receptor. As is often the case in science, the identification of the high-affinity NGF receptor came from an unexpected source. In 1986, Mariano Barbacid's laboratory discovered a gene for a tyrosine kinase receptor that was identified in colon carcinoma cells. This **oncogene** encoded a protein with a tyrosine kinase domain coupled to a sequence typically associated with non-muscle tropomyosin, a common cellular component. The tropomyosin sequence was located in place of the extracellular binding domain, leading the receptor to be named the tropomyosin receptor kinase (Trk). In noncancerous cells, the gene for this receptor, now called

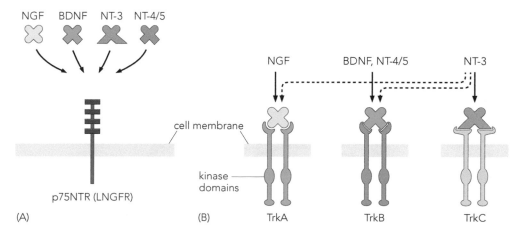

Figure 8.14 All neurotrophins bind to the p75 neurotrophin receptor, but selectively bind to specific Trk receptors. (A) The p75 neurotrophin receptor, initially called the low-affinity NGF receptor (LNGFR), was the first receptor identified that could bind to NGF. It was later found that NGF, BDNF, NT-3, and NT-4/5 all bound to this receptor with equal, low affinity. Thus, the name was changed to the p75 neurotrophin receptor (p75NTR). (B) Each neurotrophin binds to a specific Trk receptor with high affinity (solid arrows). NGF binds with high affinity to TrkA, the first Trk receptor discovered. BDNF and NT-4/5 bind with high affinity to TrkB, while NT-3 binds to TrkC with high affinity. When NT-3 is present at sufficiently high concentrations, it can also bind TrkA and TrkB (dashed arrows).

TrkA, produces a 140-kD protein that is expressed on restricted neuronal populations as well as on PC12 cells. Subsequent investigations by Luis Parada and others confirmed that TrkA is the high-affinity NGF receptor that scientists sought for so many years.

Following the discovery of TrkA, other members of the Trk family were soon identified, and different binding specificities were observed for each neurotrophin. The second Trk receptor discovered, TrkB, is the high-affinity receptor for BDNF and NT-4/5. TrkC is the high-affinity receptor for NT-3. However, NT-3 can also bind TrkA and TrkB in some contexts (Figure 8.14B). In addition to the full-length tyrosine kinase receptors, TrkB and TrkC also exist in truncated forms that lack the kinase domain.

The distribution of full-length Trk receptors on different neuronal populations matches the known growth factor preferences of the neurons; each Trk receptor is expressed on neurons known to respond to the associated neurotrophin during development or adulthood. In the 1990s, as gene knockout mice became available for the first time, studies confirmed the role of Trk receptors *in vivo*. In the PNS, the deficits found in the *TrkA*, *TrkB*, and *TrkC* knockout mice largely matched those observed in mice lacking the corresponding neurotrophin. For example, mice lacking *TrkA* had significant decreases in the number of sympathetic and DRG neurons that were comparable to the losses observed in mice lacking *Ngf*. Mice lacking *TrkB* had a decrease in nodose neurons, as did mice lacking genes for the TrkB ligands, *BDNF* and *NT-4/5* (**Figure 8.15**).

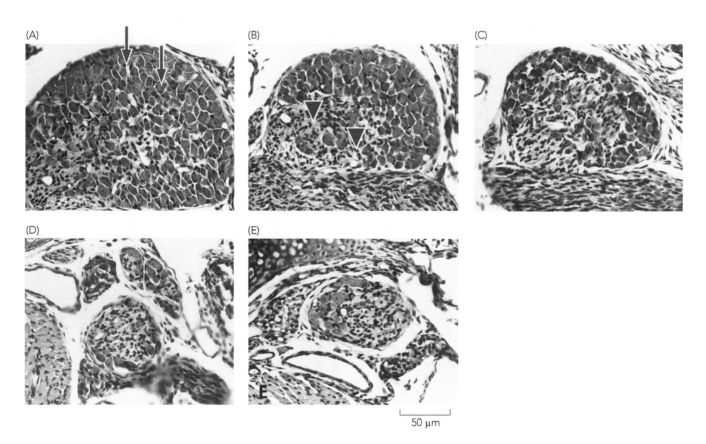

50 µm

Figure 8.15 Gene knockout mice confirm a role for Trk receptors *in vivo*. (A) A section through a nodose ganglion of a wild-type mouse reveals a large ganglion with healthy neurons throughout (arrows). Fewer healthy neurons and more dying (pyknotic neurons, arrowheads B) are observed in the nodose ganglia from mice lacking *Bdnf*. Pyknotic neurons were also common in mice lacking *Nt-4/5* (C), both *Bdnf* and *Nt-4/5* (D), or *TrkB* (E). Subpopulations of nodose neurons rely on either BDNF or NT-4/5 and therefore the *Bdnf/Nt-4/5* double knockout mouse has the fewest surviving neurons and most closely resembles the *TrkB* mutation. [From Erickson JT, Conover JC, Borday V et al. [1996] *J Neurosci* 16:5361–5371. With permission from Society of Neuroscience.]

Although the neuronal losses are generally similar in the mice lacking a given Trk receptor and a corresponding neurotrophin, subtle differences are noted in some neuronal populations, suggesting the presence of overlapping or redundant mechanisms to prevent the complete loss of a neuronal population in the absence of a single *Trk* gene.

Activation of Trk Receptors Stimulates Multiple Intracellular Signaling Pathways

The Trk receptors function similarly to other receptor tyrosine kinases. The neurotrophins are dimers, and each portion of the ligand binds to a corresponding Trk receptor monomer, causing these subunits to approach each other. Once the two receptor subunits are in close contact, they cross-phosphorylate: Each subunit phosphorylates tyrosine residues on the intracellular portion of the other subunit (**Figure 8.16**).

As more was discovered about the structure and function of Trk receptors, a model emerged to explain how neurotrophins undergo retrograde transport to the cell body. In this model, a ligand bound to the transmembrane receptor protein is taken up by the cell in a specialized transport vesicle called an **endosome**. The receptor's ligand-binding domain is enclosed within the vesicle, while its kinase domain extends outward from the vesicle surface (**Figure 8.17**). Because the ligand continues to bind the receptor, the receptor

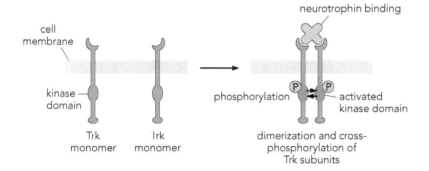

Figure 8.16 Neurotrophin binding leads to Trk dimerization and autophosphorylation. Neurotrophins exist as dimers that bind to Trk monomer subunits. The binding of the neurotrophin ligand brings the receptor subunits into close contact, so they cross-phosphorylate each other's intracellular kinase domains, thus initiating signal transduction cascades in the receptor-bearing cell.

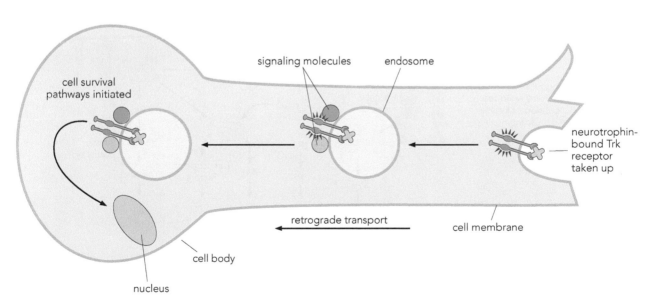

Figure 8.17 A neurotrophin and its bound receptor are retrogradely transported by an endosome. When a neurotrophin binds to its corresponding Trk receptor, the ligand-receptor complex is internalized and transported along the axon to the cell body in a transport vesicle called an endosome. The ligand remains inside the endosome, while the receptor kinase domains extend outward from the vesicle surface. Because the ligand is still bound to the receptor, the receptor remains active during transport and recruits additional signaling molecules to the endosome surface. Once the endosome complex reaches the cell body, it is able to initiate signal transduction cascades associated with cell survival. [Adapted from Miller FD & Kaplan DR [2001] *Neuron* 32:767–770.]

remains active during transport, and additional signaling molecules, such as adaptor proteins, are recruited to the endosome. Different proteins associate with endosomes at early and late stages of transport. Once the endosome complex reaches the cell body, intracellular signaling cascades are initiated to promote cell survival. While this model is currently proposed as the primary method of transporting neurotrophins to the cell body, other models suggest that the entire ligand-receptor complex does not need to reach the cell body to activate downstream signaling molecules. Transport requirements for neurotrophin signaling remain an ongoing focus of investigation.

Activation of the Trk receptor's kinase domains leads to phosphorylation of additional tyrosine residues outside the kinase domains. These phosphorylated sites then recruit other proteins that initiate intracellular signaling cascades. Among the proteins recruited are adapter proteins such as Shc (C-terminal Src-homology-2 domain) and the enzyme phospholipase C gamma (PLC$_\gamma$). These proteins, in turn, activate molecular pathways leading to various transcriptional events that regulate functions as varied as neurite outgrowth and synaptic plasticity, cell differentiation, and neuronal survival. The precise intracellular effects of neurotrophin binding depends, in part, on the cellular context and stage of development.

PLC$_\gamma$, Ras (a GTPase initially found in rat sarcoma), and phosphatidylinositol 3-kinase (PI3-K) are the three primary intracellular pathways activated by Trk receptor binding. In neurons, PLC$_\gamma$ generally influences synaptic plasticity, while Ras regulates the differentiation of cells, and PI3-K promotes cell survival. These pathways often interact through additional proteins, however, and overlapping functions are found in several neuronal contexts. For example, Ras can also influence cell survival pathways.

Figure 8.18 provides an overview of the signaling pathways activated by the binding of neurotrophins to Trk receptors. Activation of the PLC$_\gamma$

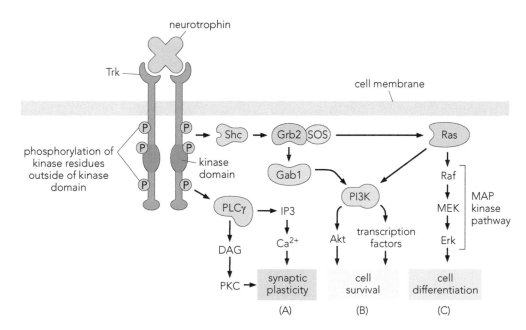

Figure 8.18 Trk phosphorylation initiates signal transduction cascades that regulate synaptic plasticity, cell survival, and cell differentiation. Upon neurotrophin binding, phosphorylation of kinase residues outside of the kinase domains recruits adaptor proteins to the receptor, thereby initiating various signal transduction cascades. Recruitment of PLC$_\gamma$ to the phosphorylated Trk receptor can activate both the IP3 and DAG pathways that influence synaptic plasticity (A). Recruitment of the Shc adaptor protein leads to the subsequent recruitment of Grb2 and ultimately to activation of GAB1 or Ras to influence the PI3 kinase pathway that promotes cell survival (B). Additionally, recruitment of Shc followed by Grb2 interacting with SOS can activate Ras and initiate the MAP kinase pathway to influence cell differentiation (C). Akt, murine thymoma viral oncogene homolog; Erk, extracellular signal-regulated kinase; GAB1, Grb2-associated binding protein 1; Grb2, Growth factor receptor-bound protein 2; MAP, Mitogen-activated protein; MEK, MAP kinase/Erk kinase; PI3-K, phosphatidylinositol 3-kinase; SOS, Son of sevenless. [Adapted from Reichardt LF [2006] *Philos Trans R Soc Lond B Biol Sci* 361(1473):1545–1564.]

pathway occurs when PLC$_\gamma$ associates with phosphorylated sites on the intracellular portion of the Trk receptor. Activation of PLC$_\gamma$ leads to the subsequent activation of the IP3 (inositol 1,4,5-triphosphate) and DAG (diacylglycerol) pathways. IP3 induces release of Ca^{2+} from intracellular stores, while DAG activates PKC (protein kinase C). Both the IP3 and DAG pathways regulate the expression of genes and proteins associated with synaptic plasticity.

Ras pathway activation is initiated with the recruitment of the adaptor protein Shc to the phosphorylated tyrosine residues on the receptor. The Shc protein recruits Grb2 (Growth factor receptor-bound protein 2), which interacts, in turn, with SOS (Son of sevenless) to activate Ras. Ras then activates the MAP (Mitogen-activated protein) kinase signaling cascade, which includes the sequential phosphorylation of Raf (rapidly accelerated fibrosarcoma kinase), MEK (MAP kinase/Erk kinase), and Erk (extracellular signal-regulated kinase). Phosphorylated Erk enters the nucleus, where it regulates various genes associated with neuronal differentiation.

The PI3-K pathway can be activated by Ras or by a sequence of adaptor proteins, including Shc, Grb2, and GAB1 (Grb2-associated binding protein 1). PI3-K then activates the protein kinase Akt (named for a murine thymoma viral oncogene homolog) or transcription factors. Activation of either influences the expression of genes associated with neuronal survival. In addition, Akt can inhibit cell death pathways, as discussed later in the chapter.

Regulated expression of Trk receptors is needed throughout development to mediate the effects of these different signaling pathways. For example, BDNF-TrkB signaling is important during formation and reorganization of synaptic connections. Therefore, TrkB must be rapidly and accurately transported to the correct cell surface when needed (see Chapter 10).

Full-Length Trk Receptors Interact with Truncated Trk Receptors and p75NTR to Influence Cell Survival

The response of a neuron to a neurotrophin can further be modified by the presence of truncated, noncatalytic receptors expressed on the same cell as the full-length Trk receptors. As noted previously, TrkB and TrkC exist in truncated, noncatalytic forms. Although the function of these receptors remains largely unclear, studies suggest truncated TrkB and TrkC may sequester their corresponding neurotrophins to increase the local concentration of those growth factors, then somehow direct the ligands to full-length receptors on the cells that require them. Alternatively, the truncated receptors may take up and sequester growth factors to prevent activation of a full-length Trk receptor. For example, it has been hypothesized that truncated TrkC could bind local NT-3 to prevent this neurotrophin from activating TrkB at the wrong stage of development (**Figure 8.19**A).

Another noncatalytic receptor, the low-affinity p75NTR, is co-expressed with Trk receptors on many neuronal populations. The function of p75NTR remained a mystery even as the functions of the various Trk receptors were being documented. The p75NTR binds all neurotrophins with equally low affinity, but because it lacks a catalytic subunit and therefore cannot transduce an intracellular signal, p75NTR's possible function in neurotrophin signaling was not readily apparent.

As scientists continued to investigate the actions of p75NTR and Trk receptors during development, a possible role for p75NTR emerged: The low-affinity receptor might help concentrate the neurotrophins at the target cell and "hand off" mature neurotrophin molecules to the high-affinity Trk receptor. This would presumably optimize signaling efficiency (Figure 8.19B). Evidence to support this hypothesis includes the observation that neurons expressing a high-affinity receptor plus p75NTR can survive and extend neurites at concentrations of neurotrophin lower than the concentrations required by neurons that express the high-affinity receptor alone.

Figure 8.19 Truncated Trk receptors and p75NTR interact with full-length Trk receptors to modify cellular responses. Several models have been proposed to explain how truncated Trk receptors and p75NTR interact with full-length Trk receptors to influence neurotrophin signaling. (A) One suggestion is that truncated TrkC receptors can take up excess NT-3 to ensure the ligand does not activate full-length TrkB receptors at the wrong time in development. (B) p75NTR may bind NGF then release it to nearby, unbound TrkA receptors to enhance the efficiency of NGF signaling.

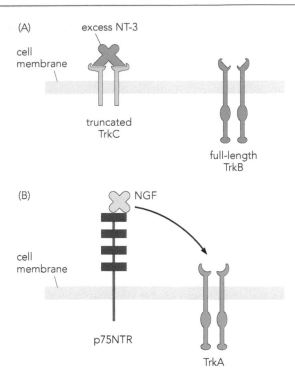

For example, *in vitro* experiments showed that DRG neurons isolated from mice lacking *p75NTR* require much more NGF to survive than DRG neurons from wild-type mice expressing both *TrkA* and *p75NTR*. Thus, under some circumstances, p75NTR appears to enhance the response of neurons to neurotrophins.

As studies of p75NTR and Trk receptors continued, an additional, surprising role for p75NTR emerged. Scientists discovered that while p75NTR enhanced Trk receptor signaling to promote neuronal survival in some contexts, it also mediated neuronal cell death in some neuronal populations. The mechanisms by which neurotrophins and p75NTR interact to mediate neuronal death during development are detailed later in the chapter.

Other Growth Factors Also Regulate Neuronal Survival and Outgrowth

The preceding descriptions of neuron-target interactions focus on the role of neurotrophins. However, these are not the only molecules that influence developing neurons. Many other growth factors—including members of the fibroblast growth factor (FGF) family, the glial-cell-line-derived neurotrophic factor (GDNF) family (GDNF, neurturin, artemin, and persephin), and ciliary neurotrophic factor (CNTF) and related cytokines—influence the survival of various neuronal populations. However, not all these proteins meet the criteria for target-derived growth factors as defined by the original neurotrophic hypothesis.

A description of the discovery and functions of CNTF is provided as just one example of a growth factor unrelated to neurotrophins. The discovery of CNTF initially appeared to reveal another classic, target-derived growth factor. However, as more was learned about CNTF it became apparent that this growth factor and its corresponding receptor had several unique traits. The history of the discovery of CNTF and its functions illustrates another example of how scientists often begin an investigation seeking to answer one question, only to uncover several new and important findings.

Ciliary Neurotrophic Factor Is Isolated Based on an Assay for Developing Ciliary Ganglion Neurons

The muscles within the ciliary body and choroid of the eye are innervated by cholinergic parasympathetic ciliary ganglion (CG) neurons. In the developing chick, about half of the CG neurons normally die between embryonic days 9 to 14 (E9–E14). Because NGF, the only identified growth factor in the 1970s, did not influence the survival of CG neurons, scientists sought to isolate the growth factor required by these developing neurons.

Cell culture methods were established for CG neurons, and it was soon found that CG survival was supported by a variety of tissues and extracts, including skeletal muscle, heart-cell-conditioned medium, whole chick embryo extracts, and eye tissue. In 1979, Silvio Varon and colleagues isolated a survival- and outgrowth-promoting factor from the ciliary body and choroid of embryonic chick eye, then focused on purifying this eye-derived factor. In 1984, extracts from 300 chick eyes yielded a partially purified protein that promoted the survival and outgrowth of embryonic CG neurons. The protein was called **ciliary neurotrophic factor** (**CNTF**). Subsequent studies found CNTF also promoted the survival and outgrowth of neurons from sympathetic and dorsal root ganglia of embryonic chick and rodents. Thus, CNTF's effects were not limited to CG neurons. CNTF was subsequently purified from heart tissue and adult rat sciatic nerve, indicating that CNTF was not limited to eye tissues. Although the initial studies suggested CNTF was a target-derived growth factor for CG neurons, as the sequence of CNTF and its tissue distribution were defined in mammals, it became apparent that CNTF did not function in the same way as neurotrophins.

For example, the neurotrophic hypothesis proposes that target-derived factors are secreted, but the CNTF protein lacked the signal sequence typical of secreted proteins. Instead CNTF's structure was typical of a cytosolic protein and therefore must be released from cells by a different mechanism. The expression patterns of CNTF were also unexpected. CNTF is not highly expressed in embryonic tissues at the time when most CNTF-responsive neurons reach their targets. In fact, CNTF expression increases in most tissues after target innervation. Of all the tissues examined at various stages of development, the strongest CNTF signal was found in the adult rat sciatic nerve. While CNTF did not fit its anticipated role as a neurotrophic factor for developing CG neurons, CNTF and related molecules are now recognized for their importance in neuronal differentiation and synaptogenesis.

As investigations of CNTF continued, it was noted that in some contexts, CNTF stimulated axonal regeneration and promoted the survival of damaged adult neurons. These observations, coupled with the findings that CNTF is a cytosolic protein expressed in non-target tissues, led to the idea that CNTF functioned primarily in adult tissues—perhaps in response to injury. This hypothesis was supported by studies in which the axons of facial motor neurons were severed (axotomized). While axotomy normally leads to degeneration of the associated facial motor neurons, application of CNTF to the severed axons prevented neuronal death.

In the rodent nervous system, CNTF not only rescued adult motor neurons injured by axotomy, but also promoted motor neuron survival when injected into a mouse model of amyotrophic lateral sclerosis (ALS), a neurodegenerative disease also known as Lou Gehrig's disease. The experimental finding that CNTF could rescue degenerating motor neurons led to clinical trials in humans with ALS. Unfortunately, systemic injections of CNTF were unable to slow the progression of the disease in humans. Among the many difficulties faced in treating motor neurons in humans is the inability to get sufficient quantities of factor across the blood—brain barrier to reach the target neurons. Another difficulty is diagnosing motor neuron disease early enough to deliver growth factors before neuronal damage is too advanced to be reversed. Researchers continue to test growth factors and delivery systems to treat degenerating motor neurons with the goal of curing ALS, or at least slowing its progression.

The CNTF Receptor Requires Multiple Components to Function

The discovery of a receptor for CNTF by Samuel Davis and colleagues in the early 1990s demonstrated that it functioned quite differently than the Trk receptors used by neurotrophins. Unlike Trk receptors, the alpha component of the receptor complex—the portion of the receptor that binds to CNTF—does not have a transmembrane domain, but instead is attached to the cell membrane by a GPI linkage (**Figure 8.20**)—a characteristic that was surprising at the time.

It is now known that the receptor for CNTF shares characteristics with other cytokine receptors. The CNTF receptor functions as a tripartite receptor, requiring three components for signaling to occur. CNTF binds to the CNTF-α receptor component, which then recruits gp130, a glycoprotein of about 130 kD (Figure 8.20). The gp130 component is also used by certain hematopoietic cytokines, such as LIF (leukemia inhibitory factor) and some interleukins (see **Box 6.1**). Once CNTF binds to CNTF-α and gp130 is recruited, LIF receptor β joins the complex. The heterodimerization of gp130 and LIF receptor β leads to tyrosine phosphorylation and intracellular signaling cascades to promote neuronal survival.

Growth Factors Unrelated to CNTF Promote Survival of Developing CG and Motor Neurons

More than 40 years after the initial identification of CNTF in eye tissue, studies seeking the target-derived growth factors required by CG neurons continue. It currently appears that two members of the GDNF family—GDNF and neurturin—are important for promoting survival of these parasympathetic neurons *in vivo*.

Investigators also continue to characterize the target-derived signals required for embryonic motor neurons. The observations that CNTF promotes the survival of developing motor neurons *in vitro* and influences motor neuron survival in adult rodents initially suggested that it might be the long-sought target-derived factor for embryonic motor neurons. However, CNTF is not expressed in muscle tissue at the correct time in development *in vivo* and therefore does not meet the criteria expected of a target-derived neurotrophic factor for motor neurons.

Early studies by Hamburger and others demonstrated that motor neuron survival was influenced by the size of available limb bud tissue, yet the identity of the limb-derived growth factor is still unknown. Several identified growth factors meet some of the necessary criteria to be considered a target-derived factor for motor neurons. Among the growth factors that

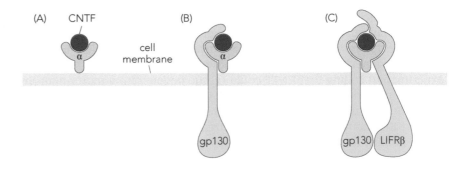

Figure 8.20 CNTF receptors function as a tripartite complex. (A) CNTF binds to the α component of the CNTF receptor complex. (B) The α component interacts, in turn, with gp130, a receptor component shared by CNTF and other cytokines. (C) A third component, the leukemia inhibitory factor receptor β subunit (LIFRβ), is then recruited to form the functional CNTF receptor. [Modified from Stahl N & Yancopoulos GD [1994] *J Neurobiol* 25:1454–1466.]

promote survival of motor neurons in at least some contexts are GDNF, BDNF, NT-4/5, IGF (insulin-like growth factor), VEGF (vascular endothelial growth factor), and cardiotrophin-1, a cytokine related to CNTF. In addition to growth factors present in skeletal muscle, many growth factors are expressed in the Schwann cells surrounding the peripheral axons of motor neurons. These are believed to be an additional source of survival-promoting factors for innervating motor neurons. The number of candidate growth factors identified for motor neurons has been referred to as "an embarrassment of riches." However, despite the large number of growth factors that can influence motor neuron survival, it remains uncertain how these factors interact to regulate motor neuron survival *in vivo* during normal development.

PROGRAMMED CELL DEATH DURING NEURAL DEVELOPMENT

As mentioned at the beginning of the chapter, the idea that cell death was a part of normal neural development was not apparent in the early twentieth century. However, as more was learned about neuron-target interactions and the role of trophic factors, it became apparent that such death was a consistent, naturally occurring phenomenon. This type of cell death, now called programmed cell death (PCD), is defined as the loss of a large number of cells in a distinct spatial and temporal pattern within a given species.

PCD, as discussed here in Chapter 8, refers to the loss of neurons that occurs normally during neural development. However, it should be kept in mind that PCD occurs not only in the nervous system, but also in a variety of animal tissues, as well as in plants, yeast, and bacteria. Thus, over the course of evolution, cells have developed and maintained mechanisms to induce cell death. That cells would expend energy to do this suggests there must be some evolutionary advantage to what may at first seem to be a wasteful process.

It is also important to note that PCD is not a result of an abnormality in the cell itself or from extrinsic damage that leads to cellular injury and necrotic cell death. Instead, PCD is an active cellular process that is regulated by intracellular signaling pathways activated by either external signals or intrinsic genetic mechanisms. Because of the unique features of PCD, scientists can distinguish types of PCD from necrotic cell death using a variety of methods (**Box 8.2**).

Box 8.2 Scientists Study Cellular Appearance and DNA Fragmentation to Distinguish Types of Cell Death

Scientists have developed many techniques to identify cells that have died by necrosis or apoptosis, a form of PCD. The term **apoptosis** is often applied inadvertently to all forms of PCD; while apoptosis is among the most common types of PCD, it is a distinct form characterized by specific cellular features. **Figure 8.21** compares the cellular changes that occur as a healthy cell undergoes necrotic or apoptotic cell death. During necrotic cell death, the dying cell begins to swell and develop cytoplasmic vacuoles. The nucleus and organelles then begin to break down, and as the cell continues to expand, the cell membrane eventually ruptures. Because macrophages become active when the cellular contents are released, an inflammatory response occurs in response to necrotic cell death.

During apoptotic cell death, the cell cytoplasm shrinks, causing the cells to appear smaller. When examined under the microscope, an apoptotic cell reveals a pyknotic nucleus with highly condensed nuclear chromatin. In the early stages of apoptosis, the cell membrane and organelles remain largely intact, but the cytoplasm and nucleus eventually break down, and the DNA becomes fragmented. The

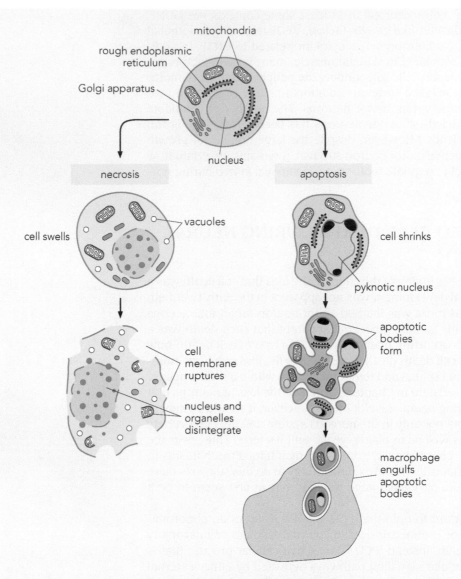

Figure 8.21 Characteristics of cell death resulting from necrosis and apoptosis. When a normal cell dies, the stages of cell death differ depending on whether the cell undergoes necrosis or apoptosis. At the onset of necrotic death, the cell begins to swell, and vacuoles form in the cytoplasm and organelles. The cell continues to enlarge as the nuclei and organelles disintegrate, until the cell membrane ultimately ruptures. The release of cell contents provokes an inflammatory response. As the cell initiates apoptotic death, its cytoplasm and nuclear chromatin begin to condense. At first, the organelles and cell membrane remain intact, even as the cell body and nucleus begin to shrink. However, the cell surface soon becomes convoluted, and the nucleus and organelles break off into discrete, membrane-bound fragments called apoptotic bodies. These apoptotic bodies are degraded and ultimately phagocytized by nearby macrophages or glial cells. [From Kerr JFR [1995] *Trends Cell Biol* 5:55–57.]

nucleus and organelles form distinct, membrane-bound fragments termed **apoptotic bodies** that are a hallmark of apoptosis. These apoptotic bodies are then phagocytized by macrophages to remove cellular debris. In some cases, glial cells adjacent to the apoptotic cells will function as macrophages to remove the apoptotic bodies. There is no inflammatory response to apoptosis because phagocytosis occurs before the dying cells can release their contents. As a result, nearby cells are not damaged by apoptotic PCD. In recent years scientists have begun to identify and characterize several subtypes of PCD.

Type 1 PCD is the apoptotic cell death described here. Type 2 PCD is characterized by the presence of lysosomes and autophagic vacuoles, whereas type 3 PCD involves swelling in the Golgi apparatus, nuclear envelope, and rough endoplasmic reticulum, as well as vacuolization of the mitochondria. The causes and purpose of these distinct forms of PCD continue to be investigated.

The characteristic changes that occur during apoptosis have been used to develop laboratory techniques that readily identify apoptotic cells. For example, DNA fragmentation can be identified as a **DNA ladder** visualized using agarose gel electrophoresis. Apoptosis breaks double-stranded chromosomal DNA into fragments. The breaks occur between nucleosomes creating fragments that are approximately 180 base pairs (bp) or multiples of 180 bp in length (that is, 360 bp, 540 bp, 720 bp, and so on). When apoptotic tissues or cells are treated to rupture the cell membrane and the resulting **lysates** are run on a gel, the fragments are separated based on size and distributed along the length of the gel, resembling the rungs of a ladder (**Figure 8.22**A). The size of the bands produced is determined by comparing them to marker standards that are run at the same time. Cells dying by necrosis do not undergo orderly DNA fragmentation, so lysates of those cells do not produce the ladder-like pattern on a gel.

To visualize apoptotic death in tissues or individual cells, the TUNEL method (terminal deoxynucleotidyl transferase dUTP nick-end labeling) can be used. The fragmented, nicked ends of DNA can be detected when tissues are incubated with an enzyme (TdT, deoxynucleotidyl transferase) and the nucleotide dUTPs (2′-deoxyuridine 5′-triphosphates). The TdT adds the dUTPs to the nicked ends of the DNA (Figure 8.22B). The dUTPs are also labeled with a marker that is attached prior to incubation or after attaching to the nicked ends of the DNA. When viewed through a microscope, the markers are visible and reveal which cells have undergone apoptosis (Figure 8.22C).

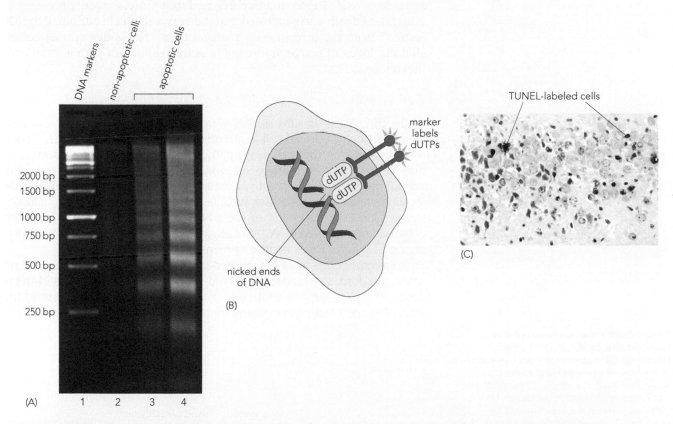

Figure 8.22 The fragmented DNA characteristic of apoptosis can be identified in cell lysates and intact cells. (A) Separation of DNA by gel electrophoresis distinguishes apoptotic and non-apoptotic cell death. During apoptosis, DNA is broken into fragments of 180 base pairs (bp) or multiples of 180 bp. When cell lysates from apoptotic cells are run on an agarose gel, DNA fragments are separated by size, forming what looks like a ladder. Lysates from non-apoptotic cells do not form such a pattern. Marker standards (1) are run on the gel to estimate the size of the bands detected in the cell lysates. Cells that do not undergo apoptotic cell death do not reveal a ladder-like pattern (2). Lysates from apoptotic cells reveal a ladder-like pattern of bands in sizes consistent with apoptotic DNA fragmentation (3 and 4). (B) The TUNEL (terminal deoxynucleotidyl transferase dUTP nick-end labeling) method allows visualization of apoptotic cells. Fragmented DNA can be visualized in tissue sections or cell cultures using the TUNEL method. When terminal deoxynucleotidyl transferase is added to tissues in the presence of dUTPs, the enzyme adds dUTPs to the fragmented, or nicked, ends of DNA. In this example, a marker that labels dUTP is subsequently added, and a chemical reaction has taken place to allow visualization of the cells through a microscope. (C) A tissue section from the hamster hippocampus illustrates apoptotic cells (brown) labeled with the TUNEL method. [(A) From Wang, Q, Wang Y-L, Wang, K et al. [2015] *Oncol Lett* 9:278–282; (C), from Xiao S-Y, Guzman H, Zhang H, et al. [2001] *Emerg Infect Dis* 7:714–721.]

Studies Reveal Cell Death Is an Active Process Dependent on Protein Synthesis

In the 1980s, scientists made several observations that suggested PCD was not merely a passive process that resulted from a lack of neurotrophic factor. *In vitro* and *in vivo* studies demonstrated that blocking RNA or protein synthesis prevented cell death that would normally occur. The results indicated that some proteins must be required to initiate cell death.

One such study from the lab of Eugene Johnson, Jr., utilized the established cell culture methods for dissociated sympathetic neurons. The researchers first cultured the neurons for several days in the presence of NGF, then added anti-NGF antiserum to block the NGF effects and thereby induce cell death (**Figure 8.23**A). The time course of these experiments corresponded to the stages at which neurons would contact target cells and compete for NGF *in vivo*.

The researchers found that cell death typically occurred within 24 to 48 hours following NGF removal (Figure 8.23B). However, cell death was prevented when the protein synthesis inhibitor cycloheximide was added to the cultures lacking NGF (Figure 8.23C). Thus, by preventing the formation of new proteins, the cells were able to survive even in the absence of NGF. These studies confirmed that proteins were necessary to initiate cell death and suggested that the survival effects attributed to NGF resulted from the inhibition of such proteins. These discoveries helped shift the focus of research to explore active cell death pathways in cells that undergo PCD.

Cell Death Genes Are Identified in *C. Elegans*

At the time these *in vitro* studies were being conducted, other scientists were starting to unravel cell death mechanisms in other systems. It soon became apparent that cell death pathways are shared across species. As noted in previous chapters, *C. elegans* provides several advantages as an animal model. The comparatively small number of cells found in specific locations helps scientists track what becomes of individual cells, including neurons. In *C. elegans*, 131 of the 1090 somatic cells undergo PCD. The cells that are removed by PCD are always the same cells, and death always occurs at specific times in development.

Several genes were discovered in the 1990s that prevent cell death when mutated. Researchers inferred that the wild-type alleles of these genes would cause cell death under normal conditions, supporting the emerging idea that some forms of cell death were active, genetically

Figure 8.23 Programmed cell death requires the production of new proteins. Sympathetic neurons harvested at the time of normal target innervation were used to test whether protein synthesis was necessary to induce cell death. (A) When cultured in the presence of NGF, dissociated sympathetic neurons survived and extended long processes. (B) When NGF was withdrawn and antibodies to NGF were added to remove any remaining growth factor, the neurites degenerated and the neurons began to die within 24 to 48 hours. (C) However, if inhibitors of protein synthesis such as cycloheximide were added during the first 24 to 48 hours following NGF removal, the cells survived. These studies indicated that protein synthesis was needed to activate cell death.

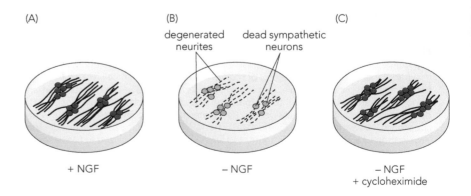

(A)

(B)
degenerated neurites   dead sympathetic neurons

(C)

+ NGF

− NGF

− NGF + cycloheximide

Table 8.1 Pro-Apoptotic or Anti-Apoptotic Genes in *C. elegans*.

C. elegans Gene	Role in PCD	Activity in Wild-Type	Loss-of-Function Phenotype	Mammalian Homologs
Ced-3	Pro-apoptotic	Cell death pathway	↓ cell death	*Caspase*
Ced-4	Pro-apoptotic	Cell death pathway	↓ cell death	*Apaf-1*
Ced-9	Anti-apoptotic	Cell survival pathway	↑ cell death	*Bcl-2*
Egl-1	Pro-apoptotic	Cell death pathway	↓ cell death	*Bim, Bad, Bid, Puma, Noxa*

controlled processes. The first genes found to regulate cell death in *C. elegans* are known as *Ced* (cell death abnormality) genes, because mutations in these genes disrupt the normal patterns of cell death. However, as other members of the *Ced* family were identified, it was recognized that wild-type alleles of some *Ced* genes promote cell survival. The *Ced* genes that promote cell death are called pro-apoptotic genes, while *Ced* genes that promote survival are called anti-apoptotic genes (**Table 8.1**).

Two pro-apoptotic genes are *Ced-3* and *Ced-4*. If either of these genes is absent, the number of cells undergoing PCD decreases. If both *Ced-3* and *Ced-4* are absent in *C. elegans*, all PCD is prevented, indicating that these genes are necessary for mediating cell death pathways. An important anti-apoptotic gene, *Ced-9*, prevents cell death that would be induced by the expression of *Ced-3* and *Ced-4*. If *Ced-9* is mutated, there is an increase in cell death. Thus, the balance of *Ced-9*, *Ced-3*, and *Ced-4* activity determines whether a cell lives or dies. A fourth gene further regulates the activity of these *Ced* gene products. *Egg-laying defective 1* (*Egl-1*), first described in 1983, is a member of the large *Egl* family of genes that plays multiple roles in *C. elegans* development that extend beyond egg-laying behavior. The *Egl-1* gene inhibits the anti-apoptotic actions of *Ced-9*, thus allowing pro-apoptotic genes to become active and induce to cell death.

Figure 8.24 illustrates how the various CED proteins interact to promote cell survival or cell death. In the absence of active EGL-1, CED9 forms a complex with CED4, preventing CED4 from interacting with CED3 and initiating the cell death pathway. Thus, when CED9 and CED4 form a complex, the cell survives. However, if EGL-1 interacts with CED9, CED9 is no longer able to complex with CED4. With CED9 inactivated by EGL-1, the pro-apoptotic CED4 can separate from the CED9/CED4 complex to interact with and activate CED3, which then induces cell death.

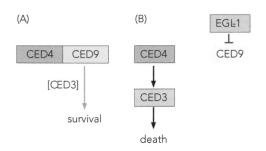

Figure 8.24 Cell death and cell survival pathways in *C. elegans* . (A) Cell survival results when the anti-apoptotic protein CED9 forms a complex with the pro-apoptotic protein CED4. This prevents CED4 from interacting with pro-apoptotic CED3 and thus stops cell death. (B) Apoptosis results if the pro-apoptotic EGL-1 inhibits anti-apoptotic CED9. If CED9 cannot maintain a complex with CED4, CED4 is free to dissociate from the CED4/CED9 complex and interact with pro-apoptotic CED3, leading to cell death.

Homologs of the *C. Elegans Ced* and *Egl* Genes Contribute to the Mammalian Apoptotic Pathway

Soon after the discovery of the pro- and anti-apoptotic genes in *C. elegans*, homologs of the *Ced* and *Egl-1* genes were identified in vertebrates. These homologs regulate pathways that are similar, though more complex. Scientists also began to explore how trophic factors interact with these pathways to regulate neuronal survival. The homolog of *Ced-3* is *Caspase-9* (Table 8.1). Caspases are cysteine proteases that cleave target proteins next to an aspartate residue. The specificity of the enzymes, involving a cysteine at the enzyme's active site that cleaves near aspartate, gives rise to the term caspase. The first mammalian caspase related to the CEDs was originally called ICE (interleukin-1β converting enzyme). This enzyme is now known as Caspase-1 and is about 28% similar to the CED3 protein.

The mammalian homolog of *Ced-4* is *Apaf-1* (apoptotic protease activating factor-1). Apaf-1 and other proteins form a complex called an **apoptosome**, a critical component of the cell death pathway. The homolog of *Ced-9* is *Bcl-2*, a member of a large family of molecules initially discovered as a part of the B-cell lymphoma-2 family of genes.

The Bcl-2 family includes multiple genes divided into two subfamilies: one anti-apoptotic, the other pro-apoptotic. The best-characterized anti-apoptotic, survival-promoting members of the Bcl-2 family are *Bcl-2* and *Bcl-xL*, while the best-known pro-apoptotic members are *Bax* and *Bak*, which interact with several related genes to promote cell death. The Bcl-2 family is further divided into three classes of proteins based on the number of Bcl-2 homology (BH) domains. The anti-apoptotic proteins, Bcl-2 and Bcl-xL, have four BH domains (BH1, BH2, BH3, and BH4), whereas the pro-apoptotic proteins Bax and Bak have three BH domains (BH1, BH2, and BH3). A second group of pro-apoptotic proteins has only the BH3 domain. These BH3-only proteins include Bad, Bim, Bid, Puma, and Noxa. The *Egl-1* gene in *C. elegans* is homologous to the BH3-only proteins in mammals. Like the effects of *Ced* gene mutations, deleting anti- or pro-apoptotic *Bcl-2*-related genes in mice alters the extent of cell death in the PNS and CNS. When the anti-apoptotic genes *Bcl-2* or *Bcl-xL* were deleted, more neurons died. Conversely, when the pro-apoptotic gene *Bax* was deleted, more neurons survived.

The cleavage of distinct pro-caspases activates the corresponding caspases that then mediate cell death pathways. One of the first cell death pathways described in mammals involves the release of cytochrome *c* from the mitochondria. When pro-apoptotic signals activate Bax or Bak, these proteins form complexes at the mitochondrial membrane that somehow induce the formation of a pore, allowing release of cytochrome *c* into the cytosol. The released cytochrome *c* then forms a complex with Apaf-1 (**Figure 8.25**, pathway 1). In the presence of ATP, Apaf-1 undergoes a conformational change that allows it to interact with pro-caspase-9, forming an apoptosome. Pro-caspase-9 is cleaved within the apoptosome, and the resulting mature caspase-9 functions as the **initiator caspase** that activates, in turn, subsequent caspases, such as caspase-3, the predominant **effector caspase** in neurons undergoing cell death.

The pathway outlined in Figure 8.25 is often called the mitochondrial pathway or the **intrinsic pathway**. This pathway can be activated by cellular stress, DNA damage, or the loss of trophic factor, such as that observed following the withdrawal of NGF from sympathetic neurons *in vitro* (see Figure 8.23). The removal of growth factor also leads to the activation of BH3-only proteins such as Bim and Puma. These pro-apoptotic

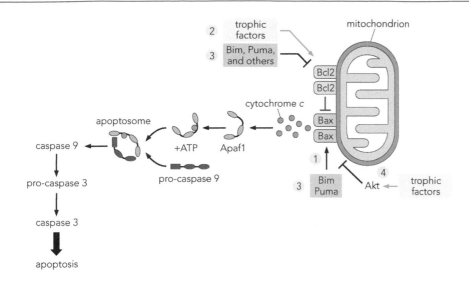

Figure 8.25 Cell death and cell survival pathways in mammals. The figure illustrates some of the ways in which mammalian pro- and anti-apoptotic pathways interact. (1) Activation of the pro-apoptotic protein Bax leads to the release of cytochrome *c* from mitochondria. Cytochrome *c* then interacts with Apaf-1, which in the presence of ATP undergoes a conformational change that allows it to interact with pro-caspase-9, a pro-apoptotic enzyme. Apaf-1, cytochrome *c*, and pro-caspase-9 form a structure called an apoptosome. Pro-caspase-9 is cleaved within the apoptosome, and the resulting caspase-9 cleaves pro-caspase-3. The activated caspase-3 then causes cell death. (2) However, if trophic factors are present, they can activate the anti-apoptotic protein Bcl-2, which in turn inhibits the activity of Bax, thereby preventing the release of cytochrome *c* from mitochondria and blocking the apoptotic pathway. (3) Other pro-apoptotic proteins such as Bim and Puma are activated upon removal of growth factors and can promote apoptosis either by activating Bax or inhibiting Bcl-2. (4) Trophic factors can also promote survival by inhibiting pro-apoptotic proteins. [Adapted from Kristiansen M & Ham J [2014] *Cell Death Differ* 21(7):1025–1035.]

proteins either activate Bax directly or inhibit the pro-survival functions of Bcl-2 and Bcl-xL (Figure 8.25, pathway 3).

Trophic factors can increase neuronal survival by inhibiting the apoptotic pathway at various points. For example, trophic factors can increase the synthesis of the Bcl-2 protein that then inhibits the activity of Bax and Bak at the mitochondrial membrane, thereby preventing the release of cytochrome *c* (Figure 8.25, pathway 2). Without cytochrome *c*, the apoptosome does not form and the caspases that induce cell death cannot be activated. In addition, when trophic factors stimulate the PI3-K pathway described earlier (see Figure 8.18), phosphorylated Akt both phosphorylates transcription factors associated with cell survival and phosphorylates and inhibits pro-apoptotic proteins such as Bim, Bad, and Bax (Figure 8.25, pathway 4). Thus, trophic factors not only stimulate survival pathways, but also actively inhibit the cell death cascade.

p75NTR and Precursor Forms of Neurotrophins Help Mediate Neuronal Death during Development

The final element of cell death pathways to consider is the activation of an **extrinsic pathway** through "death receptors." Two important observations helped elucidate how the extrinsic pathway functions in the developing nervous system. One discovery was that p75NTR is part of the tumor necrosis receptor super family (TNFRSF), a group of receptors that are often called death receptors because ligands that bind to them initiate intracellular cell death pathways. The p75NTR is also now called the TNFRSF-16. A role for p75NTR in cell death was noted in cell line studies by Dale Bredesen and colleagues. They noted that under culture conditions that induced cell death, the extent of cell death was enhanced in cells expressing p75NTR. In addition, Barde and colleagues noted that whereas NGF normally prevented cell death in retinal neurons expressing both TrkA and p75NTR, NGF induced cell death in retinal neurons that expressed only p75NTR.

The second influential discovery regarding the extrinsic pathway of cell death was the finding that different forms of neurotrophins serve distinct roles in the developing nervous system. Neurotrophins are first generated as **pro-neurotrophins**—precursor molecules of approximately 270 amino acids. Pro-neurotrophins are then cleaved to yield mature neurotrophins of roughly 120 amino acids. Initially, it was thought that only the mature forms were important during neural development. In recent years, however, roles for the unprocessed, precursor forms in regulating cell death have been identified.

Whereas the mature, processed neurotrophins bind to p75NTR with low affinity, the unprocessed pro-neurotrophins bind to p75NTR with higher affinity. In the developing nervous system, p75NTR binds to unprocessed pro-neurotrophins, often in the presence of a co-receptor called sortilin. Adaptor proteins are then recruited to the intracellular portion of the receptor complex, a region often called the **death domain**, to initiate the signaling cascades that lead to cell death (**Figure 8.26**). The exact mechanism by which this occurs is still under investigation, but in many instances the *c*-jun N-terminal kinase (JNK) pathway is involved. When NGF is withdrawn from PC12 cells, for example, the pro-survival Erk pathway is no longer activated, but instead the JNK pathway is initiated. Several targets downstream of JNK then initiate cell death. In some cases, JNK activates *c*-jun transcription factors that translocate to the nucleus to activate pro-apoptotic genes. In other contexts, JNK activates p53, which in turn activates pro-apoptotic genes such as *Bax* to initiate caspase cascades that cause cell death (Figure 8.26).

Investigators continue to explore how neurotrophins and pro-neurotrophins interact with p75NTR to influence appropriate patterns of cell survival and death. It appears that co-expression of specific Trk receptors influences how p75NTR responds to different neurotrophins. As noted, in neurons that only express p75NTR without a Trk receptor, cell death can be initiated in the presence of a pro-neurotrophin. However, when both a Trk receptor and p75NTR are present, cell survival is often enhanced upon binding of the corresponding neurotrophin. However, other experiments found that a non-corresponding neurotrophin can induce cell death in neurons expressing both a Trk receptor and p75NTR. For example, during normal development, neurons of the sympathetic chain ganglia respond to NGF but not BDNF.

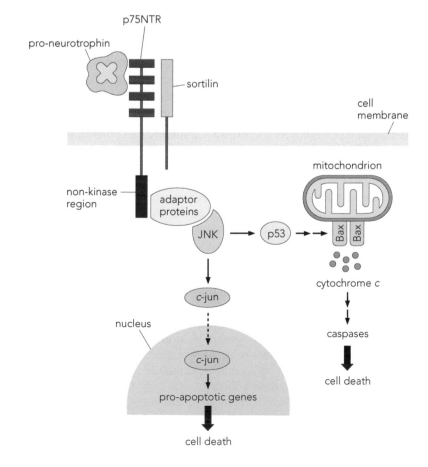

Figure 8.26 Binding of unprocessed pro-neurotrophins to the p75NTR complex activates cell death pathways. When an unprocessed pro-neurotrophin molecule binds to the extracellular region of a p75NTR-sortilin complex, adaptor proteins are recruited to the intracellular, non-kinase region of p75NTR to initiate signaling cascades such as the *c*-jun N-terminal kinase (JNK) pathway. JNK activation can induce cell death via multiple mechanisms. JNK can activate the transcription factor *c*-jun, which then translocates to the nucleus (dashed arrow) to activate pro-apoptotic genes. The JNK pathway can also activate p53, which in turn enhances the activity of Bax. The pro-apoptotic Bax causes cytochrome *c* to be released from the mitochondria, thus allowing it to form an apoptosome and activate the caspase cascade (double arrows) that leads to cell death.

If sympathetic neurons expressing TrkA and p75NTR are exposed to NGF the neurons survive. However, BDNF normally has no effect on these neurons because it does not bind to TrkA. Yet, when BDNF was experimentally overexpressed, more sympathetic neurons died, presumably from increased levels of pro-BDNF acting through the p75NTR receptor.

The combined activities of neurotrophins, pro-neurotrophins, Trk receptors, and p75NTR are clearly important in regulating the balance of cell survival and death *in vivo*. Other signals are also important, such as inhibitor of apoptosis proteins (IAP) found in the cytoplasm. IAPs inhibit the activity of caspase-3 and caspase-9 and thereby promote cell survival. These IAPs can be blocked, in turn, by the mitochondrial protein Smac/Diablo. How IAPs function in different cells remains an area of ongoing research.

Determining the pathways that regulate cell survival and death is not only important for understanding how cell death occurs under normal conditions, but may lead to methods to halt cell death in neurodegenerative disorders. Neurodegenerative diseases include Huntington's disease, Parkinson's disease, spinal muscular atrophy (SMA), amyotrophic lateral sclerosis (ALS), Alzheimer's disease and other dementias (**Box 8.3**), and some forms of narcolepsy (**Box 8.4**). The genetic and environmental triggers that cause these diseases continue to be studied, as highlighted in Boxes 8.3 and 8.4.

Box 8.3 Developing Neuroscientists: The Disease Spectrum of Amyotrophic Lateral Sclerosis and Frontotemporal Dementia

Michelle Johnson, Ph.D.

Michelle Johnson graduated from Oberlin College in 2015 where she majored in neuroscience and biology and minored in chemistry. She then completed her Ph.D. in neuroscience at Emory University working with Thomas Kukar. She is now a postdoctoral research fellow and an Institutional Research and Academic Career Development Awards (IRACDA)* scholar at Tufts University in the Department of Developmental, Molecular, and Chemical Biology.

Development of a nervous system is a complex and highly regulated process. From *in utero* to early adulthood, humans produce billions of neurons. Neurons, which are highly specialized for communication through chemical and electrical signaling, support an organism's ability to breathe, eat, move, and respond to external stimuli. As such, an organism's survival is dependent on the health of its neurons. Neurons are a naturally long-lived cell type with most surviving the length of the organism's lifespan. To survive 80+ years, neurons must withstand multiple toxic insults from injuries, infections, and trauma. It is the combination of a neuron's genes and environment that will enhance or diminish its ability to survive these insults. Regardless, all neurons will experience the insult of time, termed aging.

Healthy neurons must be adaptive cells and respond to various stressors by modulating antioxidant production, protein chaperones, neurotrophic factors, and even firing rates. However, as neurons age they accumulate a wide range of injuries from DNA lesions, increased oxidative stress, and the buildup of damaged organelles and misfolded proteins. These damages are directly associated with the physical deterioration of neurons and their connections, resulting in a breakdown in neuronal activity and function. While a level of deterioration in neurons and their connections is associated with healthy aging, progressive degeneration of neuronal networks is not. Termed neurodegeneration, this pathological process causes specific structures within the nervous system to atrophy or waste away. Neurodegeneration tends to begin in one region of the nervous system. However, overtime it progresses, leading to an overall loss of nervous tissue and function. Given the range of insults neurons accumulate, aging is the primary risk factor for most cases of neurodegeneration. Even so, other risk factors have been described including genetic mutation, chemical exposure, nutrition, head trauma, alcohol consumption, and infection.

Pathogenic neurodegeneration manifests clinically as neurodegenerative disease (ND). There are multiple types of NDs including Alzheimer's disease (AD), Parkinson's disease (PD, see **Box 10.3**), Amyotrophic Lateral Sclerosis (ALS), and Frontotemporal Dementia (FTD). While these diseases all share common hallmarks of disease including increased oxidative stress, proteasomal impairment, mitochondrial dysfunction, and protein aggregation, they differ in their clinical presentation. Interestingly, a subset of

ALS and FTD cases share genetic, neuropathological, and clinical markers of disease even though they target different neuronal populations (motor neurons and cortical neurons, respectively). Specifically, ALS is a progressive ND that affects upper and lower motor neurons, leading to limb spasticity, speech difficulties, muscle wasting, and weakness. In contrast, FTD targets cortical neurons within frontal and temporal lobes of the brain. Accordingly, FTD is characterized by alterations in either behavior and personality (frontal lobe degeneration) or language comprehension and communication (temporal lobe degeneration). In the last 20 years, reports of patients with ALS receiving a secondary diagnosis of FTD has increased substantially. It is now estimated that up to 50% of ALS patients will develop some level of cognitive impairment resembling FTD. Conversely, it is estimated that 30% of patients with a primary diagnosis of FTD will develop some level of motor impairment. This connection has led some researchers to propose that ALS and FTD are not two distinct diseases but instead represent a disease spectrum where pure ALS and pure FTD exist on either end. As such, the underlying triggers of disease for ALS and FTD may be similar.

Data suggests that toxic aggregation of misfolded proteins into intracellular structures termed inclusions (a hallmark of disease for ALS and FTD) can trigger neuropathology. In the mid-2000s researchers found that postmortem brain tissue from a subset of ALS and FTD cases contained intracellular protein aggregates of TAR (transactive response) DNA binding protein 43 (TDP-43). The amount of cytoplasmic protein aggregation was directly correlated with toxicity. TDP-43 is an RNA/DNA binding protein involved in splicing, RNA biogenesis, transcription, and RNA trafficking.

Following these reports, aggregates composed of another RNA/DNA binding protein with similar cellular functions, fused in sarcoma (FUS), were also identified. This sparked the hypothesis that ALS and FTD pathology may be due to dysfunction in RNA biogenesis, trafficking, and metabolism. Understanding why these proteins accumulate in both ALS and FTD is an active area of research.

One known reason that these proteins aggregate is genetic mutation. Causative mutations were identified in the genes that code for the aggregating proteins: *TARDBP* (TAR DNA Binding Protein) and *FUS*. While 50+ mutations have been identified in both *TARDBP* and *FUS*, other mutations have been identified in genes involved in additional cellular functions including autophagy (*SQSTM1*, Sequestosome-1; *VCP*, Valosin Containing Protein; *TBK1*, TANK binding protein 1; *UBQLN2*, Ubiquilin 2) and mitochondria function (*CHCHD10*, Coiled-coil-helix-coiled-coil helix domain containing 10). In fact, the most common cause of familial ALS and FTD are mutations in the gene *C9orf72* (Chromosome 9 Open Reading Frame 72), a gene that codes for a multifunctional protein involved in autophagy, cellular stress response, axon growth, and cytoskeletal formation.

Altogether, the interrelated neuropathology and genetics of ALS and FTD may hold the key to understanding why neurons are degenerating and how neuronal survival can be maintained so patients can be effectively treated.

*Institutional Research and Academic Career Development Awards (IRACDA) are postdoctoral Career Development Awards (K12) sponsored by the National Institutes of Health, National Institute of General Medical Sciences (NIGMs).

Box 8.4 A Neurologist's Perspective: Narcolepsy: A Neurodegenerative Disorder of the Sleep-Wake Cycle

Erica Grazioli, D.O.

Dr. Erica Grazioli majored in biology at Gannon University before attending medical school at the Lake Erie College of Osteopathic Medicine. She completed her neurology residency at the University of Pittsburgh Medical Center Hamot and a fellowship in multiple sclerosis at the University of Buffalo Jacobs Neurological Institute.

Normal sleep and wakefulness are controlled by interactions of many brain regions, neuropeptides, hormones, and the environment. Narcolepsy is a rare disorder that results from a loss of neurons in the lateral hypothalamus. The lost neurons normally produce the neuropeptide orexin, also called hypocretin, as a product of the prepro-orexin gene. Orexin has an excitatory effect in the central nervous system and increases the activity of brain regions involved in wakefulness such as the raphe nuclei, locus coeruleus, and tuberomammillary nucleus. Orexin typically promotes and stabilizes wakefulness.

When orexin-producing neurons are lost in narcolepsy, elements of sleep intrude during wakefulness and elements of wakefulness intrude into sleep. This leads to excessive daytime sleepiness and disturbed nighttime sleep. The degree of sleepiness in narcolepsy can be so severe that one dozes off with little or no warning leading to "sleep attacks." Other symptoms seen in narcolepsy include hallucinations as one is falling asleep or waking up, sleep paralysis, and cataplexy, a loss of muscle tone during waking hours often triggered by strong emotion.

During normal rapid eye movement (REM) sleep we have loss of muscle tone which keeps us from acting out our dreams. In episodes of sleep paralysis, this REM atonia carries over briefly into the wake period so that a person is waking but is unable to move for up to several minutes. Cataplexy too is a manifestation of the normal muscle atonia of sleep intruding into waking hours. Cataplexy often manifests as partial weakness affecting the facial muscles causing ptosis (drooping eyelid) or interruption of facial expressions, such as smiling. Cataplexy can affect larger muscle groups as well, leading one to collapse from leg weakness.

There are two types of narcolepsy. Narcolepsy type 1 (NT1) is narcolepsy with cataplexy. Narcolepsy type 2 (NT2) is narcolepsy without cataplexy. Evaluation of narcolepsy typically includes a type of sleep study called a multiple sleep latency test or MSLT. The MSLT is a daytime test that follows an overnight polysomnogram ("sleep study") and consists of a series of scheduled opportunities to nap. During each nap period the sleep latency (time to sleep onset) and any transitions into REM sleep are recorded.

The diagnosis of NT1 (narcolepsy with cataplexy) requires both of the following:

Daily periods of irrepressible need to sleep or daytime lapses into sleep occurring for at least three months and

One or both of the following:

Episodes of cataplexy and a mean sleep latency of less than or equal to ($\leq$) 8 minutes and two or more sleep onset REM sleep periods (SOREMPs) on MSLT. A SOREMP (within 15 minutes of sleep onset) on a preceding nocturnal polysomnogram may replace one of the SOREMPs on the MSLT.

Low cerebrospinal fluid (CSF) orexin concentration

NT2 can be diagnosed when a person presents with other symptoms of narcolepsy but does not have cataplexy or low CSF orexin. The diagnosis of NT2 requires an overnight polysomnogram followed by an MSLT that demonstrates a mean sleep latency of $\leq$ 8 minutes and two or more SOREMPs.

The development of narcolepsy has been liked to both genetic and autoimmune factors. Genes of the HLA (Human leukocyte antigen) gene complex found on chromosome 6 are associated with NT1, specifically, the DQB1*0602 haplotype (group of alleles inherited from a parent). This haplotype is presents in 95% of patients with NT1. However, only about 25% of monozygotic twins are concordant for narcolepsy, suggesting an important role for environmental and autoimmune factors in addition to genetics. Furthermore, DQB1*0602 is found in approximately 20–30% of the general population, but only a small percentage of those with the allele develop narcolepsy.

The role of environmental triggers is supported by the observation that the onset of narcolepsy appears greatest in the spring, suggesting an immune process triggered by winter infection. Furthermore, individuals in several European countries developed NT1 soon after receiving Pandemrix, a 2009 H1N1 influenza vaccine. The risk to develop narcolepsy following this vaccine was greatest in those who also had DQB1*0602. However, the exact mechanism underlying this increased risk of developing NT1 remains unclear. Other H1N1 vaccines, including those using the same adjuvant (a substance that increases immune response), did not increase the incidence of NT1. One proposed mechanism is that there was a secondary cross-reactivity between the human orexin receptor and influenza nucleoprotein A in the vaccine. The increased incidence of narcolepsy in specific European countries in 2010–2011 led to increased research on NT1 and NT2 that continues today, with studies focusing on the mechanisms governing neurodegeneration in specific neurons in the lateral hypothalamus, as well as the interactions of the autoimmune response, genetic predisposition, and environmental triggers in causing narcolepsy.

SUMMARY

This chapter reviewed the role of target-derived growth factors and the mechanisms by which they and other growth factors regulate survival and PCD during development. Over the long and interesting history of growth factor research, scientists have come to understand that growth factors not only promote survival by activating specific intracellular pathways but also inhibit pathways that promote cell death. A delicate balance of growth factor production, receptor expression, and signal transduction cascades determines whether a neuron lives or dies. To ensure proper formation of the nervous system, all these components must be carefully regulated to make sure the right cells are living or dying at the correct time of development.

As illustrated in many examples in this chapter, the importance of trophic factors extends beyond development. Additional roles for the neurotrophins and other growth factors continue to be discovered, and evidence now shows that such factors play multiple roles, such as regulating the expression of neurotransmitters and ion channels at various times during development and adulthood. While their role in the CNS remains less clear than in the PNS, it appears that neurotrophins are often involved in regulating aspects of cellular differentiation, synaptic plasticity, and the maintenance of synaptic connections. These latter topics are discussed in Chapter 10. Many of the signaling pathways used by neurotrophic factors may also be adapted for other developmental events. For example, in addition to interactions with sortilin, p75NTR can also interact with other co-receptors such as those for NOGO (neurite outgrowth inhibitor, also known as reticulon 4) to influence neural crest cell migration.

FURTHER READING

Adler R, Landa KB, Manthorpe M & Varon S (1979) Cholinergic neuronotrophic factors: Intraocular distribution of trophic activity for ciliary neurons. *Science* 204(4400):1434–1436.

Andreska T, Lüningschrör P & Sendtner M (2020) Regulation of TrkB cell surface expression-a mechanism for modulation of neuronal responsiveness to brain-derived neurotrophic factor. *Cell Tissue Res* 382(1):5–14.

Barde YA, Edgar D & Thoenen H (1982) Purification of a new neurotrophic factor from mammalian brain. *EMBO J* 1(5):549–553.

Barde YA, Lindsay RM, Monard D & Thoenen H (1978) New factor released by cultured glioma cells supporting survival and growth of sensory neurons. *Nature* 274:818.

Bothwell M (1995) Functional interactions of neurotrophins and neurotrophin receptors. *Annu Rev Neurosci* 18:223–253.

Bredesen DE & Rabizadeh S (1997) p75NTR and apoptosis: Trk-dependent and Trk-independent effects. *Trends Neurosci* 20:287–290.

Bueker ED (1948) Implantation of tumors in the hind limb field of the embryonic chick and the developmental response of the lumbosacral nervous system. *Anat Rec* 102(3):369–389.

Buss RR, Sun W & Oppenheim RW (2006) Adaptive roles of programmed cell death during nervous system development. *Annu Rev Neurosci* 29:1–35.

Campenot RB (2009) NGF uptake and retrograde signaling mechanisms in sympathetic neurons in compartmented cultures. *Results Probl Cell Differ* 48:141–158.

Chao MV, Bothwell MA, Ross AH, et al. (1986) Gene transfer and molecular cloning of the human NGF receptor. *Science* 232(4749):518–521.

Cohen S (2004) Origins of growth factors: NGF and EGF. *Ann NY Acad Sci* 1038:98–102.

Curtis R & DiStefano PS (1994) Neurotropic factors, retrograde axonal transport and cell signalling. *Trends Cell Biol* 4(11):383–386.

Davies AM (2003) Regulation of neuronal survival and death by extracellular signals during development. *EMBO J* 22(11):2537–2545.

Davis S, Aldrich TH, Valenzuela DM, et al. (1991) The receptor for ciliary neurotrophic factor. *Science* 253(5015):59–63.

Dekkers MP, Nikoletopoulou V & Barde YA (2013) Cell biology in neuroscience: Death of developing neurons: New insights and implications for connectivity. *J Cell Biol* 203(3):385–393.

Frade JM, Rodríguez-Tébar A & Barde YA (1996) Induction of cell death by endogenous nerve growth factor through its p75 receptor. *Nature* 383:166–168.

Hallberg P, Smedje H, Eriksson N, et al. (2019) Pandemrix-induced narcolepsy is associated with genes related to immunity and neuronal survival. *EBioMed* 40: 595–604.

Hamburger V (1939) Motor and sensory hyperplasia following limb bud transplantations in chick embryos. *Physiol Zool* 12:268–284.

Hashino E, Shero M, Junghans D, et al. (2001) GDNF and neurturin are target-derived factors essential for cranial parasympathetic neuron development. *Development* 128:3773–3782.

Hendry IA, Stockel K, Thoenen H & Iversen LL (1974) The retrograde axonal transport of nerve growth factor. *Brain Res* 68(1):103–121.

Johnson D, Lanahan A, Buck CR, et al. (1986) Expression and structure of the human NGF receptor. *Cell* 47(4):545–554.

Johnson MA, Deng Q, Taylor G, et al. (2020) Divergent FUS phosphorylation in primate and mouse cells following double-strand DNA damage. *Neurobiol Dis* 146:105085.

Kaplan DR, Hempstead BL, Martin-Zanca D, et al. (1991) The trk proto-oncogene product: A signal transducing receptor for nerve growth factor. *Science* 252(5005):554–558.

Kristiansen M & Ham J (2014) Programmed cell death during neuronal development: The sympathetic neuron model. *Cell Death Differ* 21(7):1025–1035.

Leibrock J, Lottspeich F, Hohn A, et al. (1989) Molecular cloning and expression of brain-derived neurotrophic factor. *Nature* 14:149–152.

Levi-Montalcini R (1975) NGF: An uncharted route. In *The Neurosciences: Paths to Discovery* (Worden FG, Swazney JP & Adelman G eds), pp. 245–265. MIT Press.

Levi-Montalcini R, Meyer H & Hamburger V (1954) In vitro experiments on the effects of mouse sarcomas 180 and 37 on the spinal and sympathetic ganglia of the chick embryo. *Cancer Res* 14(1):49–57.

Lindsay RM, Alderson RF, Friedman B, et al. (1991) The neurotrophin family of NGF-related neurotrophic factors. *Restor Neurol Neurosci* 2(4):211–220.

Martin DP, Schmidt RE, DiStefano PS, et al. (1988) Inhibitors of protein synthesis and RNA synthesis prevent neuronal death caused by nerve growth factor deprivation. *J Cell Biol* 106(3):829–844.

Martin-Zanca D, Hughes SH & Barbacid M (1986) A human oncogene formed by the fusion of truncated tropomyosin and protein tyrosine kinase sequences. *Nature* 319(6056):743–748.

Oppenheim RW (1996) Neurotrophic survival molecules for motoneurons: An embarrassment of riches. *Neuron* 17(2):195–197.

Pettmann B & Henderson CE (1998) Neuronal cell death. *Neuron* 20(4):633–647.

Radeke MJ, Misko TP, Hsu C, et al. (1987) Gene transfer and molecular cloning of the rat nerve growth factor receptor. *Nature* 325(6105):593–597.

Reichardt LF (2006) Neurotrophin-regulated signalling pathways. *Philos Trans R Soc Lond B Biol Sci* 361(1473):1545–1564.

Root J, Merino P, Nuckols A, Johnson M & Kukar T (2021) Lysosome dysfunction as a cause of neurodegenerative diseases: Lessons from frontotemporal dementia and amyotrophic lateral sclerosis. *Neurobiol Dis* 154:105360.

Sarkanen T, Alakuijala A, Julkunen I & Partinen M (2018) Narcolepsy associated with pandemrix vaccine. *Curr Neurol Neurosci Rep* 18(7):43.

Sendtner M, Kreutzberg GW & Thoenen H (1990) Ciliary neurotrophic factor prevents the degeneration of motor neurons after axotomy. *Nature* 345(6274):440–441.

Sendtner M, Schmalbruch H, Stockli KA, et al. (1992) Ciliary neurotrophic factor prevents degeneration of motor neurons in mouse mutant progressive motor neuronopathy. *Nature* 358(6386):502–504.

Skaper SD (2008) The biology of neurotrophins, signalling pathways, and functional peptide mimetics of neurotrophins and their receptors. *CNS Neurol Disord Drug Targets* 7(1):46–62.

Stahl N & Yancopoulos GD (1994) The tripartite CNTF receptor complex: Activation and signaling involves components shared with other cytokines. *J Neurobiol* 25(11):1454–1466.

Wislet S, Vandervelden G & Rogister B (2018) From neural crest development to cancer and vice versa: How p75[NTR] and (Pro)neurotrophins could act on cell migration and invasion? *Front Molec Neurosci* 11:244.

Zhang N, Kisiswa L, Ramanujan A, et al. (2021) Structural basis of NF-κB signaling by the p75 neurotrophin receptor interaction with adaptor protein TRADD through their respective death domains. *J Biol Chem* 297(2):100916.

Zigmond RE (2011) gp130 cytokines are positive signals triggering changes in gene expression and axon outgrowth in peripheral neurons following injury. *Front Mol Neurosci* 4:62.

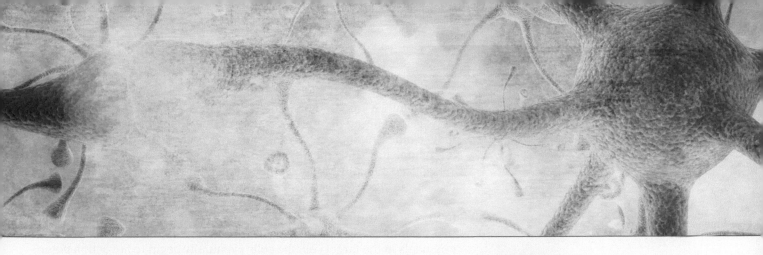

Synaptic Formation and Reorganization Part I: The Neuromuscular Junction

This chapter introduces common features of two essential aspects of neurodevelopment: the process of **synaptogenesis**—the formation of new synaptic contacts in the nervous system—and the process of **synaptic reorganization**—the subsequent strengthening or loss of a subset of these connections. The mechanisms regulating the formation and reorganization of the vertebrate **neuromuscular junction** (**NMJ**), the most studied of all the neural connections, are then detailed. Mechanisms underlying synapse formation and reorganization in the central nervous system (CNS) follow in Chapter 10.

The study of synapses, including how they initially form and later reorganize, has a long history—one often marked by lively debates and strong differences of opinion. As noted in previous chapters, neurobiologists in the late 1800s debated whether neurons communicated through a syncytial network or through connections between individual cells. Evidence ultimately demonstrated that neurons communicated through small spaces—connections now termed **synapses**. In the first half of the twentieth century, scientists engaged in another debate, this one regarding whether the communication at synapses occurred primarily through chemical or electrical signals. These differing opinions were often referred to as the "war of the soups and the sparks," with the "soups" referring to chemical signals and the "sparks" to electrical signals. As these two hypotheses were tested, it was eventually determined that most synapses utilize chemical signals in the form of **neurotransmitters** to mediate neural communication. Chapters 9 and 10 therefore focus on the development of structures associated with **chemical synapses**. Mechanisms regulating the development of electrical synapses, such as those formed by **gap junctions**, also remain an active area of study, and at some synapses, chemical and electrical signals work together to optimize neuronal communication.

Creating a functional chemical synapse is a dynamic process that takes place over an extended period and requires several steps. For example, an extending, motile growth cone must transform into a nerve terminal capable of releasing a specific neurotransmitter, and the target cell must produce the corresponding neurotransmitter receptors. Both cells also produce the many specialized proteins needed for rapid synaptic transmission. All these events occur during synaptogenesis.

DOI: 10.1201/9781003166078-9

Perhaps surprisingly, after all the effort to convert each synaptic partner into a highly differentiated cell, a subset of synaptic connections is lost during the normal course of development. Thus, rather than induce differentiation in only the subset of synaptic connections required for neural function, the nervous system instead over-produces highly specialized synaptic contacts, then eliminates a portion of them. As will be seen in the examples from the NMJ and CNS neurons, the elimination of connections depends often, though not entirely, on neural activity. The stabilization of synapses occurs in part by maintaining synapses that form functional partners. As the innervating neuron initiates action potentials in the target cell, the cells eventually begin to fire action potentials in synchrony with one another. The phrase "cells that fire together, wire together" describes one common mechanism by which cells maintain selected synaptic connections. Any target cells that do not receive sufficient input, or those that fire asynchronously with the innervating neuron, will eventually lose the synaptic connection; thus, those that are "out of sync, lose their link." This concept was originally introduced by Donald Hebb to explain synaptic changes that occur during learning or memory, and such synaptic changes are therefore referred to as Hebbian modifications or Hebbian plasticity.

There are many shared mechanisms governing synaptogenesis and reorganization in the NMJ and CNS, as well as several aspects unique to each synaptic type. This chapter begins with an overview of the synaptic structures found in synapses throughout the nervous system. Next, the specific elements found in the NMJ are described, followed by descriptions of key experiments that led to the current models of how connections at the NMJ are formed and reorganized. As with other aspects of nervous system development, many of the molecules and signaling pathways identified in synaptic development are conserved across vertebrate and invertebrate animal models.

CHEMICAL SYNAPSE DEVELOPMENT IN THE PERIPHERAL AND CENTRAL NERVOUS SYSTEMS

Numerous studies over the past century have demonstrated how chemical synaptic transmission occurs between two neurons or between a neuron and a specialized cell, such as a muscle fiber or sensory cell. In most chemical synapses, an action potential signals a **presynaptic nerve terminal** to release a neurotransmitter that diffuses across the synaptic space to bind to corresponding receptors on the surface of the **postsynaptic partner cell**, thereby initiating a response in that cell. Charles Sherrington first used the term "synapse" in 1887 to describe the small gap between communicating nerve cells. However, the chemical synapse was not visualized conclusively until the mid-1950s, when techniques for preparing tissues for electron microscopy became sufficiently advanced to provide images of nerve cells. In 1955, George Palade and Sanford Palay and Eduardo De Robertis and H. Stanley Bennett proved that neurons are separated by a well-defined **synaptic cleft**. These studies were also the first to reveal the ultrastructural organization of the synaptic elements that are studied in such detail today.

By the late twentieth century, further advances in microscopy—particularly in the areas of time-lapse and fluorescence imaging—and new methods in cellular and molecular biology led to the identification of hundreds of intracellular synaptic specializations. The intricate components of synaptic structures are currently studied at levels of detail unimaginable even 20 years ago.

Reciprocal Signaling Leads to the Development of Unique Synaptic Elements in Presynaptic and Postsynaptic Cells

As an extending nerve fiber approaches a target cell, the tip transforms from a motile growth cone to a presynaptic nerve terminal. The presynaptic terminal must now form new structures, such as **synaptic vesicles** that contain the correct neurotransmitter, protein complexes called **active zones** that direct the synaptic vesicles to precise locations along the nerve terminal membrane, and **voltage-dependent calcium (Ca^{2+}) channels** that open in response to an action potential and initiate the release of the neurotransmitter into the synaptic cleft (**Figure 9.1**).

The postsynaptic cell develops its own specialized elements, including **neurotransmitter receptors** that selectively bind neurotransmitters and **scaffold proteins** to hold the receptors in place directly across from the active zones of the presynaptic terminal (Figure 9.1). Excitatory neurons in the CNS are further characterized by an electron-dense organelle called the **postsynaptic density** (**PSD**) that helps anchor postsynaptic proteins across from the presynaptic terminal. Both pre- and postsynaptic partners produce various **adhesion molecules** to stabilize the newly formed synaptic connections. The precise alignment of pre- and postsynaptic elements ensures

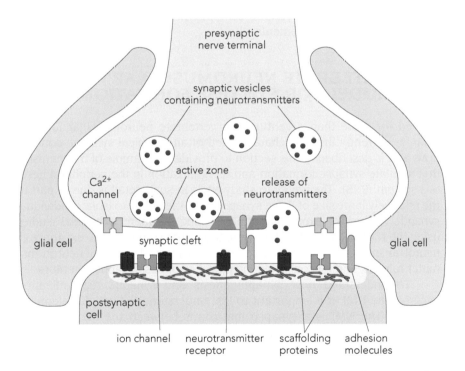

Figure 9.1 Pre- and postsynaptic cells have specialized synaptic elements and are surrounded by glial cell processes. Presynaptic nerve terminals are characterized by an abundance of synaptic vesicles. The synaptic vesicles line up at the active zones—areas of proteins along the presynaptic membrane adjacent to the postsynaptic cell. Voltage-dependent Ca^{2+} channels open in response to an action potential and initiate the release of neurotransmitter into the synaptic cleft. The neurotransmitter molecules bind to neurotransmitter receptors on the postsynaptic cell. Receptors are aligned directly across from the active zones so that neurotransmitter reaches the receptors quickly. Also located in postsynaptic cells are ion channels that open in response to neurotransmitter binding and initiate the action potential in that cell. Scaffold proteins help anchor the postsynaptic receptors and ion channels across from the presynaptic nerve terminal. Various adhesion molecules found in pre- and postsynaptic cells initiate or stabilize the synaptic contacts. Finally, glial cells surround the pre- and postsynaptic contacts to provide signals to maintain the synaptic elements and to remove cellular debris left by retracting nerve terminals. These basic synaptic elements are found at synapses throughout the nervous system.

neurotransmitters released by the synaptic vesicles reach the postsynaptic receptors as quickly as possible.

Vertebrate synapses often include glial cells. In such cases, the synapse is referred to as a **tripartite complex** because it is comprised of a presynaptic cell, a postsynaptic cell, and a glial cell (Figure 9.1). At many synapses in the central and peripheral nervous systems, the glial cells extend cellular processes that surround pre- and postsynaptic cells to help maintain synaptic elements and remove cellular debris generated during synapse elimination.

As scientists documented how synaptic elements formed during the early stages of synaptogenesis, several noted that some postsynaptic specializations began to form even before the presynaptic terminal arrived. These observations suggested that the postsynaptic cell differentiates first to provide signals needed for the differentiation of and connectivity with the presynaptic cell. However, other studies reported that the presynaptic nerve terminal produces some synaptic elements prior to contact with the postsynaptic cell, suggesting that presynaptic differentiation was independent of postsynaptic signals. In many cases, it was simply unclear whether pre- or postsynaptic elements were the first to arise. It now appears that both pre- and postsynaptic cells begin to express immature synaptic elements prior to contact, but only differentiate organized, mature synaptic elements in response to signals obtained from the synaptic partner. Therefore, both pre- and postsynaptic cells provide reciprocal cues to induce maturation of synaptic elements.

THE VERTEBRATE NEUROMUSCULAR JUNCTION AS A MODEL FOR SYNAPSE FORMATION

Studied for more than a century, the vertebrate neuromuscular junction (NMJ) is currently the best characterized of all chemical synaptic connections and is described in this section to provide an example of mechanisms that regulate synapse formation and reorganization in the peripheral nervous system (PNS). The ability to study NMJs in such detail is due in part to the relatively large size of these synapses and the ability to isolate individual synaptic regions for experimentation. Among the many influential studies using the NMJ as a model system were those that (1) proved the existence of neurotransmitters, (2) identified synaptic vesicles as the sites of neurotransmitter release, and (3) confirmed the presence of postsynaptic receptors.

To understand how synaptogenesis and synaptic reorganization occur at the NMJ, it is important to first understand the cellular elements involved. The NMJ is a synapse that forms between the axon terminal branches of presynaptic motor neuron and a postsynaptic skeletal muscle fiber. Skeletal muscles, comprised of multiple muscle fibers, are attached to bones, often via connective tissue called tendons, and are under voluntary control. Depending on the size of the particular skeletal muscle, a motor neuron can innervate up to several hundred muscle fibers by extending multiple nerve terminal branches (**Figure 9.2**). The number of muscle fibers that a single motor neuron innervates is called a **motor unit**, because stimulation of that neuron leads to contraction of all the muscle fibers it innervates. Each terminal branch contacts a specialized region on the skeletal muscle called a **motor endplate**, an oval-shaped region of a muscle cell that appears slightly elevated above the rest of the cell surface at the time of innervation. Within the endplate region, the terminal branches end as **synaptic boutons** (terminal end bulbs), where they form a synaptic connection with the muscle fiber. Myelinating Schwann cells surround the axons of motor neurons, while specialized perisynaptic Schwann cells surround the terminal branches (Figure 9.2).

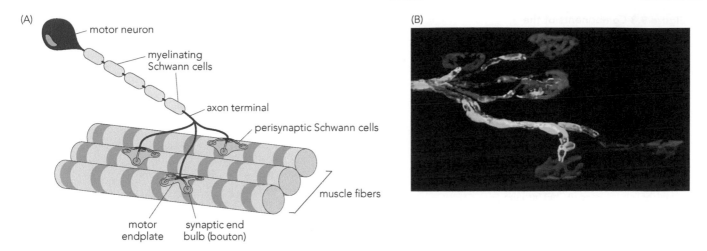

Figure 9.2 Vertebrate motor neurons innervate multiple skeletal muscle fibers. Motor neurons form synaptic connections with skeletal muscle fibers. The sites of synaptic contact are called neuromuscular junctions. (A) In adult vertebrates, each motor neuron extends a myelinated axon that branches to establish contact with several skeletal muscle fibers. Each terminal branch establishes synaptic contacts at a specialized region on the muscle surface called the motor endplate. The synapses form between the synaptic end bulbs (boutons) and the muscle cell membrane in the endplate. Terminal branches are unmyelinated but are covered by specialized Schwann cells called perisynaptic Schwann cells. (B) Elements of the NMJ of the mouse soleus muscle are identified with fluorescent labels that detect region-specific proteins. Schwann cells around the motor axon are labeled in green (S100β). The motor nerve is labeled in red (neurofilament), and motor end plate in blue (α bungatotoxin that identifies acetylcholine receptors on muscle fibers). [(B), Courtesy of Jason Newbern.]

At the NMJ, the Presynaptic Motor Axon Releases Acetylcholine to Depolarize the Postsynaptic Muscle Cell

Acetylcholine (ACh) is the neurotransmitter utilized by vertebrate motor neurons. ACh is synthesized in the presynaptic neurons by the enzyme **choline acetyltransferase (ChAT)**, then packaged into synaptic vesicles that line up at the active zones. The active zones are protein complexes located in discrete regions of the presynaptic terminal that help anchor the synaptic vesicles directly across from the nicotinic **acetylcholine receptors (AChRs)** located on the postsynaptic skeletal muscle cell. These ionotropic AChRs are ligand-gated cation channels that cluster at the crests of the **postjunctional folds**—invaginations in the skeletal muscle cell membrane (**Figure 9.3**). The localization of the AChRs to the crests of the postjunctional folds ensures that the released ACh reaches and binds to the receptors rapidly.

In response to an action potential initiated in a motor neuron, voltage-gated Ca^{2+} channels accumulated near the presynaptic active zones open, leading to an influx of Ca^{2+} ions. Ca^{2+} causes the synaptic vesicles to fuse with the neuronal membrane and release ACh into the synaptic cleft. The binding of ACh to the AChRs leads to an influx of sodium (Na^+) and other cations into the postsynaptic muscle cell and the subsequent rapid activation of voltage-gated Na^+ channels located in the troughs of the postjunctional folds (Figure 9.3). As these channels open, more sodium flows in, causing the muscle to depolarize. This postsynaptic action potential in the muscle fiber is called an **endplate potential (EPP)**. The EPP then propagates away from the endplate in both directions to initiate muscle contraction.

Each muscle fiber and myelinated axon is surrounded by a basal lamina. The synaptic cleft, however, contains the **synaptic basal lamina**, a unique structure with a different protein composition. Among the proteins found in the synaptic basal lamina is **acetylcholine esterase (AChE)**, the enzyme that breaks down ACh. Muscle cells secrete the predominant form of AChE found at the NMJ, and this enzyme becomes concentrated in the synaptic basal lamina. Because of this, any excess neurotransmitter

Figure 9.3 Components of the mature neuromuscular junction (NMJ). In the presynaptic motor neuron, acetylcholine (ACh) is synthesized by choline acetyltransferase (ChAT). The ACh is then packaged into synaptic vesicles that line up along the active zones in the presynaptic nerve terminal. The active zone proteins are concentrated opposite the postjunctional folds of the postsynaptic muscle cell. The crests of the folds contain the acetylcholine receptors (AChR), while the troughs contain voltage-gated sodium (Na^+) channels. The postsynaptic cell releases acetylcholine esterase (AChE) into the synaptic cleft, where it breaks down excess ACh. The precise arrangement of these pre- and postsynaptic elements ensures rapid and reliable neurotransmission at the NMJ. Perisynaptic Schwann cells surround the NMJ, and basal lamina covers the presynaptic terminal and muscle surface. In contrast, the synaptic basal lamina is restricted to the synaptic cleft.

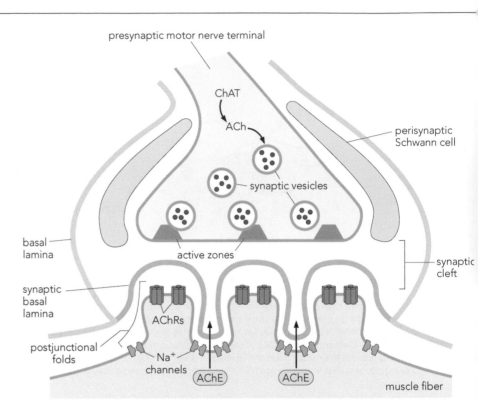

is removed quickly to prevent signaling beyond what is required to initiate muscle contraction. Several assays that detect ChAT, ACh, AChRs, and AChE both *in vitro* and *in vivo* have been extremely useful in identifying many of the developmental events that take place during synaptogenesis at the NMJ.

The Distribution of AChRs Is Mapped in Developing Muscle Fibers

Skeletal muscle tissue arises from the fusion of mesoderm-derived, spindle-shaped, mononucleated **myoblasts** (**Figure 9.4**A). During embryonic development, several myoblasts fuse to form a primary myotube that eventually extends the length of the future muscle (Figure 9.4B, C). Secondary myotubes arise soon after and fuse to the existing primary myotubes (Figure 9.4D). The myotubes continue to develop and ultimately form a mature **muscle fiber** (or **myofiber**), which is a long, cylindrical, multinucleated cell. Each myofiber is multinucleated because it arises from the fusion of the individual myoblasts (Figure 9.4E). The nuclei of these myoblasts are visualized at intervals of about 10 μm along the length of the muscle fiber. Mature human muscle fibers can be up to 30 cm long and 10 to 100 μm wide. Bundles of muscle fibers make up the mature skeletal muscle (Figure 9.4F).

As myotubes start to form, they begin to express the genes encoding AChRs. Each myotube expresses AChRs across the cell surface in a generally uniform pattern. However, some newly formed AChRs are located in small aggregates and many are found near the middle third of the cell surface—the area of the future motor endplate. This is a form of prepatterning in which an accumulation of centrally located AChRs is present prior to innervation. The receptors are still able to migrate along the muscle surface, however, and they are not restricted to the site where they first accumulate (**Figure 9.5**A).

The prepatterning of AChRs is modified once the motor nerve terminals start to contact their target myotubes. Prior to contact, motor nerve terminals begin to release some ACh from the few synaptic vesicles

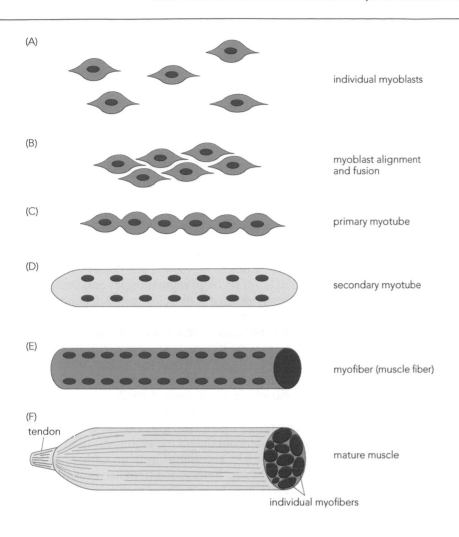

Figure 9.4 The development of skeletal muscle. (A) Individual, mononucleated myoblasts are derived from mesoderm. (B) The individual myoblasts begin to align with one another and fuse together. (C) The fusion of myoblasts continues until a primary myotube extends the length of the future muscle. (D) Secondary myotubes arise soon after the formation of primary myotubes, fusing with the primary myotubes. (E) The secondary myotubes ultimately form a cylindrical, multinucleated myofiber, also called a muscle fiber. The nuclei from the original myoblasts are located at intervals along the length of the muscle fiber. (F) Bundles of individual myofibers form a mature skeletal muscle that attaches to a bone, usually via a connective tissue bundle called a tendon. In this figure, one end of the muscle is cut away to show the organization of the myofibers.

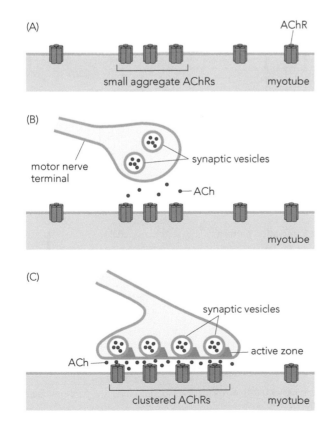

Figure 9.5 Acetylcholine receptors form prior to nerve contact and exhibit prepatterning. (A) As myotubes begin to form, they express AChRs. The AChRs are distributed in a generally uniform pattern along the length of the myotube surface, although some small aggregates of AChRs are noted. Many of these AChRs are located near the middle one-third of the muscle fiber—the site of the future endplate. This organization represents a form of prepatterning of AChRs prior to innervation. (B) As the immature motor nerve terminal approaches the myotube, the few synaptic vesicles that are present begin to release ACh that binds to nearby AChRs. (C) Once the nerve terminal reaches the myotube, pre- and postsynaptic elements begin to mature. Synaptic vesicles increase in number and line up along the newly formed active zones. The release of ACh leads to the migration of many postsynaptic AChRs, causing them to cluster directly across from the nerve terminal. Those that are not associated with the nerve terminal are eventually lost.

present (Figure 9.5B). Once the nerve terminal contacts a myotube, there is an increase in ChAT expression in the motor neurons, indicating a corresponding increase in ACh production. The ACh is packaged into synaptic vesicles, which have increased in number, and line up along the newly developed active zones to be released at the sites of muscle cell contact. As the presynaptic terminal reaches the myotube, the postsynaptic AChRs begin to migrate laterally in the muscle cell membrane to cluster directly across from the innervating nerve terminal (Figure 9.5C). Most of the remaining AChRs that lie outside the area of synaptic contact, in the extrasynaptic regions, are eventually lost.

The Density of Innervation to Muscle Fibers Changes during Vertebrate Development

Developmental changes in muscle innervation patterns were first documented in electrophysiological studies. In 1970, Paul Redfern recorded EPPs from early postnatal and adult rat diaphragm muscles and unexpectedly found that the EPPs in early postnatal muscles varied depending on the stimulus strength applied. In contrast, the adult EPPs were constant no matter what input stimulus was provided. These findings led to the hypothesis that early postnatal muscles were innervated by more than one motor axon, and the changes in EPPs correlated with how many fibers were available for stimulation. Subsequent investigations supported this hypothesis. Detailed histological and electrophysiological studies in rodents by Michael Brown and colleagues determined that muscle fibers are initially innervated by more than one motor axon, then gradually over the course of the first three postnatal weeks, all but one of the motor axons are eliminated (**Figure 9.6**A, B). For example, recordings from an early postnatal muscle fiber revealed a double EPP and innervation from two motor axons, whereas a single EPP and innervation from a single axon were characteristic of an adult muscle fiber (Figure 9.6C, D).

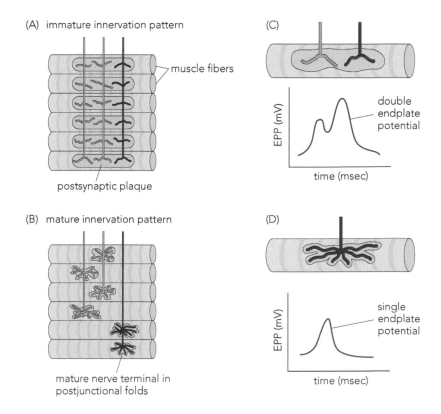

Figure 9.6 The number of motor nerve terminals that contact a muscle fiber is modified during development. Motor neurons branch to innervate multiple muscle fibers. (A) During the early stages of NMJ development, each muscle fiber is multiply innervated. The terminal branches from the different motor neurons intermingle with each other at a single plaque-like area near the central region of the muscle fiber. (B) A mature muscle fiber is innervated by only one motor neuron, but each motor neuron continues to innervate more than one muscle fiber. The site of innervation is restricted to the central region of the muscle fiber. (C) Electrophysiological recordings revealed that when early postnatal muscle fibers were innervated by two motor axons, a double endplate potential (EPP) was recorded. (D) In contrast, adult muscle fibers innervated by a single motor axon had a single endplate potential. [(C and D), Adapted from Brown M, Keynes R & Lumsden A [2001] *The Developing Brain*, Oxford University Press.]

(A) immature innervation pattern

muscle fibers

postsynaptic plaque

(B) mature innervation pattern

mature nerve terminal in postjunctional folds

(C)

EPP (mV)

double endplate potential

time (msec)

(D)

EPP (mV)

single endplate potential

time (msec)

In rodents, multiply innervated sites begin to form in most skeletal muscles around embryonic day 14, approximately one week before birth. During the first few postnatal weeks, the density of innervation decreases and each muscle fiber becomes innervated by a single motor neuron (Figure 9.6B). However, each adult vertebrate motor neuron branches to innervate more than one muscle fiber (Figure 9.6B).

At the postsynaptic site, morphological changes occur as the NMJ matures. Initially the terminal branches from different motor neurons overlap with one another near the central region of the muscle fiber (**Figure 9.7**). At this stage the morphological appearance of the postsynaptic site resembles a flattened plaque-like structure. As the system matures, the surface of the muscle fiber begins to invaginate, creating the postjunctional folds. The folds continue to elaborate until they establish what is often described as a "pretzel-like" appearance. As the postjunctional folds are maturing, excess nerve terminal contacts withdraw. The terminals of the remaining motor neurons elaborate to align with the expanding postjunctional folds and take over regions formerly occupied by the lost terminals. At the end of this maturation process, a single motor neuron innervates the muscle fiber near its central region. The area of synaptic contact represents about 0.1% of the available muscle surface. Researchers wondered how the NMJ becomes restricted to such a small region of the muscle fiber and many suspected that the muscle must produce signals that restrict where a NMJ forms.

The Synaptic Basal Lamina Is a Site of NMJ Organizing Signals

As mentioned in Chapter 7, John Langley first proposed the idea that "chemiotactic" cues from target cells induce formation of synaptic connections. His observations in 1895 were based on studies of regenerating mammalian sympathetic neurons that re-innervated their original target cells in the superior cervical ganglion. In 1907, J. Fernando Tello discovered a similar mechanism in studies of the mammalian NMJ. When Tello damaged the axons of motor neurons, he noted that the regenerated nerve fibers grew back to contact the same sites on the muscle as the original axons. These

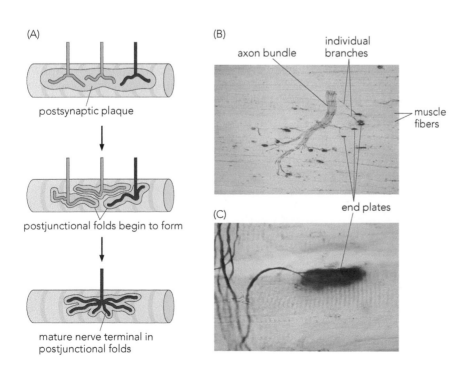

(A)

postsynaptic plaque

postjunctional folds begin to form

mature nerve terminal in postjunctional folds

(B)

axon bundle

individual branches

muscle fibers

end plates

(C)

Figure 9.7 Changes occur at the postsynaptic site as the NMJ matures. (A) The developmental sequence of changes in the nerve terminal and postjunctional folds. Initially all the nerve terminals intermingle in the plaque-like region on the muscle fiber. Postjunctional folds begin to invaginate in the muscle fiber and motor nerve terminals become segregated from one another within the expanding postjunctional folds. As the postjunctional folds mature, excess nerve contacts are lost. The postjunctional folds continue to expand and the remaining nerve terminals extend to align with the unoccupied areas in the postjunctional folds. (B) Micrograph revealing the mature innervation pattern in the rat NMJ. The individual axons leave the axonal bundle and send out terminal branches that contact a single site on a muscle fiber. The endplate region is stained to reveal the sites of contact along the individual muscle fibers. (C) An enlarged image of the endplate contacted by a single terminal branch. [(B and C), From Gorio A, Marini P & Zanoni R [1983] *Neurosci* 8:417–429.]

observations suggested that muscle-derived signals directed synapse formation to a specific region of the muscle surface, at least during the process of regeneration.

By the mid-1970s, scientists had begun to focus on identifying the putative signal directing NMJ formation. Among those searching for muscle-derived signals were Uel Jackson McMahan and colleagues. Like most labs at the time, their focus was on identifying signals arising from the postsynaptic muscle surface. This group made the surprising discovery, however, that the signal to guide regenerating motor axons back to the original synaptic site was not located in the muscle fiber, but was located in the synaptic basal lamina—the unique form of basal lamina restricted to the synaptic cleft.

The importance of the synaptic basal lamina was first noted in frogs that had developed muscle atrophy. Surprisingly, innervation remained even at sites where muscle was missing completely, indicating that the muscle was not needed to support the presynaptic connection at the NMJ (**Figure 9.8**). Because the basal lamina remained around the area of degenerated muscle fiber and the synaptic basal lamina was present in the region of the former synaptic cleft, McMahon's lab decided to examine the synaptic basal lamina as a possible source of signals.

Under normal conditions, if both motor axons and muscle fibers are damaged in an adult frog (**Figure 9.9**A), regeneration results in a new motor nerve terminal, new muscle tissue, and the formation of a fully functional NMJ (Figure 9.9B). To study the role of the synaptic basal lamina, the lab developed methods to prevent the regeneration of either the muscle tissue or the motor axons. In one set of experiments, muscles were prevented from regenerating by irradiating the muscle tissue. Although the muscle was missing, the surrounding basal lamina remained—including the synaptic basal lamina that previously adhered to the postjunctional folds on the muscle surface. Using electron microscopy and histological staining methods, the scientists confirmed that even in the absence of muscle tissue, the regenerated motor axons contacted the original synaptic sites on the synaptic basal lamina. Further, the presynaptic elements differentiated as expected, with the synaptic vesicles and active zones accumulating directly opposite the folds of the synaptic basal lamina (Figure 9.9C). In other experiments, muscle was allowed to regenerate, but the motor axons were cut and prevented from re-innervating the new muscle tissue. In these preparations, AChRs formed in the new muscle at the site of original innervation, directly below the folds of remaining synaptic basal lamina. These experiments suggested that the synaptic basal lamina also directed localization of AChRs (Figure 9.9D). Together these studies provided the first evidence that the signals regulating NMJ synaptogenesis originated in the synaptic basal lamina. Since this initial discovery, other molecules

Figure 9.8 The synaptic basal lamina persists at synaptic sites following muscle degeneration. (A) An electron micrograph shows the three cellular elements of an NMJ under normal conditions. A nerve terminal (N), perisynaptic Schwann cell (S), and muscle fiber (M) are visible. Within the nerve terminal, synaptic vesicles (SVs) are concentrated above the postjunctional folds (PJFs) of the muscle fibers. The dark staining reveals acetylcholinesterase (AChE), the enzyme that breaks down excess ACh, concentrated in the basal lamina. The synaptic basal lamina (SBL) extends into the postjunctional folds. (B) Following muscle degeneration, the synaptic basal lamina persists in the synaptic cleft and the area of former postjunctional folds (arrows). The synaptic vesicles (SV) are again concentrated above the former postjunctional folds and the perisynaptic Schwann cell (S) surrounds the nerve terminal. Thus, in the absence of muscle tissue, synaptic basal lamina persists and the presynaptic elements assemble as usual. [From Sanes JR, Marshall LM & McMahan UJ [1978] *J Cell Biol* 78: 176–198.]

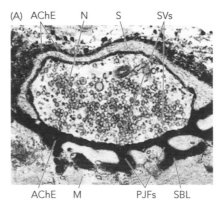

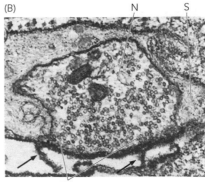

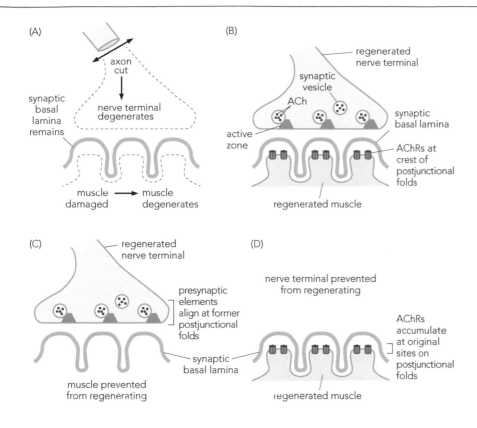

Figure 9.9 Experiments revealed the synaptic basal lamina contains signals to organize pre- and postsynaptic elements. The NMJ of a frog regenerates following damage to motor axons and muscle tissue. (A) When the innervating motor axon is cut and the target muscle is damaged, both the nerve terminal and muscle fiber degenerate. The synaptic basal lamina, however, remains. (B) Following a recovery period, both the axon and muscle fiber regenerate and pre- and postsynaptic elements assemble as usual, creating a new, fully functional synaptic connection. (C) When the muscle fiber was irradiated so that it could not regenerate, the regenerated motor nerve terminal contacted the remaining synaptic basal lamina at the original synaptic site and produced active zone proteins and synaptic vesicles above the regions of the former postjunctional folds. This suggested the synaptic basal lamina contained signals that guided the nerve terminal and induced presynaptic specializations. (D) Conversely, when the motor nerve terminal was prevented from regenerating, but the muscle fiber was allowed to regenerate, the new muscle produced AChRs concentrated in the crests of the postjunctional folds. Like the presynaptic terminal, these new postsynaptic elements were located at the site of the original synapse. This suggested that the synaptic basal lamina also contained signals that organize postsynaptic elements.

needed for both pre- and postsynaptic specializations have been found in this specialized region.

A regenerating nerve fiber growing to a target in an adult animal encounters a very different cellular environment than a newly formed nerve fiber extending in the embryo. Therefore, the signals important during regeneration may be quite different from those required during development. Researchers tested putative signals in both contexts and, as will be seen, many signals detected in models of NMJ regeneration are also present during NMJ development.

AChRs Cluster Opposite Presynaptic Nerve Terminals in Response to Agrin Released by Motor Neurons

Studies of the synaptic basal lamina first focused on signals that induce postsynaptic AChR aggregation. Two experimental methods greatly aided these investigations. The first was a myotube cell culture assay developed to map AChR distribution *in vitro*. To visualize the distribution of AChRs along the muscle surface, a snake toxin called α-bungarotoxin was conjugated to a radioactive, enzymatic, or fluorescent marker (**Figure 9.10**A). α-Bungarotoxin specifically binds to nicotinic AChRs and can be used experimentally to block neural transmission. Additionally, labeled α-bungarotoxin marks the location of AChRs on the cell surface. Studies using labeled α-bungarotoxin found that AChRs were distributed across the surface of the myotubes if motor neurons were absent (Figure 9.10B). Although some small clusters of AChRs were noted in these preparations, the distribution changed considerably when motor neurons were added. In co-cultures of motor neurons and myotubes, the AChRs began to cluster below the areas of nerve fiber contact (Figure 9.10C). This method provided a means to investigate AChR distribution in response to a variety of *in vitro* experimental manipulations, including the addition of soluble factors and local application of ACh.

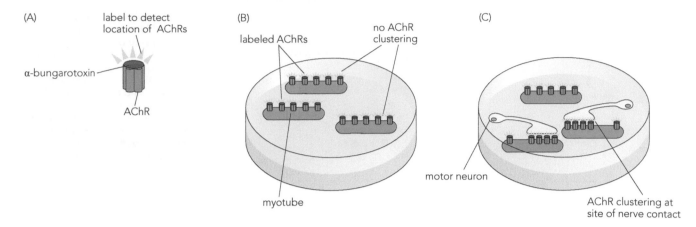

Figure 9.10 Labeled α-bungarotoxin is used to map the distribution of AChRs on myotubes *in vitro*. (A) α-Bungarotoxin binds to nicotinic AChRs and can be used as an experimental tool to inhibit neural transmission or label receptors. When α-bungarotoxin is conjugated to a radioactive, enzymatic, or fluorescent marker, the location of AChRs along a muscle can be visualized. (B) When mytotubes were grown in cell culture in the absence of motor neurons, unclustered AChRs were observed along the cell surface. (C) When motor neurons were added to the myotube cultures, AChRs clustered at sites of nerve contact.

A second important experimental approach utilized the electric organ from the ray *Torpedo californica* (**Figure 9.11**A), an animal model first used in 1772. As discovered in the 1930s, the electric organ of this fish contains a remarkably high concentration of cholinergic synapses. AChRs are so abundant in these rays that Christopher Miller described *Torpedo californica* as "essentially a swimming purified AChR." The electric organ shares many features of the NMJ and has been used in multiple studies of synaptogenesis over the years. Importantly, like muscle cells, the cells of the electric organ are covered in basal lamina, thus providing a large source of material to analyze. Without the discovery of the electric organ of *Torpedo* as an abundant source of cholinergic synapses, many of the biochemical and molecular discoveries related to the function of the NMJ would not have occurred as early as they did, if at all.

Using the *Torpedo* electric organ, the McMahon lab extracted insoluble proteins from basal lamina and tested the ability of the protein extracts to cluster AChRs in cultured chick myotubes. With this method, the lab

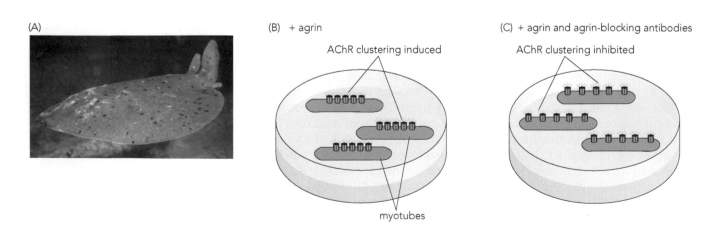

Figure 9.11 Agrin induces clustering of AChRs *in vitro*. (A) Photograph of the ray *Torpedo californica*. The electric organ of *Torpedo californica* shares many characteristics of the NMJ, including cells that are covered in basal lamina. This basal lamina provided an abundant source of material used for isolating proteins, such as agrin, that are necessary for AChR aggregation. (B) Labeled α-bungarotoxin was used to identify the location of AChRs in cultured myotubes. Addition of purified agrin to the myotube cultures induced clustering of the AChRs. (C) When antibodies that block agrin function were added to these cultures, the clustering of AChRs was inhibited. [(A), Courtesy of Daniel Gotshall/NOAA.]

eventually purified a proteoglycan they called **agrin**. When added to cultures of myotubes, purified agrin induced clustering (aggregation) of AChRs (Figure 9.11B). The clustering of AChRs in response to agrin was inhibited when antibodies to agrin were added to cell culture preparations (Figure 9.11C). Further studies indicated that agrin is concentrated in the synaptic basal lamina, produced by motor neurons, and released by the presynaptic nerve terminal. The importance of agrin in NMJ formation was confirmed in knockout mice lacking *Agrin*. In these mice, AChRs are present at normal levels but clustering is greatly reduced and stable NMJs fail to form. The mice die at birth due to respiratory failure caused by a lack of functional innervation to the muscles of respiration.

Together these findings led to what was called the **agrin hypothesis**. In this model, first described in the 1990s, agrin released by the innervating nerve terminal becomes concentrated in the synaptic basal lamina, where it signals postsynaptic AChRs to cluster opposite the innervating presynaptic nerve terminal. In the absence of agrin, such receptor clustering does not occur and a functional NMJ fails to form. The basic tenants of the agrin hypothesis have been supported, although modifications were added when a role for ACh was later discovered.

The Agrin Hypothesis Is Revised Based on Additional Observations

While the initial studies clearly demonstrated that agrin plays a major role in formation of the NMJ, other seemingly conflicting observations were also made. One observation was that agrin is also produced in muscle tissue and adjacent Schwann cells. This suggested that AChR clustering and NMJ formation could be regulated by any of the cells at the NMJ and may not require nerve-terminal-derived agrin. However, subsequent studies identified different isoforms of agrin. In mammals, the agrins are designated X, Y, or Z to indicate the splice site where amino acids are added in the different isoforms. Nerve terminals release the Z form of agrin (called Z^+ agrin), and this isoform has a much greater clustering effect than muscle-derived agrin that lacks the amino acid insert at the Z site (Z^- agrin). It is the Z^+ isoform that influences AChR clustering at the NMJ. The role of Z^- agrin in muscle is not yet known. It does not induce receptor clustering, even in the absence of Z^+ agrin, and it is localized to many other nonneural tissues, including lung and kidney. The roles of the Y and X isoforms of agrin are also unclear. Y agrin binds heparin and has been proposed to modify interactions between agrin and other molecules located at the NMJ. However, in the absence of Y agrin, AChR clustering still occurs.

Another important discovery was that the ACh released from the innervating motor nerve terminal led to dispersal of AChR away from the presynaptic terminal (**Figure 9.12**A). Studies in 2005 demonstrated that ACh appears to block the synthesis of new AChR subunits while also initiating the removal of existing AChRs by endocytosis. Thus, the innervating nerve terminal releases signals that regulate both clustering and dispersal of AChRs. This observation led to the **agrin–ACh hypothesis**, a proposal that considers the release of both agrin and ACh from the innervating nerve terminal. In this model, it is thought that neural agrin functions to stabilize the prepatterned clusters of AChRs at the central region of the muscle fiber by counteracting the effects of ACh. The nerve-derived agrin becomes concentrated in the synaptic basal lamina, where it signals to receptors located on muscle fibers, stabilizing nearby AChRs to prevent their dispersal when exposed to ACh (Figure 9.12B). In contrast, any AChRs not stabilized by agrin are lost upon exposure to ACh.

In vivo evidence supporting the agrin–ACh hypothesis was found in mice lacking genes for both agrin and ChAT—the enzyme that synthesizes

Figure 9.12 The agrin–ACh hypothesis suggests how nerve terminal signals interact to regulate dispersal and clustering of AChRs. (A) ACh released by the presynaptic nerve terminal causes AChRs to disperse away from the site of nerve contact. (B) When both ACh and agrin are released, the AChRs cluster below the nerve terminal. This suggests that the agrin acts to stabilize existing AChRs and prevent dispersal initiated by ACh.

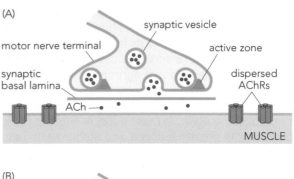

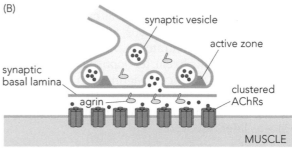

ACh. When both signals are absent, AChRs still cluster as expected. This indicates that in the absence of ACh, agrin is not required to cluster the AChRs. Further, the lack of agrin does not decrease AChR clustering because there is no ACh present to initiate dispersal of the receptors. Thus, under normal conditions, neural agrin promotes AChR clustering by preventing ACh-induced dispersal.

The Receptor Components MuSK and LRP4 Mediate Agrin Signaling

With the initial discovery of agrin, several labs began to search for a receptor on muscle fibers that would mediate agrin-induced clustering of AChRs. These investigations ultimately led researchers working with George Yancopolous to identify a new receptor tyrosine kinase in 1995. The receptor was named MuSK for <u>mu</u>scle <u>s</u>pecific <u>k</u>inase. Subsequent experiments demonstrated that MuSK was an essential component of the agrin-signaling pathway. For example, MuSK is expressed at basal levels in developing muscle cells, particularly in the central region of the muscle fiber. MuSK expression is upregulated at the time of normal innervation, as well as during re-innervation of denervated muscle. In *Musk* knockout mice, AChRs are present, but no receptor clustering is observed and functional NMJs fail to form. Like the *Agrin* knockout mice, these mice die at birth due to respiratory failure.

Even though experimental evidence indicated MuSK signaling was required for postsynaptic clustering of AChRs, scientists noted that the extracellular portion of the MuSK receptor was unable to directly bind agrin. Therefore, it was proposed that an additional signaling component must be involved. This putative component was referred to as MASC, for <u>m</u>yotube-<u>a</u>ssociated <u>s</u>pecificity <u>c</u>omponent. MASC was presumed to bind agrin before interacting with MuSK to induce clustering of AChRs.

Several labs sought this missing signaling component, and more than ten years after the discovery of MuSK, a co-receptor was identified. In 2009, low-density lipoprotein receptor-related protein 4 (LRP4) was found to bind Z^+ agrin directly. LRP4 and MuSK form a complex in the muscle cell membrane, and the interaction between the two co-receptors

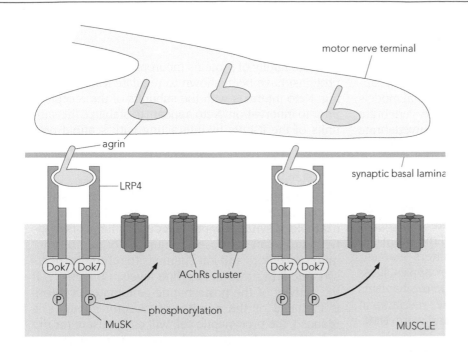

Figure 9.13 Agrin activity is mediated by the LRP4-MuSK receptor complex. Agrin released by the nerve terminal is concentrated in the synaptic basal lamina. Agrin binds to the LRP4 receptor that interacts with the receptor tyrosine kinase MuSK. The interaction between LRP4 and MuSK is further increased upon agrin binding. Agrin binding leads to an increase in MuSK phosphorylation, the recruitment of adaptor proteins such as Downstream of kinase 7 (Dok7), and the clustering of AChRs.

is increased upon binding of agrin to LRP4 (**Figure 9.13**). Agrin binding also leads to an increase in MuSK phosphorylation. Phosphorylated MuSK then recruits and phosphorylates the cytoplasmic adapter protein Downstream of kinase 7 (Dok7). Once activated, Dok7 further increases the phosphorylation of MuSK and recruits additional adapter proteins that ultimately lead to the clustering of AChRs (Figure 9.13). Among the adaptor proteins recruited by Dok7 are CRK (v-crk avian sarcoma virus CT10 oncogene homolog) and CRK-L (CRK-like) that activate Rho GTPase-binding proteins associated with the actin cytoskeleton that help stabilize AChR clustering. The requirement for LRP4 and Dok7 in NMJ formation was confirmed in mice lacking either gene. As with mice lacking *Musk*, in mice lacking either *Lrp4* or *Dok7*, aggregation of AChRs was absent, and the NMJ failed to form. The importance of Dok 7 is also revealed in congenital muscular disorders in humans. Human *Dok7* mutations are one cause of congenital myasthenic syndrome (CMS), a condition that causes muscle weakness (myasthenia) and fatigue upon exertion of skeletal muscles.

Additional proteins are also proposed to interact with MuSK to regulate postsynaptic specializations. For example, several isoforms of vertebrate Wnt have been found to interact with Frizzled-like regions on the extracellular portion of MuSK. In zebrafish, the Wnt-MuSK interaction appears to influence prepatterning and clustering of AChRs in at least some contexts. In some experiments, the addition of Wnt isoforms to cultures of myotubes increased AChR clustering even in the absence of agrin, suggesting a role for Wnts in NMJ formation. Not all studies found a direct effect of Wnts on clustering, however, and many *in vitro* and *in vivo* studies suggest that the clustering of AChRs increases when both agrin and Wnts are present. The signaling pathways involved in this additive effect are not fully established. MuSK has been found to interact with components of the canonical Wnt signaling pathway (see Figure 4.17), including Dishevelled (Dvl) and adenomatous polyposis coli (APC). This interaction may link MuSK to the cytoskeletal elements associated with stabilizing AChRs at the nerve terminal. The Wnt homolog, *Wingless* (*Wg*), is also an important component of NMJ formation in *Drosophila* (**Box 9.1**).

Box 9.1 Invertebrate Model Systems of NMJ Formation

The NMJs of *Drosophila* and *C. elegans* have provided considerable insight into the mechanisms of synaptic formation across species and both animal models share several common features with the vertebrate NMJ. Like vertebrates, studies in these invertebrate models revealed that pre- and postsynaptic elements begin to appear prior to the formation of a synaptic contact, and postsynaptic receptors cluster at the site of the arriving presynaptic nerve terminal. However, there are also many unique aspects of NMJ formation in these species. Whereas the vertebrate NMJ utilizes ACh, the fly NMJ utilizes glutamate, and the worm can use one of three different neurotransmitters—namely, ACh, glutamate, or serotonin. The examples illustrated here highlight some of the unique mechanisms regulating clustering of neurotransmitter receptors in the postsynaptic membrane of the fly and worm NMJ.

Development of the *Drosophila* NMJ

In the fly NMJ, pre- and postsynaptic elements begin to appear prior to the formation of a synaptic contact. The *Drosophila* NMJ uses the neurotransmitter glutamate that binds to the corresponding ionotropic ligand-gated glutamate receptors, the iGluRs on the postsynaptic muscle. The arrival of the presynaptic nerve terminal induces aggregation of the receptors below the terminal. This clustering depends, at least in part, on *Wingless* (*Wg*), a homolog of the vertebrate Wnt proteins. Wg is secreted by the motor neurons and binds to Frizzled (Fz) receptors on the developing muscle cells (**Figure 9.14**). In the absence of *Wg*, glutamate receptors and scaffold proteins are disrupted across the muscle surface. Additional signals from

the Neto family of proteins (neuropilin and tolloid-like proteins) have been shown to regulate iGluR clustering. Neto interacts with the subunits of the receptors prior to innervation. Neto appears to stabilize the subunits of the iGluRs by interacting with scaffold proteins in the muscle (Figure 9.14). Thus, both Fz and Neto associate with postsynaptic scaffold proteins to help stabilize the iGluR in the fly NMJ.

The synaptic elements in the fly NMJ can also be altered in response to changes in neural activity. For example, if synaptic input is reduced, the postsynaptic neuron can increase the number of neurotransmitter receptors and ion channels to return the postsynaptic firing rate to baseline levels. Conversely, if the activity of the postsynaptic glutamate receptors is inhibited or the number of receptor subunits is reduced, the presynaptic cell will compensate for the reduced muscle response by increasing neurotransmitter release. In this way, the muscle contraction remains at a stable level. This is a form of **homeostatic plasticity** in which the partner cells increase expression of synaptic elements in response to alterations in neural signaling. Homeostatic plasticity is also observed in mammalian NMJ and CNS neurons (see Chapter 10).

Development of the *C. Elegans* ACh NMJ

One of the neurotransmitters utilized at the NMJ of *C. elegans* is ACh. Among the receptors present on muscles of the worm body wall are AChRs that bind to the agonist levamisole. These levamisole-sensitive receptors reveal prepatterned clustering in the postsynaptic muscle prior to innervation. The clustering is further regulated in part by signals derived from the muscle cells. Muscle cells secrete a protein called LEV-9 (levamisole resistant-9) that interacts with LEV-10, a transmembrane muscle protein. Another secreted, muscle-derived protein called OIG-4 (one immunoglobulin domain containing protein-4) enhances this clustering. Clustering is further regulated in response to nerve-derived signals. As

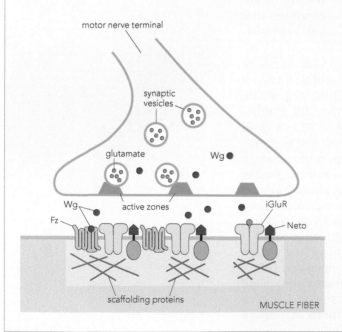

Figure 9.14 The *Drosophila* NMJ. The *Drosophila* presynaptic motor nerve terminal releases the neurotransmitter glutamate that binds to corresponding ionotropic ligand-gated glutamate receptors (iGluRs) on the muscle fiber. Clustering of the iGluRs below the nerve terminal is influenced by Frizzled (Fz) receptors and the Neto family of proteins that interact with postsynaptic scaffold proteins. Wingless (Wg) secreted by the motor nerve terminal binds to the Frizzled (Fz) receptors located on the muscle fiber. Signals from the Neto family of proteins interact with the subunits of the iGluRs prior to innervation and appear to stabilize the subunits of the iGluRs. Together, Fz and Neto associate with postsynaptic scaffold proteins to help stabilize the iGluRs in the fly NMJ.

the presynaptic motor nerve terminal approaches, the differentiating nerve fiber releases the protein MADD-4 (Map kinase-activating death domain 4). MADD-4 helps cluster the AChRs across from the presynaptic terminal. The intracellular mechanisms by which MADD-4 activates this clustering is not yet known (**Figure 9.15**).

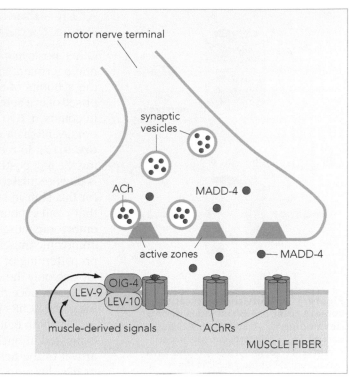

Figure 9.15 The *C. elegans* NMJ. ACh is one of the neurotransmitters utilized at the *C. elegans* NMJ. Prepatterning of the levamisole-sensitive AChRs is observed in the postsynaptic muscle fiber prior to innervation. The clustering is influenced by a protein called LEV-9 (levamisole resistant-9) secreted by muscle fibers. LEV-9 interacts with a transmembrane muscle protein, LEV-10, and another muscle-derived protein called OIG-4 (one immunoglobulin domain containing protein-4) to enhance this clustering. As the presynaptic motor nerve terminal approaches the differentiating muscle cell, it releases the protein MADD-4 (Map kinase-activating death domain 4) that further increases the clustering of the AChRs across from the presynaptic terminal.

Rapsyn Links AChRs to the Cytoskeleton

Other molecules also interact with MuSK to cluster and stabilize the AChRs so they align precisely with the presynaptic active zones. Rapsyn, first discovered in the late 1980s, co-localizes with AChRs and mice lacking rapsyn fail to cluster AChR. However, unlike *Musk-* or *Agrin*-mutant mice, the unclustered receptors of rapsyn-deficient mice remain near the central region of the muscle fiber, rather than being widely dispersed across the muscle. This suggested rapsyn is not directly influenced by the release of ACh from the nerve terminal.

Rapsyn is now known to function downstream of MuSK to help cluster AChRs—likely by interacting with other proteins to link the receptors to components of the cytoskeletal network. Several other proteins—including dystroglycan and utrophin—are thought to interact with rapsyn and cytoskeletal proteins to stabilize AChRs. Additionally, signaling through the vertebrate homolog of the *Drosophila tumorous imaginal disc 1* (*tid1*) may link MuSK to rapsyn (**Figure 9.16**).

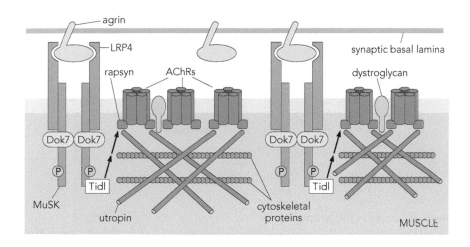

Figure 9.16 Rapsyn helps anchor AChRs to the cytoskeleton. Rapsyn interacts with AChRs, utropin, dystroglycan, and cytoskeletal proteins to anchor the AChRs to the postsynaptic site. MuSK activity appears to be linked to rapsyn by several downstream signaling pathways. In this example, the binding of agrin to the LRP4-MuSK receptor complex activates Tid1 (tumorous imaginal disc 1), which interacts, in turn, with rapsyn to stabilize the AChRs in the postsynaptic membrane.

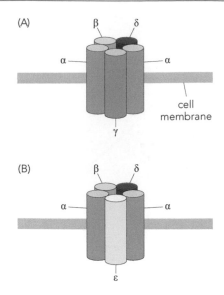

Figure 9.17 Acetylcholine receptor subunit composition changes during development. (A) Embryonic AChRs contain five subunits: two α, one β, one δ, and one γ. (B) Mature AChRs also have five subunits, but the γ subunit is not synthesized in adults. Instead, an ε subunit is synthesized.

AChR Subunits Are Synthesized in Nuclei Adjacent to the Nerve Terminal

Other postsynaptic modifications take place upon arrival of the motor nerve terminal including changes in the location of nuclei that transcribe the subunits of embryonic and adult AChRs. AChRs are pentamers comprised of five subunits. Both embryonic and adult AChRs have four subunits in common: two α, one β, and one δ. However, the fifth subunit in embryonic AChRs is a γ subunit, whereas in adult AChRs it is an ε subunit (**Figure 9.17**). In rodents, the switch in subunit composition takes place during the first postnatal weeks when the expression of embryonic γ subunits is downregulated and the expression of the ε subunits begins. The reasons for this shift in subunit composition are not entirely clear, but it appears that γ and ε subunits have different functions. For example, the γ subunit in embryonic muscle seems to be more effective in depolarizing the smaller embryonic muscle cells. The embryonic γ subunits may also influence the prepatterning of AChRs prior to the arrival of the nerve terminal. In mice lacking only the γ subunit, the normal AChR prepatterning does not occur. However, once the shift to the production of the ε subunit begins in these mice, the AChRs begin to accumulate near the central region of the muscle.

At both embryonic and adult stages, the genes for AChR subunits are expressed in nuclei concentrated adjacent to the nerve terminal. Thus, the arrival of the nerve terminal not only induces migration and clustering of AChRs, but also stimulates the migration and accumulation of nuclei that express genes for the AChR subunits. The nuclei near the site of nerve terminal contact are called **subsynaptic nuclei**, whereas remaining nuclei outside the region of innervation are called **extrasynaptic nuclei** (**Figure 9.18**). How the nuclei migrate to the site of nerve contact is not entirely clear.

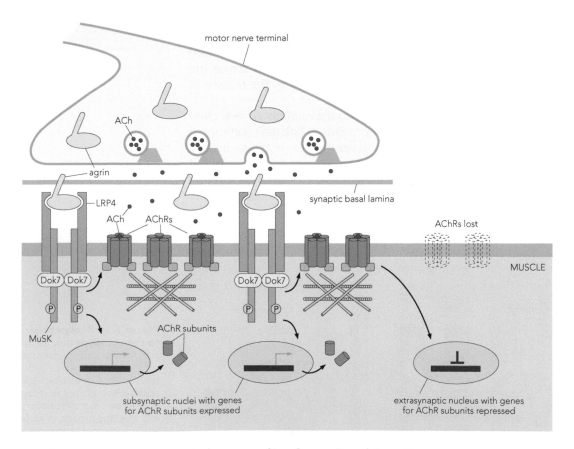

Figure 9.18 Genes for AChR subunits are selectively expressed in subsynaptic nuclei. In addition to agrin's role in clustering AChRs, activation of MuSK also contributes to the synthesis of AChR subunits in subsynaptic nuclei. In contrast, ACh released by the nerve terminal represses transcription of AChR subunit genes in the extrasynaptic nuclei, so receptors located in extrasynaptic regions of the muscle are eventually lost.

The release of ACh by the innervating nerve terminal and the resulting depolarization of the muscle fiber represses AChR subunit transcription in the extrasynaptic nuclei (Figure 9.18). This repression means that AChRs are not replaced during normal receptor turnover so receptors in the extrasynaptic regions of the muscle are eventually lost. However, AChR subunit expression is not repressed in the subsynaptic nuclei. Instead, transcription of AChR subunits continues in these regions, leading to the local production of receptors at the nerve terminal. Although the mechanisms governing the accumulation of subsynaptic nuclei at the site of nerve terminal contact are still being investigated, the local production of AChR subunits is currently thought be influenced by the agrin concentrated in the synaptic basal lamina. The agrin signals through LRP4-MuSK to help mediate gene transcription in the subsynaptic nuclei. Thus, it appears that agrin not only induces receptor clustering and prevents AChR dispersal in response to ACh, but also influences the local production of receptor subunits at the synaptic site.

Other signals may also help regulate synthesis of AChR subunits. One early candidate for regulating synthesis of AChR was isolated from spinal cord and brain extracts in the 1990s. This protein was originally termed ARIA for acetylcholine receptor-inducing activity, but was subsequently found to be Neuregulin 1 (Nrg1), a protein discussed in Chapter 6 for its roles in glial cell development. Nrg1/ARIA is produced by the motor neuron and transported down the nerve terminal where it is released. Nrg1 binds to ErbB receptors, which are located on muscle, nerve, and Schwann cells (**Figure 9.19**). Initial

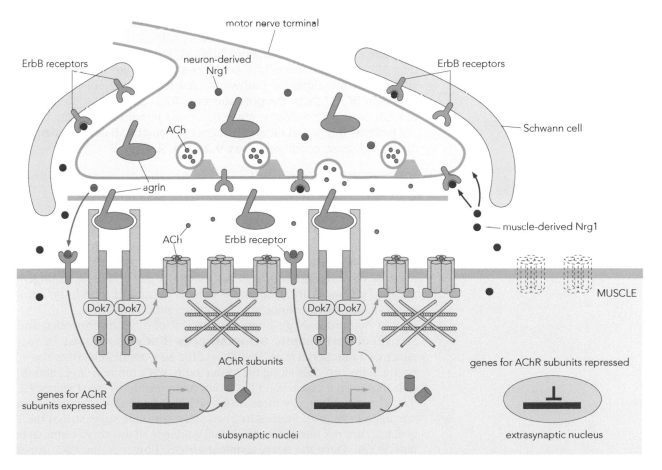

Figure 9.19 Neuregulin 1 may also influence the synthesis of AChR subunits in subsynaptic nuclei. Neuregulin 1 (Nrg1) produced by the motor neuron is believed to bind to ErbB receptors on the muscle fiber and interact with agrin to enhance the synthesis of AChR subunits in subsynaptic nuclei (orange arrows). However, the mechanism by which this occurs is still not clear. Further, Nrg1 is released by muscle and ErbB receptors are also found on motor neurons and perisynaptic Schwann cells (dark red arrows). Muscle-derived Nrg1 activates ErbB receptors on Schwann cells to promote their survival. Schwann cells, in turn, help maintain the NMJ. Thus, the Nrg1 pathway may mediate multiple aspects of NMJ formation and maintenance.

in vitro studies suggested that Nrg1 increased mRNA transcript levels for the α, δ, and ε subunits of AChR. Thus, for many years Nrg1 was thought to be the nerve-derived signal for the synthesis of AChR subunits. In recent years, however, the proposed role of Nrg1 has been revised.

Because mice lacking the gene for *Nrg1* are embryonic lethal and die prior to the formation of the NMJ, conditional knockout mice were generated in which ErbB receptors were only deleted in muscle, but not other cell types. Neuromuscular synapses still formed in these mice, suggesting that Nrg1 signaling through ErbB receptors on muscle fibers was not required for the synthesis of AChR subunits and NMJ formation. Because the ErbB receptors remained on motor neurons and Schwann cells, Nrg1 released by either motor neurons or muscle cells could have activated those receptors to stabilize the NMJ (Figure 9.19). One hypothesis is that Nrg1 may regulate the survival, proliferation, or migration of Schwann cells, which are also important for NMJ development. Therefore, loss of muscle fiber ErbB receptors would not eliminate NMJ formation.

The current working model suggests agrin and Nrg1 somehow interact to modulate transcription of the correct level of AChR subunits. Agrin appears to be the primary mediator of AChR synthesis in subsynaptic nuclei, but alone is insufficient to generate the number of receptors normally found at the NMJ. For example, one set of *in vivo* experiments found that agrin injected into adult muscle, at a site away from the active NMJ, induced ectopic postsynaptic specializations such as AChRs and their associated anchoring proteins. However, the concentration of AChRs produced was lower than at the original synapse. These findings supported the idea that additional signals are needed for full transcriptional activation of AChR subunits. Nrg1 may be one such signal.

Since the initial discoveries of agrin, MuSK, rapsyn, and ARIA/Nrg1 over 25 years ago, there have been considerable advances in our understanding of the signaling pathways used to regulate clustering and stabilization of AChRs in the postjunctional folds of the NMJ. Despite a vast amount of new knowledge, many questions remain about how the different proteins interact at different stages of normal NMJ development and in different disease conditions **(Box 9.2, Box 9.3)**.

Perisynaptic Schwann Cells Play Roles in NMJ Synapse Formation and Maintenance

The first studies of the NMJ focused on how motor neurons and muscle cells interacted to regulate formation and stabilization of pre- and post-synaptic elements. As scientists uncovered various ways that the synaptic partner cells interact with each other, it became apparent that a third element of the synapse was also necessary for successful synaptogenesis. Synaptic glial cells are now known to have important roles in synaptogenesis, synapse maintenance, and synaptic remodeling.

In mammals, Nrg1 promotes survival of all Schwann cells, and the survival of **perisynaptic Schwann cells** (**PSCs**) is required for maintenance of a functional NMJ. The PSCs of the NMJ are distinct from the myelinating Schwann cells along the motor axon. PSCs are non-myelinating glial cells that form a cap around the motor axon terminals (see Figures 9.2 and 9.8). There are typically three or four PSCs at each NMJ, depending on the size of the synapse. Studies in frogs and mammals suggest that the PSCs generally are not needed for the initial guidance of the nerve terminal to the muscle cell. Once the nerve terminal arrives, however, the PSCs appear to provide signals necessary to stabilize and maintain the synaptic connection. For example, in frogs in which PSCs were selectively ablated, the NMJ only remained functional for several hours. Over the course of a week, the nerve terminal withdrew and the NMJ was lost in the absence of the PSC.

In vitro studies further demonstrated the importance of Schwann cells in maintaining the NMJ. Frog Schwann cells grown in cell culture released factors that induced synaptogenesis when added to co-cultures of frog neurons and muscle cells. The addition of the Schwann-cell-conditioned medium led to an increase in synaptic contacts between the co-cultured motor neurons and muscle cells, as well as increased clustering of AChRs in the muscle cells. Subsequent investigations revealed that the Schwann cells release transforming growth factor β1 (TGFβ1). In both frog and rat, TGFβ1 is detected in Schwann-cell-conditioned medium and expressed in PSCs. Addition of purified TGFβ1 to co-cultures of neurons and muscle cells increased synaptic contacts and AChR clustering similar to Schwann-cell-conditioned medium. Further, these effects on NMJ formation were inhibited when TGFβ1 blocking antibodies or binding proteins were added to Schwann-cell-conditioned medium, or when the receptor for TGFβ1 was blocked with a specific kinase inhibitor. Together, these studies demonstrated the importance of Schwann-cell-derived TGFβ1 in promoting synaptogenesis. TGFβ1 was also noted to increase agrin expression in rat spinal motor neurons, suggesting a possible mechanism underlying the increased AChR clustering seen in the cultured muscle cells (**Figure 9.20**). Thus, PSCs provide essential signals that regulate multiple aspects of synaptogenesis at the NMJ.

The Synaptic Basal Lamina Concentrates Laminins Needed for Presynaptic Development and Alignment with Postjunctional Folds

As first noted in the studies by McMahon and colleagues in the 1980s, the synaptic basal lamina contains proteins that organize both presynaptic and postsynaptic specializations. While considerable progress has been made in identifying proteins that regulate development of postsynaptic elements, comparatively less is known about signals that regulate presynaptic development. This was due in part to the lack of an assay to directly assess development of specific presynaptic elements. As more presynaptic proteins have been identified, methods to label these proteins, and cell culture assays to monitor how presynaptic elements are formed and maintained in response to synaptic basal lamina-derived signals have been developed.

Like all basal lamina, the synaptic basal lamina contains collagen, entactin, and proteoglycans. Additionally, the synaptic basal lamina contains AChE, agrin, neuregulin, and specific laminin isoforms that differ from those in the extrasynaptic basal lamina. Laminins are currently among the best-characterized signals that influence presynaptic development. Several distinct isoforms of laminin are prominent in this region, including laminin-4, laminin-9, and laminin-11. As noted in previous chapters, laminins are a large family of proteins that interact through heterophilic binding. Laminin isoforms differ in the composition of various α, β, and γ chains. For example, laminins in the synaptic basal lamina differ from extrasynaptic laminins based on the composition of their β chains. In contrast, laminin isoforms concentrated in the synaptic basal lamina differ from each other based on their α-chain subunits. The chain composition of extrasynaptic laminin-1, found mainly on the muscle surface, is $\alpha 1$, $\beta 1$, and $\gamma 1$. Based on this subunit composition, laminin-1, the first laminin discovered, is now often referred to as laminin-111. Laminins concentrated in the synaptic basal lamina include laminin-9, laminin-11, and laminin-4, which differ in their α subunits. The laminin-9 chain is comprised of α4, β 2, and γ1; the laminin-11 chain is comprised of α5, β2, and γ1 subunits; and the laminin-4 chain is comprised of α2, β2, and γ1 subunits. Thus, laminin-9 is also called laminin-421, laminin-11 is called laminin-521, and laminin-4 is called laminin-221.

It now appears that each laminin in the synaptic basal lamina has a specific function to ensure proper alignment of pre- and postsynaptic elements at the NMJ (**Table 9.1**). Laminin-9 is necessary to localize the presynaptic active zones above the postjunctional folds, whereas laminin-11 participates

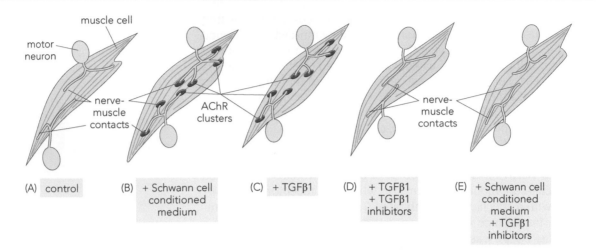

Figure 9.20 TGFβ1 released by Schwann cells supports NMJ synaptogenesis. Frog muscle cells (light pink) and motor neurons (green) were cultured together, and the number of synaptic contacts and areas with AChR clusters (dark pink) were compared in different cell culture conditions. (A) In control conditions lacking Schwann cell-conditioned medium or TGFβ1, AChR clustering was not observed, even at sites of nerve muscle contact. (B) In the presence of Schwann cell-conditioned medium, the number of nerve-muscle contacts increased and AChR clusters were observed (dark pink) at the sites of contact. (C) Similar results were observed when the cells were treated with TGFβ1. However, nerve-muscle contacts and AChR clustering decreased when TGFβ1 inhibitors were added to the cultures treated with TGFβ1 (D). (E) TGFβ1 inhibitors also decreased nerve-muscle contacts and AChR clustering when added to cultures treated with Schwann cell-conditioned medium, furthering supporting the hypothesis that TGFβ1 is released by Schwann cells to support NMJ synaptogenesis. [Adapted from Feng Z & Ko CP [2008] *J Neurosci* 28:9599–9609.]

in forming the presynaptic terminals and restricting the growth of PSCs so that their processes do not enter the synaptic cleft. In contrast, laminin-4 is needed for development of the postjunctional folds on the muscle surface.

As noted in previous chapters, laminins can bind to various combinations of integrin receptors located on neuronal cell surfaces. In the early part of the twenty-first century, researchers made the surprising discovery that the laminins containing the β2 subunit can also specifically bind to the pore-forming subunits of the subtypes of voltage-dependent Ca^{2+} channels

Table 9.1 Signals for NMJ Formation.

Signals Regulating Postsynaptic Elements	Localization	Function
Acetylcholine	Released by presynaptic motor nerve terminal	Disperses AChRs
Agrin (Z+ isoform)	Released by presynaptic motor nerve terminal	Stabilizes clusters of AChR below nerve terminal; regulates synthesis of AChR subunits in subsynaptic nuclei
LRP4	Located in muscle fiber membrane	Binds agrin
Laminin-4	Concentrated in synaptic basal lamina	Influences development of postjunctional folds
MuSK	Located in muscle fiber membrane	Associates with LRP4; mediates effects of agrin
Neuregulin	Released by motor nerve terminal and muscle	Influences survival of Schwann cells; plays a role in synthesis of AChR subunits
Rapsyn	Co-localized with AChRs in muscle	Interacts with MuSK; links AChRs to cytoskeleton
Signals Regulating Presynaptic Elements	**Localization**	**Function**
Laminin-9	Concentrated in synaptic basal lamina	Organizes active zones above postjunctional folds
Laminin-11	Concentrated in synaptic basal lamina	Restricts growth of Schwann cell processes; influences growth of presynaptic terminals
LRP4	Located in muscle fiber membrane	May influence presynaptic differentiation independent of MuSK

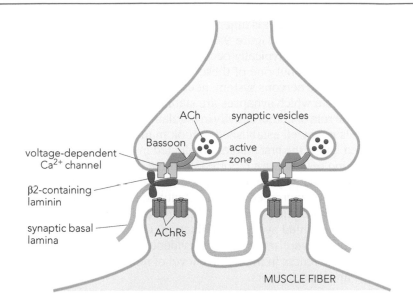

Figure 9.21 Laminins concentrated in the synaptic basal lamina bind to voltage-dependent Ca²⁺ channels in the presynaptic motor nerve terminal. Laminins containing the β2 subunit bind to the pore-forming subunits of voltage-dependent Ca²⁺ channels in the presynaptic nerve terminal. The intracellular regions of the Ca²⁺ channels also interact with the cytosolic scaffold protein Bassoon that is associated with the active zone protein complex. In this way, the laminins in the synaptic basal lamina organize the presynaptic elements above AChRs clustered at the postjunctional folds on the postsynaptic muscle fiber.

concentrated in the presynaptic motor nerve terminal. Interestingly, these laminins do *not* bind to the subtypes of voltage-dependent Ca²⁺ channels located *outside* of the nerve terminal. Thus, laminins concentrated in the synaptic basal lamina are able to selectively organize and stabilize presynaptic elements. More recently, it was observed that the intracellular domains of the voltage-dependent Ca²⁺ channels interact with proteins associated with the active zone protein complex. One such protein is Bassoon, a cytosolic scaffold protein located at the active zone complex. Thus, the synaptic basal lamina-concentrated laminins directly bind to and stabilize the presynaptic voltage-dependent Ca²⁺ channels that are linked, in turn, to the components of the active zone protein complex (**Figure 9.21**). Because active zones line up adjacent to postjunctional folds, linking voltage-dependent Ca²⁺ channels to active zone proteins ensures that Ca²⁺ influx leads to the localized release of ACh at the AChRs clustered at the top of the postjunctional folds. The importance of this laminin-voltage-dependent Ca²⁺ channel interaction was demonstrated in various knockout mice. In mice lacking genes for either β2 laminin or the presynaptic voltage-dependent Ca²⁺ channels, fewer active zones were present in the presynaptic nerve terminal.

Other signals regulate presynaptic differentiation as well. Among the candidate molecules identified to date are members of the fibroblast growth factor family (FGF). FGFs are released by muscle cells and bind to corresponding FGF receptors on the presynaptic terminal, where they can regulate aspects of synaptic vesicle clustering. LRP4 is also expressed on presynaptic terminals, where it may influence synaptic vesicle clustering, although conclusive evidence of this has not yet been found. As described in Chapter 10, other molecules regulate the differentiation of presynaptic elements in the CNS. Some of these may also play a role in NMJ development.

MODELS OF SYNAPTIC ELIMINATION IN THE NMJ

The formation of mature synaptic contacts is a prolonged and dynamic process that includes a period in which some synaptic connections are eliminated and others are strengthened. The time course over which this happens for NMJ synapses has been well documented. Initially, during embryogenesis, each motor neuron innervates more than one muscle

(A)

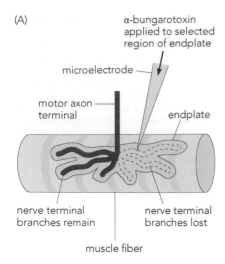

α-bungarotoxin applied to selected region of endplate

microelectrode

motor axon terminal

endplate

nerve terminal branches remain

nerve terminal branches lost

muscle fiber

(B)

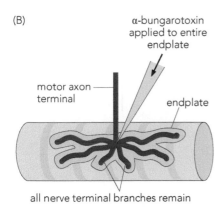

α-bungarotoxin applied to entire endplate

motor axon terminal

endplate

all nerve terminal branches remain

Figure 9.22 Differences in neural activity influence synapse elimination and stabilization. α-Bungarotoxin was used to inhibit neural transmission at the NMJ. (A) When α-bungarotoxin was injected into a small region of the motor endplate, only the nerve terminals at the site of inhibition were lost. (B) When α-bungarotoxin was applied to the entire endplate, all the nerve terminals remained. This indicated that synapse elimination only occurred when there were differences in the amount of neural transmission between the nerve terminal branches.

fiber, and each muscle fiber is innervated by multiple presynaptic terminals at a single location (see Figure 9.6). In rodents, the subsequent process of synaptic elimination typically occurs in the first three postnatal weeks. During this stage, all but one of these presynaptic terminals is lost. As in other regions of the nervous system, neural activity in the presynaptic neuron helps regulate which synapses are stabilized and which are lost.

While the role of neural activity in regulating synapse stabilization and elimination is now well established, it took many years to determine how this occurred. Since the first studies conducted in the 1970s, several seemingly conflicting results were reported. In some experiments, for example, increased neural activity increased the extent of synapse elimination. In other experiments, increased neural activity stabilized NMJ connections. In still other experiments, blocking all neural activity stabilized NMJ connections and prevented synapse elimination. Over time, these seemingly conflicting results were reconciled by evidence that synapse elimination depended on *differences* in the relative level of neural activity in presynaptic neurons and muscle fibers.

The Relative Levels of Neuromuscular Activity Determine Which Terminal Branches Remain at the Endplate

Experiments conducted by Jeff Lichtman and colleagues illustrated how differences in relative levels of neural activity influence synaptic elimination in the rodent NMJ. In one set of studies, neural transmission through AChRs was blocked at singly innervated NMJs with the AChR-binding antagonist α-bungarotoxin. The toxin was selectively applied via microelectrodes either at a restricted region of the endplate or across the entire endplate (**Figure 9.22**). When the inhibitor was applied to a limited region of the endplate, only the motor nerve terminal branches and AChRs at that site were lost. The contacts at the untreated region of the endplate remained intact. In contrast, when the entire NMJ was inhibited, no synapses were eliminated. Thus, synapses were eliminated only when there were differences in levels of neurotransmission between different terminal branches at a single endplate.

Several models have been proposed over the years to account for this effect. Most note there must be different signals released by the pre- or postsynaptic cell in response to neural activity. These are often described as "punishment" or toxic signals and "protective" or trophic signals. For example, one early model suggested that the muscle released a protease, whereas the nerve terminal released a protease inhibitor. If activity in the nerve and muscle occurred at the same time, both signals would be released at the same time. Therefore, the protease inhibitor (protective signal) would block the effects of the protease (punishment signal) and the synapse would remain (**Figure 9.23**A). If the activity of the pre- and postsynaptic cells did *not* match, however, then the signals would be released at different times. If muscle activity were greater, more "punishment" signal would be released without enough "protective" signal released from the nerve terminal to block its activity. Thus, the nerve terminal would withdraw (Figure 9.23B). If neural activity was absent in both cells, then neither signal would be released, and all synaptic contacts would remain (Figure 9.23C).

BDNF and Pro-BDNF Are Candidates for the Protective and Punishment Signals

The protective/punishment model provided a framework for continued studies of how differences in neural activity regulate synapse elimination and stabilization at the NMJ and perhaps other synapses in the PNS and CNS. Studies in the CNS and PNS indicated that some target cells release BDNF in response to neural activity, suggesting a candidate molecule for

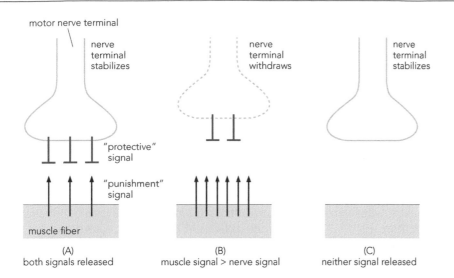

(A)
both signals released

(B)
muscle signal > nerve signal

(C)
neither signal released

Figure 9.23 Model of activity-dependent synapse elimination in the NMJ. Synapse elimination in the NMJ results when there are differences in the activity levels of the motor neuron and muscle fiber. One model to explain this effect proposed that muscle released a "punishment" signal that could damage the motor nerve terminal and initiate its withdrawal. The motor neuron could prevent nerve terminal withdrawal, however, by releasing a "protective" signal to block the effect of the muscle-derived signal. (A) When both the motor neuron and muscle are active at the same time, both the punishment and protective signals are released, and the motor nerve terminal is stabilized. (B) If the activity levels of the motor neuron and muscle cell are asynchronous, the signals are released at different times. In this example the muscle activity is greater and therefore the amount of punishment signal released is greater. Thus, the protective signal is insufficient to counteract the punishment signal and the nerve terminal withdraws. (C) If the motor neuron and muscle cell are inactive at the same time, neither signal is released. Therefore, the synapse remains.

NMJ reorganization. As noted in Chapter 8, neurotrophins are initially synthesized as precursor forms (pro-neurotrophins) that are then cleaved to mature forms. Pro-neurotrophins bind to the p75NTR-sortilin receptor complex, whereas mature neurotrophins bind to Trk receptors. Recent studies in frog and mouse suggest that different forms of BDNF serve as the punishment and protective signals at the NMJ. Based on a series of *in vitro* and *in vivo* experiments, a model emerged in which pro-BDNF is released by muscle cells in an activity-dependent manner and innervating nerve fibers express both p75NTR-sortilin and Trk B receptors. When the released pro-BDNF binds to and signals through the p75NTR-sortilin receptor complex on motor axons, those innervating nerve fibers retract. However, neural activity also initiates secretion of proteases from the nerve terminal that cleave muscle-derived pro-BDNF to its mature form. The mature BDNF then activates the TrkB receptors on the remaining nerve fibers to stabilize those synaptic connections (**Figure 9.24**). Thus, proteases released in response to neural activity in the presynaptic cells change the muscle derived pro-BDNF punishment signal mediated by the p75NTR-sortilin

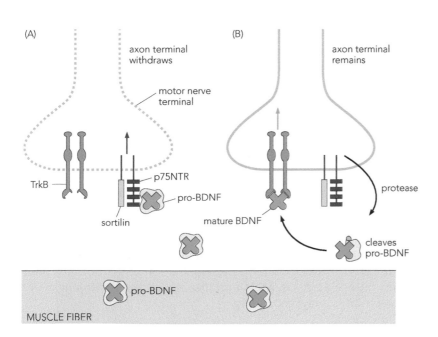

Figure 9.24 Different forms of BDNF may provide both a punishment and protective signal. (A) Pro-BDNF is released by muscle fibers in an activity-dependent manner. When pro-BDNF binds to p75NTR-sortilin receptors on innervating motor nerve terminals, it initiates signal transduction pathways that lead to nerve terminal withdrawal. (B) Neural activity causes proteases to be released by the nerve terminal. These proteases cleave the pro-BDNF to generate mature BDNF that then binds to the TrkB receptors also located on innervating motor nerve terminals. This signaling pathway stabilizes the nerve contact. Thus, the nerve-derived protease converts the muscle-derived punishment signal into a muscle-derived protective signal.

receptors to a BDNF protective signal mediated by TrkB receptors. In this model, the protease converts the "punishment" signal to a "protective" signal. Those motor axons that do not release a protective protease signal are eliminated. In this way, neural activity determines which axons respond to BDNF and therefore, which axons are stabilized.

In vivo studies confirm the importance of pro-BDNF and BDNF in differentially regulating NMJ elimination and stabilization. In one set of experiments, p75NTR knockout mice were injected with a sortilin propeptide that inhibits the binding of pro-BDNF. This approach rendered both p75NTR and sortilin receptors unable to transduce pro-BDNF signals and therefore more synapses remained. Conversely, in experiments in which TrkB signaling was inhibited, greater synaptic elimination occurred. A metalloprotease inhibitor was injected into muscle fibers to block the cleavage of pro-BDNF to mature BNDF. This prevented activation of TrkB receptors and led to more synaptic elimination. Thus, the balance of available pro-BDNF and BDNF signaling, regulated by neural activity, helps determine the extent of synaptic elimination. Researchers continue to investigate how these events are regulated at the NMJ and whether similar events occur in CNS synapses.

Perisynaptic Schwann Cells Influence the Stability of Synaptic Connections

The PSCs not only release signals to support synaptic contacts (see Figure 9.20), but also are active during synapse elimination. PSCs phagocytize and remove the cellular debris associated with eliminated axon terminals. It may be that the PSCs randomly choose terminals to remove, or they may be able to detect which synaptic contacts are less active then selectively remove only those synapses. PSCs, like other glial cells, express receptors for neurotransmitters. PSCs express both muscarinic AChRs activated by ACh and purinergic receptors activated by adenosine-5′-triphosphate (ATP). Both ACh and ATP are released during NMJ formation and elimination. Thus, PSCs have the means to monitor the activity of presynaptic terminals and influence which synapses are stabilized, and which are lost.

Box 9.2 Post-Polio Syndrome: When New Branches of Motor Nerve Terminals Can No Longer Function

During the 1940s and 1950s, one of the most feared diseases was poliomyelitis, a viral infection that causes degeneration of motor neurons in the spinal cord and brain stem. With the introduction of the first polio vaccine in 1955, polio was largely eradicated from the United States and most other countries. By the 1990s, only a few countries continued to have outbreaks of polio, though cases have emerged in several countries during the past year.

Although often thought of as a disease that always results in paralysis or death, the outcomes of a polio infection are quite variable. Some individuals experience only a few mild symptoms. Others experience muscle weakness, whereas others develop paralysis. Many of those who initially experience muscle weakness go on to recover muscle function. This recovery appears to be due to the branching of motor nerve terminals from healthy motor neurons to nearby muscle fibers that lost innervation when infected motor

neurons died. Thus, the new branches compensate for the lost motor neurons.

Post-polio syndrome (PPS) is a condition that appears three to four decades after the initial infection with the polio virus. The new symptoms include gradually progressive muscle weakness and muscle atrophy. The cause of PPS is unclear, but it appears the new connections made after the initial infection gradually become overworked and can no longer handle the increased demands. These motor neurons or their connections then begin to degenerate, causing the progressive weakness and atrophy noted by patients. Currently there is no cure for PPS and treatments rely primarily on physical and occupational therapy to maintain functional connections as long as possible. As scientists continue to study NMJ formation and elimination under normal conditions, insights into the causes and treatment of PPS may also emerge.

Box 9.3 A neurologist's Perspective: Myasthenia Gravis

Kathlyn Callenius, D.O.

Kathlyn Callenius, is a neurologist who majored in biology at the University of Minnesota before completing medical school at the Lake Erie College of Osteopathic Medicine. Her residency training in neurology at the University of Pittsburgh Medical Center Hamot included a focus on neuromuscular disorders and electromyography (EMG).

The importance of functional pre- and postsynaptic elements at the neuromuscular junction (NMJ) may not be fully appreciated unless one experiences a disruption in synaptic transmission. In some cases, interfering with even a single postsynaptic element is enough to alter daily activities, as illustrated in this example.

Imagine you get home after a long day feeling a bit more tired than usual and notice you see two identical books when attempting to read a novel that evening. Over the next couple days, you notice activities that normally would not affect you instead leave you feeling very weak. Something as simple as going up the stairs while carrying laundry is exhausting. A week later, you realize that you are having difficulty breathing. You go to the emergency room at your local hospital, and they urgently talk to you about putting in a breathing tube to make sure that you are getting enough oxygen. While they are getting things ready for the procedure, you hear the words "myasthenia gravis" being whispered around the room. The next thing you know, you are slowly waking up in the intensive care unit. You are coughing from the tube in your throat and see the machine that is helping you breathe. The neurologist comes into the room to discuss the events that led up to your current situation and the diagnosis of myasthenia gravis. How did something this serious happen so quickly?

Myasthenia gravis is an autoimmune disorder in which the body creates antibodies that target receptors in the NMJ. Although a variety of antibodies is produced, they all target receptors at the postsynaptic site, and the majority specifically block acetylcholine receptors (AChRs). The antibodies attach to the receptors preventing acetylcholine (ACh) from binding. Thus, when a person engages in any physical activity, some of the released ACh can bind to available receptors but the ACh in the synaptic cleft is forced to compete with the antibodies for access to open receptors. As a result, fewer receptors are available for ACh binding. Other antibodies block binding to MuSK or LRP4 to interfere with ACh transmission. In all cases, the disrupted transmission is perceived by patients as a feeling of increased weakness during normal activities. The longer the duration of the activity, whether lifting objects, running, looking in a certain direction, or even breathing, the weaker the person becomes as fewer and fewer receptors are bound by ACh.

There are two main categories for myasthenia gravis: ocular and generalized. Ocular myasthenia means that only the eye muscles are affected. Ocular myasthenia impairs control of eye movements so patients normally report seeing double. They may also have trouble keeping their eyelids open, and one eyelid may droop more than the other. Patients with ocular myasthenia need to be closely monitored, even if visual symptoms improve, as they can transition into generalized myasthenia, often within two years from the onset of visual symptoms.

Generalized myasthenia involves weakness in any muscle in the body, including the eye muscles and the diaphragm. Patients may report difficulties watching television or reading at the end of the day, note that climbing stairs is fatiguing, or they have more difficulty chewing certain foods, like a piece of meat. They may also have some vague complaints of shortness of breath.

Patients with myasthenia gravis should be evaluated for thymoma, a tumor of the thymus gland. The thymus gland is an active part of the immune system during childhood, but as we age, it shrinks and becomes inactive. The association between this tumor and myasthenia gravis is a paraneoplastic process, meaning the antibodies are created in reaction to the tumor. However, instead of targeting the tumor, the antibodies attack the neuromuscular junction, causing myasthenic symptoms. Many cases of myasthenia gravis caused by thymus gland tumors improve once the tumor is removed.

Treatment of cases not associated with thymoma can become a little complicated and depend on the severity of symptoms. The two main categories for treatment involve symptomatic treatments and immunosuppressant medications. Many patients are started on a medication that inhibits acetylcholinesterase (AChE) and slows the breakdown of ACh in the synaptic cleft. This gives the ACh more time and a better chance of binding to available AChRs. The medication does not stop the antibodies but helps reduce symptoms associated with muscle weakness.

The second approach involves drugs that help to fight the antibodies. These include steroids to dampen the immune response and immunomodulatory agents. Someone experiencing a myasthenic crisis and requiring intubation, such as described in the opening example, would need to be treated aggressively and would likely be given intravenous immunoglobulin (IVIG) or plasma exchange (plasmapheresis) while in the hospital. IVIG is usually a combination of antibodies from donated blood. The mechanism is not completely understood, but it can help fight against the abnormal antibodies as well as calm down the patient's immune system. This leads to symptom reduction that can last for several weeks. The second option, plasma exchange, is similar to kidney dialysis. The procedure involves filtering the patient's blood and removing the antibodies that are causing the issue. Improvements after plasma exchange can also last several weeks. Both approaches are effective but usually require subsequent treatments. Because IVIG is an infusion it is usually easier to give to patients. However, treatments depend on many factors, including patient history and response to previous treatments. While not curable, symptoms can be reduced, and treatment approaches modified as needed.

Box 9.4 A Neurologist's Perspective: Amyloid Protein Deposits Affecting Sensorimotor Nerves

Joshua A. Smith, D.O.

Joshua A. Smith, graduated from Grove City College where he majored in molecular biology. He completed medical school at the Lake Erie College of Osteopathic Medicine followed by residency in adult neurology and a fellowship in clinical neurophysiology. He is currently associate director of the neurology residency program at the University of Pittsburgh Medical Center Hamot.

When people hear the term amyloid, they often think of the accumulation of amyloid plaques in the brains of patients with Alzheimer's disease, an age-related condition that leads to dementia. However, amyloid deposits also impair sensorimotor functions in the peripheral nervous system and organs such as the liver, kidney, and heart. Damage that arises from these conditions, collectively known as **amyloidosis**, are directly related to the accumulation of amyloid protein deposits in and around the tissues. In contrast, it is still unclear if the deposits of beta amyloid (peptides derived from the amyloid precursor protein) in the brain are the cause of Alzheimer's dementia.

Amyloidosis is really a group of disorders, all of which are characterized by tissue deposition of abnormally folded proteins. The amyloid deposits form when globular, soluble proteins misfold into a beta-pleated sheet secondary structure creating insoluble, nonfunctional fibrils. These fibrils then cause organ dysfunction. The abnormally folded proteins may arise from sustained increased protein concentration, acquired protein mutations, hereditary protein mutations, or the proteolytic remodeling of a wild-type protein.

Amyloidosis is classified in a variety of ways, often by the organ involved or the source of the misfolded protein. Subtypes are usually termed A (for amyloid) followed by another letter or abbreviation to denote the protein source. For example, AL amyloidosis refers to amyloid formed by the buildup of abnormal light (L) chains (kappa or lambda light chain immunoglobulins). The abnormally folded light chains are produced by plasma cells and the resulting fibrils accumulate in tissues, organs, and nerves. AL amyloidosis is one of the more common types and patients may have a plasma cell dyscrasia (abnormality) or overt multiple myeloma (cancer of plasma cells) that requires treatment with chemotherapies. AA amyloid refers to amyloid formed by serum amyloid A proteins (A), proteins that are normally produced by the liver and found in blood. However, when serum amyloid A fragments accumulate amyloidosis results, often following systemic autoimmune conditions or chronic infections. Patients with AA amyloidosis require treatment of their underlying systemic autoimmune or chronic infectious disease.

ATTRm refers to amyloid arising from hereditary mutations (m) in the TTR (transthyretin) gene. In contrast, ATTRwt refers to amyloid deposits formed by wild-type transthyretin monomers. It is important for clinicians to distinguish the subtypes as this influences treatment options and prognosis. Among hereditary amyloidosis, there are three main subtypes: ATTRm, apolipoprotein A1-related amyloidosis, and gelsolin-related amyloidosis. Neither apolipoprotein A1- nor gelsolin-related amyloidosis have a specific treatment. ATTRm used to be treated with liver transplantation if found early in the disease course, as transthyretin is produced by the liver and transplantation stopped further mutated protein production. However, more recent treatment options are less invasive. Medications that stabilize transthyretin and prevent fibril formation have been approved for patients with cardiac disease related to amyloidosis.

Many subtypes of amyloidosis involve the peripheral nerves leading to sensory and motor deficits. This causes what is termed a classical length-dependent sensorimotor axonal peripheral polyneuropathy, meaning the longest nerves, such as those to the feet or hands, are the first to be affected. Amyloidosis-associated neuropathies tend to progress more rapidly than other common neuropathies, such as those associated with diabetes mellitus. When the small diameter fibers are involved, autonomic dysfunction often occurs and presents with decreased sweating, impaired gastric motility, orthostasis (low blood pressure upon standing up), and sexual dysfunction. When the gastrointestinal system is involved, a range of symptoms including diarrhea, malabsorption, rectal bleedings (hematochezia), bloody vomit (hematemesis), obstruction, or perforation can occur. At times focal cranial neuropathies can develop, though this is most common in the hereditary gelsolin variant.

Two medications have recently been approved for hereditary transthyretin associated amyloid neuropathy. One uses double-stranded silencing RNA and the other an antisense oligonucleotide to interfere with the translation of mRNA into transthyretin protein. There is hope that with further research these medications may treat patients' organ dysfunction beyond neuropathy.

In summary, amyloidosis is a heterogeneous group of disorders which stem from abnormal protein deposition in tissues, including peripheral nerves. Although in the past treatments were limited, more treatments are emerging, and early diagnosis and treatment can increase the duration and improve the quality of a patient's life.

SUMMARY

To establish a functional vertebrate NMJ, reciprocal signals are exchanged between pre- and postsynaptic cells. Both the presynaptic motor nerve terminal and the developing myotube express synaptic elements prior to making initial contact. Once contact is initiated, however, signals are produced to form a stable, mature synaptic connection. For example, agrin is released from the nerve terminal to bind to the LRP4 receptor that interacts with the MuSK receptor on the muscle. This leads to clustering of AChRs below the presynaptic nerve terminal and prevents ACh-induced dispersal of AChR. Several molecules, including rapsyn, help anchor the receptors in place by linking them to the cytoskeleton. Agrin, signaling through MuSK, also regulates synthesis of AChR subunits in subsynaptic nuclei.

In addition to agrin, the nerve also releases Neuregulin (Nrg1), which binds ErbB receptors that further contribute to the maintenance of the NMJ and possibly to localized synthesis of AChRs. Laminin signals in the presynaptic nerve terminal help organize the active zones to align directly above the postjunctional folds. Together, all these events result in a highly organized NMJ, with synaptic vesicles located across from the AChRs that are clustered at the crest of the postjunctional folds. The numerous molecular signals and the precise timing of events create a synapse structured to provide rapid and reliable neural transmission from motor neurons to skeletal muscle. Not all NMJ connections that form during development remain into adulthood, however. Differences in neural and muscle activity at individual NMJs regulate signals, such as BDNF, that determine which synapses are lost and which are maintained. Further, PSCs release signals to support the remaining NMJ and participate in the removal of cellular debris from the lost connections. Understanding how cellular debris is normally removed may help scientists develop methods to target accumulations of misfolded proteins concentrated at sensorimotor nerve fibers (**Box 9.4**).

Scientists continue to investigate NMJ synaptogenesis and stabilization and have already identified additional molecules, beyond those described here, that appear to interact with MuSK and other postsynaptic molecules. Given the numerous signaling pathways identified to date and the availability of well-established assays for studying NMJ differentiation and synapse elimination, scientists will continue to expand our understanding of how the vertebrate NMJ is formed and reorganized.

FURTHER READING

Axelrod D (1983) Lateral motion of membrane proteins and biological function. *J Membr Biol* 75:1–10.

Balice-Gordon RJ & Lichtman JW (1993) In vivo observations of pre- and postsynaptic changes during the transition from multiple to single innervation at developing neuromuscular junctions. *J Neurosci* 13:834–855.

Balice-Gordon RJ & Lichtman JW (1994) Long-term synapse loss induced by focal blockade of postsynaptic receptors. *Nature* 372:519–524.

Brown MC, Jansen JK & Van Essen D (1976) Polyneuronal innervation of skeletal muscle in new-born rats and its elimination during maturation. *J Physiol* 261:387–422.

Burden SJ, Sargent PB & McMahan UJ (1979) Acetylcholine receptors in regenerating muscle accumulate at original synaptic sites in the absence of the nerve. *J Cell Biol* 82:412–425.

Chen PJ, Martinez-Pena YVI, Aittaleb M & Akaaboune M (2016) AChRs are essential for the targeting of rapsyn to the postsynaptic membrane of NMJs in living mice. *J Neurosci* 36:5680–5685.

Culican SM, Nelson CC & Lichtman JW (1998) Axon withdrawal during synapse elimination at the neuromuscular junction is accompanied by disassembly of the postsynaptic specialization and withdrawal of Schwann cell processes. *J Neurosci* 18:4953–4965.

Duxson MJ, Usson Y & Harris AJ (1989) The origin of secondary myotubes in mammalian skeletal muscles: Ultrastructural studies. *Development* 107:743–750.

Feng Z & Ko CP (2008) Schwann cells promote synaptogenesis at the neuromuscular junction via transforming growth factor-beta. *J Neurosci* 28:9599–9609.

Gautam M, Noakes PG, Moscoso L, et al. (1996) Defective neuromuscular synaptogenesis in agrin-deficient mutant mice. *Cell* 85:525–535.

Ghazanfari N, Fernandez KJ, Murata Y, et al. (2011) Muscle specific kinase: Organiser of synaptic membrane domains. *Int J Biochem Cell Biol* 43:295–298.

Glass DJ, Apel ED, Shah S, et al. (1997) Kinase domain of the muscle-specific receptor tyrosine kinase (MuSK) is sufficient for phosphorylation but not clustering of acetylcholine receptors: Required role for the MuSK ectodomain? *Proc Natl Acad Sci USA* 94:8848–8853.

Glass DJ, Bowen DC, Stitt TN, et al. (1996) Agrin acts via a MuSK receptor complex. *Cell* 85:513–523.

Godfrey EW, Nitkin RM, Wallace BG, et al. (1984) Components of torpedo electric organ and muscle that cause aggregation of acetylcholine receptors on cultured muscle cells. *J Cell Biol* 99:615–627.

Gu Y & Hall ZW (1988) Characterization of acetylcholine receptor subunits in developing and in denervated mammalian muscle. *J Biol Chem* 263:12878–12885.

Hallock PT, Xu CF, Park TJ, et al. (2010) Dok-7 regulates neuromuscular synapse formation by recruiting Crk and Crk-L. *Genes Dev* 24:2451–2461.

Herbst R (2020) MuSk function during health and disease. *Neurosci Lett* 716:134676.

Je HS, Yang F, Ji Y, et al. (2012) Role of pro-brain-derived neurotrophic factor (proBDNF) to mature BDNF conversion in activity-dependent competition at developing neuromuscular synapses. *Proc Natl Acad Sci USA* 109:15924–15929.

Je HS, Yang F, Ji Y, et al. (2013) ProBDNF and mature BDNF as punishment and reward signals for synapse elimination at mouse neuromuscular junctions. *J Neurosci* 33:9957–9962.

Kummer TT, Misgeld T & Sanes JR (2006) Assembly of the postsynaptic membrane at the neuromuscular junction: Paradigm lost. *Curr Opin Neurobiol* 16:74–82.

Lieth E & Fallon JR (1993) Muscle agrin: Neural regulation and localization at nerve-induced acetylcholine receptor clusters. *J Neurosci* 13:2509–2514.

Lin S, Maj M, Bezakova G, et al. (2008) Muscle-wide secretion of a miniaturized form of neural agrin rescues focal neuromuscular innervation in agrin mutant mice. *Proc Natl Acad Sci USA* 105:11406–11411.

Liu Y, Padgett D, Takahashi M, et al. (2008) Essential roles of the acetylcholine receptor gamma-subunit in neuromuscular synaptic patterning. *Development* 135:1957–1967.

McMahan UJ & Slater CR (1984) The influence of basal lamina on the accumulation of acetylcholine receptors at synaptic sites in regenerating muscle. *J Cell Biol* 98:1453–1473.

Meier T, Masciulli F, Moore C, et al. (1998) Agrin can mediate acetylcholine receptor gene expression in muscle by aggregation of muscle-derived neuregulins. *J Cell Biol* 141:715–726.

Nishimune H (2012) Active zones of mammalian neuromuscular junctions: Formation, density, and aging. *Ann NY Acad Sci* 1274:24–32.

Nishimune H, Sanes JR & Carlson SS (2004) A synaptic laminin-calcium channel interaction organizes active zones in motor nerve terminals. *Nature* 432:580–587.

Nitkin RM, Smith MA, Magill C, et al. (1987) Identification of agrin, a synaptic organizing protein from Torpedo electric organ. *J Cell Biol* 105:2471–2478.

Redfern PA (1970) Neuromuscular transmission in new-born rats. *J Physiol* 209:701–709.

Reist NE, Magill C & McMahan UJ (1987) Agrin-like molecules at synaptic sites in normal, denervated, and damaged skeletal muscles. *J Cell Biol* 105:2457–2469.

Sanes JR & Lichtman JW (1999) Development of the vertebrate neuromuscular junction. *Annu Rev Neurosci* 22:389–442.

Sanes JR & Lichtman JW (2001) Induction, assembly, maturation and maintenance of a postsynaptic apparatus. *Nat Rev Neurosci* 2:791–805.

Sanes JR, Marshall LM & McMahan UJ (1978) Reinnervation of muscle fiber basal lamina after removal of myofibers: Differentiation of regenerating axons at original synaptic sites. *J Cell Biol* 78:176–198.

Sekijima Y (2015) Transthyretin (ATTR) amyloidosis: Clinical spectrum, molecular pathogenesis and disease-modifying treatments. *J Neurol Neurosurg Psychiatry* 86(9):1036–1043.

Shi L, Fu AK & Ip NY (2012) Molecular mechanisms underlying maturation and maintenance of the vertebrate neuromuscular junction. *Trends Neurosci* 35:441–453.

Smith MA, Yao YM, Reist NE, et al. (1987) Identification of agrin in electric organ extracts and localization of agrin-like molecules in muscle and central nervous system. *J Experimtl Biol* 132:223–230.

Son YJ & Thompson WJ (1995) Schwann cell processes guide regeneration of peripheral axons. *Neuron* 14:125–132.

Tezuka T, Inoue A, Hoshi T, et al. (2014) The MuSK activator agrin has a separate role essential for postnatal maintenance of neuromuscular synapses. *Proc Natl Acad Sci USA* 111:16556–16561.

Valenzuela DM, Stitt TN, DiStefano PS, et al. (1995) Receptor tyrosine kinase specific for the skeletal muscle lineage: Expression in embryonic muscle, at the neuromuscular junction, and after injury. *Neuron* 15:573–584.

Walsh MK & Lichtman JW (2003) In vivo time-lapse imaging of synaptic takeover associated with naturally occurring synapse elimination. *Neuron* 37:67–73.

Witzemann V, Barg B, Criado M, et al. (1989) Developmental regulation of five subunit specific mRNAs encoding acetylcholine receptor subtypes in rat muscle. *FEBS Lett* 242:419–424.

Wu H, Xiong WC & Mei L (2010) To build a synapse: Signaling pathways in neuromuscular junction assembly. *Development* 137:1017–1033.

Yang X, Arber S, William C, et al. (2001) Patterning of muscle acetylcholine receptor gene expression in the absence of motor innervation. *Neuron* 30:399–410.

Zong Y & Jin R (2013) Structural mechanisms of the agrin-LRP4-MuSK signaling pathway in neuromuscular junction differentiation. *Cell Mol Life Sci* 70:3077–3088.

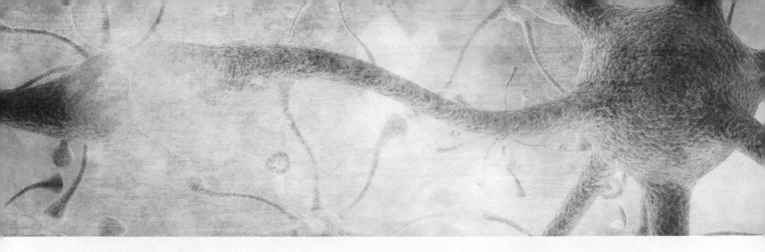

Synaptic Formation and Reorganization Part II: Synapses in the Central Nervous System

10

The formation of central nervous system (CNS) synapses begins early in embryogenesis. In many species, synapses are later modified or eliminated in response to intrinsic cues or neural activity. CNS synapses also continue to be reorganized throughout adulthood by a process called **neural plasticity**. Like the neuromuscular junction (NMJ) detailed in Chapter 9, synaptic formation, stabilization, and reorganization in the CNS rely on the exchange of reciprocal cues between presynaptic and postsynaptic partner cells.

Investigating synapse formation and reorganization in the brain and spinal cord presents more challenges than those encountered in studying the NMJ. There are many reasons why the NMJ has proven to be an advantageous experimental system including its relatively large size, experimental accessibility, structural simplicity, and use of a single neurotransmitter system (acetylcholine). In contrast, CNS synapses generally lack these same features. Typically, CNS synapses are smaller, less accessible, and more difficult to isolate from surrounding cells. In addition, CNS synapses can utilize one or more of the hundreds of identified neurotransmitters and neuropeptides. Further, the small size and vast number of neurons in the CNS make it difficult to isolate individual pre- and postsynaptic cells. One of the largest obstacles to studying specific synaptic connections in the CNS is that each neuronal cell body and dendrite is almost completely covered with the nerve terminals of synaptic partners and glial cell processes. The density of this innervation was highlighted in a 1969 study of cat spinal motor neurons in which a series of electron micrographs were compiled to illustrate the density of innervation on a single motor neuron (**Figure 10.1**A). In more recent years, fluorescent microscopy has been used to label synaptic elements and document synaptic contacts on CNS neurons under normal and experimental conditions (Figure 10.1B).

Despite the challenges of isolating individual pre- and postsynaptic regions, many synaptic specializations have been discovered in CNS neurons over the past 25 years. The number of unique synaptic elements identified in the tiny pre- and postsynaptic regions is rather astonishing. Among CNS synaptic specializations are proteins necessary for clustering synaptic

DOI: 10.1201/9781003166078-10

Figure 10.1 Neurons in the central nervous system (CNS) are contacted by multiple synaptic partners. (A) A drawing based on serial electron micrographs of a single adult cat spinal cord motor neuron reveals that the neuronal cell body is covered with multiple synapses (boutons and giant boutons) and processes from astrocytes. Each raised area on the neuron represents a point of synaptic or astrocyte contact. Dendrites and the axon also receive multiple synaptic contacts. (B) A fluorescent label specific for presynaptic terminals (red) reveals the large number of synaptic connections on the cell body and dendrites (green) of a single motor neuron. The axon is not stained in this image. [(A), From Poritsky R [1969] *J Comp Neurol* 135(4):432–451; (B), courtesy of Olaf Mundigl and Pietro de Camillo.]

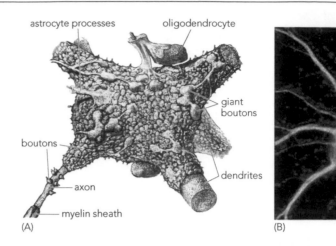

vesicles at the presynaptic active zone and proteins required for clustering neurotransmitter receptors at the postsynaptic membrane. Like the NMJ, having synaptic vesicles and neurotransmitter receptors closely associated ensures rapid and reliable synaptic transmission. Identifying synaptic elements and their multiple functions has allowed investigators to develop hypotheses to explain how synaptic formation, stabilization, reorganization, and elimination take place in the CNS.

This chapter focuses on some of the common mechanisms that influence the formation and reorganization of synaptic elements in the CNS. Many of the same signals and intracellular signaling pathways used at other stages of neural development are also used during these processes, providing further examples of how the nervous system utilizes available signals rather than creates new proteins for each developmental event. Among the molecules important for CNS synaptogenesis are cell adhesion molecules, brain-derived neurotrophic factor (BDNF), and members of the Wnt and Ephrin families. These molecules often interact with Rho GTPases or other proteins to influence cytoskeletal elements that in turn impact synapse stabilization, elimination, or remodeling.

The chapter concludes with descriptions of mechanisms that regulate synaptic reorganization and elimination in the CNS during early postnatal and adult stages. While these events take place throughout the vertebrate CNS, examples from the mammalian visual system are highlighted because development of this system has been described in previous chapters and the system remains one of the most common model systems for studies of synaptic refinement.

EXCITATORY AND INHIBITORY SYNAPSES IN THE CENTRAL NERVOUS SYSTEM

Understanding the structure and function of the different synaptic elements associated with excitatory and inhibitory synapses helps interpret the current models of CNS synaptogenesis. At **excitatory synapses,** the release of the presynaptic neurotransmitter causes the postsynaptic neuron to depolarize and thereby increases the likelihood that the neuron will fire an action potential. Excitatory synapses are usually formed between a presynaptic terminal and a postsynaptic **dendritic spine**, the small protrusion that extends from the dendritic shaft. A smaller number of excitatory synapses form between the axon terminal and the dendritic shaft. The excitatory synapses that contact dendritic spines primarily form in one of two ways. A **terminal synapse** forms when the presynaptic nerve terminal contacts a dendritic spine. Such synapses are most like those seen in the vertebrate NMJ, in which a nerve terminal contacts a muscle fiber. In

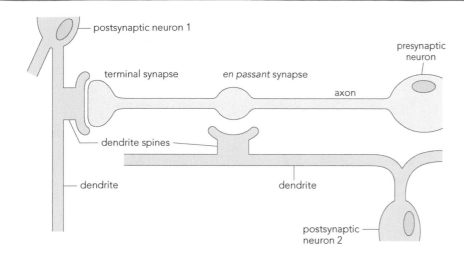

Figure 10.2 Organization of terminal and *en passant* synapses in the CNS. An excitatory terminal synapse forms between a presynaptic axon terminal and a postsynaptic dendritic spine, as shown at postsynaptic neuron 1. An *en passant* synapse occurs when presynaptic specializations form along the axon shaft and contact a dendritic spine (postsynaptic neuron 2). Inhibitory synapses are structurally similar, but typically form along the dendritic shaft.

contrast, an ***en passant* synapse** forms when a region of the axon shaft enlarges and produces presynaptic specializations that contact a dendritic spine on the postsynaptic cell (**Figure 10.2**). These are similar in morphology to the synapses that form in the invertebrate nervous system.

Glutamate is the primary excitatory neurotransmitter in the CNS. Glutamate binds to several different glutamate receptors, including those that form ion channels (ionotropic) and those that use G proteins (metabotropic). Two ionotropic receptors that have been extensively studied during synaptogenesis are the NMDA (*N*-methyl D-aspartic acid) and AMPA (alpha-amino-3-hydroxy-5-methyl-4-isoxazolepropionic acid) receptors. These names reflect the name of specific chemical agonists that bind to each receptor.

In addition to glutamate, some excitatory neurons package and release other neurotransmitters, such as acetylcholine (ACh), serotonin (5-hydroxytryptamine, 5-HT), dopamine (DA), or one of the various neuropeptides. In some cases, a single excitatory neuron will release more than one neurotransmitter. This typically involves glutamate and one of the modulatory neurotransmitters such as histamine, 5-HT, or DA. The combination of neurotransmitters helps determine whether the postsynaptic cell will fire an action potential.

Inhibitory synapses often form at the dendritic shaft or near the postsynaptic cell body, although some of these synapses are located at distal regions or dendritic spines. At inhibitory synapses, the release of the presynaptic neurotransmitter hyperpolarizes the postsynaptic neuron, decreasing the likelihood that the neuron will fire an action potential.

Most inhibitory neurons package and release GABA (gamma aminobutyric acid) or glycine—neurotransmitters that bind to corresponding receptors on the postsynaptic neuron. As with the excitatory neurons, neuromodulators may be co-released with inhibitory neurotransmitters to regulate the response of the postsynaptic cell and determine whether an action potential fires.

Many Presynaptic and Postsynaptic Elements Are Similar in Excitatory and Inhibitory Synapses

The general organization of pre- and postsynaptic elements is similar for excitatory and inhibitory synapses in the CNS. Like the presynaptic terminal of a motor neuron, the presynaptic terminals of CNS axons contain synaptic vesicles and active zone proteins. These presynaptic elements are concentrated above the corresponding neurotransmitter receptors clustered at the postsynaptic membrane. Other presynaptic elements include voltage-gated ion channels, adhesion molecules, and scaffolding proteins that help anchor the presynaptic elements at the presynaptic site (**Figure 10.3**).

Figure 10.3 Pre- and postsynaptic specializations in excitatory and inhibitory CNS synapses. (A) Presynaptic elements in an excitatory glutamatergic synapse include the glutamate-releasing synaptic vesicles, active zone proteins, voltage-gated ion channels, scaffold proteins, and adhesion molecules. The postsynaptic dendritic spine is characterized by a prominent postsynaptic density (PSD) that contains the NMDA and AMPA neurotransmitter receptors, voltage-gated ion channels, adhesion molecules, and scaffold proteins such as postsynaptic density-95 (PSD-95). (B) Similar synaptic elements are found in the presynaptic cells of an inhibitory synapse. GABA (γ-aminobutyric acid) and glycine are common inhibitory neurotransmitters released by different inhibitory neurons. Active zone proteins, voltage-gated ion channels, adhesion molecules, and scaffold proteins are also found in the presynaptic terminal of inhibitory neurons. Inhibitory synapses typically form along the dendritic shaft that lacks a prominent postsynaptic density. However, scaffold proteins concentrate at the site of synaptic contact to help anchor GABA or glycine receptors near the presynaptic elements.

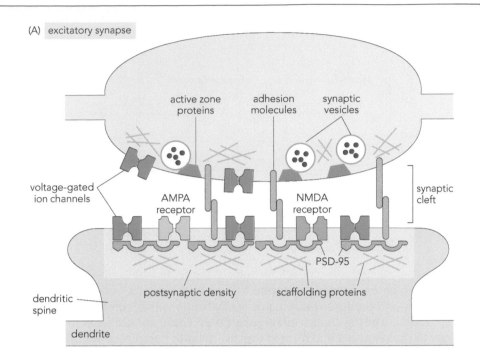

(A) excitatory synapse

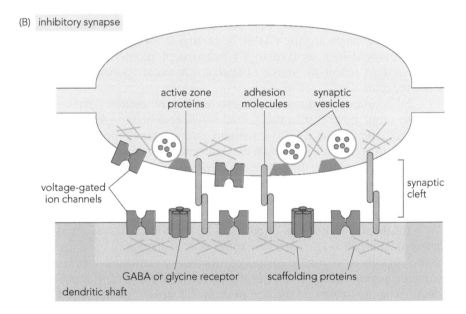

(B) inhibitory synapse

The postsynaptic specializations of CNS neurons include neurotransmitter receptors, voltage-gated ion channels, and adhesion molecules. The postsynaptic elements clustered at the site of synaptic contact are also held in place by various scaffold proteins, often with distinct proteins found in excitatory and inhibitory neurons (Figure 10.3).

Unlike the NMJ, there is no synaptic basal lamina in the synaptic cleft between CNS neurons, although there is a concentration of visible proteinaceous material. Because the synaptic cleft is smaller at CNS synapses, the pre- and postsynaptic cells are close enough that adhesion molecules, such as those described in a subsequent section, can span the cleft and stabilize connections (Figure 10.3).

The Postsynaptic Density Is an Organelle Found in Excitatory, but Not Inhibitory, Synapses

In 1959, soon after the identification of synapses by electron microscopy, E. George Gray reported structural differences in the postsynaptic regions of excitatory and inhibitory synapses. These studies revealed two types of synapses, called type I (asymmetric) and type II (symmetric), based on the appearance of electron-dense material (**Figure 10.4**). In symmetric synapses, the pre- and postsynaptic densities appear similar. With asymmetric neurons, however, a prominent electron-dense region is observed in the postsynaptic cell. This difference is due to the presence of a specialized region in excitatory neurons called the **postsynaptic density** (**PSD**). It is now understood that the asymmetric synapses are excitatory, whereas the symmetric synapses are inhibitory. Unlike excitatory synapses, inhibitory synapses do not have a prominent PSD. Instead, the postsynaptic membrane is comprised of a simpler protein composition that accomplishes similar functions, such as clustering neurotransmitter receptors (see Figure 10.3B).

The PSD of excitatory synapses is a disc-shaped, electron-dense organelle that adheres to the postsynaptic cell membrane to anchor proteins adjacent to the active zones of the presynaptic cell. Since the 1970s, scientists have isolated hundreds of proteins from this small postsynaptic region. Neurons in different regions of the CNS express different PSD proteins. Among the common PSD proteins are ion channels, cytoskeletal proteins, adhesion molecules, signaling kinases, and scaffolding proteins. The protein composition of the PSD is linked to the function of each excitatory synapse, and size of the PSD is thought to correlate with the strength of the synaptic connection. Examples of how changes in the protein composition of the PSD occur as synapses are strengthened, weakened, or reorganized during development and adulthood are described in later sections of the chapter.

Of the scaffold proteins found in postsynaptic excitatory neurons, the PSD-95 (postsynaptic density protein-95) family is one of the best characterized to date (see Figure 10.3A). The PSD-95 family contains four members: PSD-95, PSD-93, and the synapse-associated proteins-97 and -102 (SAP-97 and SAP-102, respectively). Currently, the role of PSD-95 is the best understood.

PSD-95 interacts with other proteins to cluster and anchor those proteins within the PSD. These interactions occur through the PDZ domains

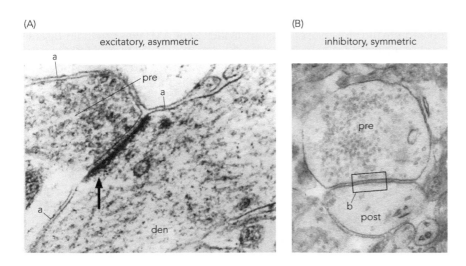

(A) excitatory, asymmetric

(B) inhibitory, symmetric

Figure 10.4 Electron micrographs reveal symmetric and asymmetric synapses. In 1959 George Gray published photomicrographs highlighting differences in the postsynaptic densities of excitatory (asymmetric) and inhibitory (symmetric) synapses. (A) The postsynaptic region of the dendrite (den) of excitatory synapse reveals an area of electron-dense material (arrow) that is not observed in other regions of the dendrite (areas labeled a) or in the presynaptic nerve terminal (pre). Thus, the synapse is asymmetric. (B) At inhibitory synapses, the regions on either side of the synaptic cleft (box, b) appear similar because the postsynaptic region does not contain any electron-dense material. Thus, inhibitory synapses are symmetrical (pre, presynaptic nerve terminal; post, postsynaptic neuron). [From Gray G [1959] *J Anat* 93:420–433.]

Figure 10.5 Adhesion molecules stabilize contacts between pre- and postsynaptic cells. Cadherins are thought to stabilize initial synaptic connections. Cadherins bind to one another through homophilic mechanisms that require calcium (Ca^{2+}). SynCAM I (synaptic cell adhesion molecule I) is a member of the IgG superfamily that also interacts through homophilic binding, but unlike cadherins, SynCAM I does not require calcium. The binding of SynCAM I molecules leads to interaction with scaffold proteins on pre- and postsynaptic neurons to stabilize cytoskeletal elements. Other nontraditional adhesion molecules also help stabilize synaptic contacts. In some neurons, ephrin B ligands on the presynaptic neuron bind to EphB receptors on the postsynaptic neuron. The cytoplasmic tails of these molecules are also thought to interact with scaffold proteins to stabilize cytoskeletal elements.

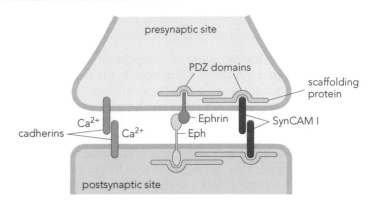

found in each of the proteins. PDZ domains were named for the three proteins in which the domains were first identified—the vertebrate P̲SD-95, the *Drosophila* d̲iscs large, and the tight junction protein (Z̲O-1). PDZ domains are about 90 amino acids in length and are among the most common protein domains known to interact with and form large complexes with other proteins. PDZ-containing proteins are particularly prominent in areas of cell–cell contact, such as synapses.

Cell Adhesion Molecules Mediate the Initial Stabilization of Synaptic Contacts

In the CNS, it appears that the pre- and postsynaptic cells first establish tentative contacts and then rely on reciprocal signals to determine whether further differentiation into a mature synaptic contact will occur. The initial tentative contacts are established through interactions with various cell adhesion molecules (CAMs) found on each putative synaptic partner. Several CAMs influence the stabilization of CNS synapses including members of the cadherin and IgG superfamilies. In the presence of calcium, cadherins interact through homophilic binding mechanisms to mediate contact and stabilization of pre- and postsynaptic partners (**Figure 10.5**). Cadherins also influence the postsynaptic clustering of glutamate and GABA receptor subunits of excitatory and inhibitory synapses, respectively.

SynCAM I (synaptic cell adhesion molecule I) is a member of the IgG superfamily expressed throughout the vertebrate CNS, particularly at excitatory synapses. The SynCAM family consists of four genes that are expressed only in vertebrates. SynCAM I is found in several tissues, but its highest level of expression is in the CNS where peak expression corresponds to periods of synaptogenesis. SynCAM I homophilic binding does not require the presence of calcium. In fact, SynCAM I was originally discovered in a search for neuronal cell adhesion molecules that did not require calcium. Among the functions associated with SynCAM I are the induction of presynaptic specializations and the stabilization of pre- and postsynaptic elements. Stabilization of synaptic connections occurs because the cytoplasmic regions of SynCAM I interact with the PDZ domains of scaffold proteins in the pre- and postsynaptic cells (Figure 10.5). The scaffold proteins help anchor the appropriate synaptic elements and stabilize the synaptic connection. Similar mechanisms are used by other members of the IgG superfamily including the nectins, a family of four cell–cell adhesion molecules. For example, nectin 3, expressed on presynaptic neurons can interact with nectin 1 expressed on postsynaptic neurons. Synapses are stabilized through the interaction of nectin's cytoplasmic tails with the PDZ domains of scaffold proteins, such as the actin-binding protein afadin. The indirect link to the cytoskeleton allows nectins to influence synaptic reorganization. Nectins also interact with cadherins and ephrin/Eph

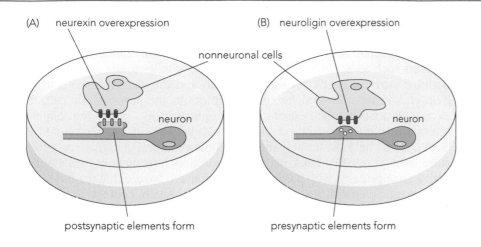

(A) neurexin overexpression

(B) neuroligin overexpression

nonneuronal cells

neuron

neuron

postsynaptic elements form

presynaptic elements form

Figure 10.6 *In vitro* **studies reveal that neurexin and neuroligin induce different synaptic elements.** Neurexin is a presynaptic protein that binds to postsynaptic neuroligin. (A) Neurexin was overexpressed in nonneuronal cells then cultured with CNS neurons. The neurons that contacted the neurexin-expressing nonneuronal cells began to accumulate postsynaptic elements such as neurotransmitter receptors and PSD proteins. (B) In contrast, when neuroligin was overexpressed in nonneuronal cells, the contacting neurons produced presynaptic elements such as active zone proteins and synaptic vesicles. Together, these experiments demonstrated that the neurexin–neuroligin interactions induce pre- or postsynaptic specializations in partner cells.

molecules located at synapses. In some CNS neurons ephrin ligands and Eph receptors help stabilize newly formed CNS synapses. As with other adhesion molecules the cytoplasmic regions of the ephrin B ligands and the EphB2 receptors interact with the PDZ domains of scaffold proteins to stabilize synaptic connections (Figure 10.5).

Neurexins and Neuroligins Also Induce Formation of Synaptic Elements and Stabilize Synaptic Contacts

Neurexins and neuroligins, first discovered in the 1990s, are a family of cell surface proteins expressed in neurons throughout the CNS. The neurexins are in the presynaptic cell, while the neuroligins are in the postsynaptic cell. The two molecules attach to each other through heterophilic binding mechanisms and signal bidirectionally so that each induces synaptic elements in the partner cell. Evidence of this bidirectional signaling was demonstrated in cell culture experiments in which nonneuronal cell lines were generated to overexpress either neurexin or neuroligin. The nonneuronal cells were then co-cultured with CNS neurons. In cultures with nonneuronal cells expressing neurexins, postsynaptic elements were induced in the neurons. In contrast, in cultures in which nonneuronal cells expressed neuroligin, presynaptic elements were induced in the neurons (**Figure 10.6**).

Each binding partner in the neurexin–neuroligin complex has an intracellular domain that interacts with the PDZ domains of local scaffold proteins. On the presynaptic side, neurexins, of which there are over 3000 potential isoforms, interact with the PDZ domains associated with active zone proteins (**Figure 10.7**). It has been proposed that the numerous

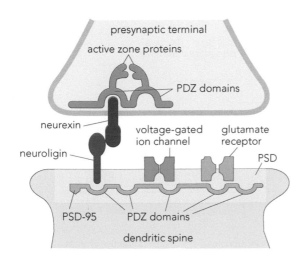

presynaptic terminal

active zone proteins

PDZ domains

neurexin

voltage-gated ion channel

glutamate receptor

PSD

neuroligin

PSD-95

PDZ domains

dendritic spine

Figure 10.7 Neurexin and neuroligin interact with PDZ domains of scaffold proteins. In the presynaptic terminal, neurexin interacts with PDZ domains of scaffold proteins that help anchor associated active zone proteins. On the postsynaptic neuron, neuroligin interacts with a PDZ domain of PSD-95. The multiple PDZ domains of PSD-95 help anchor glutamate receptors and ion channels to the postsynaptic site.

potential isoforms of neurexin help establish synaptic specificity in the CNS. However, whether this occurs has not yet been determined.

At presynaptic sites, neurexins interact with PDZ domains of scaffold proteins such as Scribble to localize synaptic vesicles to the active zones. Scribble provides a link between neurexin and the actin cytoskeleton to help cluster synaptic vesicles near receptors in the postsynaptic cell. In excitatory postsynaptic neurons, the cytoplasmic tails of the neuroligins interact with a PDZ binding domain of PSD-95. PSD-95 has multiple PDZ domains that help cluster and anchor ion channels and glutamate receptor subunits (Figure 10.7). Thus, neurexin–neuroligin binding links presynaptic active zone proteins with postsynaptic proteins to ensure that the pre- and postsynaptic elements are closely aligned at the site of contact.

One model of how CNS connections form suggests that adhesion molecules such as the cadherins are important in forming initial synaptic connections. Then proteins that interact with scaffold proteins, such as SynCAM I, EphB/ephrin B, the nectins, or neurexin/neuroligin, stabilize the pre- and postsynaptic elements. Further stabilization and maturation of synaptic elements occurs as the pre- and postsynaptic cells continue to exchange reciprocal signals.

Reciprocal Signals Regulate Pre- and Postsynaptic Development

Like the NMJ, specific pre- and postsynaptic elements are present before synaptogenesis begins. Many of the necessary proteins are therefore ready to interact with each other as soon as a synaptic contact is made. For example, prepatterning is observed in the presynaptic axons of excitatory *en passant* synapses. Initially, as an axon extends, synaptic vesicle proteins transiently accumulate at several locations along the axon. Synaptic vesicles then begin to preferentially accumulate at specific sites along the axon shaft. A number of signals likely determine where the presynaptic terminal will form. In some cases, it appears that local inhibitory signals prevent formation of presynaptic elements in certain areas of the axon. In *C. elegans*, for example, the netrin homolog Unc-6 is secreted by local guidepost cells to inhibit presynaptic specializations from forming at adjacent axonal locations. It is hypothesized that similar mechanisms operate in the vertebrate nervous system, with inhibitory cues arising from local nonneuronal cells or the postsynaptic dendrite (**Figure 10.8**).

Prepatterning of postsynaptic elements also occurs. Dendritic spines begin to accumulate postsynaptic elements prior to synaptic contact. However, contact with the presynaptic neuron is required for further maturation. The immature dendritic spine makes several transient contacts with

Figure 10.8 Pre- and postsynaptic signals determine where an excitatory synapse will form. Presynaptic specializations only form at restricted regions of the axon shaft. Inhibitory signals from the dendrite shaft and adjacent nonneuronal cells help prevent presynaptic elements from forming at other sites along the axon. The developing presynaptic site releases signals that attract and stabilize dendritic spines to the correct location.

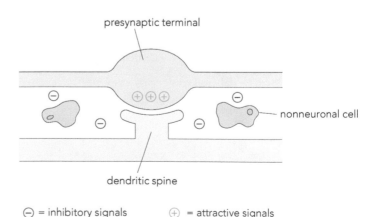

presynaptic terminal

nonneuronal cell

dendritic spine

⊖ = inhibitory signals   ⊕ = attractive signals

the presynaptic axon shaft in what appears to be an attempt by the dendrite to identify and respond to cues originating in the presynaptic axon. In at least some neuronal populations, the presynaptic neuron may provide attractive signals that direct the initial contact between the immature postsynaptic dendritic spine and the presynaptic axon (Figure 10.8).

Dendritic Spines Are Highly Motile and Actively Seek Presynaptic Partners

Dendritic spines begin as highly motile filopodia that extend from the dendritic shaft (**Figure 10.9**A). These immature spines are slender processes rich in filamentous actin (F-actin) that seem to function much like the filopodia that project from the growth cone (see Chapter 7). The dendritic filopodia eventually alter their morphology and become mature dendritic spines (Figure 10.9B–D).

The morphology of dendritic spines is characterized by the overall shape of the neck and head regions. Those identified as stubby spines have no obvious neck and a larger head, whereas thin spines have a small head and long, thin neck. Mushroom-shaped spines have a large head and a thick neck (**Figure 10.10**). The size of the spine is associated with the synaptic strength, with larger heads being stronger due, in part, to the increased space available for the postsynaptic neurotransmitter receptors.

The development of dendritic spines is a process called **spinogenesis**. In many ways, spinogenesis appears to be linked to synaptogenesis, because the number of spines is correlated with the number of synapses that form. However, there are differences that suggest the two developmental events are not directly linked. For example, dendritic spines form over a period of minutes, whereas the formation of mature synapses often takes place over a period of days or weeks. Moreover, in at least some cellular contexts, spinogenesis can occur in the absence of axonal contact. In both *Weaver* and *Reelin* mutant mice that are missing presynaptic cerebellar granule cells (see Chapter 5), the dendritic spines on postsynaptic Purkinje cells still develop and appear morphologically similar to wild-type mice. These findings suggest that spines can form independent of synaptic contact. However, it is hypothesized that there are subtle structural changes in

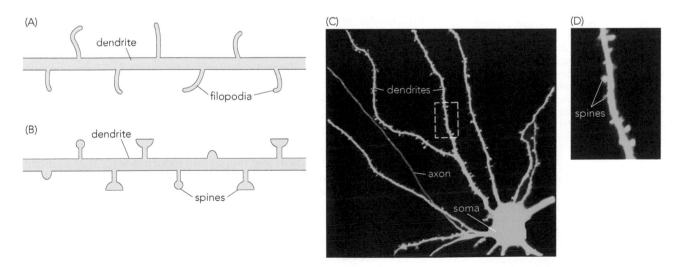

Figure 10.9 Dendritic filopodia are highly motile extensions that later form spines. (A) Dendritic filopodia are actin-rich protrusions that are highly motile prior to synapse formation. As each filopodium establishes a point of synaptic contact, it is converted to a dendritic spine (B). (C) An example of dendritic spines in a hippocampal neuron expressing green fluorescent protein. The boxed region is shown enlarged in panel D. Source: (C and D), From Tolias KF, Duman JG & Um K [2011] *Prog Neurobiol* 94:133–148.

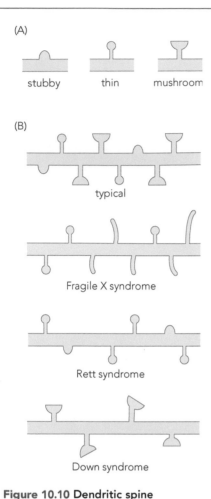

(A)

stubby   thin   mushroom

(B)

typical

Fragile X syndrome

Rett syndrome

Down syndrome

Figure 10.10 Dendritic spine morphology is altered in neurodevelopmental disorders. (A) Mature dendritic spines reveal different morphologies, which are described as stubby, thin, and mushroom-shaped. (B) Compared to the spine density and morphologies observed in typical brains, those with neurodevelopmental disorders reveal changes characteristic for each disorder. With Fragile X syndrome, spine density is variable and the spines present tend to be longer and immature, often resembling filopodia. In Rett syndrome, there is a decrease in spine density and fewer mushroom-shaped spines, whereas in Down syndrome, there are fewer spines and those present have larger, atypical spine heads. [(A), Adapted from Yuste R & Bonhoeffer T [2004] *Nat Rev Neurosci* 5:24; (B), adapted from Phillips M & Pozzo-Miller L [2015] *Neurosci Lett* 601:30–40.]

postsynaptic specializations due to the loss of reciprocal signals between pre- and postsynaptic partners.

Whether spinogenesis is directly linked to synapse formation or not, the importance of spinogenesis is noted by the large number of neurodevelopmental disorders associated with decreases in dendritic spine number and altered morphology. For example, reduced spine density and alterations in spine morphology have been noted in schizophrenia, autism, Fragile X syndrome, Rett syndrome, and Down syndrome. Studies of these disorders in mouse models or human autopsy specimens have revealed many consistent changes associated with each disorder. For example, compared to the spine density and morphology observed in typical adult brains, those with Fragile X syndrome have long spines that resemble filopodia, a morphology that suggests these spines never reach a mature state (Figure 10.10B). The density of spines is variable, at least in those specimens observed to date. In individuals with Rett syndrome, spine density is reduced and there are fewer spines exhibiting a mushroom-shaped morphology. Individuals with Down syndrome have fewer spines and those present tend to have large heads (Figure 10.10B). Studies indicate that spine density begins to decrease after 4 months of age, suggesting there is excess pruning of putative synaptic connections during early postnatal development. As discussed later in the chapter, loss of synaptic connections is a normal part of postnatal development, but in those with Down syndrome the loss of dendritic spines is much greater.

BDNF Influences Dendritic Spine Motility and Synaptogenesis

Molecules important for other aspects of neural development, such as brain-derived neurotrophic factor (BDNF), also influence dendritic spine motility and synapse formation. In mice lacking the BDNF receptor TrkB, there were fewer dendritic spines and a reduced number of synapses in the hippocampus. Other studies found that BDNF released by presynaptic neurons activated TrkB receptors on postsynaptic dendrites, increasing dendritic filopodia motility, which, in turn, increased the likelihood that a filopodium would initiate a synaptic connection.

In recent years, mechanisms by which BDNF can influence filopodia motility have been identified. BDNF influences several Rho family small GTPase signaling pathways. Rho family small GTPases, a subfamily of the Ras superfamily of small GTPases, are G proteins utilized in multiple intracellular signaling cascades and often influence actin and microtubule dynamics (see Chapter 5). Rho GTPases are activated by various guanine nucleotide exchange factors (GEFs) that convert GDP to GTP. The ability of BDNF to influence different GEFs and activate Rho GTPases provides a direct means for this neurotrophin to regulate cytoskeletal dynamics during synaptogenesis.

For example, a role for GTPase signaling pathways in synaptogenesis was observed in cultured rat hippocampal neurons where BDNF activated the GEFs, Vav2 (vav guanine nucleotide exchange factor 2), and Tiam1 (T-cell lymphoma invasion and metastasis inducing protein 1) to influence formation of dendritic spines. In other experiments, synapse formation was induced when BDNF activated the Rho family GTPase Rac1 (Ras-related C3 botulinum toxin substrate 1). Thus, by activating specific GEFs and Rho family GTPases, BDNF-TrkB signaling regulates the motility and stability of the cytoskeletal elements that influence the formation and stability of synaptic connections (**Figure 10.11**).

BDNF is also known to increase the density of postsynaptic receptors in spinal cord inhibitory neurons. The mechanisms by which changes in receptor density occur is not entirely clear, but it seems to involve signaling through the MAP kinase pathway that is activated following the binding

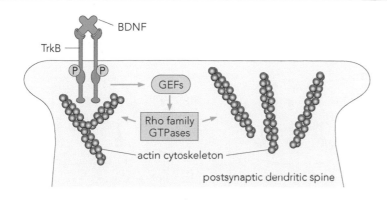

postsynaptic dendritic spine

Figure 10.11 BDNF activates Rho family small GTPases to influence cytoskeletal dynamics. BDNF released by the presynaptic terminal binds to TrkB on the postsynaptic neuron. Phosphorylated TrkB can then activate GEFs (guanine exchange factors), which stimulate activity of various members of the Rho family of small GTPases. The Rho GTPases then influence the motility of the actin cytoskeleton to regulate synapse formation and stability.

of BDNF to TrkB (see Chapter 8). Thus, BDNF signaling activates different intracellular signaling cascades to influence multiple aspects of synaptic development. Notably, TrkB can be rapidly transported to the cell surface from existing intracellular stores so BDNF signaling can occur when needed during synapse formation and reorganization.

Eph/Ephrin Bidirectional Signaling Mediates Presynaptic Development

The ephrin ligands and Eph receptors, both the EphA/ephrin A and EphB/ephrin B subclasses, influence synaptogenesis. As noted in earlier chapters, ephrin ligands bind to Eph tyrosine kinase receptors to initiate signal transduction cascades in the receptor-bearing cell through forward signaling. In addition, the binding of the Eph receptors to the membrane-bound ligand can also induce signal transduction cascades in the ligand-bearing cell through the process of reverse signaling. Both forward and reverse signaling mechanisms are used during pre- and postsynaptic development. For example, in some cellular contexts, clustering of presynaptic elements occurs when presynaptic Eph receptors are activated via forward signaling. In other contexts, presynaptic elements cluster when presynaptic ephrin ligands are activated via reverse signaling. To illustrate how these signals coordinate during synaptogenesis, examples of EphB/ephrin B signaling are highlighted.

Experiments conducted in the laboratories of Matthew Dalva and Mark Henkemeyer have contributed to current models of how ephrin signaling influences synapse formation. For example, an *in vitro* assay was developed in which ephrin B3 was expressed in nonneuronal cells. These cells were then co-cultured with hippocampal neurons. A fluorescently labeled marker for excitatory presynaptic neurons (VGLUT1—vesicular glutamate transporter 1—a protein found in the membrane of glutamate-containing synaptic vesicles) was then used to assess whether ephrin B3 induced presynaptic specializations in the hippocampal neurons. The presynaptic marker was only observed in the axons that contacted the ephrin B3-expressing nonneuronal cells. This suggested ephrin B3 mediated presynaptic differentiation of excitatory synapses by binding to endogenous EphB receptors on the hippocampal neurons (forward signaling). This finding was consistent with other studies showing EphB activation leads to presynaptic differentiation. In contrast to VGLUT1 labeling, presynaptic markers of inhibitory synapses were not detected in any of the cultures.

Postsynaptic EphB receptors signaling through presynaptic ephrin B1 and ephrin B2 ligands (reverse signaling) also influence presynaptic development by regulating the clustering of synaptic vesicles and the maturation of active zone proteins. Evidence for ephrin B-directed synaptic vesicle clustering was found in *in vitro* studies of embryonic rat cortical neurons co-cultured with nonneuronal cells expressing EphB receptors. Clustering of synaptic vesicles occurred at the sites of contact between

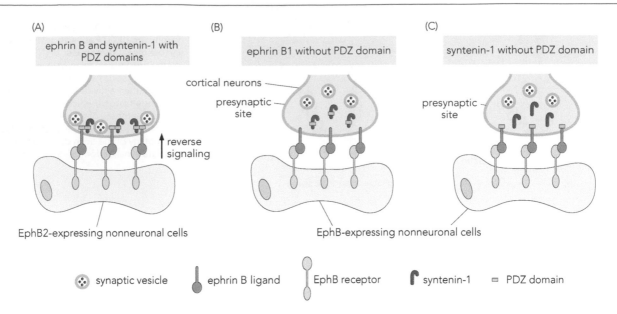

Figure 10.12 EphB receptors signal through ephrin B1 and ephrin B2 ligands to cluster presynaptic vesicles in cortical neurons. Cortical neurons were co-cultured with nonneuronal cells expressing EphB receptors. (A) When EphB-expressing nonneuronal cells contact cortical neurons expressing ephrin B1 or ephrin B2, reverse signaling leads to interaction of ephrin B PDZ domains with the PDZ domains of syntenin-1. This interaction induces clustering of synaptic vesicles at the site of contact. (B) Synaptic vesicle clustering did not occur when cortical neurons overexpressed ephrin B1 lacking the PDZ domain. (C) Similarly, when the PDZ domains of syntenin-1 were absent, the synaptic vesicles failed to cluster.

the EphB-expressing nonneuronal cells and the cortical neurons. Using a similar approach, subsequent experiments revealed that if *ephrinB1* or *ephrinB2* gene expression were reduced using the short interfering RNA method (siRNA; see Chapter 4), fewer synaptic vesicles were detected in the presence of EphB-expressing nonneuronal cells. If the expression of both *ephrinB1* and *ephrinB2* were reduced, a further decrease in synaptic vesicle accumulation was noted. In contrast, cortical neurons with decreased levels of another ligand, *ephrinB3*, did not have a corresponding decrease in synaptic vesicle clustering. Thus, only ephrin B1 and ephrin B2 appear to influence presynaptic vesicle clustering in cortical neurons.

To understand how these ephrin Bs might influence clustering of synaptic vesicles, cortical neurons expressing wild-type ephrin B1 were compared to cortical neurons expressing ephrin B1 that lacked the PDZ binding domain. In the neurons expressing the mutated form of ephrin B1, the addition of EphB-expressing nonneuronal cells now failed to influence synaptic vesicle density. This suggested that the PDZ binding domains of ephrin B1 are required to recruit the presynaptic proteins needed to cluster synaptic vesicles (**Figure 10.12**).

Syntenin-1 is a protein known to interact with ephrin B during presynaptic development. Syntenin-1 is a PDZ-containing adaptor protein involved in cellular processes that require formation of protein complexes. Syntenin-1 was first identified as a melanoma differentiation gene and has since been found to be important not only for tumorigenesis, but also for receptor trafficking and synapse formation. Syntenin-1 is located at both the presynaptic active zone and the PSD, indicating roles in assembling presynaptic and postsynaptic signaling molecules.

To investigate the function of syntenin-1 in presynaptic development, syntenin-1 lacking the PDZ domain was overexpressed in cortical neurons. In these neurons, the EphB2-expressing nonneuronal cells no longer induced synaptic vesicle clustering, similar to what was observed when the PDZ binding domain of ephrin B1 was missing (Figure 10.12C). This suggests that syntenin-1 PDZ domains interact with ephrin B1 PDZ domains to cluster the presynaptic vesicles and associated proteins at the proper site. One hypothesis is that the ephrin Bs recruit syntenin-1 to the proper presynaptic location so that syntenin-1 can then anchor synaptic vesicle proteins.

Eph/Ephrin Signaling Initiates Multiple Intracellular Pathways to Regulate the Formation of Postsynaptic Spine and Shaft Synapses

Studies in cortical and hippocampal neurons also indicate that ephrin–Eph interactions influence aspects of postsynaptic development. For example, reverse signaling initiated by presynaptic EphB receptors influences differentiation of ephrin-expressing postsynaptic neurons. These ephrins co-localize with PSD-95 and other PDZ-containing scaffold proteins located at the postsynaptic cell membrane. Interestingly, Eph–ephrin interactions appear to mediate different aspects of synaptic formation depending on the age and type of CNS neuron, as well as the specific PDZ-containing adaptor proteins utilized. For example, *in vitro* experiments suggest that postsynaptic ephrin B3 interacts with the adaptor protein GRIP1 (glutamate-receptor-interacting protein 1) to regulate development of excitatory dendritic shaft synapses whereas ephrin B3 interacts with different adaptor proteins to influence development of spine synapses.

In one series of cell culture assays, an EphB receptor was added to cultures of hippocampal neurons to initiate reverse signaling through ephrin B3. This ephrin B3 signaling induced GRIP1 clustering at the postsynaptic membrane of shaft synapses (**Figure 10.13**A). GRIP1 is an adaptor protein that interacts with both ephrin B3 and AMPA receptors. The direct interaction of ephrin B3 and GRIP1 was demonstrated when the PDZ domains of ephrin B3 were mutated then overexpressed in hippocampal neurons. The overexpression of the mutated ephrin B3 acted as a dominant negative and prevented normal interactions between the PDZ domains of ephrin B3 and GRIP1. In the absence of functional ephrin B3 PDZ domains, the addition of EphB2 to the cell cultures no longer induced GRIP1 clustering. Thus, under normal developmental conditions, Eph/ephrin B3 reverse signaling appears to recruit GRIP1 to the membrane of postsynaptic shaft synapses where it presumably helps anchor glutamate receptors below the presynaptic terminal.

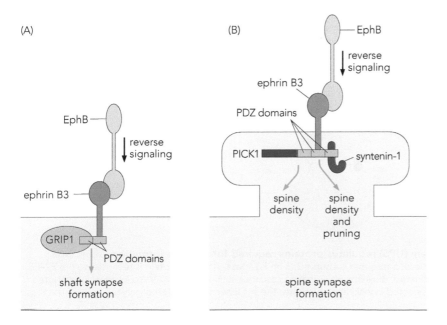

Figure 10.13 Postsynaptic ephrin B3 interacts with different adaptor proteins to regulate excitatory shaft and excitatory spine synapse development. In hippocampal neurons, EphB activates postsynaptic ephrin B3 through reverse signaling. (A) Ephrin B3 interacts with the adaptor proteins GRIP1 (glutamate-receptor-interacting protein 1) to influence formation of shaft synapses, but not spine synapses. (B) The density of spine synapses increases when activated ephrin B3 interacts with the adaptor protein PICK1 (protein interacting with C kinase-1) or with syntenin-1. Syntenin-1 can also regulate the subsequent pruning of excess synapses. Both PICK1 and syntenin-1 interact with cytoskeletal elements in the dendritic spine, thereby linking ephrin B3 signaling to dendritic spine structure and density.

In studies of hippocampal dendritic spine synapses, when presynaptic EphB receptors activated postsynaptic ephrin B3, the PDZ-containing adaptor proteins PICK1 (protein interacting with C kinase-1) and syntenin-1 were recruited to the postsynaptic cell membrane. When the PDZ domains of ephrin B3 interacted with the PDZ domains of PICK1, dendritic spine density increased. In contrast, ephrin B3 PDZ domains interacting with syntenin-1 PDZ domains influenced both spine density and the subsequent pruning of excess spines. Thus, postsynaptic modifications resulting from ephrin B signaling take place not only during synaptogenesis, but also during synapse maturation and refinement.

Evidence suggests that in some cases, Eph/ephrin signaling pathways may intersect with BDNF-TrkB signaling pathways to impact synapse formation, maturation, and refinement. Additionally, like BDNF-TrkB signaling, Eph/ephrin signaling also leads to the activation of specific GEFs and Rho GTPases to mediate cytoskeletal dynamics that in turn regulate spine motility and stabilization. Other signaling pathways regulate cytoskeletal dynamics in pre- and postsynaptic neurons. For example, the ubiquitin proteasome system (UPS) influences cytoskeletal dynamics in the presynaptic terminals of invertebrate and vertebrate synapses (**Box 10.1**).

Box 10.1 Ubiquitin-Mediated Pathways Regulate Synapse Formation

The ubiquitin proteasome system (UPS) was originally discovered in the mid-twentieth century as a system for degrading proteins. Proteins tagged with ubiquitin, a 76-amino acid peptide that attaches to other proteins, are transferred to one of the many **proteasomes** found in the nucleus or cytoplasm of the cell. Only ubiquitinated proteins enter the proteasome, where they are degraded into small peptides and released back into the cell for further use. The proteasome is often called the protein-recycling center within a cell.

In addition to the important role in recycling proteins, the UPS regulates several aspects of synaptic development. This first became apparent at the beginning of the current century in studies of invertebrates. It is now recognized that ubiquitination mediates many cellular pathways that ultimately regulate synaptic size, stabilization, and elimination by rapidly degrading various pre- and postsynaptic proteins.

The specific effect of ubiquitin on a target protein depends on the length and configuration of the attached ubiquitin chains. Enzymes called E1, E2, and E3 ubiquitin ligases are used to attach the ubiquitin chains to a target protein. First, an E1 ligase activates ubiquitin then interacts with the E2 ubiquitin carrier ligase. Next, the E2 ligase is coupled with one of the over 100 specific E3 ligases. Together, this E2–E3 complex transfers ubiquitin to the target protein (**Figure 10.14**).

One highly conserved E3 ligase was first discovered in invertebrate animal models in 2000. The ligase is called highwire (Hiw) in *Drosophila* and RPM-1 (regulator of presynaptic morphology-1) in *C. elegans*. Within five years, homologs were identified in zebrafish, where it is called Esrom, and in mammals, where it is called Phr1 (PAM (protein associated with Myc)/highwire/RPM-1) or MycBP2 (Myc binding

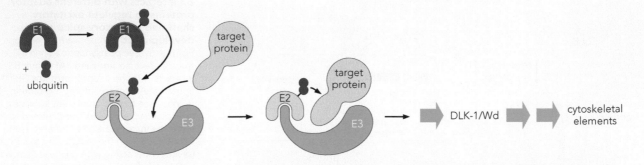

Figure 10.14 The ubiquitin proteasome system (UPS) regulates proteins required for synapse formation. Ubiquitin is activated by E1 ligase, then transferred to an E2 carrier ligase that is coupled to an E3 ligase. A target protein is recruited to the E2–E3 ligase complex. Once the target protein is ubiquitinated, downstream signaling cascades such as the DLK-1/Wd kinases are activated. By activating intracellular pathways that degrade selected cytoskeletal elements, the synapses are rapidly reorganized.

protein-2). In all these animal models, this E3 ligase is important for presynaptic development. Mutations of the E3 ligase gene led to morphological changes in the presynaptic terminals, as well as changes in the number of synaptic contacts that reached the target cell.

Interestingly, the results of a mutation in this E3 ligase gene vary with species. For example, in the *Drosophila* NMJ, loss of *Hiw* led to overgrowth of presynaptic axon terminals, whereas a mutation in *C. elegans* or vertebrates resulted in disorganized synapse morphology and fewer contacts at the correct target cell. In all cases, however, development of the correct number of functional synapses depended on the expression of the corresponding E3 ligase gene.

Scientists have now identified several downstream targets of these E3 ligases. In *C. elegans*, RPM-1 interacts with the protein DLK-1 (dual leucine zipper kinase), whereas in *Drosophila* the homologous kinase is called Wallenda (Wd). DLK-1/Wd activates other signaling pathways to ultimately regulate cytoskeletal elements to influence the motility and morphology of the presynaptic terminal.

Understanding the role of the UPS in neural development remains an active area of research. Scientists have discovered other E3 ligases that regulate the density of postsynaptic receptors, the stability of postsynaptic scaffold proteins, and the morphology of dendritic spines. Further, the role of the UPS is not limited to synaptogenesis. Other E3 ligases have been associated with neuronal migration and axonal guidance. Thus, the UPS, the protein-recycling center of a cell, has been adapted to function as a rapid, reliable, and necessary signaling system for regulating many aspects of neural development.

Wnt Proteins Influence Pre- and Postsynaptic Specializations in the CNS

The Wnt family of proteins also provides bidirectional signals to regulate development of pre- and postsynaptic elements in the CNS. During synaptogenesis, Wnt proteins activate specific Frizzled (Fz) receptors to initiate different intracellular signaling cascades that ultimately mediate pre- and postsynaptic development.

In vitro and *in vivo* studies in the vertebrate CNS revealed that Wnt7a is secreted by postsynaptic cells in the cerebellum, hippocampus, and spinal cord to guide presynaptic terminals expressing the receptor Fz5. Once the presynaptic terminal is at the correct postsynaptic location, Wnt7a then induces clustering of synaptic vesicles and active zone proteins by regulating microtubule stability in the presynaptic terminal.

The importance of this signaling pathway in presynaptic development was seen in experiments in which Wnt7a was added to cell cultures of hippocampal neurons. In these assays, the addition of Wnt7a led to the clustering of presynaptic vesicles and associated proteins whereas the addition of Wnt7a inhibitors prevented this clustering. For example, when either soluble Frizzled receptor proteins (sFRPs) or Dickkopf were added to the hippocampal neurons, the Wnt7a-induced clustering of presynaptic proteins was prevented. *In vivo* studies of mice lacking Wnt7a also found defects in the clustering of synaptic vesicles and associated presynaptic proteins, further demonstrating the need for Wnt7a in presynaptic development.

The Wnt signaling pathway used during presynaptic development is known as the **divergent canonical pathway (Figure 10.15)**. This pathway is similar to the canonical pathway (see Chapter 4), but transcription of Wnt target genes does not occur. Instead, Wnt signals locally to regulate Dishevelled (Dvl) activity. Dvl then inhibits GSK3β, leading to a decrease in phosphorylation of microtubule-associated protein 1B (MAP 1B) and changes in the location of the protein APC (adenomatous polyposis coli). APC is not only a component of the destruction complex found in

Figure 10.15 Wnt signaling through the divergent canonical pathway regulates presynaptic development. In the divergent canonical pathway, Wnt binding to the Frizzled LRP5/6 receptor complex causes Dishevelled to inhibit GSK3β. This leads to a decrease in phosphorylation of microtubule associated protein 1B (MAP 1B, not shown) and relocation of the APC protein that was attached to the distal ends of microtubules. With the removal of APC (dashed arrows), the growth of the axon is stopped and the microtubules stabilize. This enables the microtubules to attract and attach presynaptic proteins.

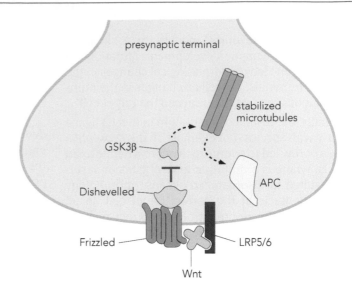

the canonical pathway, but also a binding protein that attaches to the distal (plus) end of microtubules. In this role, APC helps accumulate microtubules to the actively extending growth cone. Wnt signaling through the divergent canonical pathway leads to the removal of APC from the microtubules. This prevents further growth of the axon and leads to changes in axon terminal morphology. The changed morphology then allows the stabilized microtubules to attract and attach presynaptic proteins. To date, the Wnt divergent pathway has been associated with the formation of presynaptic terminal synapses, but it is unclear if this pathway is used at all CNS synapses.

Different Wnts Regulate Postsynaptic Development at Excitatory and Inhibitory Synapses

Wnt7a is also secreted by the presynaptic cell to influence postsynaptic specializations at excitatory synapses. Presynaptic Wnt7a influences dendritic spine formation and recruits the PSD-95 protein to the postsynaptic site (**Figure 10.16**). Thus, Wnt7a functions through a bidirectional signaling mechanism to influence the maturation of synaptic elements on both presynaptic and postsynaptic sites of excitatory synapses. However, Wnt7a does *not* appear to influence the development of postsynaptic elements at inhibitory synapses.

Another Wnt family member, Wnt5a, influences the development of inhibitory, as well as excitatory, synapses by activating different downstream signaling pathways. At excitatory synapses, Wnt5a clusters the PSD-95 protein in the dendritic spine. In at least some neuronal populations, Wnt5a functions in an **autocrine** manner—that is, the same postsynaptic cell releases and then binds Wnt5a. Although the receptor used by Wnt5a in this context is still unknown, Wnt5a appears to activate the JNK signal cascade to induce PSD-95 clustering (Figure 10.16). At inhibitory synapses, Wnt5a induces clustering of GABA$_A$ receptors by activating the CamKII (calmodulin-dependent kinase II) signaling pathway. Thus, Wnt5a can differentially regulate the development of postsynaptic specializations in excitatory and inhibitory synapses.

Glial Cells Contribute to CNS Synaptogenesis

Like the Schwann cells at the NMJ, glial cells in the CNS regulate some aspects of synapse formation. There is often a delay between the time that CNS axons reach their target cells and the time that synapses form. This

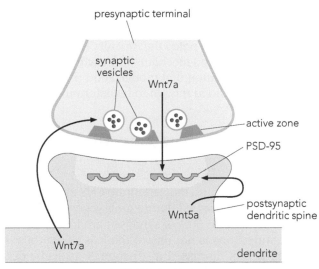

presynaptic terminal

synaptic vesicles

Wnt7a

active zone

PSD-95

postsynaptic dendritic spine

Wnt5a

Wnt7a

dendrite

excitatory synapse

Figure 10.16 Wnt proteins regulate development of presynaptic and postsynaptic elements. Postsynaptic Wnt7a regulates the accumulation of active zone proteins and synaptic vesicles in the developing presynaptic terminal. The presynaptic terminal also releases Wnt7a to recruit PSD-95 into the PSD, leading to further maturation of the postsynaptic site. In addition, Wnt5a released by the postsynaptic dendrite functions in an autocrine manner to help cluster the PSD-95 within the PSD.

period generally coincides with the production of glial cells. Therefore, researchers have long suspected that glial cells release molecules that contribute to synapse formation.

Among the first studies demonstrating the importance of glial-derived signals in CNS synaptogenesis were those using cultures of retinal ganglion cells (RGCs). *In vitro* studies in the 1990s found that astrocytes induced synaptic elements and helped establish new synaptic connections between the cultured RGCs. However, this effect did not require cell contact between astrocytes and neurons, because astrocyte-conditioned medium also induced synaptogenesis when added to cultures of RGCs.

Subsequent studies by Ben Barres and colleagues identified thrombospondin-1 (TSP-1) and thrombospondin-2 (TSP-2) as the proteins produced by astrocytes. There are five members of the TSP glycoprotein family, all of which are produced by multiple cell types in the body to mediate a variety of cell–cell and cell–matrix interactions. In the CNS, TSP-1 and TSP-2 are produced by astrocytes during periods of synaptogenesis.

In vitro studies revealed that synaptogenesis induced by TSP-1 or TSP-2 is similar to that observed with astrocyte-conditioned media. The importance of these proteins was further noted *in vivo*. Mice lacking both *TSP-1* and *TSP-2* form about 30% fewer CNS synapses than do wild-type mice. However, mice lacking only one *TSP* gene did not reveal any changes in synaptic number, indicating that these proteins have redundant functions.

How TSPs mediate synaptogenesis is still not entirely clear. A receptor for TSPs, α2δ1, is expressed on RGCs and other CNS neurons. This receptor has a binding site for the drug gabapentin—a medication used to treat epilepsy and other neural conditions. In the presence of gabapentin, the α2δ1 receptor is blocked, and synapse formation is prevented, indicating that TSPs normally act through this receptor to induce synapse formation (**Figure 10.17**). TSPs alone, however, do not induce a fully functional synapse, suggesting that other factors are necessary for synapse maturation.

Other growth factors associated with regulating aspects of CNS synaptogenesis, such as BDNF and ephrins, are detected in astrocytes and their corresponding receptors are found on dendritic spines. The other CNS glial cells—oligodendrocytes and microglia—also appear to impact aspects of synaptogenesis. For example, studies indicate microglia release the cytokine tumor necrosis factor α (TNFα) that can strengthen newly formed

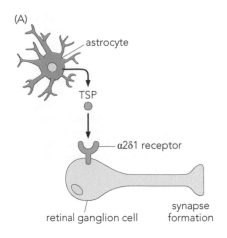

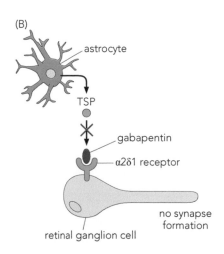

Figure 10.17 Thrombospondins released by astrocytes induce synapse formation. (A) Thrombospondins (TSP) released by astrocytes bind to the α2δ1 receptor on retinal ganglion cell neurons to induce the formation of synapses between RGCs. Both TSP-1 and TSP-2 are released by astrocytes. (B) The α2δ1 receptor also binds the drug gabapentin. When gabapentin is present, TSPs cannot bind to the receptor, so new RGC synapses cannot form. These studies indicated the necessity of α2δ1 receptor activation in mediating the effects of TSPs on synaptogenesis.

synapses. Thus, several factors produced by glial cell populations have the potential to interact and impact synaptogenesis and synaptic maintenance in the CNS. However, it is also noted that not all CNS synapses are tripartite and thus the importance of and mechanisms by which astrocytes and other glia contribute to the formation and modification of synaptic connections in development and adulthood remain to be determined.

SYNAPSE ELIMINATION AND REORGANIZATION IN THE CNS

During postnatal development, the vertebrate CNS eliminates many of the synaptic connections established during embryogenesis. In some regions of the vertebrate CNS, up to 50% of synapses are lost. The elimination of these synaptic connections is not due to the loss of neurons themselves. Programmed cell death, discussed in Chapter 8, typically occurs during embryonic development, whereas synapse elimination typically begins during early postnatal development. In contrast to normal vertebrate CNS development, a widespread loss of synaptic connections does not take place in the invertebrate CNS.

Much like the process of synapse elimination in the vertebrate NMJ, the loss of synaptic connections in the vertebrate CNS appears to be due, in part, to differences in the levels of neuronal activity between pre- and post-synaptic partners. The relative level of neural activity between neighboring axons, as well the timing of action potential firing in the postsynaptic cell often determines whether a synaptic connection remains.

While evidence for the importance of neuronal activity in regulating synaptic connections has accumulated for over 60 years, the mechanisms by which neural firing patterns regulate such connectivity are still not fully understood. In some cases, growth factors such as BDNF are released in response to neural activity that may help maintain synapses. Other growth factors proposed to help stabilize connections, at least transiently, include ciliary neurotrophic factor (CNTF), leukemia inhibitory factor (LIF), and glial cell line-derived neurotrophic factor (GDNF). Presumably, the synapses that do not have access to adequate levels of these growth factors are the ones that are lost. However, whether the levels of these growth factors are directly regulated by neural activity levels remains to be established.

The Vertebrate Visual System Is a Popular Model to Study Synapse Elimination and Reorganization

Much of what is known about the processes underlying synapse elimination and reorganization in the vertebrate CNS was first discovered in the visual system. The visual system is a convenient model system to study synapse formation and elimination for several reasons. In many vertebrate animal models, the visual system develops postnatally and the visual pathways are generally accessible. Further, the organization of the pathways allow for manipulations to test how synapses are formed and reorganized. In the visual system of mammals that rely on binocular vision, for example, several CNS regions become segregated into layers, with each cellular layer receiving synaptic inputs from either the right or left eye. Such segregation is noted in each **lateral geniculate nucleus (LGN)** of the thalamus, a target of the RGCs, and in the **primary visual cortex (V1)**, the target of LGN axons.

The role of neural activity in shaping synaptic connections in the vertebrate visual system has been studied since the mid-twentieth century. As outlined in the next section, in some regions of the mammalian visual system, synaptic stabilization, elimination, and reorganization are influenced by spontaneous embryonic neural activity, while in other regions these processes are driven by visual stimuli. In both cases, changes in the number of dendritic contacts and postsynaptic receptors are influenced by the amount of neural stimulation received.

Spontaneous Waves of Retinal Activity Stabilize Selected Synapses in LGN Layers

In mammalian visual systems, retinal ganglion cells project to both the superior colliculi and LGN. Each LGN is organized into cellular layers. The number of layers varies with species, ranging from two in rodents to six in primates. In all these animal models, the RGCs from both eyes initially project axons to postsynaptic neurons in all LGN layers. These overlapping innervation patterns then gradually segregate. For example, in the ferret, axons from the temporal side of the retina project mainly to the ipsilateral LGN and make synaptic connections in layers A1 and C1. In contrast, most axons from the nasal half of the retina cross to the contralateral LGN to contact cells in layers A and C (**Figure 10.18**).

Neural activity generated in the retina is necessary for synaptic reorganization and segregation to occur in the LGN. If tetrodotoxin (TTX), a drug that blocks action potentials, is infused into the eyes during the period that RGC axons normally segregate into different LGN layers, synaptic reorganization fails to occur. Instead, the axons of RGCs from each eye remain spread across all LGN layers. Interestingly, the process of synaptic reorganization in the LGN takes place before the photoreceptor cells of the retina are fully developed and functional. Thus, visual stimulation is not required for this form of synaptic reorganization. In the late 1980s, scientists proposed that RGCs produce spontaneous action potentials that in turn stimulate neurons in the LGN. In the 1990s, Carla Shatz and colleagues conducted a series of experiments using electrophysiological recordings, calcium imaging, and axonal labeling methods to document the patterns of spontaneous retinal activity in the visual system of the ferret. In the ferret, the formation of the synaptic layers in the LGN occurs from postnatal day 1 (P1) through P21, and eye opening and the onset of vision begin at P30, thus providing a system to study the developmental changes in synaptic connections between RGCs and postsynaptic LGN neurons before and after vision begins.

The Shatz lab demonstrated that RGCs produce spontaneous electrical activity that spreads quickly to adjacent RGCs in waves. The waves do not spread across the entire retina during this short period, but are propagated across only a limited segment. At the completion of the wave, the previously active RGCs remain silent for up to two minutes. After one wave is inactivated, another wave begins at a new, apparently random

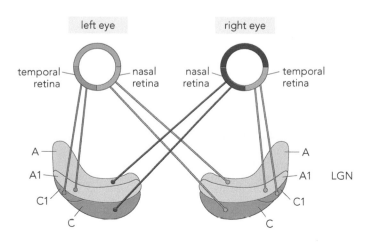

Figure 10.18 Retinal ganglion axons contact distinct layers within the lateral geniculate nucleus (LGN). Retinal axons that project from the temporal (T) or nasal (N) side of the retina become segregated into specific cellular layers in the LGN. In the ferret, temporal axons project mainly to the ipsilateral LGN to contact neurons in layers A1 and C1. Most axons from the nasal half of the retina reach the contralateral LGN to contact neurons in layers A and C. [Adapted from Sharma J & Sur M [2014] In *Biology and Diseases of the Ferret* [3rd ed, Fox JG & Marini RP eds], pp. 711–734. Wiley and Sons, Inc.]

site in the retina. Therefore, RGCs throughout the retina become active at different times. Additionally, the waves produced in each eye begin at different times and in different retinal locations. As a result, the stimulation that reaches the postsynaptic neurons in the various LGN layers is not synchronous. This asynchronous stimulation creates variable amounts of postsynaptic stimulation to each LGN neuron that ultimately influences axon projection patterns to each layer.

Similar results were found in studies of patterning in the mouse superior colliculus. Using optogenics (see Chapter 1), Zhang and colleagues were able to control the activity of RGCs in each eye. When the RGCs were stimulated so activity in each eye was asynchronous, axonal segregation in the different layers of the superior colliculus was even greater than in normal conditions. Conversely, if both eyes were stimulated so synchronous activity resulted, the layer-specific segregation was blocked.

The onset and patterning of retinal waves under normal conditions continues to be explored. One recent study suggests the waves also influence the neurons in the superior colliculus that are important for forward motion detection. This would help establish pathways for neural functions needed soon after eye opening.

Competition between Neurons Determines Which Synaptic Connections Are Stabilized

Repeated stimulation of a postsynaptic LGN neuron by RGCs causes that LGN neuron to fire action potentials. Some of the resulting LGN action potentials will occur at the same time as action potentials from some RGC inputs. When such synchronous firing occurs, the synaptic connections between those RGC axons and the LGN neuron becomes strengthened. However, when the firing patterns from RGC inputs and an LGN neuron are asynchronous, those presynaptic connections withdraw (**Figure 10.19**).

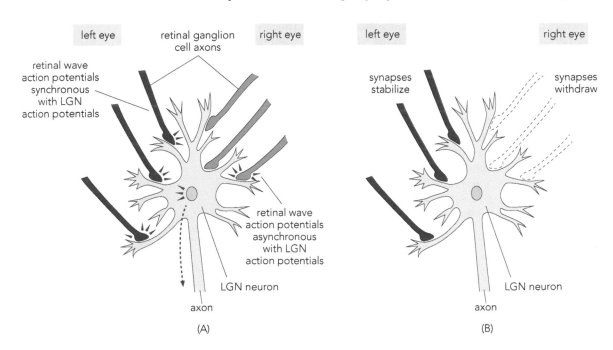

Figure 10.19 Synchronous activity determines which synapses stabilize in the lateral geniculate nucleus (LGN). Retinal ganglion cells (RGCs) produce spontaneous waves of electrical activity that stimulate target neurons in the LGN. Because each eye produces waves at different times, the stimulation to a target neuron varies. (A) The wave of action potentials across RGCs in the left eye is sufficient to activate a neuron in the LGN and that neuron begins to fire action potentials at the same time as the left RGCs. Thus, the LGN neuron action potentials become synchronous with those of the retinal wave. The retinal wave action potentials from the right eye however are asynchronous with the action potentials of the LGN neuron. (B) Because synchronous firing did not occur between the RGC inputs from the right eye and the LGN neuron, the contacts from the right eye are withdrawn. However, the synchronous firing between RGC inputs from the left eye and the LGN neuron allow those RGC synapses to stabilize.

This is an example of the concept of Hebbian plasticity in which "cells that fire together, wire together; those out of sync, lose their link."

As in the NMJ, it appears that differences in the level of neural stimulation determine which synaptic connections are stabilized and which are lost. As noted in Chapter 9, muscle fibers initially receive multiple synaptic inputs. When neural activity was blocked at restricted regions along the postsynaptic muscle fiber, the synaptic connections at that region withdrew. However, if the entire NMJ was blocked, synaptic connections remained. Thus, only when differences in the levels of neural activity were present did the synapses reorganize (see Figures 9.22 and 9.23).

Similar results were found in the studies of RGC projections to the LGN. Drugs that inhibit cAMP were injected into the eyes of ferrets during the period of synaptic reorganization in the LGN. The available levels of cAMP influence the activity of retinal waves. Thus, when the drug was injected into both eyes, neither eye produced retinal waves. Because the RGC activity in each eye was equally inhibited, there were no differences in neural activity across the LGN layers, and therefore the axons of the RGCs spread throughout all the layers of the LGN. When the drug was injected into only one eye, however, the RGC axons from the treated eye withdrew connections in the LGN, while the axons from the untreated eye maintained connections with neurons in all layers of the LGN. Thus, when differences in neural activity are present, the postsynaptic neurons receiving the greater amount of stimulation—in this case, those innervated by the untreated eye—form stable synaptic connections. Under normal conditions, the spontaneous temporal and spatial differences in retinal wave activity influence the synaptic organization in LGN cell layers. What is still unclear is whether the retinal waves provide cues to set up initial patterning in the LGN or whether the waves are used to establish the final LGN layers.

Early Visual Experience Establishes Ocular Dominance Columns in the Primary Visual Cortex

Presynaptic neural activity also influences synaptic organization in the primary visual cortex. Whereas neurons in the LGN stabilize synaptic connections independent of visual stimulation, the neurons in the primary visual cortex require visual stimulation to maintain synaptic connections.

Studies by David Hubel and Torsten Wiesel begun in the 1960s, for which they won the Nobel Prize in Physiology or Medicine in 1981, were among the first to detail the role of neural activity in establishing mature synaptic connections. As with the LGN, the layers of the primary visual cortex in several mammalian animal models are initially innervated by connections from both eyes. The RGCs project to neurons in the LGN, and those LGN neurons project to the primary visual cortex. In animal models that rely on binocular vision, such as cats and some monkeys, the inputs originating from both eyes initially overlap in layer IV of the primary visual cortex. However, during early postnatal development the projections segregate into distinct regions known as **ocular dominance columns (ODCs)**, which arise from alternating projections originating in the right and left eyes (**Figure 10.20**).

Under normal developmental conditions, visual experience shapes the patterning of the ODCs during early postnatal development. This occurs because each eye views a given visual stimulus from a slightly different angle. Thus, any visual stimulus will activate a different portion of each retina, and it is this difference in stimulation that ultimately leads to the formation of alternating columns in layer IV of the primary visual cortex (Figure 10.20A).

Hubel and Wiesel conducted a series of studies to determine how these ODCs formed. Using radiolabeled neuronal tracing methods and

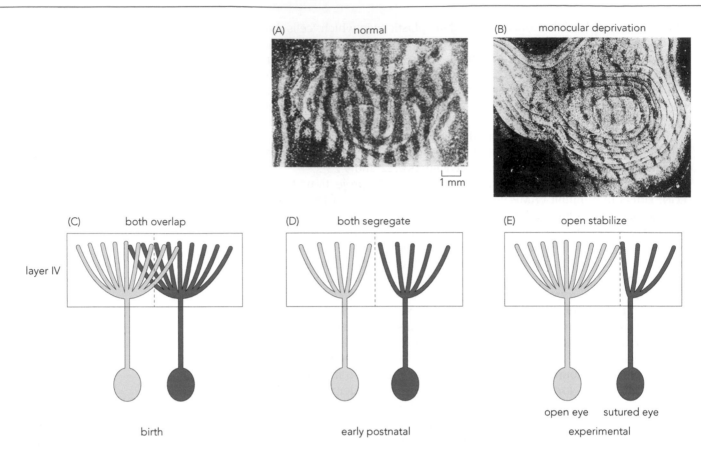

Figure 10.20 Studies reveal activity-dependent changes in synaptic organization of the primary visual cortex. (A) Layer IV of the primary visual cortex of the macaque monkey is normally organized into alternating columns representing input from each eye. In this example, one eye was injected with a radioactive tracer (light bands). The dark bands represent the areas receiving input from the eye that was not injected with the tracer. Each eye projects to alternating columns of equal size. (B) Experiments were conducted in which one eye was sutured shut so that only one eye received visual stimulation during postnatal development. When the radioactive tracer was injected into the open eye, it was found that the layer IV ocular dominance columns were larger (light bands) than those of the sutured eye (dark bands). This was found to be due to stabilization of existing contacts from the open eye and the loss of some connections from the sutured eye. (C–E) Schematic representation of the changes that occur. (C) At the time of birth, axonal projections from both eyes intermingle in layer IV of the visual cortex. (D) During postnatal development, the eyes normally segregate into ocular dominance columns, such that each eye is represented by alternating columns. (E) When one eye is sutured shut, the activity from the open eye becomes sufficient to maintain initial contacts. Thus, the columns for the open eye appear expanded compared to the sutured eye. Source: (A and B), From Hubel DH, Wiesel TN & LeVay S [1977] *Philos Trans R Soc Lond Biol* 278:377–409.

electrophysiological recordings, the researchers discovered that if one eye was sutured shut during the early postnatal period, the ODCs associated with the open eye became wider than under normal conditions. In contrast, the ODCs of the closed eye became smaller (Figure 10.20B). Over time, most of the connections from the sutured eye were lost.

The expansion of the size of the ODCs associated with the open eye could suggest that visual stimulation increased the number of synaptic connections that formed and stabilized. However, Hubel and Wiesel found that the increased size of a cortical column was due to the maintenance of connections that would normally have been eliminated during development, rather than the formation of new connections. These results indicated that because the sutured eye no longer received visual stimulation, it was no longer able to provide competing neural activity to the ODC neurons it normally innervated. Therefore, the preexisting connections from the other, open eye were able to stabilize (Figure 10.20C–E). Like the studies in the LGN, when TTX was injected into both eyes to prevent activation of the RGCs, reorganization of ODCs did not occur. Together, these experiments were among the first to demonstrate the importance of differing levels of neural activity in regulating the stabilization of synaptic connections.

Another major finding from these studies was that the effect was reversible during a limited period in development. If the sutured eye was re-opened and the previously opened eye sutured shut while synaptic connections were still forming, the columns associated with the previously shut eye widened while the columns from the previously open eye became smaller. Thus, synaptic reorganization in response to visual stimulation was plastic during a limited period of postnatal development. This window of plasticity is now called a **critical period**—a time in neural development during which synaptic reorganization can take place. After this period, changes to synaptic connections are less likely to occur and any changes that do take place are much more limited compared to changes that occur prior to the close of the critical period. In the studies by Hubel and Wiesel, for example, when the sutured and opened eyes were reversed later in development or in adulthood, the width of the cortical columns no longer reversed as they did during the critical period.

The molecular mechanisms underlying activity-dependent changes in CNS synaptic connections are beginning to be elucidated. It appears that the expression of many proteins used during the initial formation of synaptic contacts, such as BDNF and ephrins, are also influenced by neuronal activity. The expression of these proteins can then modify synaptic structure and may further enhance neural transmission by influencing the number of receptors anchored at the postsynaptic membrane.

Homeostatic Plasticity Contributes to Synaptic Activity

An additional form of plasticity contributes to the structural and functional changes in synapses that accompany normal postnatal development. **Homeostatic plasticity** occurs when pre- and postsynaptic cells modify synaptic elements and synaptic output in response to changes in the overall level of neural activity. This form of plasticity occurs in the CNS and NMJ of vertebrates and invertebrates (see also Box 9.1) and has become a particularly active area of research in the past decade. Whereas Hebbian modifications take place within a period of minutes, homeostatic changes occur over a period of hours to days. Hebbian and homeostatic plasticity appear to balance one another to maintain stable levels of neural firing. Maintaining action potential firing patterns is necessary to sustain a stable functional network of neurons as other synaptic changes, such as those associated with learning and memory, take place.

Examples of the principles of homeostatic changes are illustrated in **Figure 10.21**. The desired baseline level of neural firing and the expression of postsynaptic neurotransmitter receptors and ion channels are shown in the first panel. If the overall level of neural stimulation is then lowered over an extended period, the postsynaptic cell will compensate for the reduced

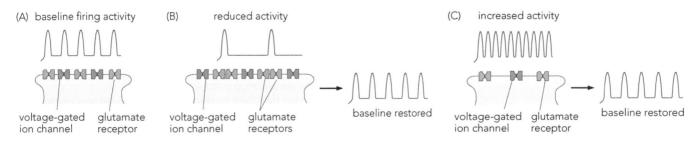

Figure 10.21 Homeostatic plasticity ensures stable neural firing patterns. (A) Neurons have a desired baseline level of neural firing activity. Each postsynaptic neuron expresses a given number of neurotransmitter receptors and ion channels. (B) If neural activity is reduced for a prolonged period, the postsynaptic neurons will increase the number of postsynaptic neurotransmitter receptors and ion channels to restore neural firing patterns to baseline levels. (C) If neural activity is increased over a prolonged period, the postsynaptic cells will decrease expression of neurotransmitter receptors and ion channels to restore neural activity to baseline levels.

activity by increasing the number of neurotransmitter receptors and ion channels expressed. Conversely, if neural stimulation is maintained at a high level, homeostatic changes take place to reduce the levels of postsynaptic receptors and ion channels. In both examples, these changes allow the neurons to adjust to an altered environment and return firing patterns to the baseline level. This form of homeostatic plasticity, also called **synaptic scaling**, therefore maintains the desired Hebbian modifications, but prevents further widespread changes in synaptic connectivity that would be detrimental to nervous system function.

Other structural changes are observed during homeostatic plasticity. As noted, the synapse size and spine density are correlated with synaptic strength. If the synaptic signaling is altered experimentally through sensory deprivation or decreasing the number or function of postsynaptic neurotransmitter receptors, subsequent changes in dendritic spines will occur. In response to decreased activity, the number of spines increases and the size of the associated PSDs also increase, providing additional space for postsynaptic elements to accumulate. In addition, the presynaptic neurons increase synaptic vesicle release and the size of the active zone in response to sensory deprivation to further offset the effects of reduced input.

Homeostatic changes also occur at inhibitory synapses (**Box 10.2**). In the visual system, experimentally induced lesions made at restricted sites of the retina led to a decrease in synaptic activity initially, but over a period of 48 hours, neural activity returned to baseline as the size of dendritic spines on the excitatory pyramidal neurons of the visual cortex increased and the number of spines on inhibitory neurons decreased. These results suggest that inhibitory neurons help restore baseline firing by reducing their inhibitory signaling as neural activity decreases.

Studies further indicate that the amount of synaptic elimination during postnatal development decreases in conditions of sensory deprivation, presumably in an effort to reduce the extent of synaptic connections lost.

Intrinsic and Environmental Cues Continue to Influence Synapse Organization at All Ages

It is rather difficult to establish when development of the nervous system ends. This book focuses on the events that occur from neural induction through the initial stages of synaptogenesis and synaptic reorganization. However, changes in synaptic connections continue throughout the lifespan of an animal. While some neuronal populations exhibit the greatest extent of synaptic refinement during early postnatal stages, others continue to modify synaptic connections until puberty. In many vertebrate species, the changes in hormone levels that accompany puberty are thought to limit the ability of some synapses to further modify connections. Thus, hormonal changes stabilize or "crystalize" synapses, often by regulating the stability and organization of postsynaptic elements such as neurotransmitter receptors. In humans, for example, the ability to acquire new language skills or a second language begins to decrease following puberty. Examples of synapse crystallization are also seen in male songbirds. Once testosterone levels begin to increase during puberty, the birds lose the ability to learn new songs due to an inability to reorganize synaptic connections in the CNS regions associated with song learning.

In other areas of the nervous system, ongoing changes in synaptic connections occur throughout adulthood. For example, the neural processes underlying learning and memory require changes in synaptic connections. Numerous studies in the hippocampus, the area of the brain most associated with learning and memory, provide examples of how cellular mechanisms regulate the selective strengthening and weakening of synapses in response to various types of stimulation. At excitatory synapses, strengthened synapses have an increased dendritic spine size and more AMPA receptors in the postsynaptic membrane. These changes allow for enhanced neural transmission and are the basis of the process called long-term potentiation (LTP). The opposite changes are seen as synapses weaken. Weakened synapses lose dendritic spines and have fewer AMPA receptors in the postsynaptic membrane, leading to reduced efficiency of neural transmission. These changes are part of the process called long-term depression (LTD). Recent experiments suggest that pro-BDNF signaling through the p75NTR receptor may reduce spine number and weaken synaptic transmission in at least some CNS neurons, reminiscent of the differential effects of BDNF-TrkB and pro-BDNF-p75NTR signaling in the NMJ (see Figure 9.24).

Box 10.2 Developing Neuroscientists: Changes in Ion Channel Expression Following the Onset of Hearing

Briana Carroll, Ph.D.

Briana Carroll graduated from Oberlin College in 2011 with a major in neuroscience. She completed a Ph.D. at Florida State University with Richard Hyson, investigating developmental changes in the auditory system following hearing onset. She is currently a postdoctoral fellow at the University of Chicago where she studies cortico-cortical and cortico-thalamo-cortical pathways.

Synaptic plasticity refers to changes in the connections made between two neurons. Another form of plasticity occurs during development known as homeostatic plasticity, which involves changes in pre- and postsynaptic elements in a single neuron. Homeostatic intrinsic plasticity occurs within a single neuron and refers to changes in the expression or biophysical properties of the ion channel proteins that characterize a given neuron. By altering the expression or properties of ion channels, the behavioral characteristics of a neuron are also altered. In many cases, changes in the number of synaptic inputs, or their firing patterns, contribute to intrinsic plasticity. The chick auditory system has been used as a model system for understanding the normal development of CNS connections and the effects of external stimuli on both synaptic and intrinsic neuronal properties.

The auditory nerve fibers of the eighth cranial nerve (CN VIII) connect to hair cells in the inner ear (see Chapter 6) and terminate in the cochlear nuclei in the brainstem. The chick cochlear nuclei are divided into three regions: nucleus magnocellularis (NM), nucleus angularis (NA), and nucleus laminaris (NL). NM and NA receive input from the ipsilateral CN VIII while NL receives bilateral input from the NM (shown ipsilaterally in **Figure 10.22**). The neurons from cochlear nuclei send ascending branches to the superior olivary nuclei (SON), which are found on each side of the brainstem. In addition, descending fibers from SON provide inhibitory connections to the three regions of the cochlear nuclei. This descending inhibitory firing is activated in response to loud or high-frequency sounds and is thought to help shape the response of the ascending signals by suppressing auditory neuron output. Thus, the SON neurons provide a type of gain control to suppress auditory input when necessary.

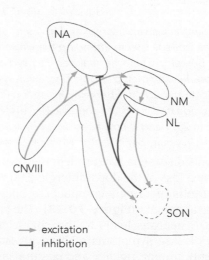

Figure 10.22 The cochlear nuclei in the chick auditory brainstem receive excitatory and inhibitory input. The cochlear nuclei of the chick auditory brainstem are divided into the nucleus angularis (NA), nucleus magnocellularis (NM), and nucleus laminaris (NL). Excitatory input originates from the eighth cranial nerve (CN VIII). These nuclei send ascending projections to the superior olivary nuclei (SON). The neurons in the SON also send descending, inhibitory fibers to all three regions of the cochlear nuclei. These inhibitory fibers are activated in response to loud or high-frequency sounds to suppress auditory output when needed. Ion channel expression increases after the onset of hearing to allow this inhibitory control to occur.

If the intrinsic plasticity of SON neurons is important for maturation of the gain control, then the properties of ion channels in SON neurons would be expected to change following the onset of hearing. To investigate these possibilities the properties of SON neurons were evaluated at different developmental time points prior to and after hearing onset.

The ion channel properties of single SON neurons were evaluated using the patch-clamp recording technique. Over the course of normal development, the neurons became more excitable following the onset of hearing, indicating greater ion channel expression. In a subset of recordings, drugs that target sodium and potassium channels were applied. Altering the conductance of these channels had little effect prior to hearing onset, further suggesting that ion channel expression was more limited prior to the onset of hearing. Together, the studies suggest that an increased expression of ion channels facilitates the increased excitability observed in SON neurons following hearing onset.

It is still unclear whether hearing onset itself triggers increased ion channel expression. Whatever the mechanism, increased ion channel expression is an important developmental step that provides mature neurons in the SON the ability to exert gain control in the lower NM, NA, and NL auditory nuclei.

In addition to the normal changes in synaptic connections that take place during neural development and adulthood, there are several conditions that lead to a loss or disorganization of synapses. For example, altered synaptogenesis may underlie neurodevelopmental disorders such as autism and schizophrenia and neurodegenerative disease such as Parkinson's disease (**Box 10.3**).

Box 10.3 Neurologist's Perspective: Characteristics, Treatment, and Possible Causes of Parkinson's Disease

Gustavo Buitron Carvajal, M.D.

Gustavo Buitron Carvajal, completed his associate's degree at De Anza College before attending the University of California at Berkeley where he majored in psychology. After graduation, he completed a post-baccalaureate premedical certificate through the Berkeley Extension Program. Dr. Carvajal graduated from the Indiana University School of Medicine in 2021 and is now a resident physician in neurology at the University of Pittsburgh Medical Center Hamot.

Parkinson's disease is a neurodegenerative condition caused by a loss of the dopaminergic (DA) neurons of the substantia nigra pars compacta, a body of nuclei of the basal ganglia (**Figure 10.23**). The progressive loss of DA neurons leads to the characteristic Parkinson's disease symptoms of bradykinesia (slowed movement), tremor, rigidity, and postural instability.

To understand the way in which DA neuron degeneration leads to the symptoms of Parkinson's disease, it is helpful to first review the basic anatomical structures and their functional roles in healthy individuals. The basal ganglia, composed of several collections of nuclei in the deeper structures of the brain, are responsible for the complex sequence of movements we carry out daily. These subcortical structures include the striatum (comprised of the caudate nucleus and putamen), globus pallidus externa (GPe), globus pallidus interna (GPi), subthalamic nucleus (STN), substantia nigra pars compacta (SNc), and substantia nigra pars reticulata (SNr). Other structures outside the basal ganglia that contribute to the control of movement are the primary motor cortex and the thalamus (Figure 10.23B).

These structures work as a highly interconnected system creating two major neural circuits to control movement: the direct pathway and the indirect pathway. Synaptic neurotransmission in both pathways is mediated either by glutamate for excitatory effects or GABA for inhibitory effects.

In the direct pathway, excitatory projections from the primary motor cortex reach the striatum which in turn sends inhibitory projections to the GPi and SNr. Inhibitory projections from these regions reach the thalamus, which then sends excitatory projections to the primary motor cortex. With the indirect pathway, the GPe and STN serve as intermediate points, with the GPe sending inhibitory projections to the STN which then sends excitatory projections to the GPi/SNr that, in turn, send inhibitory projections to the thalamus. Completing the direct and indirect pathways are the connections between the dopamine neurons of the SNc and the spiny projection neurons (SPNs) of the striatum. SPNs express either excitatory dopamine D1

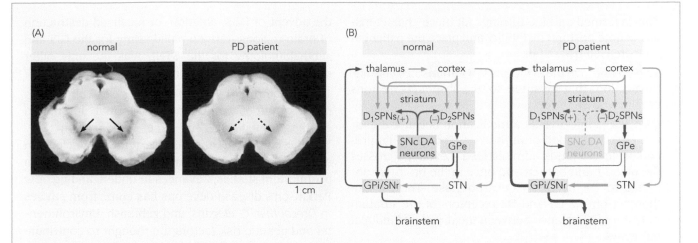

Figure 10.23 Parkinson's disease symptoms are caused by the degeneration of dopamine neurons in the substantia nigra and accompanying changes in synaptic transmission. (A) In a normal brain, the dopamine neurons appear dark in the substantia nigra pars compacta (solid arrows). These neurons are missing in patients with Parkinson's disease (PD; dashed arrows). Both panels show coronal sections of postmortem brains. (B) Under normal conditions, the direct pathway increases movement while the indirect pathway decreases movement through a series of excitatory (green) and inhibitory (red) projections. Dopamine from neurons in the substantia nigra pars compacta (SNc) binds to excitatory D1 receptors and inhibitory D2 receptors on spiny projection neurons (SPNs) in the striatum to balance the activity of the direct and indirect pathways. In Parkinson's disease, the loss of dopamine neurons in the substantia nigra pars compacta (SNc) causes an imbalance so that there is an over-activation of the indirect pathway (thicker arrows) at the expense of the direct pathway (thinner arrows), leading to an overall decrease in movement. **Direct pathway**: Excitatory projections from the primary motor cortex to the striatum which send inhibitory projections to the GPi and SNr which in turn send inhibitory projections to the thalamus. The thalamus then sends excitatory projections to the primary motor cortex. **Indirect pathway**: Excitatory projections from the primary motor cortex to the striatum which sends inhibitory projections to the GPe, the GPe sends inhibitory projections to the STN which provides excitatory inputs to the GPi/SNr that in turn send inhibitory projections to the thalamus. Dopamine D1 receptors, D1; dopamine D2 receptors, D2; Globus pallidus externa, GPe; globus pallidus interna, GPi; SPN, spiny interneurons; subthalamic nucleus, STN; substantia nigra parts compacta, SNc; substantia nigra pars reticulata, SNr. [Luo L [2021] *Principles of Neurobiology*, 2nd ed, CRC Press.]

receptors (direct pathway) or inhibitory dopamine D2 receptors (indirect pathway; Figure 10.23B).

Under normal conditions, activation of the direct pathway increases movement while activation of the indirect pathway decreases movement. In Parkinson's disease, the degeneration of the SNc causes an imbalance in the motor activity contributed by each pathway thereby leading to over-activation of the indirect pathway at the expense of the direct pathway, and thus an overall decrease in movement (hypokinesis).

Patients with Parkinson's disease develop hypokinetic symptoms that interfere with activities of daily living. Typically, the progressive hypokinetic sequelae of bradykinesia, tremor, rigidity, and postural instability begins between the fourth and seventh decades of life, with males afflicted more than females.

Bradykinesia leads to a progressive diminution in how fast and how much motor function is executed. For example, if a patient taps his or her index finger and thumb repetitively, one will notice a decrease in the number of taps (speed) and a decrease in the distance between fingers (amplitude) with each successive attempt. Similarly, when watching a patient write, one will notice their print transition from normal sized to smaller sized print over time. This is termed micrographia. Other forms of bradykinesia are hypomimia, a reduction in facial expression and hypophonia, a reduction in vocal volume over time.

The tremor of a patient with Parkinson's disease is a rhythmic oscillation of the hands, with a frequency of 5 to 7 Hertz. This tremor occurs at rest and appears as if an object is being maneuvered between the first and second digits in a pill-rolling manner.

Rigidity reflects an increased muscle tone during passive movement of the limbs that is independent of direction or velocity. If one passively moves the arm of a patient and observes a smooth and steady (sustained) increase in his or her muscle tone throughout the manipulation, the resistance is termed lead-pipe rigidity. Alternatively, if in the same exercise, one observes an intermittent, ratchet-like increase in muscle tone, this form of resistance is termed cogwheel rigidity.

When walking, a patient with Parkinson's disease often leans his or her upper body forward, has the upper extremities flexed at the elbows and wrist with a loss of swing, and flexes the lower extremities at the hips and knees using short steps close to each other, termed a shuffling gait. On redirection of the body, the patient moves slowly and does not twist the torso.

This is termed en-bloc turning. All these characteristics denote postural instability and place the patient at an increased risk of falls.

To help improve the symptoms of Parkinson's disease, medications that substitute for the absent DA neurons are prescribed.

DA itself is infective as a medication because it cannot cross the blood-brain barrier. Therefore, its precursor, levodopa, is administered. Levodopa crosses the blood-brain barrier, and once in the brain, is converted to DA by DOPA decarboxylase. The DA can then act on the D1 and D2 receptors of the striatum to restore the balance between the direct and indirect pathways.

Carbidopa-levodopa is a single medication with two pharmacological agents: a DA decarboxylase inhibitor and the precursor to DA. Carbidopa inhibits DOPA decarboxylase and prevents the conversion of levodopa to DA in the peripheral circulation. This has a twofold effect. The first effect is to prevent activation of DA receptors outside of the central nervous system. This stops the unwanted levodopa side effects of nausea and vomiting (emesis), which arise from action on the area postrema, a structure located outside the central nervous system at the fourth ventricle. The mitigation of the side effects of levodopa is reflected in the medication's brand name (Sinemet), which stems from the Latin, "without emesis." The second effect of carbidopa is to allow for more levodopa to cross the blood-brain barrier where it can be converted to DA.

In the peripheral circulation, levodopa also is converted to 3-O-methyldopa (3-OMD) by catechol-O-methyl transferase (COMT). To prevent this, a third medication that inhibits COMT is used. By stopping the conversion of levodopa to 3-OMD, more levodopa persists in the peripheral circulation and ultimately reaches the central nervous system where it is converted to DA.

The fourth type of medication used for Parkinson's disease are DA agonists that cross the blood-brain barrier and act directly on the D1 and D2 receptors of the striatum.

When symptoms of Parkinson's disease progress and are refractory to medications, surgical intervention such as deep brain stimulation (DBS) is an option. With DBS, electrodes are placed into selected targets of the brain to stimulate, or modulate, their activity and alleviate symptoms. For example, stimulation inhibits the GPi preventing inhibitory signals from reaching the thalamus, thereby allowing thalamocortical activity so movement can proceed. Stimulation of the STN has similar downstream effects. Prior to the advent of DBS, ablation, or localized destruction of brain regions (termed pallidotomy for the GPi and subthalamotomy for the STN), was the only other means of reducing symptoms.

The symptoms of Parkinson's disease were recognized centuries before the first formal description of "the shaking palsy" by James Parkinson in 1817. Yet, the etiology of the disease remains unclear. In addition to mammalian animal models and human postmortem studies, advances in understanding how Parkinson's disease develops has come from studies in *Drosophila*, *C. elegans*, and zebrafish. Environmental and genetic risk factors are thought to contribute to the development of Parkinson's disease, and most cases appear to involve changes in synaptic vesicle, mitochondrial, or proteasomal function. About 95% of the Parkinson's disease cases are sporadic, meaning they arise from unknown causes; however, familial forms caused by genetic mutations also exist, with over 20 associated genes identified to date. Epigenetic changes such as hypo- and hypermethylation of genes associated with Parkinson's disease are also thought to contribute to some forms of Parkinson's disease.

The aggregation of α synuclein in the presynaptic cell, particularly in inclusion bodies called Lewy bodies, is seen in many cases of Parkinson's disease. Although not all patients have Lewy bodies, aggregates of α synuclein are common. Normally, α synuclein, a member of the synuclein family of proteins, is involved in synaptic vesicle transport, neurotransmitter release, and presynaptic terminal size. In Parkinson's disease, α synuclein becomes misfolded into amyloid fibrils (Box 9.4) that interfere with DA uptake and release from synaptic vesicles. Increases in α synuclein are also thought to interfere with mitochondrial function. The UPS system used to tag proteins for degradation normally helps remove the damaged mitochondria by a pathway involving the kinase PINK1 (PTEN-induced kinase 1; PTEN, phosphatase and tensin homolog) and the UPS associated enzyme Parkin.

Recent mammalian studies suggest that the cells most susceptible to degeneration in Parkinson's disease originate from the same progenitor populations. Thus, early developmental events may contribute to later onset of this disease. Although not a neurodevelopmental disorder, there is increasing evidence to suggest that susceptibility to Parkinson's begins during embryogenesis.

SUMMARY

Remarkable progress has been made in understanding the mechanisms regulating synaptogenesis in the CNS, particularly in identifying signaling molecules that induce the formation of pre- and postsynaptic specializations and mediate synaptic stabilization.

Initial synaptic contacts in the CNS are often induced by interactions between cell adhesion molecules (CAMS). PDZ-containing proteins that interact with local scaffold proteins then help stabilize select connections. In both the NMJ and CNS reciprocal signals are exchanged between synaptic partners. These signals lead to the induction of active zone proteins and the clustering of synaptic vesicles and ion channels in the presynaptic terminal. Reciprocal signals also mediate the clustering and stabilization of neurotransmitter receptors and ion channels at the postsynaptic membrane. Additionally, glial cells provide signals that help form and maintain synapses. Perhaps surprisingly, some of the proteins known to have essential roles in the formation of the NMJ do not appear to function similarly during CNS synaptogenesis. For example, although agrin and neuregulin are detected in many CNS neurons, they are not required for synapse formation. Moreover, rapsyn, which is the protein used for clustering acetylcholine receptors (AChR) at the NMJ, is not used in the CNS, even at central cholinergic synapses. Thus, the identification of a synaptic protein in one region of the nervous system does not always indicate the same function at another synaptic site. In the CNS, BDNF/TrkB, Eph/ephrin, and Wnt/Fz signaling pathways are among the many identified to date that induce the clustering of synaptic specialization in pre- and postsynaptic neurons.

Scientists continue to study how individual connections are selected for elimination during neural development. In most cases, synapse elimination does not appear to result from any inherent defect in the connections that are lost and in many cases synapse elimination does not lead to a net decrease in synaptic effectiveness. Instead, the remaining connections become more efficient at stimulating the postsynaptic cells.

CNS synapses are often modified throughout adulthood. This plasticity can be influenced by a combination of environmental factors and neural experience. In some cases, intrinsic or extrinsic factors lead to disordered synaptogenesis. One long-term goal of studying synaptogenesis is to develop therapeutic interventions to promote or restore malformed synaptic connections.

FURTHER READING

Ahmad-Annuar A, Ciani L, Simeonidis I, et al. (2006) Signaling across the synapse: A role for Wnt and Dishevelled in presynaptic assembly and neurotransmitter release. *J Cell Biol* 174(1):127–139.

Andreska T, Lüningschrör P & Sendtner M (2020) Regulation of TrkB cell surface expression-a mechanism for modulation of neuronal responsiveness to brain-derived neurotrophic factor. *Cell Tissue Res* 382(1):5–14.

Aoto J, Ting P, Maghsoodi B, et al. (2007) Postsynaptic ephrinB3 promotes shaft glutamatergic synapse formation. *J Neurosci* 27(28):7508–7519.

Beekman JM & Coffer PJ (2008) The ins and outs of syntenin, a multifunctional intracellular adaptor protein. *J Cell Sci* 121(9):1349–1355.

Biederer T, Sara Y, Mozhayeva M, et al. (2002) SynCAM, a synaptic adhesion molecule that drives synapse assembly. *Science* 297(5586):1525–1531.

Blumenfeld H (2021) Chapter 16: The basal ganglia. In *Neuroanatomy through Clinical Cases* (Blumenfeld H ed), pp. 740–793. Oxford University Press.

Calabrese B, Wilson MS & Halpain S (2006) Development and regulation of dendritic spine synapses. *Physiology* 21:38–47.

Carrasco MA, Castro P, Sepulveda FJ, et al. (2007) Regulation of glycinergic and GABAergic synaptogenesis by brain-derived neurotrophic factor in developing spinal neurons. *Neurosci* 145(2):484–494.

Carroll BJ, Bertram R & Hyson RL (2018) Intrinsic physiology of inhibitory neurons changes over auditory development. *J Neurophysiol* 119(1):290–304.

Cherra SJ & Jin Y (2015) Advances in synapse formation: Forging connections in the worm. *WIREs Dev Biol* 4(2):85–97.

Christopherson KS, Ullian EM, Stokes CC, et al. (2005) Thrombospondins are astrocyte-secreted proteins that promote CNS synaptogenesis. *Cell* 120(3):421–433.

Church FC (2021) Treatment options for motor and non-motor symptoms of Parkinson's disease. *Biomolecules* 11(4):612.

Davis GW (2013) Homeostatic signaling and the stabilization of neural function.*Neuron* 80(3):718–728.

de Agustín-Durán D, Mateos-White I, Fabra-Beser J & Gil-Sanz C (2021) Stick around: Cell-cell adhesion molecules during neocortical development. *Cells* 10(1):118.

DeFelipe J (2015) The dendritic spine story: An intriguing process of discovery.*Front Neuroanat* 9:14.

Farias GG, Alfaro IE, Cerpa W, et al. (2009) Wnt-5a/JNK signaling promotes the clustering of PSD-95 in hippocampal neurons. *J Biol Chem* 284(23):15857–15866.

Flavell SW & Greenberg ME (2008) Signaling mechanisms linking neuronal activity to gene expression and plasticity of the nervous system.*Annu Rev Neurosci* 31:563–590.

Ge X, Zhang K, Gribizis A, et al. (2021) Retinal waves prime visual motion detection by simulating future optic flow. *Science* 373(6553):eabd0830.

Giachello CN, Montarolo PG & Ghirardi M (2012) Synaptic functions of invertebrate varicosities: What molecular mechanisms lie beneath. *Neural Plast* 2012:670821.

Giagtzoglou N, Ly CV & Bellen HJ (2009) Cell adhesion, the backbone of the synapse: "Vertebrate" and "invertebrate" perspectives. *CSH Perspect Biol* 1(4):a003079.

Goodman CS & Shatz CJ (1993) Developmental mechanisms that generate precise patterns of neuronal connectivity. *Cell* 72:77–98.

Graf ER, Zhang X, Jin SX, et al. (2004) Neurexins induce differentiation of GABA and glutamate postsynaptic specializations via neuroligins. *Cell* 119(7):1013–1026.

Hamilton AM & Zito K (2013) Breaking it down: The ubiquitin proteasome system in neuronal morphogenesis. *Neural Plast* 2013:196848.

Hruska M & Dalva MB (2012) Ephrin regulation of synapse formation, function and plasticity. *Mol Cell Neurosci* 50(1):35–44.

Keck T, Keller GB, Jacobsen RI, et al. (2013) Synaptic scaling and homeostatic plasticity in the mouse visual cortex in vivo. *Neuron* 80(2):327–334.

Keith D & El-Husseini A (2008) Excitation control: Balancing PSD-95 function at the synapse. *Front Mol Neurosci* 1:4.

Khammari A, Arab SS & Ejtehadi MR (2020) The hot sites of α-synuclein in amyloid fibril formation. *Sci Rep* 10(1):12175.

McClelland AC, Hruska M, Coenen AJ, et al. (2010) Trans-synaptic EphB2-ephrin-B3 interaction regulates excitatory synapse density by inhibition of postsynaptic MAPK signaling. *Proc Natl Acad Sci USA* 107(19):8830–8835.

McClelland AC, Sheffler-Collins SI, Kayser MS & Dalva MB (2009) Ephrin-B1 and ephrin-B2 mediate EphB-dependent presynaptic development via syntenin-1. *Proc Natl Acad Sci USA* 106(48):20487–20492.

Mulligan KA & Cheyette BN (2012) Wnt signaling in vertebrate neural development and function. *J Neuroimmune Pharmacol* 7(4):774–787.

Munno DW & Syed NI (2003) Synaptogenesis in the CNS: An odyssey from wiring together to firing together. *J Physiol* 552(1):1–11.

Opperman KJ & Grill B (2014) RPM-1 is localized to distinct subcellular compartments and regulates axon length in GABAergic motor neurons. *Neural Dev* 9:10.

Park TJ & Curran T (2014) Neurobiology: Reelin mediates form and function. *Curr Biol* 24(22):R1089–1092.

Penn AA, Riquelme PA, Feller MB & Shatz CJ (1998) Competition in retinogeniculate patterning driven by spontaneous activity. *Science* 279(5359):2108–2112.

Petzoldt AG & Sigrist SJ (2014) Synaptogenesis. *Curr Biol* 24(22):R1076–1080.

Sanes JR & Zipursky SL (2020) Synaptic specificity, recognition molecules, and assembly of neural circuits. *Cell* 181(3):536–556.

Scheiffele P (2003) Cell–cell signaling during synapse formation in the CNS. *Annu Rev Neurosci* 26:485–508.

Sharma J & Sur M (2014) The ferret as a model for visual system development and plasticity. In *Biology and Diseases of the Ferret*, 3rd ed. (Fox JG & and Marini RP eds), pp. 711–734. John Wiley and Sons, Inc.

Sheng M & Kim E (2011) The postsynaptic organization of synapses. *Cold Spring Harb Perspect Biol* 3(12):a005678.

Srinivasan E, Chandrasekhar G, Chandrasekar P, et al. (2021) Alpha-synuclein aggregation in Parkinson's disease. *Front Med* 8:736978.

Südhof TC (2018) Towards an understanding of synapse formation. *Neuron* 100(2):276–293.

Tian X & Wu C (2013) The role of ubiquitin-mediated pathways in regulating synaptic development, axonal degeneration and regeneration: Insights from fly and worm. *J Physiol* 591:3133–3143.

Tomorsky J, Parker P, Doe CQ & Niell CM (2020) Precise levels of nectin-3 are required for proper synapse formation in postnatal visual cortex. *Neural Dev* 15(1):13.

Xu NJ, Sun S, Gibson JR & Henkemeyer M (2011) A dual shaping mechanism for postsynaptic ephrin-B3 as a receptor that sculpts dendrites and synapses. *Nat Neurosci* 14:1421–1429.

Yogev S & Shen K (2014) Cellular and molecular mechanisms of synaptic specificity. *Annu Rev Cell Dev Biol* 30:417–437.

Zhang J, Ackman JB, Xu HP & Crair MC (2011) Visual map development depends on the temporal pattern of binocular activity in mice. *Nature Neurosci* 15(2):298–307.

Ziv NE & Smith SJ (1996) Evidence for a role of dendritic filopodia in synaptogenesis and spine formation. *Neuron* 17:91–102.

GLOSSARY

α-bungarotoxin A snake venom toxin that competitively binds to and blocks nicotinic acetylcholine receptors.

acetylcholine (ACh) Neurotransmitter released by a motor neuron at the neuromuscular junction in vertebrates and some invertebrates. It is also used in the vertebrate CNS, as an excitatory or modulatory neurotransmitter, and in the autonomic nervous system. In some invertebrates, it is the major excitatory neurotransmitter of the CNS.

acetylcholine esterase (AChE) Enzyme that catalyzes the breakdown of acetylcholine. At the neuromuscular junction, an isoform of the enzyme is secreted by muscle cells and becomes concentrated in the synaptic basal lamina.

acetylcholine receptor (AChR) Receptor for the neurotransmitter acetylcholine. The nicotinic AChRs (nAChRs) are nonselective cation channels; they are the postsynaptic receptor at the vertebrate neuromuscular junction and function as excitatory receptors at some CNS synapses. The metabotropic AChRs (muscarinic AChRs or mAChRs) are G-protein-coupled receptors that play a modulatory role.

action potential An elementary unit of nerve impulses that axons use to convey information across long distances. It is all-or-none, regenerative, and propagates unidirectionally in the axon.

activin Protein of the transforming growth factor β (TGFβ) superfamily. One of its many roles during embryogenesis is the induction of mesoderm.

adherens junction Protein complex that forms connections between cells by linking the actin filaments of one cell to those of an adjacent cell. Examples include the sites of attachment between apical progenitor cells and the ventricular surface of the developing neural tube.

adhesion molecule Protein located on a cell surface that mediates binding between cells or between a cell and the extracellular matrix.

agrin Protein secreted by motor neurons that induces aggregation of acetylcholine receptors in the muscle.

agrin–ACh hypothesis Idea that both agrin and acetylcholine (ACh), released by an innervating motor nerve terminal, regulate the clustering and dispersal of acetylcholine receptors, respectively. Agrin is thought to stabilize the receptors in the central region of a muscle fiber to prevent them from being displaced upon exposure to ACh.

agrin hypothesis Idea that agrin released by an innervating motor nerve terminal becomes concentrated in the synaptic basal lamina to induce the clustering of acetylcholine receptors in the muscle directly opposite the nerve terminal.

anencephaly A severe type of neural tube defect that results when the anterior portion of the neural tube fails to close; the cranial portions of the brain and the back of the skull fail to fuse. The condition is fatal.

animal pole The portion of the egg that has the least amount of yolk, located opposite the vegetal pole. It is the site of origin for cells that form ectoderm and mesoderm.

anterior neural border (ANB) In fish, the anterior-most region of the neural ectoderm that becomes a signaling center that secretes molecules to influence the patterning of forebrain regions. Called the anterior neural ridge in other vertebrates.

anterior neural ridge (ANR) The anterior-most region of the neural ectoderm that becomes a signaling center that secretes molecules to influence the patterning of forebrain regions in mice, chick, and other vertebrates. Called the anterior neural border in fish.

anterior–posterior (anteroposterior) axis Refers to the main body axis that extends from head to tail and the neuraxis that extends from the brain to the spinal cord. Also called the rostral–caudal axis.

anterior visceral endoderm (AVE) In mice, a population of extraembryonic cells that lies below the anterior neural plate. The AVE serves as a signaling center to pattern anterior portions of the neural tube as well as to specify cells that form the anterior neural ridge. Similar to the hypoblast in other species.

apical abscission The shedding (abscission) of the primary cilium of a neural progenitor cell prior to migration from the apical (ventricular) surface of the neural tube.

apoptosis A form of programmed cell death characterized by distinct cellular changes, including the fragmentation of DNA. A naturally occurring developmental process initiated by extracellular signals or intrinsic genetic mechanisms; mediated by intracellular signaling pathways involving enzymes called caspases.

apoptosome Heptamer of Apaf1 proteins that forms on activation of the intrinsic apoptotic pathway; it recruits and activates initiator caspases that subsequently activate downstream effector caspases to induce apoptosis.

apoptotic body Discrete, membrane-bound structure containing condensed fragments of organelles and the nucleus; buds off from a cell undergoing apoptosis.

astrocyte Star-shaped glial cell of the vertebrate central nervous system that plays a variety of roles in neuronal development, synaptic maintenance, and neuronal communication.

astrotactin A glycoprotein found on the surface of migrating neurons in the cerebellum and cerebral cortex; facilitates attachment of the neurons to radial glial cells.

autocrine A type of cellular communication in which a cell releases a signaling molecule that then binds to a corresponding receptor on the surface of that same cell.

autonomic nervous system (ANS) The collected parts of the nervous system that regulate the function of internal organs, including the contraction of smooth and cardiac muscles and the activities of glands.

axon A long, thin process of a neuron, it often extends far beyond the soma and propagates and transmits signals to other neurons or muscle at its presynaptic terminals.

axon terminal The distal, slightly enlarged end of an axon that forms the presynaptic connection at a synapse.

Bergmann glia Specialized radial glia found in the cerebellum that extend multiple, parallel cell processes which help guide cerebellar granule cells from the external to the internal granule cell layer.

blastocoel The central, fluid-filled cavity of an early-stage embryo.

blastocyst The early-stage, pre-implantation embryo of mammals. The structure consists of blastomeres surrounding a large blastocoel cavity.

blastoderm A group of cells (blastomeres) that forms above the central cavity (blastocoel) or yolk of the early-stage embryo.

blastodisc The flat, disc-shaped, early-stage embryo that forms two cell layers (epiblast and hypoblast) above the blastocoel cavity; characteristic of birds, reptiles, and some mammals.

blastomere A cell produced in the zygote during cell division.

blastopore The opening on the outer surface of an early-stage embryo through which cells migrate during gastrulation.

blastula A spherical-shaped, early-stage embryo consisting of blastomeres surrounding the blastocoel cavity; characteristic of amphibians and similar to the blastocyst in mammals. Often used to refer to embryos at similar stages of development in multiple species.

bone morphogenetic protein (BMP) Protein of the transforming growth factor β (TGFβ) superfamily; isoforms play multiple roles in embryogenesis, including induction of epidermal ectoderm and patterning along the dorsal–ventral axis of the neural tube.

Cajal–Retzius (CR) cell Cell located in the marginal zone of the developing neocortex and hippocampus; secretes the protein Reelin.

canonical pathway The established signaling cascade for a given molecule. When referring to Wnt signaling, the Wnt/β-catenin pathway that leads to the transcription of target genes.

caspase Intracellular protease that is involved in mediating the intracellular events of apoptosis.

CDK inhibitory protein A group of small proteins that halt progression of the cell cycle at different stages by blocking the activity of specific cyclin-dependent kinases (CDKs). Primarily involved at the gap 1 (G1) and synthesis (S) phases.

cell autonomous Of a gene, acting in the cell that produces the gene product.

cell cycle The sequence of steps through which a cell progresses prior to cell division. The stages include gap 1 (G1), synthesis (S), gap 2 (G2), and mitosis (M).

cell fate The final cell type established from a precursor cell during the course of development.

cell fate determination The process by which a cell becomes committed to forming a particular cell type. Once determined, no further developmental or experimental manipulations can alter the type of cell that forms.

cell line culture A group of homogenous cells maintained *in vitro* that descend from a single cell and therefore have the same genotype and phenotype.

cell marker A protein or gene expressed in a limited number of cell types and used to distinguish the identity of a particular cell.

cellular blastoderm In *Drosophila*, the stage of development when the nuclei located around the cortex of the blastoderm become separated by cell membranes to create individual cells. Forms from the syncytial blastoderm.

cellular determination See *cell fate determination*.

cellular differentiation The process of acquiring the characteristics of a particular type of cell. Cellular differentiation follows cellular determination.

central domain (C-domain) The region of the growth cone that lies closest to the neurite shaft and contains microtubules and organelles. Also called the central mound.

central mound The region of the growth cone that lies closest to the neurite shaft and contains microtubules and organelles. Also called the central domain (C-domain).

central nervous system (CNS) The brain and spinal cord in vertebrates; the brain and nerve cord in some invertebrates.

cephalic Relating to the head.

Cerberus A secreted protein present in the organizer and anterior visceral endoderm that blocks the activity of bone morphogenetic proteins and Wnt. In the anterior neural tube, Cerberus acts to prevent the posteriorization of forebrain regions.

chemical synapse A site at which information is transferred from a neuron to another cell across a synaptic space (cleft), where presynaptic nerve terminal specializations release chemical neurotransmitters and postsynaptic specializations specifically bind the neurotransmitter.

chemoaffinity hypothesis The concept that there are matching proteins or "chemical tags" located on the surface of axonal growth cones and their target cells. Originally proposed by Roger Sperry to describe how the axons of retinal ganglion cells reached their target cells in the optic tectum.

chemotropism The directed extension of a growth cone up or down a chemical gradient.

choice point An intersection of two cellular surfaces on which an elongating axon has the potential to extend.

choline acetyltransferase (ChAT) Enzyme needed to synthesize the neurotransmitter acetylcholine.

chordamesoderm A band of axial mesoderm that forms the notochord and related structures in vertebrate embryos.

chordin One of the first neural inducers identified; promotes formation of neural tissue by blocking the epidermal-inducing actions of bone morphogenetic proteins (BMPs).

choroid plexus The specialized epithelial cells and network of capillaries of the ventricles. The capillaries are covered by ependymal cells that produce cerebral spinal fluid.

ciliary neurotrophic factor (CNTF) A cytokine of the gp130 family of proteins that promotes survival and neurite outgrowth of several neuronal populations in the developing and adult nervous system. Initially identified in the chick eye.

co-linearity The relative location of a *Hox* gene in its cluster to its expression along the anterior–posterior axis; *Hox* genes located at the 3′ end of the cluster are expressed earlier and in more anterior regions than those at the 5′ end of the cluster.

commissure A bundle of nerve fibers that connects the two sides of the brain or spinal cord.

conditional knockout mouse The process of disrupting a gene in a specific spatiotemporal pattern or an animal in which a gene has been disrupted in a specific spatiotemporal pattern. The most common strategy for generating conditional knockouts in mice uses Cre/*loxP*-based recombination. It usually involves inserting a pair of *loxP* elements in introns that flank (an) essential exon(s) of a gene of interest. The gene of interest is only disrupted in cells in which Cre is active or in cells derived from a progenitor in which Cre was active.

contralateral Of the other side of the midline. For example, a contralateral axonal projection is an axon that crosses the midline and terminates on the side of the nervous system opposite the soma.

cortical hem A derivative of the telencephalon roof plate comprised of neuroepithelial cells that serves as a transient signaling center for dorsal–ventral patterning of neocortical regions.

cortical plate (CP) A transient layer of cells in the developing neocortex that forms between the marginal zone and preplate. The cells ultimately form layers II through VI of the adult neocortex.

cranial neural crest Neural crest cells that originate from the dorsal surface of the neural tube between the midbrain and rhombomere 6. Gives rise to structures associated with the head and neck, including cranial ganglia and skeletal structures.

critical period A sensitive period during development when experience plays an important role in shaping the wiring properties of the brain.

cyclin Protein that periodically rises and falls in concentration in step with the eukaryotic cell cycle. Cyclins activate crucial protein kinases (called cyclin-dependent protein kinases or CDKs) and thereby help control progression from one stage of the cell cycle to the next.

cyclin-dependent kinase (CDK) Protein kinase that has to be complexed with a cyclin protein in order to act. Different cyclin-CDK complexes trigger different steps in the cell division cycle by phosphorylating specific target proteins.

cytokine Extracellular signal protein or peptide that acts as a local mediator in cell–cell communication.

delaminate To separate into different layers. When referring to neural crest cells, the process by which the cells separate from the neuroepithelium of the neural tube.

dendrite Neuronal process that is typically branched with tapered ends; receives synaptic input.

dendritic spine A small protrusion on a dendrite of certain neurons that receives synaptic input from a partner neuron. The thin spine neck creates chemical and electrical compartments for each spine such that it can be modulated independently from neighboring spines.

dermamyotome A segment of a somite that later gives rise to dermis, skeletal muscle, and vascular tissue.

Dickkopf (Dkk) A secreted protein that binds to the LRP5 and LRP6 subunits of the Wnt receptor complex to block Wnt signaling. In the anterior neural tube, Dkk acts to prevent the posteriorization of forebrain regions.

diencephalon One of the five secondary brain vesicles; arises from the posterior region of the prosencephalon and forms structures that include the thalamus, hypothalamus, and optic cup.

diphtheria toxin A bacterial-derived protein synthesis inhibitor.

divergent canonical pathway A signaling cascade other than the one previously established for a given molecule. When referring to Wnt signaling, the pathway that leads microtubule stabilization.

DNA (deoxyribonucleic acid) Long double-stranded chains of nucleotides. The nucleotides consist of the sugar deoxyribose, a phosphate group, and one of four nitrogenous bases: adenine (A), cytosine (C), guanine (G), or thymidine (T).

DNA ladder The pattern of DNA fragments characteristic of apoptotic cells, as visualized by gel electrophoresis.

dominant negative A type of mutation in which the resulting gene product interferes with the function of the normal gene product.

dorsal blastopore lip (DBL) The dorsal margin of the opening (blastopore) that forms on the outer surface of the amphibian embryo. This region of the blastopore becomes a signaling center critical for directing ectoderm to a neural fate.

dorsal–ventral axis (dorsoventral axis) The body axis that runs from the back (Latin, *dorsum*) to the belly (Latin, *venter*).

downstream In a signal transduction pathway, molecules that are activated after receptor binding or after activation of an earlier step in the pathway.

dynamic microtubule In a growth cone, a motile microtubule (a hollow tube comprised of the protein tubulin) that extends past the stable microtubules of the central domain to influence the direction of growth cone extension.

ectoderm Embryonic epithelial tissue that is the precursor of the epidermis and nervous system.

ectopic Arising or placed at a location where a tissue is not normally found.

effector caspase Apoptotic caspases that catalyze the widespread cleavage events during apoptosis that kill the cell.

electroporation A procedure in which DNA containing a transgene is introduced into cells by applying electrical current to facilitate the transfer of negatively charged DNA molecules into the cells. In animals, this can be achieved by placing a micropipette containing the DNA near the cells of interest and applying electrical current.

embryonic shield In zebrafish, the organizer region that is analogous to the dorsal blastopore lip of the amphibian.

embryonic stem (ES) cell Pluripotent cell derived from early embryos that can be propagated indefinitely *in vitro* and that can give rise to all cell types of an embryo *in vivo*.

endoderm The innermost of the three germ cell layers; gives rise to the gut and related structures.

endosome A membrane-enclosed organelle produced by endocytosis. It carries newly internalized extracellular materials and transmembrane proteins.

endplate potential (EPP) Depolarization produced in a postsynaptic muscle cell by acetylcholine released from a presynaptic motor neuron in response to an action potential.

enhancer A short segment of DNA that, when bound by transcription factors, increases the likelihood that a gene will be transcribed.

en passant **synapse** Synaptic connection that forms when presynaptic elements concentrate along the axon shaft rather than at the axon terminal.

enteric ganglia The clusters of neurons that form the enteric nervous system, a division of the autonomic nervous system. Enteric ganglia are associated with the gastrointestinal tract and regulate digestion rather independently of the rest of the autonomic nervous system.

ependymal cell A type of glial cell that lines the ventricles and spinal canal.

Eph receptor Receptor tyrosine kinase that binds ephrin with its extracellular domains. Two Eph receptor subtypes, the EphA and EphB receptors, typically bind ephrin As and ephrin Bs, respectively, but this specificity is not absolute. They can also serve as ligands during reverse signaling.

ephrin Cell surface protein that acts as a ligand for Eph receptors. The ephrin family consists of two subfamilies: ephrin As are attached to the extracellular face of the plasma membrane by GPI, and ephrin Bs are transmembrane proteins. They can also serve as receptors during reverse signaling.

epiblast The outer of the two layers of the blastodisc that gives rise to all future cell layers (ectoderm, mesoderm, and endoderm). It is roughly equivalent to the animal pole of the amphibian blastula.

epiboly The movement and thinning of cell layers. In zebrafish, several stages of development are distinguished based on the percentage of the yolk that is surrounded by the blastoderm, reported as percent of epiboly.

epigenetic Molecular modifications to DNA and chromatin, such as DNA methylation and various forms of post-translational modification of histones. They do not modify the DNA sequence but can alter gene expression.

excitatory synapse Site where the release of a presynaptic neurotransmitter depolarizes the postsynaptic cell, thereby increasing the likelihood that the cell will fire an action potential.

explant A piece of tissue placed in cell culture.

extraembryonic tissue Tissues that contribute to the maintenance of an embryo, but that do not become part of the embryo proper.

extrasynaptic nucleus In the multinucleated muscle fiber (myofiber), a nucleus that is not located below the innervating motor nerve terminal.

extrinsic pathway Pathway of apoptosis triggered by extracellular signal proteins binding to cell surface death receptors.

fasciclin A cell surface glycoprotein of the IgG superfamily expressed on axons. An ortholog of neural cell adhesion molecule (NCAM) that relies on homophilic binding and often functions to direct the growth of other axons.

fasciculation The bundling of axons; the tendency for growing axons to bind to one another to form a fascicle.

fibroblast growth factor (FGF) A member of a family of secreted growth factors that act as morphogens to pattern early embryos during development.

filopodium Slender process arising from the growth cone of a neurite that actively samples the local environment to select the most favorable substrate. Helps to direct, or steer, the growth cone to a particular pathway.

floor plate A wedge-shaped region of glial cells that forms at the ventral surface of the neural tube; serves as a signaling center for neuronal patterning and axonal guidance.

folic acid Vitamin B_9, or folate.

follistatin An activin-binding protein that was one of the first neural inducers identified; promotes formation of neural tissue by blocking the epidermal-inducing actions of bone morphogenetic proteins (BMPs).

G1 (gap 1) phase Phase of the eukaryotic cell division cycle, between the end of mitosis and the start of DNA synthesis.

G2 (gap 2) phase Phase of the eukaryotic cell division cycle, between the end of DNA synthesis and the beginning of mitosis.

ganglion Cluster of neurons located in the peripheral nervous system. Plural **ganglia**.

gap gene In *Drosophila* development, a gene that is expressed in specific broad regions along the anteroposterior axis of the early embryo, and which helps designate the main divisions of the insect body.

gap junction Communicating channel-forming cell–cell junction present in most animal tissues that allows ions and small molecules to pass from the cytoplasm of one cell to the cytoplasm of the next.

gastrulation The process by which an embryo is transformed from a ball of cells into a structure with three distinct layers: ectoderm, mesoderm, and endoderm.

gene A segment of DNA that carries the instructions for how and when to make specific RNAs and proteins.

gene knockout mouse A mouse generated using genetic engineering to inactivate a specific gene. It is usually achieved by homologous recombination in embryonic stem cells to create a mutation in the target gene. The resulting mutant mouse is called a knockout mouse for that particular gene.

gene regulation The process by which a cell determines which genes will be expressed (turned on) and which will be repressed (turned off).

genetic manipulation The process of altering gene expression in a cell or organism.

germ cell layer One of the three primary tissue layers (endoderm, mesoderm, and ectoderm) of an animal embryo; also called germ layer.

glia Nonneuronal cells of the nervous system, they play essential roles for the development and function of neurons.

gliogenesis The process of generating new glial cells, the nonneuronal cells of the nervous system.

gp130 cytokine A member of a family of peptides that signal through a common receptor system that includes the gp130 subunit. Ciliary neurotrophic factor and leukemia inhibitory factor are among the members of this family of cytokines.

granule cell layer The innermost layer of the adult cerebellum that contains granule neurons as well as interneurons.

growth cone A dynamic structure at the tip of a developing neuronal process, it enables the extension of the process and guides its direction.

Hensen's node In birds, a group of cells at the anterior (cephalic) end of the primitive streak. Analogous to the dorsal blastopore lip of amphibians and node of mammals.

holoprosencephaly (HPE) A congenital disorder in which the two hemispheres fail to fully separate during embryonic development. The resulting neurological and craniofacial deficits range from mild to severe.

homeobox A 180-base-pair segment of DNA that is found in genes such as those of the *Hox* family. Disruption of this sequence leads to transformations of one body part into another.

homeodomain The 60 amino acid protein product encoded by the homeobox; common to *Hox* and other transcription factors.

homeostatic plasticity The modification of pre- and postsynaptic elements in response to changes in the overall level of neural activity; used to maintain a stable level of neural firing.

homologous recombination Exchange of nucleotide sequences between two identical or highly similar DNA molecules. It occurs naturally in certain cells due to its role in specific biological processes, such as in germ-line cells during meiotic crossing over. It is also used experimentally for genome engineering, such as the generation of knockout and knock-in alleles.

***Hox* code** The combination of *Hox* genes used to define the development of regions along the anterior–posterior axis.

***Hox* gene** A member of a family of evolutionarily conserved genes that are arranged in genomes in clusters and encode homeodomain-containing transcription factors. *Hox* genes define the anterior–posterior body axis of vertebrates and most invertebrates and contribute to neuronal fate at later stages of development.

hyperdorsalized embryo A *Xenopus* embryo experimentally treated with lithium chloride that displays an expansion of dorsal tissues, including an increase in the size of the nervous system.

hyperplasia An increase in the size of a tissue.

hypoblast The inner of the two layers of the blastodisc. Lying just above the yolk, it is eventually displaced by the definitive endoderm and ultimately forms extraembryonic tissues. It is roughly equivalent to the vegetal pole of the amphibian blastula.

hypodermis The outermost cell layer of the *C. elegans* embryo; equivalent to the ectoderm of other species.

hypoplasia A decrease in the size of a tissue.

imaginal disc Pockets of tissue found in the larval stages of some insects. The tissues are segmentally organized and attach to the inside of the larval epidermis. Although the discs play no role in larval development, they later evert during metamorphosis to form adult structures of the head, thorax, legs, and wings.

immunocytochemistry A staining method that uses antibodies to visualize the distributions of proteins in cells. See *immunohistochemistry*.

immunohistochemistry A staining method that uses antibodies to visualize the distributions of proteins in fixed tissues. The most common form uses sequential application of two antibodies: a primary antibody that binds the protein of interest and a fluorescence- or enzyme-conjugated secondary antibody that binds to the primary antibody. Protein distribution can be visualized by fluorescence or a color substrate produced by the enzyme conjugated to the secondary antibody.

immunolabeling The process of identifying antigens with fluorescent or enzymatic markers to map protein distribution in cells or tissues. See *immunohistochemistry*.

inhibitory synapse Site where the release of a presynaptic neurotransmitter hyperpolarizes the postsynaptic cell, thereby decreasing the likelihood that the cell will fire an action potential.

initiator caspase Apoptotic caspases that begin the apoptotic process, thus activating the effector caspases.

***in situ* hybridization** A method for determining mRNA distribution in tissues by hybridizing labeled gene-specific nucleic acid probes to fixed histological sections or whole-mount tissues.

interkinetic nuclear migration In the developing nervous system, the movement of the nucleus of a neural progenitor cell through its cytoplasmic processes. The nucleus travels from the ventricular (apical) surface of the neural tube to the pial (basal) surface then back to the ventricular surface, where the cell divides.

intermediate target A tissue that helps direct axons toward their final target cells by transiently expressing guidance cues to which those axons respond.

intrinsic pathway Pathway of apoptosis activated from inside the cell in response to stress or developmental signals; depends on the release into the cytosol of mitochondrial proteins normally resident in the mitochondrial intermembrane space.

ipsilateral Of the same side of the midline. For example, an ipsilateral axonal projection is an axon that does not cross the midline and therefore terminates on the same side of the nervous system as the soma.

isthmic organizer (IsO) A signaling center located between the mesencephalon and metencephalon (the isthmus) that is important for the development of the future midbrain, pons, and cerebellum.

isthmus The region of the central nervous system located between the mesencephalon and metencephalon; also called the midbrain–hindbrain border (MHB).

labeled pathway hypothesis Concept that axons express particular surface molecules ("labels") that serve as directional cues for later extending axons.

lamellipodium A veil-like, motile area of a growth cone that contains a meshwork of filamentous (fibrillar) actin (F-actin) and subunits of globular actin (G-actin); located between the filopodia.

lateral geniculate nucleus (LGN) A thalamic nucleus that receives visual input from retinal ganglion cell axons and sends output to the primary visual cortex.

lateral inhibition The process by which neighboring cells are prevented from adopting identical fates through cell–cell interactions, such as those mediated by Notch/Delta.

ligand A molecule that binds to its receptor.

lissencephaly A brain disorder in which the outer surface of the brain is characterized by a smooth surface lacking typical gyri and sulci. The disorder results from defects in neuronal migration that cause cortical neurons to stack up below one another.

lysate The material produced when cell membranes are broken apart.

marginal zone (MZ) In the developing neocortex, the layer of cells closest to the pial surface that contains the Cajal–Retzius cells; forms layer I of the adult cortex.

medial–lateral axis The body axis that extends from the midline to the sides.

mesencephalon One of the three primary brain vesicles that forms the midbrain region of the neural tube; gives rise to the superior and inferior colliculi (tectum) and midbrain tegmentum.

mesoderm The middle of the three germ cell layers that gives rise to muscle, bone, connective tissues, and the cardiovascular and urogenital systems.

metencephalon One of the five secondary brain vesicles; forms the cerebellum and pons.

microglia A glial cell that functions as the resident immune cell of the nervous system. It engulfs damaged cells and debris. It influences the stability of synaptic connections under some circumstances.

midbrain–hindbrain border (MHB) The region of the central nervous system located between the mesencephalon and metencephalon; also called the isthmus.

molecular layer The outermost of the three layers in the adult cerebellum. It contains various interneurons as well as the elaborate dendritic tress of the Purkinje cells.

morphogen Diffusible signal molecule that can impose a pattern on a field of cells by causing cells in different places to adopt different fates.

motor endplate A specialized, oval-shaped region of skeletal muscle that receives innervation from the terminal branches of a motor neuron.

motor unit A motor neuron and the set of muscle fibers it innervates.

M phase The mitosis phase of the cell cycle; results in two daughter cells.

multipotent cell Cells that can develop into more than one specific cell type, but cannot develop into all cell types of the body.

muscle fiber A muscle cell; a long cylindrical multinucleated cell arising from myotubes; also called a myofiber.

myelin The sheath surrounding some axons that is formed by the cellular processes of oligodendrocytes or Schwann cells. It improves the speed of action potential conduction.

myelencephalon One of the five secondary brain vesicles; gives rise to the medulla.

myoblast Mononucleated, undifferentiated muscle precursor cell. A skeletal muscle cell is formed by the fusion of multiple myoblasts.

myofiber A muscle cell; a long cylindrical multinucleated cell arising from myotubes; also called a muscle fiber.

neocortex Outer region of the cerebral cortex, consisting of six cell layers.

neural cell adhesion molecule (NCAM) A glycoprotein expressed on the surfaces of neurons and neurites; cells adhere to one another via homophilic binding mechanisms.

neural crest cell Cell originating along the dorsal neural tube in the vertebrate embryo. These cells migrate to give rise to a variety of tissues, including neurons and glia of the peripheral nervous system, pigment cells of the skin, and the bones of the face and jaws.

neural fold The raised area at the lateral regions of the neural plate that curl over to form the neural tube.

neural groove The central indentation along the length of the neural plate.

neural induction The process by which external signals promote the formation of neural tissue from ectoderm.

neural keel In zebrafish, the solid group of cells that forms at the midline of the neural plate. It later gives rise to the neural rod.

neural plasticity Changes of the nervous system in response to experience and learning.

neural plate The layer of ectodermal cells overlaying the notochord that invaginates and gives rise to the neural tube during neurulation.

neural rod In zebrafish, the solid central region that forms from the neural keel and separates from the surface epidermal ectoderm. The central cells later migrate away in order to form the hollow center of the neural tube.

neural tube The hollow tube that is the embryonic precursor to the brain and spinal cord. Comprised of neuroepithelial cells, it is derived from neural plate ectoderm.

neural tube defect A birth defect resulting from failure of the neural tube to close in the cephalic (brain) or spinal cord regions.

neuraxis The main axis established by the central nervous system (the brain and spinal cord); also called neuroaxis.

neurite An axon or dendrite; often used to describe neural processes extending in cell culture when it is not clear if the process is an axon or dendrite.

neuroblast A neuronal progenitor cell.

neurogenesis The process of generating new neurons.

neuromere Transient segment of the developing neural tube that provides a restricted area for further development; called prosomere in the developing forebrain and rhombomere in the developing hindbrain.

neuromuscular junction (NMJ) The synapse between a motor neuron's presynaptic terminals and a skeletal muscle cell.

neuron An electrically excitable cell that receives, integrates, propagates, and transmits information as the working unit of the nervous system.

neuronal progenitor Precursor cell that has the capacity to develop into a restricted number of neuronal types, but has not yet fully differentiated.

neuronal specificity The process by which progenitor (precursor) cells are directed to express specific genes and adopt a particular neuronal fate. Can also refer to the mapping of axonal projections to the correct target cells.

neurotransmitter Molecule that is stored in synaptic vesicles (or dense-core vesicles in the case of neuropeptides) in the presynaptic terminals, is released into the synaptic cleft triggered by presynaptic depolarization, and activates ionotropic or metabotropic receptors on a postsynaptic target cell.

neurotransmitter receptor A protein that binds and responds to a specific chemical (neurotransmitter) released by a presynaptic cell.

neurotrophic hypothesis Proposal that target tissues release limited quantities of survival-promoting proteins at the time of innervation to ensure that only the correct number of innervating neurons survive.

neurotrophin A family of secreted signaling proteins that regulate the survival, morphology, and physiology of target neurons through binding to specific receptors on those neurons. Mammalian neurotrophins include nerve growth factor (NGF), brain-derived neurotrophic factor (BDNF), neurotrophin-3 (NT-3), and neurotrophin-4/5 (NT-4/5).

node A group of cells at the anterior (cephalic) end of the primitive streak in mammals; analogous to the dorsal blastopore lip of amphibians and Hensen's node of birds.

noggin One of the first neural inducers identified; promotes formation of neural tissue by blocking the epidermal-inducing actions of bone morphogenetic proteins (BMPs).

notochord In the embryo, a midline, mesodermal structure located along the ventral surface of the neural plate and neural tube. It extends from the mesencephalon to the spinal cord, where it serves as a source of signals for patterning ventral regions of the nervous system. It is later incorporated into the vertebral column of most vertebrates.

nucleation The formation of a small complex of approximately three G-actin monomers; the first step in action polymerization.

ocular dominance column (ODC) Preference for receiving and/or representing visual input from one eye over the other eye. In the primary visual cortex of some mammals, such as cats and monkeys, cells in the same vertical columns share the same ocular dominance, thus producing ocular dominance columns.

oligodendrocyte A glial cell in the CNS that wraps axons with its cytoplasmic extension to form myelin sheath.

oligodendrocyte precursor cell (OPC) Cells that originate in the ventricular zone and migrate to sites where they differentiate into mature oligodendrocytes.

ommatidium A unit of an insect compound eye composed of photoreceptor, pigment, and supporting cells. Each eye is comprised of several hundred ommatidia.

oncogene An altered gene whose product can act in a dominant fashion to help make a cell cancerous. Typically, an oncogene is a mutant form of a normal gene (proto-oncogene) involved in the control of cell growth or division.

optogenetics A method of using light to selectively alter neuronal activity in cells genetically modified to express a light sensitive protein such as an opsin. The particular protein expressed determines whether light activates, inhibits, or modifies neuronal activity.

organizer A group of cells that induces formation of neural tissue; refers to the dorsal blastopore lip (Spemann's organizer) in amphibians, Hensen's node in birds, and the node in mammals.

ortholog Gene or protein from different species that are similar in sequence because they are descendants of the same gene in the last common ancestor of those species.

otocyst The embryonic inner ear, also called the otic vesicle.

pair-rule gene In *Drosophila* development, a gene expressed in a series of regular transverse stripes along the body of the embryo and which helps to determine its segments.

pallium The dorsal region of the telencephalon that gives rise to the cerebral cortex.

paralog Genes or proteins that are similar in sequence because they are the result of a gene duplication event occurring in an ancestral organism. Those in two different organisms are less likely to have the same function than are orthologs.

peripheral domain (P-domain) The distal, leading edge of a growth cone consisting of lamellipodia and filopodia.

peripheral nervous system (PNS) Collections of neurons (ganglia) that lie outside of the brain and spinal cord.

perisynaptic Schwann cell (PSC) Specialized, non-myelinating Schwann cell that surrounds the terminal branches of a motor neuron; a component of the neuromuscular junction tripartite complex.

periventricular heterotopia (PH) A brain disorder in which neurons fail to migrate out of the ventricular zone. Instead, the neurons, which appear to be highly differentiated, cluster in nodules near the ventricles.

pioneer axon The first axon to extend in the direction of a target tissue. It expresses guidance cues along its axon shaft to help guide subsequent axons toward a target tissue.

placode In vertebrates, an ectodermal thickening that gives rise to some cranial ganglia and sensory structures.

postjunctional fold Invagination in the cell membrane of skeletal muscle at the site of synaptic contact.

postsynaptic After, or downstream of, a synapse.

postsynaptic cell A cell that receives and responds to input from another neuron (the presynaptic neuron); a cell specialized for neurotransmitter reception, located after the synaptic cleft.

postsynaptic density (PSD) An electron-dense organelle found in excitatory synapses of the central nervous system; helps anchor certain proteins of the postsynaptic cell across from the presynaptic terminal.

prechordal plate Located at the anterior end of the epiblast, the region where surface ectoderm tightly adheres to the underlying endoderm.

precursor cell Cell with the ability to give rise to a limited number and type of specialized cells, in a state between a stem cell and differentiated cell; also called a progenitor cell.

preplate (PP) A transient layer of cells in the developing neocortex located below the pial surface; contains the first cells to migrate out of the ventricular zone. As additional cells leave the ventricular zone, the preplate separates into two layers—the marginal zone and subplate.

presynaptic cell A cell specialized to release a signal to initiate a response in a partner cell (the postsynaptic cell); the cell located before the synaptic cleft.

presynaptic nerve terminal A structure at the end (or along the trunk) of an axon that is specialized for releasing neurotransmitters onto target cells.

primary cell culture Cells harvested from an animal and maintained *in vitro*.

primary cell layers The three tissue layers (endoderm, mesoderm, and ectoderm) that form during the process of gastrulation; also called germ layers or germ cell layers.

primary cilium A single, non-motile cilium that projects from the cell membrane as a separate compartment and selectively transports proteins, including those associated with signaling pathways.

primary visual cortex (V1) The visual cortical area that receives direct input from the lateral geniculate nucleus.

primitive streak The axial thickening in ectoderm that is the site of indentation and migration for cells during gastrulation in birds, reptiles, and some mammals; similar in function to the blastopore of amphibians and embryonic shield of fish.

progenitor cell Cell with the ability to give rise to a limited number and type of specialized cells, in a state between a stem cell and differentiated cell; also called a precursor cell.

programmed cell death (PCD) A form of cell death in which a cell kills itself by activating an intracellular death program.

proneural cluster (PNC) A group of cells expressing proneural genes. Each cell in the cluster has the potential to form a neural precursor.

proneural gene Gene that provides a cell with the potential to form a neural precursor.

pro-neurotrophin Precursor protein that is cleaved to yield the mature form of a neurotrophin; thought to play a role in cell death pathways.

prosencephalon One of the three primary brain vesicles; forms the forebrain region of the neural tube and brain.

prosomeres Transient segments (neuromeres) in the embryonic forebrain that provide restricted regions for further cellular development. Numbered from 1–6, beginning at the mesencephalon and progressing to the telencephalon.

proteasome Large protein complex in the cytosol with proteolytic activity that is responsible for degrading proteins that have been marked for destruction by ubiquitination or by some other means.

Purkinje cell layer The middle cell layer in the adult cerebellum, it is characterized by a single layer of the Purkinje neuron cell bodies; also contains cerebellar interneurons.

pyknotic A shrunken nucleus with condensed chromatin which is characteristic of cells dying by apoptosis.

radial glial (RG) cell Progenitor cell in the ventricular zone that extends processes to the ventricular and pial surfaces. The processes provide surfaces for neuronal migration.

receptor A protein that binds and responds to a specific signaling molecule.

Reelin A secreted glycoprotein that influences neuronal migration and synaptic connections in the nervous system; first identified in Cajal–Retzius cells, where it helps direct the migration of cells entering the cortical plate.

resonance hypothesis Proposal that target cells establish connections with innervating axons based on matching electrical activity.

retinal progenitor cell (RPC) Multipotent cell that can give rise to any of the six neuronal populations, as well as the Müller glia, of the retina.

retinoblastoma protein (Rb) Tumor suppressor protein involved in the regulation of cell division; mutated in the cancer retinoblastoma, as well as in many other tumors. Its normal activity is to regulate the eukaryotic cell cycle by binding to and inhibiting the E2F proteins, thus blocking progression to DNA replication and cell division.

retinoic acid (RA) A vitamin A derivative.

retinula cell (R cell) A photoreceptor cell of the *Drosophila* ommatidium.

retrograde From the axon terminal to the cell body.

rhombencephalon One of the three primary brain vesicles that forms the hindbrain region of the neural tube.

rhombomere Transient segment (neuromere) in the embryonic hindbrain that provides a restricted region for further cellular development.

Ribonucleic acid (RNA) A chain of nucleotides that contain the sugar ribose, a phosphate group, and one of the four bases: adenine (A), cytosine (C), guanine (G), or uracil (U).

roof plate A wedge-shaped region of glial and neuroepithelial cells that forms on the dorsal surface of the neural tube; serves as a signaling center for neuronal patterning.

rostral–caudal (or rostrocaudal) axis Of a body axis, from head to tail.

sacral neural crest Neural crest cells that originate from the dorsal surface of the neural tube posterior to somite 28 and give rise to parasympathetic and enteric ganglia.

satellite cell Glial cell of the vertebrate peripheral nervous system that surrounds neuronal cell bodies. Thought to function similarly to astrocytes of the central nervous system.

scaffolding protein Protein that anchors other proteins to a specific location in the cell to facilitate protein interactions. Also called scaffold protein.

Schwann cell A glial cell in the PNS that wraps axons with its cytoplasmic extension to form myelin sheath.

sclerotome A segment of a somite that later gives rise to cartilage and bone of the axial skeleton and rib cage.

secreted Frizzled-related protein (sFRP) Small, soluble protein that sequesters Wnt to inhibit it from signaling through the Frizzled receptors. In the anterior neural tube, these proteins act to prevent the posteriorization of forebrain regions.

segment polarity gene In *Drosophila* development, a gene involved in specifying the anteroposterior organization of each body segment.

segmentation gene Gene expressed by subsets of cells in the embryo that refines the pattern of gene expression so as to define the boundaries and ground plan of the individual body segments.

sensory organ progenitor (SOP) In *Drosophila*, a precursor cell whose asymmetric divisions give rise to different cells (a socket cell, hair cell, sheath cell, and sensory neuron) in the external sensory organ.

short interfering RNA (siRNA) Synthetic double-stranded RNAs that bind to complementary mRNA and transiently block gene transcription.

signal transduction Process by which signals originating outside of a cell are conveyed to cytoplasmic components or the nucleus to influence cell behavior.

somal translocation A form of radial migration in which the neuronal cell body travels through its cytoplasmic process toward the pial surface; often used by the earliest generated neurons in the cerebral cortex.

somite One of the paired blocks of mesoderm tissue found alongside the neural tube; comprised of dermamyotome and sclerotome segments.

Sonic hedgehog (Shh) A morphogen that determines cell fate by regulating the expression of specific transcription factors in many developmental contexts. For instance, floor-plate-derived Shh is responsible for determining the different fates of neuronal progenitors located at different positions along the dorsal–ventral axis of the ventral spinal cord. It is also used as a midline attractant for commissural axons.

Spemann's organizer A group of cells in the dorsal blastopore lip of amphibians that induces formation of neural tissue and organizes the body axis; discovered by Hans Spemann and Hilde Mangold.

S phase The phase of the cell cycle during which DNA is synthesized.

spina bifida A neural tube defect of the spinal cord that results from incomplete closure of the neural tube and overlying vertebrae. The outcomes range from mild to severe, depending on the degree of spinal cord exposure.

spinogenesis The formation of spines on the dendrites of neurons.

stem cell Undifferentiated cell that can continue dividing indefinitely, producing daughter cells that can either commit to differentiation or remain a stem cell (in the process of self-renewal).

subpallium The ventral region of the telencephalon that gives rise to structures below (deep to) the cerebral cortex such as the basal ganglia and amygdala.

subplate (SP) A transient layer of cells in the developing neocortex that is created when the preplate is divided into two layers by the cells of the cortical plate; located between the cortical plate and intermediate zone.

subsynaptic nuclei In the multinucleated muscle fiber (myofiber), nuclei that are clustered below the innervating motor nerve terminal.

subventricular zone (SVZ) A layer of proliferative cells in the neocortex that forms between the ventricular and intermediate zones. It continues to produce neurons and glial cells after the ventricular zone is gone; it persists in regions near the lateral ventricles into adulthood.

sulcus limitans A longitudinal groove along the inner surface of the neural tube that provides an anatomical landmark separating the alar and basal plates; it extends from the midbrain to the spinal cord.

superplate An aggregate of cells below the pial surface in the cerebral cortex of mice with the reeler mutation; contains Cajal–Retzius cells and the neurons that would normally form the subplate.

synapse Communicating cell–cell junction that allows signals to pass from a nerve cell to another cell. In a chemical synapse, the signal is carried by a diffusible neurotransmitter. In an electrical synapse, a direct connection is made between the cytoplasms of the two cells via gap junctions.

synaptic basal lamina The specialized extracellular matrix found in the synaptic cleft between a motor nerve terminal and a muscle fiber. Has a unique protein composition that distinguishes it from other basal lamina.

synaptic bouton The slightly enlarged region of an axon terminal at the site of synaptic contact.

synaptic cleft A 20–100 nm gap that separates the presynaptic terminal of a neuron from its target cell.

synaptic reorganization The strengthening or loss of subsets of existing neuronal connections.

synaptic scaling The modification of postsynaptic elements in response to changes in the overall level of neural activity; used to maintain a stable level of neural firing; a form of homeostatic plasticity.

synaptic vesicle A small, membrane-enclosed organelle (typically about 40 nm in diameter) enriched at the presynaptic terminal. Each vesicle is filled with neurotransmitters and, upon stimulation, fuses with the plasma membrane to release neurotransmitters into the synaptic cleft.

synaptogenesis The formation of new synaptic contacts in the nervous system.

syncytial blastoderm In *Drosophila*, the stage of development when the nuclei of a fertilized egg have migrated to the cortex of the blastoderm and share a continuous cytoplasm.

syncytium A group of cells in which the cytoplasm is continuous.

tangential migration Mode of neuronal migration in which cells travel on existing cells or axons along routes parallel to the ventricular surface.

telencephalon The anterior-most of the five secondary brain vesicles; gives rise to the cerebral cortex, hippocampus, basal ganglia, basal forebrain nuclei, olfactory bulb, and lateral ventricles.

temporal identity factors (TIFs) Transcription factors expressed in a sequential manner that together influence the fate of a cell. Also called temporal transcription factors.

terminal synapse The type of synaptic connection that forms between a presynaptic nerve terminal and a postsynaptic cell.

tissue manipulation In experimental embryology, the process of surgically removing, adding, or relocating a piece of tissue to assess the impact of that alteration on subsequent development.

topographic map The consistent, primarily invariant projection of axons from one region of the nervous system to their target cells in another region.

transcription factor A DNA-binding protein that regulates transcription of target genes.

transcription factor code The combination of transcription factors in a precursor cell that leads to the expression or repression of genes that determine the subtype of cell that forms.

transcriptome The expressed RNAs (transcripts) in a tissue or cell.

transforming growth factor β (TGFβ) Large family of structurally related secreted proteins that act as hormones and local mediators to control a wide range of functions in animals, including during development.

tripartite complex The three cells that comprise the vertebrate synapse (a presynaptic neuron, a postsynaptic cell, and a glial cell).

tropism To grow toward. When referring to viruses, the preferential targeting to certain cells, tissues, or organisms.

trunk neural crest Neural crest cells that originate from the dorsal surface of the neural tube between somites 8 and 28 to give rise to dorsal root ganglia, portions of the sympathetic and parasympathetic ganglia, and adrenal chromaffin cells.

vagal neural crest Neural crest cells that originate from the dorsal surface of the neural tube from the posterior hindbrain to somite 7 to give rise to some cranial ganglia as well as sympathetic, parasympathetic, and enteric ganglia of the trunk.

vegetal pole The portion of the egg that has the most yolk; located opposite the animal pole; site of origin for endoderm.

ventral nerve cord (VNC) An invertebrate CNS structure posterior to the brain. It is analogous to the vertebrate spinal cord.

ventralized embryo A *Xenopus* embryo experimentally treated with UV light that develops without a nervous system or other dorsal structures.

ventricular zone (VZ) A transient single cell layer of neuroepithelial cells adjacent to the lumen of the neural tube; the site of proliferation for the first cells generated in the central nervous system.

visceral endoderm In the mouse embryo, the cell layer that becomes displaced by definitive endoderm; gives rise to extraembryonic tissues.

voltage-dependent calcium (Ca2+) channel An ion channel that allows selective passage of Ca^{2+} and whose conductance is regulated by the membrane potential.

zona limitans intrathalamica (ZLI) A signaling center in the neural tube located at the junction of the prechordal and notochordal regions, thought to influence development of the diencephalon.

zygote A fertilized egg cell.

INDEX

Page numbers in *italics* indicate a figure and page numbers in **bold** indicate a table on the corresponding page.

T - #0203 - 111024 - C378 - 280/210/17 - PB - 9780367749903 - Gloss Lamination